Magnetofluid Dynamics

Lazăr Dragoș

Professor at the Department of Mathematics and Mechanics, Bucharest University

Magnetofluid Dynamics

Editura Academiei
București
România

Abacus Press
Tunbridge Wells, Kent
England
1975

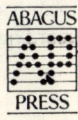

First edition published in Romanian in 1969 under the title
Magnetodinamica fluidelor by
EDITURA ACADEMIEI ROMÂNE,
Calea Victoriei, 125, București, România

Second edition in English, representing the revised
and enlarged translation of the Romanian edition
under the joint imprints
of
EDITURA ACADEMIEI ROMÂNE
and
ABACUS PRESS,
Abacus House, Speldhurst Road, Tunbridge Wells, Kent, England

Translated from Romanian by
Vasile Zoița
Institute of Physics, Bucharest
Translation Editor
Dr. John Hammel
Chelsea College, London University

ISBN 0 85626 016 9

PRINTED IN ROMANIA

Contents

Part 2

THE MICROSCOPIC THEORY

Preface

This work represents the English translation of my book "Magnetodinamica fluidelor" published by Editura Academiei in Romanian in 1969. In its present form this book is however very different from its first edition. First of all it is organized into two well defined parts: the macroscopic and the microscopic theory respectively. Then some new chapters (e.g., chapters 9, 13, 14) have been introduced and the chapters on the theory of MHD generators and thin airfoil theory have been completely altered. All the chapters and the bibliography have been revised and completed. In its present form this work contains some unpublished results.

This book is based on the author's lectures given from 1962 onwards to the fourth and fifth year students of the Fluid Mechanics Section of the Faculty of Mathematics and Mechanics of the Bucharest University. However the book exceeds those lectures by far.

The first chapter postulates the electromagnetic field equations, the expression of force for the continuous media at rest and derives (after Minkowski) their expressions for moving media (the fluids investigated in magnetodynamics are moving continuous media). I wanted to make this introduction since most of the books on magnetofluid dynamics do not specify the conditions for which the electromagnetic force has the expression $\mathbf{J}\times\mathbf{B}$ *(which is usually utilized), the structural (or constitutive) equations of the field in moving media reduce themselves to* $\mathbf{D}=\varepsilon\mathbf{E}$, $\mathbf{B}=\mu\mathbf{H}$, $\mathbf{J}=\sigma(\mathbf{E}+\mathbf{V}\times\mathbf{B})$, *the energy equation which has an extremely complex form for moving media reduces itself to the simple form one generally uses, etc.*

After a short introduction in the kinematics of fluids, chapter 2 presents the magnetofluid dynamic equations in integral form and then derive the differential equations for continuous motion. One section is devoted to fluid thermodynamics and the energy equation. By writing the differential equations in dimensionless form one is able to indicate, among others, in what conditions one may use the simplified form of the equations (neglecting the displacement currents, the convection current in Ohm's law and the electric force term). One section deals with the conservation theorems and another with the existence and uniqueness theorems.

The classical problems of flow between parallel plates and through ducts are presented in Chapter 3: the Poiseuille-Hartmann problem (with Hall effect), the Couette problem and the problem of flow through rectangular ducts.

The theoretical model of the MHD generator in the most general conditions (finite electrodes, external magnetic field applied in the electrode region only, the MHD interaction) is the subject of Chapter 4. The fluid is assumed to be incompressible and the motion to be steady and plane. The mathematical problem consists in integrating a system of second-order partial differential equations (the coefficients being discontinuous functions) with matching conditions along some jump lines and with mixed boundary conditions. Approximate series solutions which can be numerically calculated are given. The Hall effect is also considered.

The following three chapters are devoted to the theory of thin airfoils: the perfectly conducting fluid theory (Chapter 5), the finite resistivity theory (Chapter 6) and the theory of motion with Hall effect (Chapter 7). Using the linearized theory the general solutions are derived and the lift is calculated pointing out the magnetic field influence. For the particular case of nonconducting fluids the results of classical aerodynamics are recovered.

In Chapter 8 the principal results on the flow of viscous fluids past a flat plate with and without incidence are presented. The integral equations of the problem are asymptotically integrated and the classical solutions are recovered as particular cases.

The next three chapters are devoted to the theory of waves : the first two deal with weak waves and the third with strong (shock) waves. Chapter 9 presents the (linear) analytical theory of perturbations and the solution of the Cauchy problem for perfectly conducting fluids and Chapter 10 the geometrical (non-linear) theory. In the latter one can also find the classical results on the (Friedrichs) wave-front diagrams and the simple wave motion. Finally in Chapter 11 the equations of shock phenomena are derived, their properties are discussed and their solution is presented. One section is devoted to the stability problem.

The second part of the book has the purpose of deriving the microscopic equations of magnetofluid dynamics. The electromagnetic field equations are derived in Chapter 12 from the Hamilton principle and the macroscopic equations are obtained on averaging. Chapter 13 is devoted to the kinetic theory of homogeneous gases, to the derivation of Boltzmann's equation and the conservation equations. In Chapter 14 the conservation equations of (ionized) gas mixtures and Ohm's law are derived. Various types of plasmas are considered.

We have especially insisted upon the mathematical model of magnetofluid dynamics in various specific cases since the corresponding boundary value problems are one of the most complex. Indeed since the magnetic field extends all over the space the problems do not close upon themselves within the domains occupied by the fluid but are connected to other problems (other equations) from the adjacent regions. That is why I attempted to formulate as completely as possible the boundary value problems.

I do not pretend to have presented in this book all the aspects of magnetofluid dynamics (as a matter of fact this would be too great a task for any single research worker). I only intended to write an introduction to this science and to present some of its most significant results. This work also contains some of the author's results like those concerning the MHD generator theory (the whole Chapter 4), those regarding the thin airfoil theory (Chapters 6 and 7 completely and part of Chapter 5), those on the motion of viscous fluids past the flat plate (Chapter 8) and the results on the propagation of small perturbations and the solution of the Cauchy problem (Chapter 9).

I have attempted to make up for some omissions in the work by presenting an extensive bibliography. The references were grouped together according to the chapters, the first papers being noted as they were mentioned in the text; I have then included (in chronological order) some other references relevant to the corresponding chapter. Works of general character are presented (in chronological order) in the General Bibliography and are indicated in the text by the letter G followed by a number. I have adopted a chronological presentation in order to obtain a better image on the progress of the various fields of research.

The references are indicated by one number if the paper quoted belongs to the bibliography of that chapter and by two numbers (the first indicating the chapter and the second the paper) if it belongs to another chapter.

I have written this book following the advice of Prof. C. Iacob and for this I would like to express my gratitude. I would like to thank Drs. N. Marcov, D. Homentcovschi and L. Dinu for their help in writing this book.

The author would welcome any critical observations regarding the book.

Bucharest
22 Oct. 1973

L. DRAGOŞ

Introduction

Magnetofluid dynamics deals with the study of the motion of electrically conducting fluids in the presence of magnetic fields. In such a problem the magnetic field influences the fluid motion (this influence is expressed mathematically by including the electromagnetic force in the equations of motion) and the fluid motion changes in turn (through Ohm's law) the magnetic field. Therefore an interaction of the electromagnetic field with the fluid motion takes place and this interaction determines the simultaneous consideration of the fluid mechanics equations and the electromagnetic field equations. Then magnetofluid dynamics represents a synthesis of two classical sciences: fluid mechanics and electromagnetic field theory. At the same time, in order to study the motion of compressible fluids, one should take into account their thermodynamic state and thus the object of this new science turns out to be the study of the phenomena related to fluid motion in all their complexity.

From the historical point of view the first investigations in this field are due to Faraday (1836). As an effect of the action of tides the Thames River flows in both directions and thus has salt waters which are electrically conducting. Faraday attempted to detect the electric field induced by the motion of the conducting waters in the Earth's magnetic field by connecting two electrodes situated near the two banks of the river. But due to the very low conductivity of the fluid the results were hardly discernible and Faraday gave up the idea.

A second important stage in the development of this science is represented by the experiments of Hartmann and Lazarus in 1937 who pointed out the magnetic field influence on the motion of a fluid. Mercury flowing through a tube in the presence of a strong magnetic field was used as conducting fluid. Although the experimental results confirmed the theoretical studies no practical reason for continuing these investigations existed at that time.

The first genuine magnetofluid dynamic problems have been posed by astrophysics. So in 1940 the Swedish scientist H. Alfvén proved that new waves, unknown to both fluid mechanics and electromagnetism can be propagated through a conducting fluid in a magnetic field.

But the first important impetus in the development of this science was the discovery of the plasma and its production on the laboratory scale. As it is now well known, the plasma state represents the manner in which matter exists at high temperatures. A plasma is electrically conducting and has the property of maintaining its overall charge neutrality. At sufficiently high temperatures (of the order of thousands of degrees) matter cannot organize itself into electrically neutral entities (even if it has been so organized the entities would be destroyed) since the thermal energy per particle is high enough to break up these entities through frequent collisions. In such conditions the electric field created by various groups cannot capture enough particles to ensure neutralization. Therefore plasma is a mixture of positively charged, negatively charged and maybe neutral particles. In a magnetic field the charged particles of a plasma are oriented after the field lines of force. Plasma is consequently a compressible, continuous (or discrete), electrically conducting medium. It is obvious that the study of the motion of such a medium represents the object of magnetofluid dynamics. Various other fields of investigation (plasma physics, plasma chemistry, etc.) deal with other aspects of this state of matter.

A widespread belief that most of the matter in the Universe is in the plasma state exists at the present time. Especially interstellar, stellar and solar matter is supposed to exist in such a form. It follows that any astrophysical and astronautical investigations should be based on this new science. The cosmical flight of a satellite represents a motion of a body through a conducting medium in the presence of a magnetic field. But we have a magnetoaerodynamic problem even in the hypersonic-velocity aerodynamics since the gas in front of the moving body is heated by friction and becomes a conductor.

Under normal conditions the air is a weakly conducting medium since it is ionized by radioactive substances and cosmic radiation, but its electrical conductivity does not exceed 0.1 mho cm^{-1} (for the conditions in which a satellite flight takes place). The terrestrial magnetic field is also very weak. At large distances its distribution may be assumed to be that of a dipole of magnetic moment 8×10^{25}CGS units situated at the centre of the Earth. However, by an artificial increase of the gas conductivity near the satellite body and an additional magnetic field properly oriented, one can produce magnetodynamic effects and use them even in this case.

The feasibility of two major projects related to the vital energy problem depends on the results of magnetofluid dynamics: they regard the achievement of controlled thermonuclear reactions and the direct conversion of thermal energy to electrical energy by means of the magnetohydrodynamic generator. The former implies investigations on the stability and turbulence of plasmas and the latter implies studies on the motion of ionized gases through ducts.

These are only a few reasons which prove the necessity of further developing the science of magnetofluid dynamics.

Part 1

The macroscopic theory

1

Electromagnetic field theory

1.1 Electromagnetic field equations for a medium at rest

1.1.1 Maxwell's equations

The Maxwell equations – a synthesis of the results of research carried out by Coulomb, Ampère, Faraday and Maxwell – represent the foundation of the electromagnetic field theory. In fact, these equations represent axioms in which the essence of the electromagnetic field is contained as much as Newton's laws contain the essence of classical mechanics.

In order to formulate these equations [1], [2], we shall characterize the electromagnetic field by the following vectors: electric field intensity $\boldsymbol{E}(t,\boldsymbol{x})$, electric induction $\boldsymbol{D}(t, \boldsymbol{x})$, magnetic field intensity $\boldsymbol{H}(t, \boldsymbol{x})$, magnetic induction $\boldsymbol{B}(t, \boldsymbol{x})$ and density of the conduction current (due to electron motion) $\boldsymbol{J}(t, \boldsymbol{x})$. The electromagnetic field is a state of reality characterized by certain relationships between the above mentioned vectors. These relationships (equations) will be also called laws.

The first is given by what is often called Faraday's electromagnetic induction law (established experimentally in 1831). This can be stated as follows: *the electromotive force induced in a closed conductor loop Γ* (the electric field circulation) *is equal to the negative of the time derivative of the magnetic induction flux through any surface bounded by the conductor loop.* Mathematically, this is written as:

$$\int_{\Gamma} \boldsymbol{E} \cdot \mathrm{d}\boldsymbol{x} = -\frac{\partial}{\partial t}\int_{\Sigma} \boldsymbol{B} \cdot \boldsymbol{n}\, \mathrm{d}a \tag{1.1.1}$$

where $\boldsymbol{n}$ is the unit vector positively orientated, normal to the surface Σ (The right-hand rule for the positive normal: if the right-hand fingers are placed in the direction of travel around the perimeter Γ of the surface the positive normal is indicated by the thumb of the right hand) (Figure 1).

The second relationship is obtained from the magnetic circuit law (an extension of Ampère's experimental results). This is stated in the following way: *the magnetic* *field circulation along the closed contour Γ is equal to the total current flux through any surface Σ bounded by Γ.* This can be written as follows:

$$\int_{\Gamma} \boldsymbol{H} \cdot \mathrm{d}\boldsymbol{x} = \int_{\Sigma} (\boldsymbol{J} + \partial_t \boldsymbol{D}) \cdot \boldsymbol{n} \mathrm{d}a \tag{1.1.2}$$

where

$$\partial_t = \partial/\partial t$$

The term $\partial_t \boldsymbol{D}$ which has the dimensions of a current density (like $\boldsymbol{J}$) is called the displacement current density. It was pointed out by Maxwell by means of physical

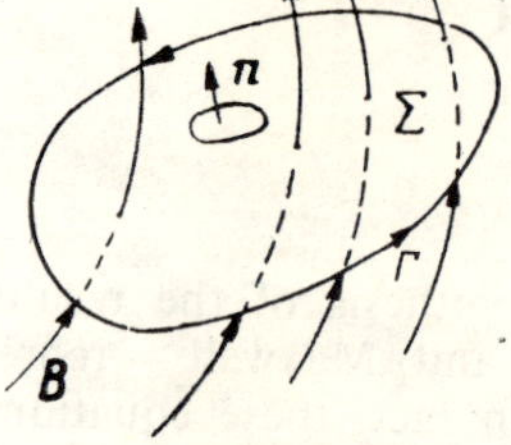

Fig. 1

considerations. Its presence is required by the fact that equations following from (1.1.2) should have an invariant character (Section 2.9).[1]

The last two field equations are derived from the following relationships:

$$\int_{\partial D} \boldsymbol{B} \cdot \boldsymbol{n} \, \mathrm{d}a = 0 \tag{1.1.3}$$

$$\int_{\partial D} \boldsymbol{D} \cdot \boldsymbol{n} \, \mathrm{d}a = \int_{D} \rho^{(q)} \mathrm{d}v = q \tag{1.1.4}$$

where D is a region of space where the electromagnetic field is present, ∂D is the surface bounding this region, $\boldsymbol{n}$ is the outward normal, $\rho^{(q)}$ is the electrical charge density and q the total charge in D. One denotes by da and dv the surface and volume elements, respectively. Equation (1.1.3) represents the induction conservation law while (1.1.4) represents the electric charge conservation law.

If $\boldsymbol{E}$ and $\boldsymbol{B}$ are class C^1 field functions one may use the Stokes formula in (1.1) and get:

$$\int_{\Sigma} (\mathrm{curl}\, \boldsymbol{E} + \partial_t \boldsymbol{B}) \cdot \boldsymbol{n} \, \mathrm{d}a = 0 \tag{1.1.1'}$$

The integrand being a continuous function and the integral being zero for any surface Σ, it follows that the integrand is zero everywhere. Therefore:

$$\operatorname{curl} \boldsymbol{E} + \partial_t \boldsymbol{B} = 0 \tag{1.1.5}$$

The argument is classical: if there was a point at which the integrand of (1.1.1) was, for instance, positive, then it would be also positive in the vicinity of this point: for Σ included in this vicinity the equality (1.1.1) would not be valid.
One also has from equation (1.1.2):

$$\operatorname{curl} \boldsymbol{H} = \boldsymbol{J} + \partial_t \boldsymbol{D} \tag{1.1.6}$$

and from (3) and (4):

$$\operatorname{div} \boldsymbol{B} = 0 \tag{1.1.7}$$

$$\operatorname{div} \boldsymbol{D} = \rho^{(q)} \tag{1.1.8}$$

In equations (1.1.3) and (1.1.4) one applies the Gauss formula (A.30), valid for class C^1 field functions. It should be kept in mind that one may use the differential forms (1.1.5)–(1.1.8) of the Maxwell equations only for the case of class C^1 field functions.
From (1.1.6) and (1.1.8), the following *continuity equation* follows:

$$\operatorname{div} \boldsymbol{J} + \partial_t \rho^{(q)} = 0 \tag{1.1.9}$$

1.1.2 Structural equations

The Maxwell equations express the essence of the electromagnetic field and therefore have a general validity. However, the characteristics of the electromagnetic field in various material media depend on the structure of these media, their thermodynamic state, history, etc. Therefore one should complete the field equations with relationships between the field vectors and the characteristic parameters of the respective medium. We shall call these relationships *structural equations.* Within the macroscopic theory – which forms the basis of the first part of this book – these relationships can only be experimentally proved. From the mathematical point of view, it is obvious that Maxwell's equations are not sufficient to determine the unknown quantities $\boldsymbol{E}$, $\boldsymbol{D}$, $\boldsymbol{H}$, $\boldsymbol{B}$, $\boldsymbol{J}$, and $\rho^{(q)}$.

The simplest media are the homogeneous and isotropic ones. In the linear approximation we may use for these media the following equations:

$$\boldsymbol{D} = \varepsilon \boldsymbol{E}, \ \boldsymbol{B} = \mu \boldsymbol{H} \tag{1.1.10}$$

where ε and μ are characteristic parameters (ε is the *electrical permitivity* and μ the *magnetic permeability*). In general, ε and μ could depend on the thermodynamic state of the medium, i.e.

$$\varepsilon = \varepsilon(\rho, T), \quad \mu = \mu(\rho, T) \tag{1.1.10'}$$

where ρ is the mass density and T the temperature. For a very large class of problems met in magnetofluid dynamics ε and μ may be considered constant. In vacuum, equations (1.1.10) become:

$$\boldsymbol{D} = \varepsilon_0 \boldsymbol{E}, \quad \boldsymbol{B} = \mu_0 \boldsymbol{H} \tag{1.1.10''}$$

where ε_0 and μ_0 are constant.

The third structural equation establishes a relationship between the current density $\boldsymbol{J}$ and the electric field intensity $\boldsymbol{E}$. This is *Ohm's law:*

$$\boldsymbol{J} = \sigma \boldsymbol{E} \tag{1.1.11}$$

where σ is the *electrical conductivity* of the medium. This law is also valid only for linear, homogeneous and isotropic media. It has an empirical character as well, and, like equations (1.1.10), has only a local validity. As we have seen, $\boldsymbol{J}$ represents the conduction current density, that is the current due to the motion of the free charges: free electrons in metals, ions in electrolytes and ionized gases. It seems obvious that the conduction current density should depend on the electric field intensity $\boldsymbol{E}$, i.e. that one should have a relationship of the form: $\boldsymbol{J} = \boldsymbol{J}(\boldsymbol{E})$. Actually, $\boldsymbol{J}$ is also determined by the magnetic field intensity. This will be shown in the second part of this book where the generalized Ohm's law will be derived from the microscopic point of view. The magnetic field effect in Ohm's law is called the Hall effect. For weakly ionized gases, one may use equation (1.1.11) to the first approximation.

As shown above, parameter σ characterizes the electrical conductivity of the medium. The smaller σ is, the higher is the electric field needed to produce an electric current and, conversely, the higher σ, the smaller the electric field needed to produce charge displacement. The media for which $\sigma \simeq 0$ will be called *insulators*, while those for which $\sigma \simeq \infty$ will be called *perfect conductors*. Obviously, there are no perfect insulators or conductors. Some typical values of σ are given in what follows [1]:

copper	5.8005×10^7	conductivity units (perfect conductor)
mercury	0.1044×10^7	conductivity units (perfect conductor)
sea water	3–5	conductivity units (good conductor)
wet earth	10^{-2} – 10^{-3}	conductivity units (poor conductor)
paraffin	10^{-14}–10^{-16}	conductivity units (insulator)

The quantity σ^{-1} is called *electrical resistivity* since Ohm's law expresses the fact that the electrical charges come up against the resistance of the medium: the higher σ^{-1}, the higher is the electric field needed to induce motion of the charges, and conve sely. A perfect conductor medium has zero resistivity.

It is obvious that, generally speaking, σ may also depend on the thermodynamic state of the medium, i.e., $\sigma = \sigma(\rho, T)$. In MFD (magnetofluid dynamics) σ may be considered constant to a good approximation. For anisotropic media structural equations (1.1.10) and (1.1.11) will be replaced by the tensor equations:

$$D = \underline{\boldsymbol{\varepsilon}} \cdot E,\ B = \underline{\boldsymbol{\mu}} \cdot H,\ J = \underline{\boldsymbol{\sigma}} \cdot E \tag{1.1.12}$$

where $\underline{\boldsymbol{\varepsilon}}$, $\underline{\boldsymbol{\mu}}$ and $\underline{\boldsymbol{\sigma}}$ are tensors.

1.1.3 Some consequences for the field equations in an isotropic and homogeneous medium

Taking into account equations (1.1.10) and (1.1.11) the Maxwell equations become:

$$\operatorname{curl} \boldsymbol{E} = -\mu \partial_t \boldsymbol{H},\ \operatorname{curl} \boldsymbol{H} = (\sigma + \varepsilon \partial_t \boldsymbol{E}) \tag{1.1.13}$$

$$\operatorname{div} \mu \boldsymbol{H} = 0,\ \operatorname{div} \varepsilon \boldsymbol{E} = \rho^{(q)} \tag{1.1.14}$$

From equation (1.1.13) one obtains six equations for the six unknown quantities E_i and H_i. The first equation of (1.1.14) is a further restriction on the magnetic field, while the second will give the charge density $\rho^{(q)}$. As a matter of fact, using the continuity equation (1.1.9) and Ohm's law, it follows that:

$$\partial_t \rho^{(q)} + \frac{\sigma}{\varepsilon} \rho^{(q)} = 0 \Rightarrow \rho^{(q)} = \rho_0^{(q)} \exp\left(-\frac{t}{\tau}\right),\ \tau = \frac{\varepsilon}{\sigma} \tag{1.1.15}$$

where $\rho_0^{(q)}$ is the initial charge density. It follows — from the last equation — that the charge density in a region D of a conducting medium ($\sigma \neq 0$) decreases exponentially, independent of the applied field; the smaller τ, the faster the decay. Particularly, if the initial charge distribution is zero, it will remain so for ever (independent of the field). It follows that the *free charges tend to accumulate on surfaces and curves*. Indeed, let us suppose that at the initial moment all the free charge of a conductor is concentrated inside a region D bounded by surface ∂D, the outside charge density being zero. According to equation (1.1.15), the charge density inside D decays in time without increasing at the outside of D. Since on the other hand the total charge is conserved (last Maxwell equation), it follows that charges accumulate on ∂D. The parameter τ (having dimensions of time) is called the *relaxation time*. For sea water $\tau = 2 \times 10^{-10}$ s while for fused quartz (a very good insulator) $\tau = 10^6$ s.

A second remark is that that from equations (1.1.13) and (1.1.14) the existence of a fundamental, free electromagnetic field follows. This field propagates as electromagnetic waves of velocity

$$c_m = \frac{1}{\sqrt{\varepsilon\mu}} \tag{1.1.16}$$

in an infinite medium.

Indeed, if one sets $\boldsymbol{J} \equiv 0$, $\rho^{(q)} \equiv 0$ and applies the *curl* operator to equation (1.1.13), one would have:

$$\left(\Delta - \varepsilon\mu\frac{\partial^2}{\partial t^2}\right)\left(\boldsymbol{E}, \boldsymbol{H}\right) = 0, \quad \Delta = \frac{\partial^2}{\partial x_1^2} + \frac{\partial^2}{\partial x_2^2} + \frac{\partial^2}{\partial x_3^2}$$

In vacuum, the electromagnetic field propagates itself with velocity:

$$c = \frac{1}{\sqrt{\varepsilon_0\mu_0}} \tag{1.1.16'}$$

Finally, taking into account equation (A.24), one gets from the Maxwell and structural equations:

$$\begin{aligned}\operatorname{div}(\boldsymbol{E} \times \boldsymbol{H}) &= \boldsymbol{H}\cdot\operatorname{curl}\boldsymbol{E} - \boldsymbol{E}\cdot\operatorname{curl}\boldsymbol{H} = -\boldsymbol{E}\cdot\partial_t\boldsymbol{D} - \boldsymbol{H}\cdot\partial_t\boldsymbol{B} - \boldsymbol{E}\cdot\boldsymbol{J}\\ &= -\frac{1}{2}\partial_t(\boldsymbol{D}\cdot\boldsymbol{E} + \boldsymbol{B}\cdot\boldsymbol{H}) - \boldsymbol{E}\cdot\boldsymbol{J}\end{aligned}$$

Integrating over some region D it follows:

$$-\frac{\partial}{\partial t}\int_D \frac{1}{2}(\boldsymbol{D}\cdot\boldsymbol{E} + \boldsymbol{B}\cdot\boldsymbol{H})\,\mathrm{d}v = \int_{\partial D} \boldsymbol{E}\times\boldsymbol{H}\cdot\boldsymbol{n}\,\mathrm{d}a + \int_D \boldsymbol{E}\cdot\boldsymbol{J}\,\mathrm{d}v \tag{1.1.17}$$

This relationship may be interpreted as the energy balance equation if one considers:

$$w = \frac{1}{2}(\boldsymbol{D}\cdot\boldsymbol{E} + \boldsymbol{B}\cdot\boldsymbol{H}) \tag{1.1.17'}$$

as being the electromagnetic energy per unit volume. It follows, from equation (1.1.17), that the decrease in energy is due to the outward energy radiation and Joule heating (first and second term in the RHS, respectively). Such an interpretation is no longer valid for moving media.

1.2 The electromagnetic field equations for moving media

1.2.1 The Lorentz group

Let us specify some results in connection with the Lorentz group. These may be also found, for instance, in reference [3]. The Lorentz group establishes the relationship between the coordinates x, y, z, t of an event in the reference frame S and the coordinates x', y', z', t' of the same event in another reference frame S'

moving uniformly with respect to S with velocity v_0 along the Ox axis. This is written mathematically as follows:

$$x' = \frac{1}{\theta}(x - v_0 t), \; y' = y, \; z' = z \tag{1.2.1}$$

$$t' = \frac{1}{\theta}\left(t - \frac{v_0 x}{c^2}\right)$$

where

$$\theta = \sqrt{1 - \frac{v_0^2}{c^2}}$$

and c is the velocity, of light.

The Lorentz group may also be written as:

$$x'_\alpha = c_{\alpha\beta} x_\beta, \; \alpha = 1, 2, 3, 4 \tag{1.2.1'}$$

where the following notations:

$$x_1 = x, \; x_2 = y, \; x_3 = z, \; x_4 = ict$$

$$(c_{\alpha\beta}) = \begin{pmatrix} \dfrac{1}{\theta} & 0 & 0 & \dfrac{iv_0}{c\theta} \\ 0 & 1 & 0 & 0 \\ 0 & 0 & 1 & 0 \\ -\dfrac{iv_0}{c\theta} & 0 & 0 & \dfrac{1}{\theta} \end{pmatrix}$$

and the summation convention with respect to β have been used.

A first necessary remark is that:

$$\det(c_{\alpha\beta}) = 1 \tag{1.2.2}$$

a fact which has deep implications: the volume element is invariant, the tensor densities become tensors, etc.

The four-dimensional space (x, y, z, t) is called the *Minkowski universe* (space), and a point in this space is called an *event*. The metric of this space is given by the equation:

$$ds = \sqrt{-dx_\alpha dx_\alpha} = c\sqrt{1 - \frac{v^2}{c^2}}\,dt \overset{\text{def}}{=} c\,d\tau \tag{1.2.3}$$

the elements ds and $d\tau$ being invariant with respect to the Lorentz group.

A quantity A_α which for a change of the reference system defined by (1.2.2) transforms itself according to the equations:

$$A'_\alpha = c_{\alpha\beta} A_\beta \tag{1.2.4}$$

will be called a *quadrivector*, while a quantity $A_{\alpha\beta}$ which transforms itself according to:

$$A'_{\alpha\beta} = c_{\alpha\gamma} c_{\beta\delta} A_{\gamma\delta} \tag{1.2.5}$$

will be called a *quadritensor* (all Greek subscripts take the values 1, 2, 3, 4). The choice of these names can be easily understood based on the analogy of (A.7) and (A.8).

An exemple of a quadrivector is the generalized velocity quadrivector defined by:

$$U_\alpha = \frac{\mathrm{d}x_\alpha}{\mathrm{d}\tau}, \qquad \alpha = 1, 2, 3, 4 \tag{1.2.6}$$

One can easily verify that:

$$U_\alpha U_\alpha = -c^2, \quad U_\alpha \frac{\mathrm{d}U_\alpha}{\mathrm{d}\tau} = 0 \tag{1.2.6$'$}$$

1.2.2 The covariant form of the Maxwell equations

Let us denote by subscripts the components of the $\boldsymbol{B}$, $\boldsymbol{E}$, $\boldsymbol{H}$, $\boldsymbol{D}$ and $\boldsymbol{J}$ vectors and introduce in the universe the following quadritensors:

$$\begin{gathered}
(F_{\alpha\beta}) = \begin{array}{l}\beta\rightarrow\\ \alpha\downarrow\end{array}\begin{pmatrix} 0 & -B_3 & B_2 & \dfrac{\mathrm{i}}{c}E_1 \\ B_3 & 0 & -B_1 & \dfrac{\mathrm{i}}{c}E_2 \\ -B_2 & B_1 & 0 & \dfrac{\mathrm{i}}{c}E_3 \\ -\dfrac{\mathrm{i}}{c}E_1 & -\dfrac{\mathrm{i}}{c}E_2 & -\dfrac{\mathrm{i}}{c}E_3 & 0 \end{pmatrix} \\
(H_{\alpha\beta}) = \begin{pmatrix} 0 & -H_3 & H_2 & \mathrm{i}cD_1 \\ H_3 & 0 & -H_1 & \mathrm{i}cD_2 \\ -H_2 & H_1 & 0 & \mathrm{i}cD_3 \\ -\mathrm{i}cD_1 & -\mathrm{i}cD_2 & -\mathrm{i}cD_3 & 0 \end{pmatrix}
\end{gathered} \tag{1.2.7}$$

One can easily verify that equations (1.1.5) and (1.1.7) may be concentrated (in the universe) in the form:

$$F_{\beta\gamma,\alpha} + F_{\gamma\alpha,\beta} + F_{\alpha\beta,\gamma} = 0 \tag{1.2.8}$$

where:

$$F_{\beta\gamma,\alpha} = \partial F_{\beta\gamma}/\partial x_\alpha, \ \ldots$$

while equations (1.1.6) and (1.1.8) become:

$$H_{\alpha\beta,\,\alpha} = J_\beta, \qquad \beta = 1, 2, 3, 4 \tag{1.2.9}$$

where J_β is the quadrivector $(J_1, J_2, J_3, \mathrm{i}c\rho^{(q)})$. The LHS term in (1.2.9) represents the divergence of tensor $H_{\alpha\beta}$. Equations (1.2.8) may also be reduced to a divergence form if one introduces the tensor $F^*_{\alpha\beta}$ *dual* to $F_{\alpha\beta}$ that is:

$$F^*_{\alpha\beta} = \frac{1}{2}\delta_{\alpha\beta\gamma\delta}F_{\gamma\delta} \tag{1.2.10}$$

where $\delta_{\alpha\beta\gamma\delta}$ is the Levi-Civita tensor (A.11). With this notation equations (1.2.8) become:

$$F^*_{\alpha\beta,\,\alpha} = 0, \qquad \beta = 1, 2, 3, 4 \tag{1.2.8'}$$

The covariant form (1.2.8′) and (1.2.9) of the Maxwell equations *proves that these equations are invariant with respect to the Lorentz group.*

Some more remarks should be made regarding the above equations. The first refers to the structure of dual tensors. So, if one formally writes:

$$(F_{\alpha\beta}) = \left(-\boldsymbol{B}, \frac{\mathrm{i}}{c}\boldsymbol{E}\right), \ (H_{\alpha\beta}) = (-\boldsymbol{H}, \mathrm{i}c\boldsymbol{D})$$

then one has the following expressions for the dual tensors

$$(F^*_{\alpha\beta}) = \left(\frac{\mathrm{i}}{c}\boldsymbol{E}, -\boldsymbol{B}\right), \ (H^*_{\alpha\beta}) = (\mathrm{i}c\boldsymbol{D}, -\boldsymbol{H})$$

The second remark refers to the structure of the equations. Since the tensors $F^*_{\alpha\beta}$ and $H_{\alpha\beta}$ are antisymmetric, it follows:

$$F^*_{\alpha\beta,\,\alpha\beta} \equiv 0, \ H_{\alpha\beta,\,\alpha\beta} \equiv 0 \quad (\Rightarrow J_{\beta,\,\beta} = 0)$$

The first identity shows that equations (1.2.8′) are not independent, and the second one that the continuity equation (1.1.9) is a consequence of equations (1.2.9).

1.2.3 Field transformations

From the previous section it follows that if $F_{\alpha\beta}$ and $H_{\alpha\beta}$ are quadritensors and J_β is a quadrivector, then the Maxwell equations will be invariant with respect to the Lorentz group. As a matter of fact, the actual situation is the other way round: the Maxwell equations are invariant with respect to the Lorentz group *(this being the fundamental postulate of the relativity theory)* so that $F_{\alpha\beta}$ and $H_{\alpha\beta}$ are quadritensors and J_β is a quadrivector. From equation (1.2.5′) it follows that if J_β is a quadrivector it will have, in the S′ reference system, the following components:

$$J_1' = \frac{1}{\theta}(J_1 - \rho^{(q)}v_0),\ J_2' = J_2,\ J_3' = J_3$$
$$\rho^{(q)'} = \frac{1}{\theta}\left(\rho^{(q)} - \frac{v_0}{c^2}J_1\right) \qquad (1.2.11)$$

or, in the non-relativistic approximation, ($v_0^2/c^2 \simeq 0$),

$$J_1' = J_1 - \rho^{(q)}v_0,\ J_2' = J_2,\ J_3' = J_3$$
$$\rho^{(q)'} = \rho^{(q)} \qquad (1.2.11\ )$$

In deriving the last equation, the fact that J_1 is of the order of magnitude of v_0 (as it follows from the first equation (1.2.11) was taken into account. Therefore, if an observer in S' measures the density $\rho^{(q)'}$ and the current J_1', an observer in S will find the current J_1' increased by $\rho^{(q)}v_0$. This term represents the so-called convection current (due to the motion of S'). In the three-dimensional space equations (1.2.11′) are written as follows:

$$\boldsymbol{J}' = \boldsymbol{J} - \rho^{(q)}\boldsymbol{v}_0 \qquad (1.2.11'')$$

In order to derive the expressions for the field vectors in the moving frame S' one uses the equations (1.2.5′) and (1.2.7). For the $F_{\alpha\beta}$ tensor one has:

$$E_1' = E_1,\ \theta E_2' = E_2 - v_0B_3,\ \theta E_3' = E_3 + v_0B_2$$
$$B_1' = B_1,\ \theta B_2' = B_2 + c^{-2}v_0E_3,\ \theta B_3' = B_3 - c^{-2}v_0E_2 \qquad (1.2.12)$$

In the non-relativistic approximation these become:

$$\boldsymbol{E}' = \boldsymbol{E} + \boldsymbol{v}_0 \times \boldsymbol{B}$$
$$\boldsymbol{B}' = \boldsymbol{B} - \frac{1}{c^2}\boldsymbol{v}_0 \times \boldsymbol{E} \qquad (1.2.13)$$

Expressions (1.2.13) are fundamental in the Minkowski theory of the electromagnetic field. First of all they show that the electromagnetic field has an objective existence and should be regarded as an entity. Its division into an electric and a magnetic field is relative (it depends on the reference system). Indeed, if $\boldsymbol{E}$ and $\boldsymbol{B}$ have given values, in the S frame one can find another, inertial, reference frame so that the electromagnetic field will have either only the electric field form (if the velocity of the frame S' is given by equation $\boldsymbol{v}_0 \times \boldsymbol{E} = c^2\boldsymbol{B}$) or only the magnetic field form (if the velocity of the S' frame is given by equation $\boldsymbol{v}_0 \times \boldsymbol{B} = -\boldsymbol{E}$). Conversely, if the field is, for instance, electrostatic (due to some electric charges only) in the S' frame, an electromagnetic field described by:

$$\boldsymbol{B} = c^{-2}\boldsymbol{v}_0 \times \boldsymbol{E}', \ \boldsymbol{E} = \boldsymbol{E}'$$

will exist in the S frame. A second necessary remark regarding equations (1.2.13) is that the last term is of the order of the relativistic correction (v_0^2/c^2) only for the case that $\boldsymbol{E}$ is of the order of magnitude of $\boldsymbol{v}_0$. This is the case for instance in MFD in those problems in which the electric field is due only to the fluid motion in the magnetic field. In the next chapter one will see that in this case $E = 0\,(v_0B)$ so that the term $c^{-2}\boldsymbol{v}_0 \times \boldsymbol{E}$ may be neglected.

Similarly, it follows (for the case $v_0^2/c^2 \simeq 0$):

$$\boldsymbol{H}' = \boldsymbol{H} - \boldsymbol{v}_0 \times \boldsymbol{D}$$

$$\boldsymbol{D}' = \boldsymbol{D} + \frac{1}{c^2}\boldsymbol{v}_0 \times \boldsymbol{H} \tag{1.2.14}$$

To conclude, it follows that only the electromagnetic field has as objective existence, this fact being expressed by the invariance property of the Maxwell equations with respect to the Lorentz group. The way the $\boldsymbol{E}$, $\boldsymbol{D}$, $\boldsymbol{H}$, $\boldsymbol{B}$ and $\boldsymbol{J}$ vectors characterize the electromagnetic field changes from one frame to another.

1.2.4 The Maxwell equations for moving media

The extension of Maxwell's equations from rest to moving media is due to Minkowski [4]. The inclusion of this subject in this place is required by the fact that the fluids studied in MFD are conducting media moving in the presence of a magnetic (or electromagnetic) field.

In what follows Minkowski's argument is presented. Let us consider a continuous medium in motion. Let $\boldsymbol{V}(t, \boldsymbol{x})$ be the velocity (with respect to the frame S) of the particle P of position vector $\boldsymbol{x}$ at time t and let S' be a reference frame with its origin at P and velocity $\boldsymbol{V}$. The S' frame may be considered inertial with respect to S at any time, so that the transition from frame S' to frame S will be performed by the equations given by the Lorentz group (where $\boldsymbol{v}_0$ will be replaced by $\boldsymbol{V}$). The particle P is at rest with respect to frame S' (frame S' is called the proper frame of particle P) while its neighbourhood – in a quasi-static state. But in such conditions the Maxwell equations in differential form are valid (they are locally valid). Since the Maxwell equations are invariant with respect to the Lorentz group it follows that at any moment one may locally use the form (1.1.5)–(1.1.8) of these equations. Therefore, *the differential form of Maxwell's equations are valid for moving media too*. However, the integral form (1.1.1)–(1.1.4) is changed since for moving media the time derivative may no longer be permuted with the integral (the manyfold Σ is made up of moving material particles, thus being a function of t). This matter will be discussed in Section 1.6.

1.2.5 Structural equations for moving media

Structural equations of the form (1.1.10), (1.1.11) are valid only in the local reference frame. Thus, one has:

$$\boldsymbol{D}' = \varepsilon \boldsymbol{E}', \ \boldsymbol{B}' = \mu \boldsymbol{H}', \ \boldsymbol{J}' = \sigma \boldsymbol{E}' \tag{1.2.15}$$

Using the transformation equations (1.2.13) and (1.2.14) one gets the following equations with respect to frame S:

$$\boldsymbol{D} + \frac{1}{c^2}\boldsymbol{V} \times \boldsymbol{H} = \varepsilon(\boldsymbol{E} + \boldsymbol{V} \times \boldsymbol{B}) \tag{1.2.16}$$

$$\boldsymbol{B} - \frac{1}{c^2}\boldsymbol{V} \times \boldsymbol{E} = \mu(\boldsymbol{H} - \boldsymbol{V} \times \boldsymbol{D}) \tag{1.2.17}$$

$$\boldsymbol{J} - \rho^{(q)}\boldsymbol{V} = \sigma(\boldsymbol{E} + \boldsymbol{V} \times \boldsymbol{B}) \tag{1.2.18}$$

By eliminating $\boldsymbol{B}$ of (1.2.16) using equation (1.2.17) and $\boldsymbol{D}$ of (1.2.17) using equation (1.2.16), it follows that:

$$\left(1 - \frac{V^2}{c^2} k_e k_m\right) \boldsymbol{D} = \varepsilon\left(1 - \frac{V^2}{c^2}\right) \boldsymbol{E} + (k_e k_m - 1)\frac{\boldsymbol{V} \times \boldsymbol{H}}{c^2} - \varepsilon(k_e k_m - 1)\left(\boldsymbol{E} \cdot \frac{\boldsymbol{V}}{c}\right)\frac{\boldsymbol{V}}{c}$$

$$\left(1 - \frac{V^2}{c^2} k_e k_m\right) \boldsymbol{B} = \mu\left(1 - \frac{V^2}{c^2}\right) \boldsymbol{H} - (k_e k_m - 1)\frac{\boldsymbol{V} \times \boldsymbol{E}}{c^2} - \mu(k_e k_m - 1)\left(\boldsymbol{H} \cdot \frac{\boldsymbol{V}}{c}\right)\frac{\boldsymbol{V}}{c}$$

where

$$k_e = \frac{\varepsilon}{\varepsilon_0}, \quad k_m = \frac{\mu}{\mu_0}$$

Or, neglecting terms of the order of V^2/c^2:

$$\begin{aligned} D &= \varepsilon E + (\varepsilon\mu - \varepsilon_0\mu_0) V \times H \\ B &= \mu H - (\varepsilon\mu - \varepsilon_0\mu_0) V \times E \end{aligned} \tag{1.2.19}$$

Therefore, for moving media (in the non-relativistic approximation) structural equations have the form (1.2.19) and (1.2.18). Fortunately, those media met in practical problems of MFD readily satisfy the condition:

$$\varepsilon\mu \simeq \varepsilon_0\mu_0 \tag{1.2.20}$$

so that equations (1.2.19) may be used in the form:

$$D = \varepsilon E, \quad B = \mu H \tag{1.2.19'}$$

One will call *simple media* those satisfying the equation (1.2.20) with constant ε and μ. One can see that even for this case the electromagnetic field cannot be determined independently of motion (the particle velocity field appears in Ohm's law (1.2.18)). On the other hand, the motion cannot be determined independently of the electromagnetic field.

1.3 The force density

A conducting material medium (solid, fluid) in motion, subject to an electromagnetic field, is influenced by the presence of this field. In order to write down the equations of motion of the medium (according to the Newtonian view) one should express the field action by a force density. This requirement is not so easily satisfied and, as a matter of fact, it has not been, to this day, expressed in such a definitive way like, for instance, the Maxwell equations. An interesting solution to this effect is due to A. Timotin [5]. For the purpose of this work, the classical results are good enough. The electromagnetic field was defined by four vectors, ***E***, ***D***, ***H*** and ***B*** connected by the Maxwell equations. The physical nature of these vectors can be pointed out by the action of the field on material bodies [1]. The expression of this action is a postulate of electrodynamics. In its simplest form, the electromagnetic force density is given by:

$$f^{(em)} = \rho^{(q)} E + J \times B \tag{1.3.1}$$

and is called the *Lorentz force*. In fact, equation (1.3.1) represents the macroscopic form of the force produced by the electromagnetic field on a charge q of velocity v in vacuum (Section 12.1). Equation (1.3.1) is based on the Coulomb law, for the electric field case, and the Biot and Savart law for the magnetic field case. These are both experimentally established laws. For an inductive development of the electromagnetic field theory ([6], [7] for instance) one starts with these laws.

In order to prove the validity of equation (1.3.1) for moving media, one should notice that this represents the three-dimensional part of the following covariant form:

$$f_\alpha^{(em)} = J_\beta F_{\beta\alpha}, \qquad \alpha = 1, 2, 3, 4 \tag{1.3.2}$$

with J_β and $F_{\beta\alpha}$ defined in the previous section. The fourth component of the *force quadrivector* (1.3.2) has the following expression:

$$f_4^{(em)} = \frac{i}{c} \boldsymbol{J} \cdot \boldsymbol{E} \tag{1.3.2'}$$

and gives the density of the work performed by the field on the medium. During the motion of the medium, the first three components of the force quadrivector will give the time change of the momentum per unit volume, while the fourth component will give the rate of change of the mechanical energy of the same unit volume.

In order to point out the *conservation laws* we shall try to write down the force density using the divergence of some tensor which will naturally come out from Maxwell's equations and the force density (1.3.1). This makes it possible for some part of the action performed through (1.3.1) upon the material medium as long range action to be transformed in action at the boundary of the material medium.

1.3.1 The electromagnetic stress tensor

Let $T_{ij}^{(em)}$ be the electromagnetic stress tensor given by:

$$T_{ij}^{(em)} = D_i E_j + B_i H_j - \frac{1}{2}(\boldsymbol{D} \cdot \boldsymbol{E} + \boldsymbol{B} \cdot \boldsymbol{H})\delta_{ij} \tag{1.3.3}$$

Let us compute the divergence of this tensor. Marking the derivatives by commas and taking into account equations (1.1.7) and (1.1.8) one has:

$$T_{ij,i}^{(em)} = \rho^{(q)} E_j + D_i E_{j,i} + B_i H_{j,i} - \frac{1}{2}(\boldsymbol{D} \cdot \boldsymbol{E} + \boldsymbol{B} \cdot \boldsymbol{H})_{,j}$$

But

$$D_i E_{j,i} - \frac{1}{2}(D_i E_i)_{,j} = D_i(E_{j,i} - E_{i,j}) + \frac{1}{2}(D_i E_{i,j} - D_{i,j} E_i)$$

and similarly,

$$B_i H_{j,i} - \frac{1}{2}(B_i H_i)_{,j} = \ldots$$

It follows that:

$$T^{(\mathrm{em})}_{ij,\,i} = \rho^{(q)}E_j + (\operatorname{curl} \boldsymbol{E} \times \boldsymbol{D})_j + (\operatorname{curl} \boldsymbol{H} \times \boldsymbol{B})_j \qquad (1.3.4)$$

$$+ \frac{1}{2}(D_i E_{i,\,j} - E_i D_{i,\,j} + B_i H_{i,\,j} - H_i B_{i,\,j})$$

Now, taking into account the Maxwell equations (1.1.5) and (1.1.6), equation (1.3.4) becomes:

$$\operatorname{div} \underline{\boldsymbol{T}}^{(\mathrm{em})} = \rho^{(q)}\boldsymbol{E} + \boldsymbol{J} \times \boldsymbol{B} + \partial_t(\boldsymbol{D} \times \boldsymbol{B}) + \boldsymbol{f}^{(\mathrm{em})}_* \qquad (1.3.5)$$

The physical explanation of (1.3.5) is straightforward. The term $\rho^{(q)}\boldsymbol{E} + \boldsymbol{J} \times \boldsymbol{B}$ is the Lorentz force acting upon the free charges of density $\rho^{(q)}$ and electric current of density $\boldsymbol{J}$. The term $\boldsymbol{f}^{(\mathrm{em})}_*$ represents the force acting upon the dielectrics and is due to polarization phenomena. In vacuum, or for simple media $\boldsymbol{f}^{(\mathrm{em})}_* \equiv 0$.

The term $\partial_t \boldsymbol{g}$ where,

$$\boldsymbol{g} = \boldsymbol{D} \times \boldsymbol{B} \qquad (1.3.6)$$

should be interpreted as the *electromagnetic momentum* and $\underline{\boldsymbol{T}}^{(\mathrm{em})}$ as the *electromagnetic stress tensor*. Indeed, integrating equation (1.3.5) upon a fixed region of space D, one obtains:

$$\int_{\partial D} \boldsymbol{n} \cdot \underline{\boldsymbol{T}}^{(\mathrm{em})} \, da = \int_D \boldsymbol{F}^{(\mathrm{em})} \, dv + \frac{\partial}{\partial t}\int_D \boldsymbol{g} \, dv, \qquad (1.3.7)$$

or

$$\int_{\partial D} \boldsymbol{n} \cdot \underline{\boldsymbol{T}}^{(\mathrm{em})} da = \frac{d}{dt}(\boldsymbol{G} + \boldsymbol{p}) \qquad (1.3.8)$$

where

$$\boldsymbol{F}^{(\mathrm{em})} = \boldsymbol{f}^{(\mathrm{em})} + \boldsymbol{f}^{(\mathrm{em})}_*, \quad \boldsymbol{G} = \int_D \boldsymbol{g} \, dv, \quad \boldsymbol{p} = \int_D \boldsymbol{V} dm$$

$\boldsymbol{p}$ is the mechanical momentum (whose change is determined by the resulting force). The electromagnetic momentum is attached to the space and, for a moving medium, is not connected with the medium.

In moving media the interpretation of equation (1.3.5) is more difficult. However the expression for the electromagnetic stress tensor remains valid for invariance reasons to be presented in a subsequent section.

For a medium whose electric permitivity and magnetic permeability depend on ρ, T and the field strength, the stress tensor contains an additional pressure due to electromagnetic striction effects. Thus for a *perfect gas*, the stress tensor becomes (Section 2.3):

$$T^{(\mathrm{em})}_{ij} = D_i E_j + B_i H_j - \frac{1}{2}\left(\boldsymbol{D} \cdot \boldsymbol{E} + \boldsymbol{B} \cdot \boldsymbol{H} - \rho E^2 \frac{\partial \varepsilon}{\partial \rho} - \rho H^2 \frac{\partial \mu}{\partial \rho}\right)\delta_{ij} \qquad (1.3.9)$$

and the ponderomotive force:

$$\boldsymbol{F}^{(\mathrm{em})} = \rho^{(q)}\boldsymbol{E} + \boldsymbol{J} \times \boldsymbol{B} + \frac{1}{2}(D_i\nabla E_i - E_i\nabla D_i + B_i\nabla H_i - H_i\nabla B_i)$$

$$+ \nabla\left(\frac{1}{2}\rho E^2 \frac{\partial \varepsilon}{\partial \rho} + \frac{1}{2}\rho H^2 \frac{\partial \mu}{\partial \rho}\right) \tag{1.3.10}$$

The last terms in the RHS of equation (1.3.9):

$$\boldsymbol{D} \cdot \boldsymbol{E} + \boldsymbol{B} \cdot \boldsymbol{H} - \rho E^2 \frac{\partial \varepsilon}{\partial \rho} - \rho H^2 \frac{\partial \mu}{\partial \rho}$$

are interpreted as an electromagnetic pressure (Section 2.3).

1.3.2 The electromagnetic energy-momentum quadritensor

It can be easily shown that the electromagnetic stress tensor (1.3.3) represents the three-dimensional part of the following *energy-momentum quadritensor* (Minkowski):

$$T^{(\mathrm{em})}_{\beta\alpha} = H_{\beta\gamma}F_{\gamma\alpha} + \frac{1}{4}H_{\gamma\delta}F_{\gamma\delta}\delta_{\beta\alpha} \tag{1.3.11}$$

The ponderomotive force density is the divergence of this quadritensor. Indeed, taking into account the Maxwell equations (1.2.8) and (1.2.9), it follows that:

$$T^{(\mathrm{em})}_{\beta\alpha,\,\beta} = J_\gamma F_{\gamma\alpha} + H_{\beta\gamma}F_{\gamma\alpha,\,\beta} + \frac{1}{4}(H_{\gamma\delta}F_{\gamma\delta})_{,\,\alpha}$$

but

$$H_{\beta\gamma}F_{\gamma\alpha,\,\beta} = H_{\gamma\delta}F_{\delta\alpha,\,\gamma} = \frac{1}{2}H_{\gamma\delta}(F_{\delta\alpha,\,\gamma} + F_{\alpha\gamma,\,\delta})$$

$$= -\frac{1}{2}H_{\gamma\delta}F_{\gamma\delta,\,\alpha} = -\frac{1}{4}(H_{\gamma\delta}F_{\gamma\delta})_{,\,\alpha} + \frac{1}{4}(H_{\gamma\delta,\,\alpha}F_{\gamma\delta} - H_{\gamma\delta}F_{\gamma\delta,\,\alpha})$$

so that,

$$T^{(\mathrm{em})}_{\beta\alpha,\,\beta} = J_\gamma F_{\gamma\alpha} + \frac{1}{4}(H_{\gamma\delta,\,\alpha}F_{\gamma\delta} - H_{\gamma\delta}F_{\gamma\delta,\,\alpha})$$

The first term in the RHS of the last equation is the Lorentz force (1.3.2) and the second, the force $\boldsymbol{f}^{(\mathrm{em})}_*$. Indeed, one could easily check up [1], [9], [G. 24] that, for a simple medium the last term is zero in the proper frame (and therefore is zero in any frame). One may therefore write:

$$T^{(em)}_{\beta\alpha,\beta} = F^{(em)}_{\alpha}, \qquad \alpha = 1, 2, 3, 4. \tag{1.3.12}$$

This equation proves that the expressions of the electromagnetic stress tensor and ponderomotive forces remain valid for moving media too. This will be of interest —in the present work — especially for simple media. Using equations (1.3.11) and (1.2.7) one gets:

$$\begin{aligned}(T_{14}, T_{24}, T_{34}) &= -\frac{i}{c}\boldsymbol{E}\times\boldsymbol{H}\\ T_{44} &= \frac{1}{2}(\boldsymbol{D}\cdot\boldsymbol{E}+\boldsymbol{B}\cdot\boldsymbol{H}) = w\end{aligned} \tag{1.3.13}$$

so that from (1.3.12) and (1.3.2′) one obtains:

$$\frac{\partial w}{\partial t} + \operatorname{div}(\boldsymbol{E}\times\boldsymbol{H}) + \boldsymbol{J}\cdot\boldsymbol{E} = 0 \tag{1.3.14}$$

The interpretation of this equation is given in (1.1.17). From the theory of relativity it is known that the fourth component given by the quadri-dimensional form of the conservation laws gives the energy balance.

1.4 Dimensions and units

The quantities used to characterize the electromagnetic field have no meaning without the introduction of a system of units. Comparing the geometrical and physical quantities with quantities of the same nature taken as standards, one obtains numbers which express the magnitude of these quantities. Only by measurement can the actual phenomena be translated into numbers which enter the equations and the results of mathematical operations gain a real meaning. In the following the dimensions and units of the quantities introduced so far will be established.

The fundamental quantities of mechanics are: length (L), time (T) and mass (M). By adopting as units for these quantities the *meter* (m), *second* (s) and *kilogramme* (kg) one obtains the so-called MKS system. Both the electromagnetic field theory and thermodynamics introduce a new quantity. Taking as a fundamental quantity the electric charge (Q) and the *coulomb* (C) as unit, one obtains the International System (SI) suggested by Giorgi in 1901 and adopted by the International Electrotechnical Commission in 1935. This is the system used in this work. From the definition of charge density one has:

$$[\rho^{(q)}] = QL^{-3}, \text{ Cm}^{-3} \text{ as unit}$$

Therefore, equations (1.8) and (1.6) give:

$$[D] = QL^{-2}, \text{ Cm}^{-2} \text{ as unit}$$

$$[J] = QL^{-2}T^{-1}, \text{ Am}^{-2} \text{ as unit}$$

$$[H] = QL^{-1}T^{-1}, \text{ Am}^{-1} \text{ as unit}$$

where A is the symbol for the *ampère*, the unit for the current intensity:

$$I = \int_{\Sigma} \boldsymbol{J} \cdot \boldsymbol{n} \, da$$

The dimensions of I are QT^{-1}. Lately, the ampère has been used as a fundamental unit in electrical engineering instead of the coulomb (which becomes a derived unit).

In the present system, force has the dimension MLT^{-2} and as unit the *newton* (N), energy has the dimension ML^2T^{-2} and as unit the *joule* (J) and power ML^2T^{-3} as dimension, and the *watt* (W) as unit. Obviously, all these units are derived from the MKS system and the dimensions from the laws of mechanics. From the expression of the volume force density (3.1) it follows:

$$[E] = MLT^{-2}Q^{-1}, \text{ JC}^{-1}\text{m}^{-1} = \text{Vm}^{-1} \text{ as unit}$$

$$[B] = MT^{-1}Q^{-1}, \text{ Vsm}^{-2} = \text{T as unit}$$

where V stands for the *volt* (the electrical potential φ unit defined by: $\boldsymbol{E} = -\text{grad}\varphi$) and T stands for the *tesla* (not to be mistaken for the time unit). From the structural equations one obtains:

$$[\varepsilon] = M^{-1}L^{-3}T^2Q^2, \text{ CV}^{-1}\text{m}^{-1} = \text{Fm}^{-1} \text{ as unit}$$

$$[\mu] = MLQ^{-2}, \text{ VsA}^{-1}\text{m}^{-1} = \text{Hm}^{-1} \text{ as unit}$$

$$[\sigma] = M^{-1}L^{-3}TQ^2, \quad \frac{\text{C}^2\text{s}}{\text{m}^2\text{kg}} \times \frac{1}{\text{m}} = \frac{1}{\text{ohm}} \times \frac{1}{\text{m}} = \frac{\text{mho}}{\text{m}} \text{ as unit}$$

where: F stands for the *farad*, H for the *henry*. In the expression for conductivity the resistivity unit ohm appeaıs as well as its inverse the mho.

In what follows, some numerical values of the fundamental constants in the SI system are given [1]:
electron charge $q_e = 1.60 \times 10^{-19}$C
electrical permitivity of vacuum $\varepsilon_0 = 8.854 \times 10^{-12} \simeq (36\pi)^{-1} \times 10^{-9}$Fm^{-1}
magnetic permeability of vacuum $\mu_0 = 4\pi \times 10^{-7} \simeq 1.257 \times 10^{-6}$ Hm^{-1}
light velocity in vacuum $c = 2.998 \times 10^8 \simeq 3 \times 10^8$ ms^{-1}

1.5 Boundary conditions

1.5.1 Discontinuity surfaces

Let us consider two material media (1) and (2) separated by a moving surface $\mathscr{S}$. The electromagnetic field is present in both medium (1) and (2) but due to the structural equations it manifests itself differently in the two media (the parameters ε, μ and σ are different). Therefore the surface will be a discontinuity surface for the quantities which characterize the field and the motion. In what follows we shall define the equations connecting the field and motion quantities on the two sides of surface $\mathscr{S}$.

Let us first define the displacement speed of some discontinuity surface. To this end, let us consider the surface at time t, $\mathscr{S}(t)$ and at time $t + \Delta t$, $\mathscr{S}(t + \Delta t)$. Let $P(\boldsymbol{x})$ be a point on $\mathscr{S}(t)$, $\boldsymbol{n}$ the normal in P at $\mathscr{S}(t)$ positively orientated towards $\mathscr{S}(t + \Delta t)$ and P' $(\boldsymbol{x} + \Delta \boldsymbol{x})$ the point of intersection of $\boldsymbol{n}$ with surface $\mathscr{S}(t + \Delta t)$. We define by the scalar quantity

$$d = \lim_{\Delta t \to 0} \frac{|\boldsymbol{PP'}|}{\Delta t} \tag{1.5.1}$$

the displacement (or drift) speed of surface $\mathscr{S}$. The vector form of this quantity is $\boldsymbol{d} = d\boldsymbol{n}$.

If $G(t, \boldsymbol{x}) = 0$ is the equation of surface $\mathscr{S}(t)$ then

$$d = -\frac{\partial_t G}{|\text{grad}\, G|} \tag{1.5.1'}$$

Indeed, noticing that $\Delta \boldsymbol{x} = |\boldsymbol{PP'}|\boldsymbol{n}$, from equation

$$G(t + \Delta t, \boldsymbol{x} + |\boldsymbol{PP'}|\boldsymbol{n}) = 0$$

(which expresses the fact that P' lies on $\mathscr{S}(t + \Delta t)$), we find that

$$\frac{\partial G}{\partial t}\Delta t + \frac{\partial G}{\partial x_i}|\boldsymbol{PP'}|n_i = 0$$

From this equation we derive the ratio of equation (1.5.1), and, on taking the limit with respect to $\Delta t \to 0$, equation (1.5.1') is derived.

1.5.2 The material derivative

By a material surface we mean a surface made up of material particles and thus a surface moving and distorting itself with the medium. It will be shown in Section 2.1 that such a surface is conserved as the locus of the same particles during the motion. At the same place, we shall prove that for the case of a field A of class C^1 we have the following equation (2.1.16′) for such a surface Σ:

$$\frac{\mathrm{d}}{\mathrm{d}t}\int_{\Sigma} \boldsymbol{A}\cdot\boldsymbol{N}\mathrm{d}a = \int_{\Sigma} \boldsymbol{a}\cdot\boldsymbol{N}\,\mathrm{d}a + \int_{\partial\Sigma} \boldsymbol{A}\times\boldsymbol{V}\cdot\mathrm{d}\boldsymbol{x} \tag{1.5.2}$$

where $\boldsymbol{N}$ is the positive normal, $\boldsymbol{V}$ the material particle velocity and

$$\boldsymbol{a} = \partial_t\boldsymbol{A} + \boldsymbol{V}\operatorname{div}\boldsymbol{A} \tag{1.5.2′}$$

The equation (1.5.2) is derived from (2.1.16′) using the Stokes theorem.

Let us now consider a plane surface Σ, orthogonal to $\mathscr{S}$, which contains elements of both medium (1) and (2) (Figure 2).

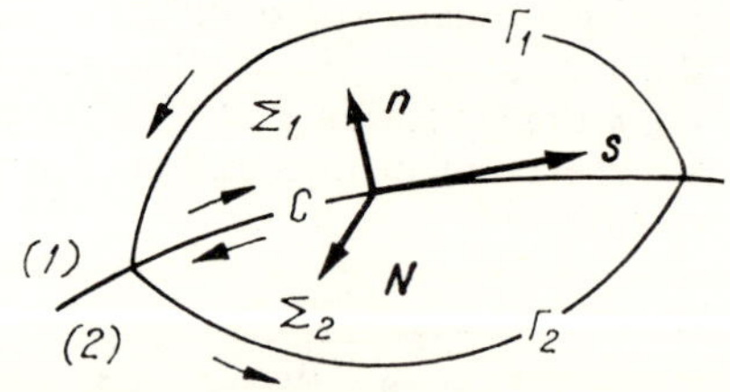

Fig. 2

We denote by Σ_1 (boundary Γ_1) the part of Σ contained in medium (1) and by Σ_2 (boundary Γ_2) the part contained in (2). Then denote by $\boldsymbol{n}$ the unit vector normal to $\mathscr{S}$ and consideı it positively orientated towards medium (1). Let $C = \mathscr{S} \cap \Sigma$ and $\boldsymbol{s}$ be its unit vector ($\boldsymbol{N} = \boldsymbol{s}\times\boldsymbol{n}$). We may apply equation (2) for each of the surface Σ_1 and Σ_2. Marking by 1 and 2 the limit values of A on C from Σ_1 and Σ_2 respectively, we have:

$$\begin{aligned}\frac{\mathrm{d}}{\mathrm{d}t}\int_{\Sigma_1} \boldsymbol{A}\cdot\boldsymbol{N}\,\mathrm{d}a &= \int_{\Sigma_1} \boldsymbol{a}\cdot\boldsymbol{N}\mathrm{d}a + \int_{\Gamma_1} \boldsymbol{A}\times\boldsymbol{V}\cdot\mathrm{d}\boldsymbol{x} + \int_C \boldsymbol{A}_1\times\boldsymbol{d}\cdot\boldsymbol{s}\,\mathrm{d}s\\ &= \int_{\Sigma_1} \boldsymbol{a}\cdot\boldsymbol{N}\mathrm{d}a + \int_{\partial\Sigma_1} \boldsymbol{A}\times\boldsymbol{V}\cdot\mathrm{d}\boldsymbol{x} + \int_C \boldsymbol{A}_1\times(\boldsymbol{d}-\boldsymbol{V}_1)\cdot\boldsymbol{s}\,\mathrm{d}s\end{aligned}$$

$$\frac{\mathrm{d}}{\mathrm{d}t}\int_{\Sigma_2} \boldsymbol{A}\cdot\boldsymbol{N}\,\mathrm{d}a = \int_{\Sigma_2} \boldsymbol{a}\cdot\boldsymbol{N}\,\mathrm{d}a + \int_{\partial\Sigma_2} \boldsymbol{A}\times\boldsymbol{V}\cdot\mathrm{d}\boldsymbol{x} - \int_C \boldsymbol{A}_2\times(\boldsymbol{d}-\boldsymbol{V}_2)\cdot\boldsymbol{s}\,\mathrm{d}s$$

The line integrals over $\partial\Sigma_1$ and $\partial\Sigma_2$ can be converted into surface integrals. Adding the above expressions we get:

$$\frac{d}{dt}\int_{\Sigma_1+\Sigma_2} \boldsymbol{A}\cdot\boldsymbol{N}\,da = \int_{\Sigma_1+\Sigma_2}\{\boldsymbol{a} + \operatorname{curl}(\boldsymbol{A}\times\boldsymbol{V})\}\cdot\boldsymbol{N}\,da + \int_C [\boldsymbol{A}\times(\boldsymbol{d}-\boldsymbol{V})]\cdot\boldsymbol{s}\,ds \tag{1.5.3}$$

where

$$[\boldsymbol{A}\times(\boldsymbol{d}-\boldsymbol{V})] = \{\boldsymbol{A}_1\times(\boldsymbol{d}-\boldsymbol{V}_1)\} - \{\boldsymbol{A}_2\times(\boldsymbol{d}-\boldsymbol{V}_2)\}$$

In general, for any function φ we write

$$[\varphi] = \varphi_1 - \varphi_2 \tag{1.5.4}$$

Let us consider now a material domain D divided into two regions D_1 and D_2 by a discontinuity surface $\mathscr{S}$. Let $\boldsymbol{N}$ be the external normal to D (Figure 44). If $\boldsymbol{A}$ is a class C^1 vector in D_1 and D_2, then, from the Gauss formula we get:

$$\int_{D_1} \operatorname{div}\boldsymbol{A}\,dv = \int_{\Sigma_1}\boldsymbol{A}\cdot\boldsymbol{N}\,da - \int_S \boldsymbol{A}_1\cdot\boldsymbol{n}\,da$$

$$\int_{D_2} \operatorname{div}\boldsymbol{A}\,dv = \int_{\Sigma_2}\boldsymbol{A}\cdot\boldsymbol{N}\,da + \int_S \boldsymbol{A}_2\cdot\boldsymbol{n}\,da$$

where Σ_1 and Σ_2 are the parts of ∂D situated in (1) and (2), respectively. Adding the above two equations, we get:

$$\int_{D_1+D_2} \operatorname{div}\boldsymbol{A}\,dv = \int_{\partial D}\boldsymbol{A}\cdot\boldsymbol{N}\,da - \int_S [\boldsymbol{A}]\cdot\boldsymbol{n}\,da \tag{1.5.5}$$

1.5.3 The integral form of Maxwell's equations for moving media

Let us consider a material curve Γ and a material surface Σ bounded by Γ. Let $\boldsymbol{N}$ be the normal to Σ, positively orientated (with respect to the positive direction along Γ). It is obvious that Maxwell equations (1.1.1), (1.1.2) are not valid for this case. They will be replaced by the following equations:

$$\int_\Gamma \boldsymbol{E}'\cdot d\boldsymbol{x} = -\frac{d}{dt}\int_\Sigma \boldsymbol{B}\cdot\boldsymbol{N}\,da \tag{1.5.6}$$

$$\int_\Gamma \boldsymbol{H}'\cdot d\boldsymbol{x} = \frac{d}{dt}\int_\Sigma \boldsymbol{D}\cdot\boldsymbol{N}\,da + \int_\Sigma \boldsymbol{J}'\cdot\boldsymbol{N}\,da \tag{1.5.7}$$

where E', H' and J' were defined in Section 1.2 (these are vector fields in the proper system) and the time derivative in Section 2.1. In order to prove equation (1.5.6), we apply the Stokes formula to the first integral and notice that, taking into account the Maxwell equations (1.1.5) and (1.1.7) (valid for moving media, too), we have

$$\operatorname{curl} \boldsymbol{E}' = \operatorname{curl}(\boldsymbol{E} + \boldsymbol{V} \times \boldsymbol{B}) = -\{\partial_t \boldsymbol{B} + \operatorname{curl}(\boldsymbol{B} \times \boldsymbol{V}) + \boldsymbol{V} \operatorname{div} \boldsymbol{B}\}$$

Using equation (2.1.16′) we get the same result for the RHS of equation (1.5.6). One can similarly prove equation (1.5.7).

The other two Maxwell equations for moving media are

$$\int_{\partial D} \boldsymbol{B} \cdot \boldsymbol{N}\, da = 0 \quad \text{(non-existence of magnetic charges)} \tag{1.5.8}$$

$$\int_{\partial D} \boldsymbol{D} \cdot \boldsymbol{N}\, da = \int_D \rho^{(q)}\, dv \quad \text{(existence of electric charges)} \tag{1.5.9}$$

where D is a (moving) material medium and ∂D its boundary.

We postulate that equations (1.5.6)—(1.5.9) are also valid for the case of material manyfolds crossed by discontinuity surfaces. As a matter of fact this is the most general form of Maxwell's equations.

1.5.4 Jump conditions

In order to derive the jump conditions across a surface which separates two media (1) and (2) we have to use the integral form of Maxwell's equations (the differential form is valid only for the case of continuous fields). In this respect we shall use equation (1.5.6) for the case of a contour Γ, as in Figure 2. The RHS surface integral is written using equation (3). Now, taking the limit $\Gamma_1 \to C$, $\Gamma_2 \to C$, we shall notice that the surface integral of equation (1.5.3) vanishes (if the integrand is a bounded function). Therefore, we get

$$\int_C [\boldsymbol{E}'] \cdot \boldsymbol{s}\, ds = \int_C [\boldsymbol{B} \times (\boldsymbol{d} - \boldsymbol{V})] \cdot \boldsymbol{s}\, ds$$

Since C is a curve of arbitrary length it follows that

$$[\boldsymbol{E}'] \cdot \boldsymbol{s} = [\boldsymbol{B} \times (\boldsymbol{d} - \boldsymbol{V})] \cdot \boldsymbol{s}$$

and then

$$[\boldsymbol{E} + \boldsymbol{d} \times \boldsymbol{B}] \cdot \boldsymbol{s} = 0 \tag{1.5.10}$$

If $\mathscr{S}$ is a material surface then equation (1.5.10) becomes

$$[\boldsymbol{E} + \boldsymbol{V} \times \boldsymbol{B}] \cdot \boldsymbol{s} = 0 \tag{1.5.11}$$

and for the case of fixed surface

$$[\boldsymbol{E}] \cdot \boldsymbol{s} = 0 \tag{1.5.12}$$

Equations (1.5.10)—(1.5.12) express the continuity conditions of the tangential component of the electric field in the proper system attached to Σ.

In a similar way we get from (1.5.7)

$$[\boldsymbol{H} - \boldsymbol{d} \times \boldsymbol{D}] \cdot \boldsymbol{s} = 0 \tag{1.5.13}$$

supposing that no surface currents or surface electric charges exist on $\mathscr{S}$. If this is not true, that is, if a surface density $\boldsymbol{J}_s$ and a surface density $\rho_s^{(q)}$ may be defined, then ($\boldsymbol{J}'$ is not bounded across $\mathscr{S}$) from equation (1.5.7) we find

$$[\boldsymbol{H} - \boldsymbol{d} \times \boldsymbol{D}] \cdot \boldsymbol{s} + (\boldsymbol{J}_s - \rho_s^{(q)}\boldsymbol{V}) \cdot \boldsymbol{N} = 0 \tag{1.5.14}$$

For the case of a material surface condition (1.5.13) becomes

$$[\boldsymbol{H} - \boldsymbol{V} \times \boldsymbol{D}] \cdot \boldsymbol{s} = 0 \tag{1.5.15}$$

and, for a fixed surface:

$$[\boldsymbol{H}] \cdot \boldsymbol{s} = 0 \tag{1.5.16}$$

Let us now consider a region D across $\mathscr{S}$. Using equation (1.5.5) for the case $\boldsymbol{A} = \boldsymbol{B}$ (magnetic induction) and equation (1.5.8) it follows:

$$\int_S [\boldsymbol{B}] \cdot \boldsymbol{n} \, \mathrm{d}a = 0$$

since div $\boldsymbol{B} = 0$ within both D_1 and D_2. Since S is an arbitrary surface, it follows that

$$[\boldsymbol{B}] \cdot \boldsymbol{n} = 0 \tag{1.5.17}$$

Similarly, from equation (1.5.9), we have:

$$[\boldsymbol{D}] \cdot \boldsymbol{n} = 0 \tag{1.5.18}$$

or

$$[\boldsymbol{D}]\cdot\boldsymbol{n} = \rho_s^{(q)} \qquad (1.5.18')$$

supposing that a surface charge density exists.

The following condition is useful for problems in which the conductivity of surface $\mathscr{S}$ is given:

$$[\boldsymbol{J}]\cdot\boldsymbol{n} = 0 \qquad (1.5.19)$$

derived from equation (1.1.9) as (1.5.18) is derived from equation (1.1.8). As a matter of fact, this condition is not independent of the ones above as much as equation (1.1.9) is not independent of equations (1.1.4)–(1.1.8). Particularly, if $\mathscr{S}$ is an insulating wall and if the fluid fills the region D_1, from equation (1.5.19) it follows:

$$\boldsymbol{J}_1\cdot\boldsymbol{n} = 0 \qquad (1.5.20)$$

Some interesting considerations on the jump conditions are due to D. Homentcovschi [9]. Their derivation from the differential form of the field equations using the theory of distributions is particularly interesting.

Note: If the two media separated by surface $\mathscr{S}$ are characterized by the same constants ε and μ then conditions (1.5.10) and (1.5.13) may be replaced by

$$[\boldsymbol{E}]\cdot\boldsymbol{s} = 0 \qquad (1.5.21)$$

$$[\boldsymbol{H}]\cdot\boldsymbol{s} = 0 \qquad (1.5.22)$$

Indeed, from (1.5.10) we find:

$$[\boldsymbol{E}]\cdot\boldsymbol{s} + d\mu[\boldsymbol{H}]\cdot\boldsymbol{N} = 0 \qquad (\boldsymbol{N} = \boldsymbol{s}\times\boldsymbol{n}) \qquad (1.5.23)$$

and from equation (1.5.13)

$$[\boldsymbol{H}]\cdot\boldsymbol{s} - d\varepsilon[\boldsymbol{E}]\cdot\boldsymbol{N} = 0 \qquad (1.5.24)$$

Writing these conditions for the case when the tangential vector $\boldsymbol{s}$ is replaced by the tangential vector $\boldsymbol{N}$, we get:

$$[\boldsymbol{E}]\cdot\boldsymbol{N} - d\mu[\boldsymbol{H}]\cdot\boldsymbol{s} = 0 \qquad (1.5.25)$$

$$[\boldsymbol{H}]\cdot\boldsymbol{N} + d\varepsilon[\boldsymbol{E}]\cdot\boldsymbol{s} = 0 \qquad (1.5.26)$$

From equations (1.5.23) and (1.5.26) equation (1.5.21) follows, and from equation (1.5.24) and (1.5.25), equation (1.5.22). These conclusions are no longer valid for the case when $d^2\varepsilon\mu = 1$, i.e., for the case when the discontinuity surface is propagating with the velocity of light.

2

Magnetofluid dynamics

2.1 Elements of the kinematics of continuous media

2.1.1 Definitions and axioms

Let $\mathcal{M}$ be some continuous medium (liquid or gas) and $\mathcal{D}_0$ and $\mathcal{D}_t$ the domains filled by this at the initial time and time t, respectively; $\mathcal{D}_0$ and $\mathcal{D}_t$ are domains of the Euclidian space $\mathscr{E}_3$. If we denote by $\boldsymbol{x}^0$ the position vector of some material particle P at the initial time and by $\boldsymbol{x}$ the position vector (of the same particle) at time t then the particle motion is described by

$$\boldsymbol{x} = \boldsymbol{x}(t, \boldsymbol{x}^0) \tag{2.1.1}$$

The vector function (2.1.) given over the whole domain $\mathcal{D}_0$ defines the motion of medium $\mathcal{M}$. We shall assume this function to be continuous (as well as its derivatives up to the third order) and by definition

$$J \stackrel{\text{def}}{=} \frac{\partial(x_1, x_2, x_3)}{\partial(x_1^0, x_2^0, x_3^0)} \neq 0 \tag{2.1.2}$$

over the whole domain $\mathcal{D}_0$ (except maybe for some singular points, curves or surfaces). From the theorem of implicit functions it follows that the transformation (2.1.1) has an inverse (the inverse function also being of class C^2). For a given $\boldsymbol{x}^0$ and variable t equation (2.1.1) defines the trajectory of particle P, and for a given t and variable (within $\mathcal{D}_0$) $\boldsymbol{x}^0$, equation (2.1.1) defines a *continuous transformation* of $\mathcal{D}_0$ into $\mathcal{D}_t$.

Due to the continuity property of transformation (2.1.1) it follows that the material manifolds (curves, surfaces and domains) are conserved during motion. That is to say that the material points making up a domain $\mathcal{D}_0$ at the initial time will lie at time t within a domain $\mathcal{D}_t$ given by transformation (2.1.1). No point external to $\mathcal{D}_0$ will get inside $\mathcal{D}_t$ and no point of $\mathcal{D}_0$ will get outside $\mathcal{D}_t$. The same statement is also true for surfaces and curves. The fact that the material particles always remain distinct from one another is atested by the existence of the inverse transformation (i.e., hypothesis (2.1.2)). This hypothesis thus expresses the property of impenetrability of matter.

As a matter of fact (2.1.2) states that no domain $\mathscr{D}_0$ of finite volume can be transformed into a domain $\mathscr{D}_t$ of zero volume. From this point of view, equation (2.1.2) expresses the property of the indestructibility (or continuity) of matter.

The coordinates of vector $\boldsymbol{x}^0\,(x_i^0)$ are called *Lagrangian* (or *material*) coordinates, while the coordinates of vector $\boldsymbol{x}\,(x_i)$ are called *Eulerian* (or *spatial*) coordinates. Any function $F(t, \boldsymbol{x})$ of the spatial coordinates may also be regarded (due to the transformation (2.1.1)) as a function $F(t, \boldsymbol{x}^0)$ of the material coordinates (and *vice-versa*). $F(t, \boldsymbol{x})$ represents the quantity F defined for the particle which at time t is situated at $\boldsymbol{x}$, while $F(t, \boldsymbol{x}^0)$ is the quantity F defined at time t for that particle which was initially situated at $\boldsymbol{x}^0$. It is obvious that the derivatives

$$\partial_t F \stackrel{\text{def}}{=} \frac{\partial F}{\partial t} \stackrel{\text{def}}{=} \frac{\partial F(t, \boldsymbol{x})}{\partial t}, \quad \dot{F} \stackrel{\text{def}}{=} \frac{dF}{dt} \stackrel{\text{def}}{=} \frac{\partial F(t, \boldsymbol{x}^0)}{\partial t} \tag{2.1.3}$$

are distinct from one another. The second one, called the *material derivative* of F, gives the rate of change of F for a given particle, while the first gives the rate of change of F as determined by an observer situated at point $\boldsymbol{x}$. It is obvious that the velocity $\boldsymbol{V}$ of some particle is the material derivative of the position vector

$$\boldsymbol{V} = \frac{\mathrm{d}\boldsymbol{x}}{\mathrm{d}t}(t, \boldsymbol{x}^0) \tag{2.1.4}$$

that is, a function of the material variables.

In fluid mechanics, the state of motion of the medium is characterized by pressure, velocity and density as functions of the Eulerian variables

$$p = p(t, \boldsymbol{x}), \quad \boldsymbol{V} = \boldsymbol{V}(t, \boldsymbol{x}), \quad \rho = \rho(t, \boldsymbol{x}) \tag{2.1.5}$$

Having determined these unknowns the problem of finding the trajectories comes to the integration of the system:

$$\frac{\mathrm{d}\boldsymbol{x}}{\mathrm{d}t} = \boldsymbol{V}(t, \boldsymbol{x}) \tag{2.1.6}$$

with initial conditions $\boldsymbol{x}(t = 0) = \boldsymbol{x}^0$.

The acceleration is the time derivative of the velocity of a certain particle, that is,

$$\boldsymbol{a} = \frac{\mathrm{d}\boldsymbol{V}}{\mathrm{d}t} = \frac{\partial \boldsymbol{V}}{\partial t} + (\boldsymbol{V}\cdot\nabla)\boldsymbol{V} \tag{2.1.7}$$

where we used the fact that velocity is expressed as a function of the Eulerian coordinates.

Equation (2.1.7) represents a particular case of the equation

$$\dot{F} = \partial_t F + (\boldsymbol{V}\cdot\nabla)F \tag{2.1.7'}$$

which relates the material derivatives to the spatial one.

2.1.2 The Lagrange criterion

The following criterion is due to Lagrange: *a necessary and sufficient condition fol the surface Σ of equation $f(t, \boldsymbol{x}) = 0$ to be a material surface (a locus of materiar points) is that*

$$\dot{f} = 0 \tag{2.1.8}$$

We know that if a surface represents a locus of material particles at a given time, it will remain the locus of the same particles at any other time.

In order to prove Lagrange's criterion, let us consider a surface Σ at time t and at a subsequent time $t + \Delta t$. By subtraction, we get

$$\partial_t f + U_i f,_i = 0, \quad f,_i \overset{\text{def}}{=} \partial f / \partial x_i$$

$\boldsymbol{U}$ being the velocity of a point P of the surface. Taking into account this formula we find that the displacement velocity (1.5.1′) of surface Σ is given by the equation

$$U_n = \boldsymbol{U} \cdot \boldsymbol{n} = \frac{U_i f,_i}{|\mathrm{grad}\, f|} = -\frac{\partial_t f}{|\mathrm{grad}\, f|}$$

Let now $\boldsymbol{V}$ be the velocity of the fluid particle which at time t was situated at P. It follows that

$$V_n = \boldsymbol{V} \cdot \boldsymbol{n} = \frac{V_i f,_i}{|\mathrm{grad}\, f|}$$

and finally

$$V_n - U_n = \frac{\dot{f}}{|\mathrm{grad}\, f|}$$

This equation proves the initial statement (Σ is a material surface when and only when $V_n = U_n$).

2.1.3 Material derivatives

(a) Let Γ_t be a material curve and $\mathrm{d}\boldsymbol{x}$ its line element. Since

$$\frac{\mathrm{d}}{\mathrm{d}t}(\mathrm{d}x_i) = \frac{\mathrm{d}}{\mathrm{d}t}\left(\frac{\partial x_i}{\partial x_j^0}\,\mathrm{d}x_j^0\right) = \frac{\partial V_i}{\partial x_j^0}\,\mathrm{d}x_j^0 = \frac{\partial V_i}{\partial x_k}\,\mathrm{d}x_k$$

it follows that

$$\frac{\mathrm{d}}{\mathrm{d}t}(\mathrm{d}\boldsymbol{x}) = (\mathrm{d}\boldsymbol{x} \cdot \nabla)\boldsymbol{V} \tag{2.1.9}$$

whence

$$\frac{d}{dt}(ds^2) = 2\,dx \cdot \underline{D} \cdot dx, \quad D_{ij} \overset{\text{def}}{=} \frac{1}{2}(V_{i,j} + V_{j,i}) \tag{2.1.9'}$$

this equation expressing the fact that the necessary and sufficient condition for the motion to be rigid is to have $\underline{D} = 0$.

Differentiating the following (obvious) equation

$$\frac{\partial x_i}{\partial x_j^0}\frac{\partial x_j^0}{\partial x_l} = \delta_{il}$$

one also obtains

$$\frac{\partial x_i}{\partial x_j^0}\frac{d}{dt}\left(\frac{\partial x_j^0}{\partial x_l}\right) = -\frac{\partial x_j^0}{\partial x_l}\frac{d}{dt}\left(\frac{\partial x_i}{\partial x_j^0}\right) = -\frac{\partial x_j^0}{\partial x_l}\frac{\partial V_i}{\partial x_j^0} = -\frac{\partial V_i}{\partial x_l}$$

and then

$$\frac{d}{dt}\left(\frac{\partial x_k^0}{\partial x_l}\right) = -\frac{\partial V_i}{\partial x_l}\frac{\partial x_k^0}{\partial x_i} \tag{2.1.10}$$

(b) Let us prove the following theorem due to Euler:

$$\frac{dJ}{dt} = J \operatorname{div} V \tag{2.1.11}$$

To this end, we shall denote by C_{ij} the algebraic complement of the element $\partial x_i/\partial x^0$ in the expansion of the determinant J. We have

$$C_{ij}\partial x_k/\partial x_j^0 = J\delta_{ki}$$

and therefore

$$\frac{dJ}{dt} = C_{ij}\frac{d}{dt}\left(\frac{\partial x_i}{\partial x_j^0}\right) = C_{ij}\frac{\partial V_i}{\partial x_k}\frac{\partial x_k}{\partial x_j^0} = J\frac{\partial V_i}{\partial x_i}$$

The summation with respect to $i\,(i = 1, 2, 3)$ comes from the fact that the derivative of determinant J is the sum of three determinants J_i (J_i being obtained from J by the differentiation of line i).

If the volume remains constant during the motion (incompressible fluid) we get from equation (2.1.11)

$$\operatorname{div} V = 0 \tag{2.1.11'}$$

(c) Let now Σ_t be a material surface determined by equation $\boldsymbol{x}^0 = \boldsymbol{x}^0(\lambda_1, \lambda_2)$ at the initial time and by equation $\boldsymbol{x} = \boldsymbol{x}(t, \boldsymbol{x}^0) = \boldsymbol{x}(t, \lambda_1, \lambda_2)$ at time t and let

$$da_i^0 = \delta_{ijk}\frac{\partial x_j^0}{\partial \lambda_1}\frac{\partial x_k^0}{\partial \lambda_2}d\lambda_1\,d\lambda_2, \quad da_i = \delta_{ijk}\frac{\partial x_j}{\partial \lambda_1}\frac{\partial x_k}{\partial \lambda_2}d\lambda_1\,d\lambda_2$$

be the orientated area elements ($da_i = n_i da$) at the two times. A simple calculation [G.24] gives:

$$da_i = J \frac{\partial x_k^0}{\partial x_i} da_k^0 \qquad (i = 1, 2, 3) \tag{2.1.12}$$

Now, taking into account equations (2.1.10), (2.1.11) and (2.1.12) we get

$$\frac{d}{dt}(da_l) = \frac{d}{dt}\left(J \frac{\partial x_k^0}{\partial x_l} da_k^0\right) = (\operatorname{div} V)\, da_l - \frac{\partial V_i}{\partial x_l} da_i \tag{2.1.13}$$

The above equations make it possible for us to derive the expressions of the material derivatives of the line, surface and volume integrals which will be used in the present work.

(a′) Let, for instance, $F(t, \boldsymbol{x})$ be a function defined on the fluid particles and Γ_t a material curve. We have the following equation:

$$\frac{d}{dt}\int_{\Gamma_t} F\, dx_i = \int_{\Gamma_t}\left(\frac{dF}{dt} dx_i + F \frac{\partial V_i}{\partial x_j} dx_j\right) \tag{2.1.14}$$

The proof of this formula is straightforward if one turns to material variables in the first integral and uses equation (2.1.9). With material variables the curve Γ_t becomes the fixed curve Γ_0 so that the derivative may be taken inside the integral.

For the case $F = A_i$, A_i being the coordinates of the vector $\boldsymbol{A}$, equation (2.1.14) gives

$$\begin{aligned}\frac{d}{dt}\int_{\Gamma_t} \boldsymbol{A}\cdot d\boldsymbol{x} &= \int_{\Gamma_t}\left(\frac{\partial A_i}{\partial t} + V_j \frac{\partial A_i}{\partial x_j} + A_j \frac{\partial V_j}{\partial x_i}\right) dx_i \\ &= \int_{\Gamma_t}\left\{\frac{\partial A_i}{\partial t} + V_j\left(\frac{\partial A_i}{\partial x_j} - \frac{\partial A_j}{\partial x_i}\right) + \frac{\partial}{\partial x_i}(A_j V_j)\right\} dx_i \qquad (2.1.14') \\ &= \int_{\Gamma_t}\{\partial_t \boldsymbol{A} + \operatorname{curl} \boldsymbol{A} \times \boldsymbol{V} + \operatorname{grad}(\boldsymbol{A}\cdot \boldsymbol{V})\}\cdot d\boldsymbol{x}\end{aligned}$$

Particularly, if Γ_t is a closed curve, (2.1.14′) becomes

$$\frac{d}{dt}\oint_{\Gamma_t} \boldsymbol{A}\cdot d\boldsymbol{x} = \oint_{\Gamma_t}(\partial_t \boldsymbol{A} + \operatorname{curl} \boldsymbol{A} \times \boldsymbol{V})\cdot d\boldsymbol{x} \tag{2.1.14''}$$

(b′) For the case of a material domain D_t we have the following equation:

$$\frac{d}{dt}\int_{D_t} F(t, \boldsymbol{x})\, dv = \int_{D_t}\{\partial_t F + \operatorname{div}(F\boldsymbol{V})\}\, dv \tag{2.1.15}$$

To prove this relationship, we shall use the method presented above. Using equation (2.1.1), we go to material variables and in this way the domain D_t goes to the fixed domain D_0. Then using Euler's theorem we get:

$$\frac{\mathrm{d}}{\mathrm{d}t}\int_{D_t} F\,\mathrm{d}v = \int_{D_0} \frac{\mathrm{d}}{\mathrm{d}t}(FJ)\,\mathrm{d}v^0 = \int_{D_t} (\dot{F} + F\,\mathrm{div}\,V)\,\mathrm{d}v$$

which proves equation (2.1.15).

(c′) Finally, for the case of a material surface Σ_t we get

$$\frac{\mathrm{d}}{\mathrm{d}t}\int_{\Sigma_t} F\,\mathrm{d}a_i = \int_{\Sigma_t} \left\{\dot{F}\,\mathrm{d}a_i + F\left((\mathrm{div}\,V)\,\mathrm{d}a_i - \frac{\partial V_j}{\partial x_i}\mathrm{d}a_j\right)\right\} \qquad (2.1.16)$$

where we used equation (2.1.13). For the case $F = A_i$ equation (2.1.16) gives

$$\frac{\mathrm{d}}{\mathrm{d}t}\int_{\Sigma_t} A\cdot n\,\mathrm{d}a = \int_{\Sigma_t} \{\dot{A} + A\,\mathrm{div}\,V - (A\cdot\nabla)V\}\cdot n\,\mathrm{d}a \qquad (2.1.16')$$

$$= \int_{\Sigma_t} \{\partial_t A + \mathrm{curl}\,(A \times V) + V\mathrm{div}\,A\}\cdot n\,\mathrm{d}a$$

also using (A.25). From the last equation we have the following criterion due to Zorawski [1], [2]: the necessary and sufficient condition for the flux of a vector A through some material surface to be constant in time is to have

$$\partial_t A + \mathrm{curl}\,(A \times V) + V\mathrm{div}\,A = 0 \qquad (2.1.17)$$

2.2 Equations of conservation of mass and momentum

2.2.1 The mass conservation equation (the continuity equation)

We shall derive in this section the first two equations to be satisfied by a fluid in motion. These equations follow from the principles of conservation of mass and momentum. Let us take into consideration those fluid particles which at the initial time are situated within the domain D_0. At time t, these particles will occupy the domain D obtained from D_0 through the transformation (2.1.1). By writing that the mass of these particles remains constant, we have:

$$\frac{\mathrm{d}}{\mathrm{d}t}\int_{D} \rho(t, x)\,\mathrm{d}v = 0 \qquad (2.2.1)$$

$\rho\ (t, \boldsymbol{x})$ being the specific mass (a positive numerical function). If the motion is *continuous*, one may use equation (2.1.15). Since D is arbitrary, it follows that

$$\partial_t \rho + \operatorname{div}(\rho \boldsymbol{V}) = 0 \tag{2.2.2}$$

which represents the differential form of the equation of mass conservation (the same as in classical fluid dynamics).

Equation (2.2.2) may be written as follows:

$$\rho \frac{\mathrm{d}}{\mathrm{d}t}\left(\frac{1}{\rho}\right) = \operatorname{div} \boldsymbol{V} \tag{2.2.2$'$}$$

Multiplying equation (2.2.2′) by J and taking into account equation (2.1.11) it follows that:

$$\frac{\mathrm{d}}{\mathrm{d}t}(\rho J) = 0 \Rightarrow \rho J = \rho^0 \tag{2.2.2$''$}$$

Equation (2.2.1) proves that mass is an integral invariant of the transformation (2.1.1).

2.2.2 The momentum conservation equation

In order to derive the equation of momentum conservation we shall notice that the actions which determine the change of momentum are of mechanical and electromagnetic nature. The actions of mechanical nature are divided into mass actions (long range actions) and surface actions (contact actions). We denote by $\boldsymbol{F}$ the mechanical action per unit mass and by $\boldsymbol{T}_n$ the action on unit area (of normal $\boldsymbol{n}$). From the stress theory (e.g. [G. 28], [G. 29]) it follows that $\boldsymbol{T} = \boldsymbol{n}.\underline{\boldsymbol{T}}$, $\underline{\boldsymbol{T}}$ being the mechanical stress tensor. The actions of electromagnetic nature are long range interactions and are represented by $\boldsymbol{F}^{(em)}$ (force per unit volume) defined in Section 1.3. Finally, the principle of momentum conservation applied to the fluid mass within D gives

$$\frac{\mathrm{d}}{\mathrm{d}t}\int_D \rho \boldsymbol{V} \mathrm{d}v = \int_D \rho \boldsymbol{F}\, \mathrm{d}v + \int_{\partial D} \boldsymbol{n} \cdot \underline{\boldsymbol{T}} \mathrm{d}a + \int_D \boldsymbol{F}^{(em)}\, \mathrm{d}v \tag{2.2.3}$$

For the case of *continuous* motion we may use equations (2.1.15) and (A.32) in the first and third integral, respectively. Since D is arbitrary it follows:

$$\partial_t(\rho \boldsymbol{V}) + \operatorname{div}(\rho \boldsymbol{V}\boldsymbol{V}) = \rho \boldsymbol{F} + \operatorname{div} \underline{\boldsymbol{T}} + \boldsymbol{F}^{(em)} \tag{2.2.4}$$

which may be written for each component

$$\partial_t(\rho V_i) + \operatorname{div}(\rho V_i \boldsymbol{V}) = \rho F_i + T_{ji,\,j} + F_i^{(em)}, \qquad i = 1, 2, 3 \tag{2.2.4$'$}$$

Using the electromagnetic stress tensor (1.3.5) equation (2.2.4) takes the following conservative form:

$$\partial_t(\rho \boldsymbol{V} + \boldsymbol{g}) + \operatorname{div}(\rho \boldsymbol{V}\boldsymbol{V} - \underline{\boldsymbol{T}}) = \rho \boldsymbol{F} \tag{2.2.5}$$

$$\underline{\boldsymbol{T}} \overset{\text{def}}{=} \underline{\boldsymbol{T}} + \underline{\boldsymbol{T}}^{(\text{em})}$$

Finally, using equation (2.2.2), we get from equation (2.2.4) and (2.2.5)

$$\rho \dot{\boldsymbol{V}} = \rho \boldsymbol{F} + \operatorname{div} \underline{\boldsymbol{T}} + \boldsymbol{F}^{(\text{em})} \tag{2.2.4''}$$

$$\rho \dot{\boldsymbol{V}} + \partial_t \boldsymbol{g} = \operatorname{div} \underline{\boldsymbol{T}} + \rho \boldsymbol{F} \tag{2.2.5'}$$

These equations become explicit only if we add the structural equation (of mechanical nature) of the fluid. For a Newtonian fluid [G. 28], [G. 29] this has the form:

$$\underline{\boldsymbol{T}} = -p\underline{\boldsymbol{U}} + \underline{\boldsymbol{\tau}} \tag{2.2.6}$$

$$\underline{\boldsymbol{\tau}} = \tilde{\lambda}(\operatorname{div} \boldsymbol{V})\underline{\boldsymbol{U}} + 2\tilde{\mu}\underline{\boldsymbol{D}}$$

p being the mechanical pressure, $\underline{\boldsymbol{U}}$ the unit tensor of the second order, $\underline{\boldsymbol{\tau}}$ the viscosity tensor, and $\tilde{\lambda}$ and $\tilde{\mu}$ the viscosity coefficient. Generally we shall admit Stokes' hypothesis ($3\tilde{\lambda} + 2\tilde{\mu} = 0$).

Substituting equation (2.2.6) into equation (2.2.4′) we find

$$\rho \dot{\boldsymbol{V}} = \rho \boldsymbol{F} - \nabla p + \frac{\tilde{\mu}}{3} \nabla \operatorname{div} \boldsymbol{V} + \tilde{\mu} \Delta \boldsymbol{V} + \boldsymbol{F}^{(\text{em})} \tag{2.2.7}$$

In this equation, the electromagnetic force will be expressed in terms of the electromagnetic field according to the equations of Section 1.3. On the other hand, the electromagnetic field is determined by Maxwell's equations (1.1.5)–(1.1.7) and the structural equations (1.2.18), (1.2.19). But structural equations also depend on $\boldsymbol{V}$, so that we have to consider all these equations (the continuity, momentum, Maxwell and structural equations) as making up a system with the following unknowns: ρ, p, $\boldsymbol{V}$, $\boldsymbol{E}$, $\boldsymbol{D}$, $\boldsymbol{H}$, $\boldsymbol{B}$, and $\boldsymbol{J}$. From these equations one may eliminate $\rho^{(q)}$ using equation (1.1.8.). In the case of an incompressible fluid (using (2.1.11′), from (2.2.2′) it follows that $\rho \equiv \rho^0$) the above-mentioned system is sufficient to determine all the unknown quantities. For the compressible fluid, another equation of state and the energy equation (which will be dealt with in the following section) should be added to these equations.

Due to the presence of velocity in the field equations (structural equations), the electromagnetic force in the momentum equation cannot be determined independently of $\boldsymbol{V}$. This fact expresses the interaction between field and motion. For a fluid at rest, this interaction disappears. The electromagnetic force is determined independently of the fluid state. In this way, the mechanical equilibrium condition is reduced to that of the classical fluid equilibrium (in which F is replaced by $F + F^{(em)}$).

2.2.3 Simple media

If the given medium satisfies conditions (1.2.20), then the electromagnetic force is reduced to the Lorentz force. It will be shown in Section 2.4 that the displacement current $\partial_t D$ in the Maxwell equations, the convection current $\rho^{(q)} V$ in Ohm's law and the electric force $\rho^{(q)} E$ in the equations of motion represent negligible terms compared with the other terms. With these simplifying assumptions, equation (2.2.7) becomes (after substituting $J = \text{curl}\, H$):

$$\rho \dot{V} = \rho F - \nabla p + \frac{\tilde{\mu}}{3} \nabla \,\text{div}\, V + \tilde{\mu} \Delta V + \text{curl}\, H \times \mu H \tag{2.2.8}$$

By applying the operator *curl* to Ohm's law written in the form

$$\text{curl}\, H = \sigma(E + V \times \mu H) \tag{2.2.9}$$

and using Maxwell's equations (1.1.7), (1.1.5) we find

$$\partial_t H = \text{curl}\,(V \times H) + \eta \Delta H, \quad \eta \overset{\text{def}}{=} 1/\sigma\mu \tag{2.2.10}$$

This is the *magnetic induction* equation. This equation is important since it contains only the two unknowns V and H. Together with (2.2.8), the continuity equation and equation (1.1.7) make up a system which is sufficient for the solution of the incompressible fluid problem.

Using the identity (A.26′) equation (2.2.8) may also be written as follows:

$$\rho \dot{V} = \rho F - \nabla P + \frac{\tilde{\mu}}{3} \nabla \,\text{div}\, V + \tilde{\mu} \Delta V + (\mu H \cdot \nabla) H \tag{2.2.8′}$$

where

$$P = p + \frac{1}{2} B \cdot H$$

Also, using (A.25), equation (2.2.10) becomes

$$\partial_t H = (H \cdot \nabla) V - (V \cdot \nabla) H - H \,\text{div}\, V + \eta \Delta H \tag{2.2.10′}$$

2.2.4 Boundary (mechanical) conditions

Let us first consider the fluid to be at contact with a fixed or moving rigid surface Σ. If the fluid viscosity is negligible, this will slip along the surface Σ, so that this may be considered a material surface. If $f(t, \boldsymbol{x}) = 0$ is the equation of surface Σ, then, according to the Lagrange criterion, we have

$$\partial_t f + (\boldsymbol{V} \cdot \nabla) f = 0$$

for $f = 0$, whence,

$$\boldsymbol{V} \cdot \boldsymbol{n}\big|_\Sigma = - \frac{\partial_t f}{|\mathrm{grad}\, f|} = \boldsymbol{U} \cdot \boldsymbol{n} \tag{2.2.11}$$

$\boldsymbol{U}$ being the velocity field of the surface (the transport velocity). From equation (2.2.11) it follows that

$$\boldsymbol{v} \cdot \boldsymbol{n}\big|_\Sigma = 0, \quad \boldsymbol{v} \overset{\text{def}}{=} \boldsymbol{V} - \boldsymbol{U} \tag{2.2.12}$$

$\boldsymbol{v}$ being the relative (with respect to the surface) velocity of the fluid. If surface Σ is fixed, we have obviously:

$$\boldsymbol{V} \cdot \boldsymbol{n}\big|_\Sigma = 0 \tag{2.2.13}$$

If the fluid is viscous, it will adhere to the wall, so that

$$\boldsymbol{v}\big|_\Sigma = 0 \tag{2.2.14}$$

For a fixed wall, we have

$$\boldsymbol{V}\big|_\Sigma = 0 \tag{2.2.15}$$

For the free surfaces of the fluid or for the separation surfaces of two fluids, the condition of continuity of the stress vector is fulfilled (according to Lamb). For inviscid fluids this condition is reduced to the continuity of pressure. To this, the condition that the respective surface is a material surface (the kinematic condition) is added.

The condition to be satisfied by the electromagnetic field are given in Section 1.5.

2.3 The energy equation

2.3.1 Elements of thermodynamics of conducting fluids

While the first equations of magnetofluid dynamics had already been defined during 1950—1955 [3], the energy equation has been satisfactorily derived by Pai only in 1957 [4]. However, as noticed by Goldstein [5], and especially by Boa Theh Chy

[6], the equation derived by Pai is only valid for constant ε and μ. Although this is the case of interest in the problems to be studied in the later chapters, we shall give here (following Chy) the general form of the energy equation, this being particularly important from a theoretical point of view.

To this end, we shall first estimate the rate of work done by the external forces on the mass of the fluid. For the fluid occupying a domain D, we have [6]:

$$\frac{\mathrm{d}\mathscr{L}}{\mathrm{d}t} = \int_D \rho F_i V_i \,\mathrm{d}v + \int_D n_j \mathscr{T}_{ji} V_i \,\mathrm{d}a = \int_D \{\rho F_i V_i + (\mathscr{T}_{ji} V_i),_j\} \,\mathrm{d}v \tag{2.3.1}$$

Noticing that

$$\mathscr{T}_{ji} = -P\delta_{ji} + \tau_{ji} + D_j E_i + B_j H_i, \quad P \overset{\mathrm{def}}{=} p + p_{\mathrm{em}} \tag{2.3.2}$$

and taking into account equations (2.2.5′) and (2.2.2′), we have

$$\begin{aligned} &\rho F_i V_i + (T_{ji} V_i),_j \\ &= (\rho \dot{V}_i + \partial_t g_i) V_i - P V_{i,i} + \Phi + (E_i D_j + H_i B_j) V_{i,j} \\ &= \rho \frac{\mathrm{d}}{\mathrm{d}t}\left(\frac{1}{2} V^2\right) - \rho P \frac{\mathrm{d}}{\mathrm{d}t}\left(\frac{1}{\rho}\right) + \Phi + \boldsymbol{V} \cdot \partial_t \boldsymbol{g} + \boldsymbol{E} \cdot (\boldsymbol{D} \cdot \nabla)\boldsymbol{V} + \boldsymbol{H} \cdot (\boldsymbol{B} \cdot \nabla)\boldsymbol{V} \end{aligned} \tag{2.3.3}$$

where

$$\Phi = \tau_{ji} V_{i,j} = \tau_{ji} D_{ij} = \underline{\boldsymbol{\tau}} : \underline{\boldsymbol{D}} \tag{2.3.4}$$

Using equation (2.2.6), it follows that Φ is a positively definite form

$$\Phi = 2\tilde{\mu} D_{ij} \{D_{ij} - \frac{1}{3}(\operatorname{div} \boldsymbol{V})\delta_{ij}\}$$

$$= 4\tilde{\mu}\{(D_{12}^2 + D_{23}^2 + D_{31}^2) + \frac{1}{6}((D_{11} - D_{22})^2 + (D_{22} - D_{33})^2 + (D_{33} - D_{11})^2)\} \geqslant 0$$

Noticing now that, on account of the continuity equation, we have

$$\int_D \rho \frac{\mathrm{d}}{\mathrm{d}t}\left(\frac{1}{2} V^2\right) \mathrm{d}v = \frac{\mathrm{d}}{\mathrm{d}t} \int_D \frac{1}{2} \rho V^2 \,\mathrm{d}v, \tag{2.3.5}$$

and taking into account the identity [6]

$$\boldsymbol{V} \cdot \partial_t \boldsymbol{g} + \boldsymbol{E} \cdot (\boldsymbol{D} \cdot \nabla)\boldsymbol{V} + \boldsymbol{H} \cdot (\boldsymbol{B} \cdot \nabla)\boldsymbol{V}$$

$$= \operatorname{div}(\boldsymbol{E}' \times \boldsymbol{H}') + \boldsymbol{J}' \cdot \boldsymbol{E}' + \rho \left\{ \boldsymbol{E} \cdot \frac{\mathrm{d}}{\mathrm{d}t}\left(\frac{\boldsymbol{D}}{\rho}\right) + \boldsymbol{H} \cdot \frac{\mathrm{d}}{\mathrm{d}t}\left(\frac{\boldsymbol{B}}{\rho}\right) \right\}$$

it follows that

$$\frac{\mathrm{d}\mathscr{L}}{\mathrm{d}t} = \frac{\mathrm{d}}{\mathrm{d}t}\int_D \frac{1}{2}\rho V^2\,\mathrm{d}v + \int_D \Phi\,\mathrm{d}v + \int_{\partial D} \boldsymbol{E}' \times \boldsymbol{H}' \cdot \boldsymbol{n}\,\mathrm{d}a + \int_D \boldsymbol{J}' \cdot \boldsymbol{E}'\,\mathrm{d}v$$

$$+ \int_D \left\{ -P\frac{\mathrm{d}}{\mathrm{d}t}\left(\frac{1}{\rho}\right) + \boldsymbol{H}\cdot\frac{\mathrm{d}}{\mathrm{d}t}\left(\frac{\boldsymbol{B}}{\rho}\right) + \boldsymbol{E}\cdot\frac{\mathrm{d}}{\mathrm{d}t}\left(\frac{\boldsymbol{D}}{\rho}\right)\right\}\mathrm{d}v \qquad (2.3.6)$$

This equation is interpreted as follows: the work done on the fluid by external forces is transformed into kinetic energy of the fluid, irreversible work performed in viscous dissipation, energy radiated away from the fluid, irreversible work done against the fluid electrical resistivity (Joule loss), reversible work performed in setting up the magnetohydrodynamic field (the last term).

In order to use thermodynamic principles, we must consider a differential fluid element in the quasi-static state. For this, equation (2.3.6) yields the following relationship (for the unit mass of the fluid):

$$\mathrm{d}\mathscr{L} + \mathrm{d}Q_r = \mathrm{d}\mathscr{L}_{ir} + \mathrm{d}\mathscr{L}_r \qquad (2.3.7)$$

$\mathrm{d}\mathscr{L}$ being the work done on the fluid element in time $\mathrm{d}t$ during a small change, $\mathrm{d}Q_r = -(1/\rho)\,\mathrm{div}\,(\boldsymbol{E}' \times \boldsymbol{H}')\,\mathrm{d}t$ the energy radiated into the element, $\mathrm{d}\mathscr{L}_{ir} = (1/\rho)(\Phi + \boldsymbol{J}' \cdot \boldsymbol{E}')\mathrm{d}t$ the energy dissipated into heat by viscosity and electrical resistivity, and $\mathrm{d}\mathscr{L}_r = -P\mathrm{d}(1/\rho) + \boldsymbol{E}\cdot\mathrm{d}(\boldsymbol{D}/\rho) + \boldsymbol{H}\cdot\mathrm{d}(\boldsymbol{B}/\rho)$ the reversible work done on the fluid; in a quasi-static transformation the change in the kinetic energy is zero.

Equation (2.3.7) expresses the fact that the energy input from mechanical work and radiation is equal to the energy spent in dissipation and in freezing-in the magnetic field, and it represents the basic relationship of the thermodynamics of electrically conducting fluids. Indeed, according to the first law of thermodynamics, the change in the total internal energy $\mathrm{d}\mathscr{U}$ for a transition from one state to another is equal to the sum of the heat received $\mathrm{d}Q$, the mechanical work done on the element and the internally radiated energy:

$$\mathrm{d}\mathscr{U} = \mathrm{d}Q + (\mathrm{d}\mathscr{L} + \mathrm{d}Q_r) \qquad (2.3.8)$$

But $\mathrm{d}Q + \mathrm{d}Q_r$ is the total heat received by the fluid element from the external source and the internal dissipations. Therefore, using the second law of thermodynamics:

$$\mathrm{d}Q + \mathrm{d}\mathscr{L}_{ir} = T\mathrm{d}\mathscr{S} \qquad (2.3.9)$$

$\mathscr{S}$ being the total entropy per unit mass and T the absolute temperature. From (2.3.7), (2.3.8) and (2.3.9) one obtains the fundamental equation of the thermodynamics of conducting fluids

$$T\mathrm{d}\mathscr{S} = \mathrm{d}\mathscr{U} + P\,\mathrm{d}\left(\frac{1}{\rho}\right) - \boldsymbol{E}\cdot\mathrm{d}\left(\frac{\boldsymbol{D}}{\rho}\right) - \boldsymbol{H}\cdot\mathrm{d}\left(\frac{\boldsymbol{B}}{\rho}\right) \qquad (2.3.10)$$

2.3.2 The perfect gas

The perfect gas is defined by the following *equation of state*:

$$p = \rho R T \tag{2.3.11}$$

and the following *caloric equation*:

$$U = c_v(T - T_0) + U_0 \tag{2.3.12}$$

where R is the universal gas constant (its value being experimentally determined as 8.3 Jdeg^{-1} mol^{-1}, and c_v the specific heat per unit mass at constant volume. These equations were experimentally determined through the works of Boyle-Mariotte and Gay-Lussac. They follow from the kinetic theory (Section (13.3)) through the fundamental hypothesis of the microscopic motion.

In the absence of the electromagnetic field, the fundamental thermodynamic equation (2.3.10) becomes:

$$T\mathrm{d}S = \mathrm{d}U + p\,\mathrm{d}\left(\frac{1}{\rho}\right) \tag{2.3.13}$$

S being the mechanical entropy and U the internal energy of mechanical nature. From (2.3.11), (2.3.12) and (2.3.13) it follows that

$$\mathrm{d}S = c_v\frac{\mathrm{d}T}{T} - R\frac{\mathrm{d}\rho}{\rho} \tag{2.3.14}$$

and then

$$S = S_0 - R\ln\left(\frac{\rho}{\rho_0}\right)\left(\frac{T_0}{T}\right)^{c_v/R} \tag{2.3.15}$$

Denoting by c_p the specific heat per unit mass at constant pressure we have $c_p = c_v + R$, whence

$$\frac{c_v}{R} = \frac{1}{\gamma - 1}, \quad \gamma \overset{\text{def}}{=} \frac{c_p}{c_v} \tag{2.3.15$'$}$$

Introducing the free energy $F(F \overset{\text{def}}{=} U - TS)$ we have

$$F = F_0 - \left(S_0 - \frac{R}{\gamma - 1}\right)(T - T_0) + RT\ln\left(\frac{\rho}{\rho_0}\right)\left(\frac{T_0}{T}\right)^{\frac{1}{\gamma - 1}} \tag{2.3.16}$$

F_0 being the internal energy in the reference state $(T = T_0, \rho = \rho_0)$.

As a matter of fact, if one uses p and ρ as independent variables (instead of T and ρ as in (2.3.15)) from (2.3.14) it follows:

$$S = S_0 + c_v\ln\left(\frac{p}{p_0}\right)\left(\frac{\rho_0}{\rho}\right)^{\gamma} = c_v\ln\frac{p}{\rho^{\gamma}} + C \tag{2.3.17}$$

C being a constant. Taking ρ and S as independent variables then from equation (2.3.17) we have

$$p = p(\rho, S) = p_0\left(\frac{\rho}{\rho_0}\right)^{\gamma} \exp\left(\frac{S - S_0}{c_v}\right) \tag{2.3.18}$$

Similarly, we get $\rho = \rho(p,S)$.

A characteristic quantity is the sound velocity. This is denoted by a and is defined thermodynamically by the formula

$$a = \sqrt{\left(\frac{\mathrm{d}p}{\mathrm{d}\rho}\right)_{S=\text{const.}}} \tag{2.3.19}$$

From the second law of thermodynamics, it follows that in the case of *non-dissipating fluids* the isentropic motions ($\mathrm{d}S = 0$) coincide with the adiabatic ones ($\mathrm{d}Q = 0$). But for adiabatic motions it follows from the first law

$$\mathrm{d}U = -p\,\mathrm{d}\left(\frac{1}{\rho}\right) \tag{2.3.20}$$

Using the caloric equation (2.3.12), equation (2.3.20) becomes

$$c_v \mathrm{d}T|_{S=\text{const}} = p\,\mathrm{d}\rho/\rho^2 \tag{2.3.21}$$

Finally, from the equation of state (2.3.11), it follows that

$$\mathrm{d}p|_{S=\text{const}} = RT\mathrm{d}\rho + \rho R\,\mathrm{d}T|_{S=\text{const}} = \gamma RT\mathrm{d}\rho \tag{2.3.22}$$

To conclude

$$a^2 = \gamma RT = \gamma\frac{p}{\rho} \tag{2.3.23}$$

One last remark refers to the adiabatic motions with simple relationships between temperature and density (on the one hand) and pressure and density (on the other hand). Indeed, from equation (2.3.21), it follows that

$$T/T_0 = (\rho/\rho_0)^{\gamma-1} \tag{2.3.24}$$

and from the equation of state

$$p/p_0 = (\rho/\rho_0)^{\gamma} \tag{2.3.25}$$

Equation (2.3.25) is used for the adiabatic and reversible processes (slow transformations of non-dissipating fluids).

2.3.3 The (electrically-conducting) perfect gas

If the gas is also electrically conducting, we shall denote by $\mathscr{F}$ its free energy ($\mathscr{F} \overset{\text{def}}{=} U - TS$). Using equation (2.3.10), one obtains

$$d\mathscr{F} = -\mathscr{S}\,dT - P\,d\left(\frac{1}{\rho}\right) + \boldsymbol{E}\cdot d\left(\frac{\boldsymbol{D}}{\rho}\right) + \boldsymbol{H}\cdot d\left(\frac{\boldsymbol{B}}{\rho}\right) \qquad (2.3.26)$$

$\mathscr{F}$ being a state function. This equation shows that the thermodynamic state of the system is characterized by the function $\mathscr{F} = \mathscr{F}(\rho, T, \boldsymbol{D}, \boldsymbol{B})$. Since

$$d\mathscr{F} = \frac{\partial \mathscr{F}}{\partial \rho}\,d\rho + \frac{\partial \mathscr{F}}{\partial T}\,dT + \frac{\partial \mathscr{F}}{\partial D_i}\,dD_i + \frac{\partial \mathscr{F}}{\partial B_i}\,dB_i$$

we obtain

$$\mathscr{S} = -\frac{\partial \mathscr{F}}{\partial \rho}, \quad P = \boldsymbol{D}\cdot\boldsymbol{E} + \boldsymbol{B}\cdot\boldsymbol{H} + \rho^2\frac{\partial \mathscr{F}}{\partial \rho} \qquad (2.3.27)$$

$$E_i = \rho\frac{\partial \mathscr{F}}{\partial D_i}, \quad H_i = \rho\frac{\partial \mathscr{F}}{\partial B_i} \qquad (2.3.28)$$

A function from which the thermodynamic state of a fluid is deduced is called a *characteristic function* The free energy $\mathscr{F}$ is such a characteristic function.

If the gas is characterized by the structural equations

$$\boldsymbol{D} = \varepsilon(\rho, T)\boldsymbol{E}, \qquad \boldsymbol{B} = \mu(\rho, T)\boldsymbol{H} \qquad (2.3.29)$$

from equation (2.3.28) it follows that

$$\mathscr{F} = F(\rho, T) + \frac{1}{2\rho}\left(\frac{D^2}{\varepsilon} + \frac{B^2}{\mu}\right) \qquad (2.3.30)$$

where F is the mechanical free energy, which for perfect gases has the form (2.3.16). Once the free energy of the system is defined, the state functions of the fluid follow from equation (2.3.27)

$$\mathscr{S} = S + S_{\text{em}} = S + \frac{1}{2\rho}\left(E^2\frac{\partial \varepsilon}{\partial T} + H^2\frac{\partial \mu}{\partial T}\right) \qquad (2.3.31)$$

$$P = p + p_{\text{em}} = \rho RT + \frac{1}{2}(\boldsymbol{D}\cdot\boldsymbol{E} + \boldsymbol{B}\cdot\boldsymbol{H}) - \frac{\rho}{2}\left(E^2\frac{\partial \varepsilon}{\partial \rho} + H^2\frac{\partial \mu}{\partial \rho}\right) \qquad (2.3.32)$$

and from $\mathscr{U} = \mathscr{F} + T\mathscr{S}$

$$\mathscr{U} = U + U_{\text{em}} = U + \frac{1}{2\rho}\left\{\frac{D^2}{\varepsilon}\left(1 + \frac{T}{\varepsilon}\frac{\partial \varepsilon}{\partial T}\right) + \frac{B^2}{\mu}\left(1 + \frac{T}{\mu}\frac{\partial \mu}{\partial T}\right)\right\} \tag{2.3.33}$$

Finally let us define the *specific enthalpy* $\mathscr{I}$ by the formula

$$\mathscr{I} = \mathscr{U} + \frac{P}{\rho} = \left(U + \frac{p}{\rho}\right) + \left(U_{\text{em}} + \frac{p_{\text{em}}}{\rho}\right) = I + I_{\text{em}} \tag{2.3.34}$$

Then we get the following equation

$$I = I_0 + \frac{\gamma R}{\gamma - 1}(T - T_0) = I_0 + c_p(T - T_0)$$

(2.3.34′)

$$I_{\text{em}} = \frac{1}{\rho}\left\{\frac{D^2}{\varepsilon}\left(1 + \frac{T}{2\varepsilon}\frac{\partial \varepsilon}{\partial T} - \frac{\rho}{2\varepsilon}\frac{\partial \varepsilon}{\partial \rho}\right) + \frac{B^2}{\mu}\left(1 + \frac{T}{2\mu}\frac{\partial \mu}{\partial T} - \frac{\rho}{2\mu}\frac{\partial \mu}{\partial \rho}\right)\right\}$$

From equation (2.3.31) it follows that, if ε and μ are constant, the gas entropy is reduced to the mechanical entropy.

From equation (2.3.32), it follows that the pressure of a perfect gas is made up of three terms: the mechanical pressure ρRT, the electromagnetic pressure $(1/2)\,(\boldsymbol{D}\cdot\boldsymbol{E}+\boldsymbol{B}\cdot\boldsymbol{H})$ expressed by means of the factor δ_{ij} in the Maxwell stress tensor and finally (the last term) the pressure due to the magnetostriction effects (this term is also zero for constant ε and μ).

2.3.4 The energy equation

Let us again consider the fluid contained within the domain D. Its total energy is

$$\int_D \left(\frac{1}{2}V^2 + \mathscr{U}\right)\rho\, \mathrm{d}v$$

Obviously, the change in the total energy comes from the work done by mass and surface stress, from the internally radiated electromagnetic energy and from the energy due to the heat flow from outside the fluid. Denoting as usually by $\boldsymbol{n}$ the external normal to ∂D, we have

$$\frac{\mathrm{d}}{\mathrm{d}t}\int_D \left(\frac{1}{2}V^2 + \mathscr{U}\right)\rho\, \mathrm{d}v = \int_D \rho \boldsymbol{F}\cdot \boldsymbol{V}\mathrm{d}v + \int_{\partial D} (\boldsymbol{n}\cdot \underline{T})\cdot \boldsymbol{V}\mathrm{d}a$$

$$- \int_{\partial D} \boldsymbol{E}' \times \boldsymbol{H}'\cdot \boldsymbol{n}\,\mathrm{d}a - \int_{\partial D} \boldsymbol{q}\cdot \boldsymbol{n}\,\mathrm{d}a \tag{2.3.35}$$

For continuous motions, one may use equation (2.3.6). Also taking into account the continuity equation we have

$$\frac{\mathrm{d}}{\mathrm{d}t}\int_D \rho\,\mathscr{U}\,\mathrm{d}v = \int_D \rho\,\frac{\mathrm{d}\mathscr{U}}{\mathrm{d}t}\,\mathrm{d}v$$

so that (2.3.33) is reduced to (D being arbitrary)

$$\rho\dot{\mathscr{U}} = \Phi + \boldsymbol{J}'\cdot\boldsymbol{E}' + \rho\left\{-P\frac{\mathrm{d}}{\mathrm{d}t}\left(\frac{1}{\rho}\right) + \boldsymbol{E}\cdot\frac{\mathrm{d}}{\mathrm{d}t}\left(\frac{\boldsymbol{D}}{\rho}\right) + \boldsymbol{H}\cdot\frac{\mathrm{d}}{\mathrm{d}t}\left(\frac{\boldsymbol{B}}{\rho}\right)\right\} - \nabla\cdot\boldsymbol{q} \tag{2.3.36}$$

Taking into account also the fundamental equation (2.3.10), equation (2.3.36) reduces to the following

$$\rho T\dot{\mathscr{S}} = \Phi + \boldsymbol{J}'\cdot\boldsymbol{E}' - \nabla\cdot\boldsymbol{q} \tag{2.3.37}$$

This is the differential form of the energy equation. The Fourier hypothesis is used for the heat flow vector $\boldsymbol{q}$

$$\boldsymbol{q} = -\varkappa\nabla T \tag{2.3.38}$$

where $\varkappa$ is a coefficient characterizing the thermal conductivity of the fluid. Using Ohm's law ($\boldsymbol{J}' = \sigma\boldsymbol{E}'$) from equation (2.3.37) we obtain

$$\rho T\dot{\mathscr{S}} = \Phi + (1/\sigma)J'^2 + \nabla\cdot(\varkappa\nabla T),\quad \boldsymbol{J}' = \boldsymbol{J} - \rho^{(q)}\boldsymbol{V} \tag{2.3.39}$$

This equation is different from the energy equation for the non-conducting fluids due to the presence of the electromagnetic entropy in $\mathscr{S}$ as well as the presence of the term $(1/\sigma)J'^2$ due to the electrical resistivity of the fluid (Joule loss). For an ionized gas, the dependence of ε and μ with the temperature may be neglected (due to the very high temperature of the gas), so that $\mathscr{S} = S$.

The term $(1/\sigma)\,J'^2$ vanishes in the case of perfectly conducting fluids, as well as in those problems whose equations may be linearized.

We shall see in the following section that the convection current $\rho^{(q)}\boldsymbol{V}$ and the displacement current in the Maxwell equations are of the order of V_0^2/c^2 (i.e., of the order of the relativistic correction), V_0 being a characteristic velocity of the motion. Neglecting these currents in equation (2.3.39) we may use to a good approximation $\boldsymbol{J}' \approx \boldsymbol{J} \approx \operatorname{curl}\boldsymbol{H}$.

The following theorem is a result of the above remarks [6.7], [6.15]: *the adiabatic motion of an inviscid electrically conducting fluid is isentropic if one of the following conditions is valid:* (i) *the fluid is perfectly conducting;* (ii) *one may use linearized equations;* (iii) *there exists a potential A such that $\boldsymbol{H} = \operatorname{grad} A$.*

To conclude, let us now consider the energy equation for a simple medium (ε and μ constant, $\varepsilon\mu \approx \varepsilon_0\mu_0$). Taking into account equations (2.3.31)–(2.3.33), as well as the continuity equation (2.2.2′) the fundamental equation (2.3.10) becomes

$$T\mathrm{d}\mathscr{S} = \mathrm{d}U + p\,\mathrm{d}\left(\frac{1}{\rho}\right) = \mathrm{d}I - \frac{\mathrm{d}p}{\rho} = \mathrm{d}U + \frac{p}{\rho}\,\mathrm{div}\,\boldsymbol{V}, \quad (\mathrm{d}\mathscr{S} = \mathrm{d}S) \tag{2.3.40}$$

Therefore, the energy equation is reduced to one of the following forms

$$\rho\dot{I} = \dot{p} + \Phi + (1/\sigma)J'^2 + \nabla\cdot(\varkappa\nabla T) \tag{2.3.41}$$

$$\rho\dot{U} = -\,p\nabla\cdot\boldsymbol{V} + \Phi + (1/\sigma)J'^2 + \nabla\cdot(\varkappa\nabla T) \tag{2.3.42}$$

for the compressible fluids and to

$$\rho\dot{U} = \Phi + (1/\sigma)J'^2 + \nabla\cdot(\varkappa\nabla T) \tag{2.3.43}$$

for the incompressible fluids. If we use the fact that for a perfect gas both the enthalpy I and the internal energy U are expressed as functions of T, then equations (2.3.41)–(2.3.43) can be used to determine the temperature T. These equations are different from the corresponding ones in classical fluid mechanics only by the presence of the term $(1/\sigma)\,J'^2$.

In order to determine the temperature T, we also need the temperature T or the heat flow on the frontier $\partial\mathscr{D}$ bounding the fluid. Particularly, if the frontier is thermally insulating then the heat flow vanishes:

$$\boldsymbol{q}\cdot\boldsymbol{n}\Big|_{\partial\mathscr{D}} = -\,\varkappa\frac{\mathrm{d}T}{\mathrm{d}n}\Big|_{\partial\mathscr{D}} = 0 \tag{2.3.44}$$

Also, for the solution of an initial-conditions problem we should obviously know the initial temperature of the fluid.

2.4 Magnetofluid dynamics equations in dimensionless variables

The equations of magnetofluid dynamics contain various *characteristic parameters* introduced by the structural equations. These are of mechanical ($\tilde{\mu}$), thermodynamic (c_v, c_p, $\varkappa$) and electromagnetic (ε, μ, σ) nature. The external conditions in which the process takes place also introduce some characteristic quantities, e.g. a velocity V_0 (when the fluid has some given motion and one has to study this motion in other conditions), a magnetic induction $\boldsymbol{B}_0$ (the external induction in the enviroment in which the fluid motion takes place), a characteristic length L_0, etc. Only some combinations of these parameters (not all of them) are important. To point out these combinations we shall write the system of equations in dimensionless variables. The theoretical foundation of this process is the Pi-theorem [7].

Let us suppose we have to find the solution of a problem in which a characteristic length L_0, a density ρ_0, a velocity V_0, an induction B_0 and a temperature T_0 are given. Taking these as basic quantities one can easily check up (taking into account the analysis of Section 1.4) that the fundamental quantities (length, mass, time, charge and temperature) become derived units. To conclude, all the variables of magnetofluid dynamics may be referred to L_0, ρ_0, V_0, B_0 and T_0. Let us now suppose that the fluid is a simple medium which obeys the perfect gas law equations (2.3.11) and (2.3.12). Indicating by asterisks the dimensionless variables, these are defined in the following way

$$x_i = L_0 x_i^* \ (i = 1, 2, 3), \ V_0 t = L_0 t^*$$

$$\rho = \rho_0 \rho^*, \ p - p_0 = \rho_0 V_0^2 p \ , \ \boldsymbol{V} = V_0 \boldsymbol{V}^*, \ T = T_0 T^*, \ L_0 \boldsymbol{F} = V_0^2 \boldsymbol{F}^*$$

$$\boldsymbol{B} = B_0 \boldsymbol{B}^*, \ \boldsymbol{E} = V_0 B_0 \boldsymbol{E}^*, \ \mu L_0 \boldsymbol{J} = B_0 \boldsymbol{J}^*, \ \mu L_0 V_0 \rho^{(q)} = B_0 \rho^{(q)*} \tag{2.4.1}$$

assuming that the characteristic parameters remain constant. Using the analysis of Section 1.4, one may check up that the asterisked variables are abstract numbers. With these variables the magnetofluid dynamic system is written as follows:

$$\partial_{t*} \rho^* + \operatorname{div}_* (\rho^* \boldsymbol{V}^*) = 0 \tag{2.4.2}$$

$$\rho^* \frac{d\boldsymbol{V}^*}{dt^*} = \rho^* \boldsymbol{F}^* - \nabla_* p^* + \frac{1}{Re} \left(\frac{1}{3} \nabla_* \operatorname{div}_* \boldsymbol{V}^* + \Delta_* \boldsymbol{V}^* \right) \tag{2.4.3}$$

$$+ Rh(\rho^{(q)*} \boldsymbol{E}^* + \boldsymbol{J}^* \times \boldsymbol{B}^*)$$

$$\operatorname{curl}_* \boldsymbol{E}^* = - \partial_{t*} \boldsymbol{B}^*, \tag{2.4.4}$$

$$\operatorname{div}_* \boldsymbol{B}^* = 0 \tag{2.4.5}$$

$$\operatorname{curl}_* \boldsymbol{B}^* = \boldsymbol{J}^* + Rc \partial_{t*} \boldsymbol{E}^*, \tag{2.4.6}$$

$$Rc \operatorname{div}_* \boldsymbol{E}^* = \rho^{(q)*} \tag{2.4.7}$$

$$\boldsymbol{J}^* = Rm(\boldsymbol{E}^* + \boldsymbol{V}^* \times \boldsymbol{B}^*) + \rho^{(q)*} \boldsymbol{V}^* \tag{2.4.8}$$

the last equation being the form of Ohm's law.

The equation of state (2.3.11) becomes

$$1 + \gamma M^2 p^* = \rho^* T^* \tag{2.4.9}$$

and the energy equation (2.3.41)

$$\rho^* \frac{dT^*}{dt^*} = (\gamma - 1)M^2 \left(\frac{dp^*}{dt^*} + \frac{1}{Re} \tau_{ij}^* \frac{\partial V_j^*}{\partial x_i^*} + \frac{Rh}{Rm} J'^{*2} \right) + \frac{1}{RePr} \Delta_* T^* \tag{2.4.10}$$

where

$$\tau_{ij}^* = -\frac{2}{3} \frac{\partial V_k^*}{\partial x_k^*} \delta_{ij} + \frac{\partial V_i^*}{\partial x_j^*} + \frac{\partial V_j^*}{\partial x_i^*} \tag{2.4.10'}$$

The following notations have been used:

$$Re = \frac{L_0 V_0}{\nu_0}, \qquad \nu_0 = \frac{\tilde{\mu}}{\rho_0} \tag{2.4.11}$$

where Re is Reynolds' number (known from classical fluid mechanics);

$$Rm = \frac{L_0 V_0}{\eta} (= \sigma \mu L_0 V_0), \quad \eta = \frac{1}{\sigma \mu} \tag{2.4.12}$$

Rm being the magnetic Reynolds' number (due to the analogy with Re);

$$Rh = \frac{V_A^2}{V_0^2} \left(= \frac{B_0^2}{\rho_0 \mu V_0^2} \right) \stackrel{\text{def}}{=} \frac{1}{A^2}, \qquad V_A = \frac{B}{\sqrt{\mu \rho_0}} \tag{2.4.13}$$

where V_A is the Alfvén velocity, A Alfvén's number and Rh the magnetic pressure number;

$$M^2 = \frac{V_0^2}{a_0^2} \left(= \frac{V_0^2}{\gamma R T_0} \right), \qquad \gamma = \frac{c_p}{c_v} \tag{2.4.14}$$

M being Mach's number (in the reference motion);

$$Pr = \frac{c_p \tilde{\mu}}{\varkappa} \tag{2.4.15}$$

and Pr Prandtl's number.

In the above equations the number

$$Rc = \frac{V_0^2}{c_m^2}, \qquad c_m = \frac{1}{\sqrt{\varepsilon\mu}} \tag{2.4.16}$$

appears, where c_m is the light velocity in the fluid equation (2.1.1). Since we assume that the medium obey the condition $\varepsilon\mu \simeq \varepsilon_0\mu_0$, it follows that this velocity is approximately equal to the light velocity in vacuum and thus Rc is very small (of the order of the relativistic correction). This remark justifies the neglect of the displacement current in equation (2.4.6) (the last term). The neglect of the terms $\rho^{(q)*}\boldsymbol{E}^*$ in equation (2.4.3) and $\rho^{(q)*}\boldsymbol{V}^*$ in equation (2.4.8) may be also justified taking into account equation (2.4.7).

To conclude, we shall also write down in dimensionless form the expression of the total tensor (2.2.5). We get

$$\mathscr{T}_{ij} = \rho_0 V_0^2 \mathscr{T}_{ij}^* \tag{2.4.17}$$

where

$$\mathscr{T}_{ij}^* = -\left\{\frac{p_0}{\rho_0 V_0^2} + p^* + \frac{1}{2} Rh(RcE^{*2} + B^{*2})\right\}\delta_{ij} + \frac{1}{Re}\tau_{ij}^*$$

$$+ Rh(RcE_i^*E_j^* + B_i^*B_j^*)$$

It is obvious that the terms in the expression of the electromagnetic stress tensor containing the electric field are also of the order of the relativistic correction and thus negligible.

Other considerations on this matter may be found in [8], [9] and [G. 12].

2.5 The conservation theorems

2.5.1 The equation of motion

Let us first consider the equation of motion in the form

$$\partial_t \boldsymbol{V} + (\boldsymbol{V}\cdot\nabla)\boldsymbol{V} = \boldsymbol{F} - \frac{1}{\rho}\operatorname{grad} p + \frac{1}{\rho}\boldsymbol{F}^{(\mathrm{em})} \tag{2.5.1}$$

and suppose the mechanical forces are conservative ($\boldsymbol{F} = \operatorname{grad} \Pi$) and that the fluid is *barotropic**.

---

* A fluid is called barotropic if it is characterized by an equation of state of the form $f(\rho, p) = 0$. The perfect gas in adiabatic motion is such a fluid.

Using the relationship (A.26′), equation (2.5.1) becomes

$$\partial_t V + \operatorname{curl} V \times V = \operatorname{grad}\left(\Pi - \int \frac{dp}{\rho} - \frac{1}{2} V^2 \right) + \frac{1}{\rho} F^{(em)} \tag{2.5.2}$$

Let now Γ_t be a closed material curve. From (2.1.14′) and (2.5.2) it follows that

$$\frac{d}{dt}\int_{\Gamma_t} V \cdot dx = \int_{\Gamma_t} \frac{1}{\rho} F^{(em)} \cdot dx \tag{2.5.3}$$

a relationship between the velocity circulation and the mechanical work done by the electromagnetic force $F^{(em)}$ along the curve Γ_t. This relationship will be used in Section 5.5. In classical hydrodynamics, relationship (2.5.3) represents Thomson's theorem: *the velocity circulation along some material curve is constant* (and equal to its initial value). In this case the circulation is an integral invariant of the transformation (2.1.1). In magnetofluid dynamics this invariant is changed at any instant by the electromagnetic force circulation.

Let us apply the *curl* operator to equation (2.5.2). We obtain:

$$\partial_t \Omega + \operatorname{curl}(\Omega \times V) = \operatorname{curl}\frac{1}{\rho} F^{(em)}, \qquad \Omega \overset{\text{def}}{=} \operatorname{curl} V \tag{2.5.4}$$

If we now consider some material surface Σ_t and use formula (2.1.16′), from equation (2.5.4) it follows:

$$\frac{d}{dt}\int_{\Sigma_t} \Omega \cdot n \, da = \int_{\Sigma_t} \left(\operatorname{curl}\frac{1}{\rho} F^{(em)} \right) \cdot n \, da = \int_{\Gamma_t} \frac{1}{\rho} F^{(em)} \cdot dx \tag{2.5.5}$$

where Γ_t is a material curve bounding Σ_t. The above equation expresses the fact that the vortex flow through a material surface Σ_t is also determined by the electromagnetic force circulation (this result could also be obtained directly from equation (2.5.3). In classical hydrodynamics this equation represents the vortex conservation theorem. However, in magnetofluid dynamics there exist some expressions which are conserved during the motion [10].

Of more interest is however the solution of equation (2.5.4). In order to find this we shall use equation (2.1.25) and the continuity equation to obtain

$$\partial_t \Omega + \operatorname{curl}(\Omega \times V) = \frac{d\Omega}{dt} - (\Omega \cdot \nabla) V - \frac{\Omega}{\rho}\frac{d\rho}{dt}$$

In this way, equation (2.5.4) becomes

$$\frac{\mathrm{d}}{\mathrm{d}t}\left(\frac{\boldsymbol{\Omega}}{\rho}\right)-\left(\frac{\boldsymbol{\Omega}}{\rho}\cdot\nabla\right)\boldsymbol{V}=\frac{1}{\rho}\operatorname{curl}\left(\frac{1}{\rho}\boldsymbol{F}^{(\mathrm{em})}\right) \tag{2.5.4'}$$

Using equations (2.1.10), it follows that

$$\left\{\frac{\mathrm{d}}{\mathrm{d}t}\left(\frac{\Omega_i}{\rho}\right)-\frac{\Omega_l}{\rho}\frac{\partial V_i}{\partial x_l}\right\}\frac{\partial x_k^0}{\partial x_i}=\frac{\mathrm{d}}{\mathrm{d}t}\left(\frac{\Omega_j}{\rho}\frac{\partial x_k^0}{\partial x_j}\right)$$

so that equation (2.5.4′) yields

$$\frac{\Omega_j}{\rho}\frac{\partial x_k^0}{\partial x_j}=\frac{\Omega_k^0}{\rho_0}+\int_0^t\frac{1}{\rho}\left(\operatorname{curl}\frac{1}{\rho}\boldsymbol{F}^{(\mathrm{em})}\right)_i\frac{\partial x_k^0}{\partial x_i}\,\mathrm{d}t \tag{2.5.6}$$

Multiplying equation (2.5.6) by ρ_0 $(= J\rho)$, we finally obtain:

$$J\Omega_l=\left\{\Omega_k^0+\int_0^t\frac{\rho_0}{\rho}\left(\operatorname{curl}\frac{1}{\rho}\boldsymbol{F}^{(\mathrm{em})}\right)_i\frac{\partial x_k^0}{\partial x_i}\,\mathrm{d}t\right\}\frac{\partial x_l}{\partial x_k^0},\quad (l=1,2,3) \tag{2.5.7}$$

This equation expresses the way in which the electromagnetic force contributes to vortex creation. Even if at some given time ($t = 0$) the motion is taking place without vortices, during the subsequent times vortices will be created by the electromagnetic force action. In its absence (i.e. in the case of the classical fluid) from equation (2.5.7), Lagrange's theorem follows

$$J\boldsymbol{\Omega}=(\boldsymbol{\Omega}^0\cdot\nabla_0)\boldsymbol{x},\quad \nabla_0\overset{\mathrm{def}}{=}\left(\frac{\partial}{\partial x_i^0}\right) \tag{2.5.7'}$$

Equation (2.5.7′) expresses the fact that *if the motion of a classical fluid is initially vortexless it will remain so during all subsequent times.* The potential motion hypothesis in classical fluid mechanics is based on this theorem (and the Bernoulli equation).

Since in magnetofluid dynamics this irrotational motion is not conserved, we may not assume that the motion is potential during a finite time interval.

2.5.2 The magnetic induction equation

Let us first consider the magnetic induction equation (2.2.10) for the hypothesis of perfectly conducting fluid

$$\partial_t\boldsymbol{H}=\operatorname{curl}(\boldsymbol{V}\times\boldsymbol{H}). \tag{2.5.8}$$

On account of Zorawski's criterion (2.1.17), equation (2.5.8) expresses the fact that the magnetic induction is frozen in the fluid (all the particles situated at a given time on a fluid line will be displaced, always remaining on the same line). This result *is known as the theorem of frozen-in field lines.* Using the same way as in equation (2.5.4') — (2.5.7) from equation (2.5.8) it follows:

$$JH = (H^0 \cdot \nabla_0)x \tag{2.5.9}$$

This equation expresses the fact that if the fluid particles were at some time $t = 0$ in motion in the absence of the magnetic field, then this property is conserved for the subsequent times. Therefore, *the motion as such, in the absence of an external magnetic field, cannot induce a magnetic field.*

Let us now consider the magnetic induction equation for the (simple medium) fluid with electrical resistivity.

$$\partial_t H = \operatorname{curl}(V \times H) + \eta \Delta H \tag{2.5.10}$$

For media at rest, equation (2.5.10) gives

$$\partial_t H = \eta \Delta H \tag{2.5.10'}$$

This is a *diffusion-like equation*: due to the electrical resistivity, the magnetic field diffuses into the field. Equation (2.5.10') is of the parabolic type, like the heat equation. It is known from the theory of partial differential equations that the time variation of the solution of this equation is described by an exponential damping time $t_0 = L_0^2 \eta^{-1} = L_0^2 \sigma\mu$, L_0 being a characteristic length of the definition domain. For the laboratory-scale conductors (small L_0) the damping time is short (e.g., $t_0 \approx 10$ s for a copper sphere of 1 m diameter). On the other hand, for cosmic-scale conductors these times may be very long. For instance, for the Earth (according to the calculations of Elsasser [11]) it would be of approximately 15.000 years, and for the Sun approximately 16.000 million years. Since this time interval is much larger than the estimated age of the Sun, it follows that the present solar magnetic field comes to a large extent from the initial magnetic field (created at the same time as the Sun). For further remarks on this subject, one may look up [G.2] and [G.13] and the bibliography indicated therein.

The coefficient η is called the diffusion coefficient since its magnitude determines the variation due to diffusion of the magnetic field. We have previously shown that the first term on the right-hand side of equation (2.5.10) represents the field change due to the fluid motion. This term will therefore be called the convection term and the second the diffusion term. The ratio of the magnitudes of these term is equal to Rm. Indeed, using the equations in dimensionless variables (2.4.8), (2.4.6) and (2.4.5) we get

$$\partial_{t*} B^* = \operatorname{curl}_* (V^* \times B^*) + (Rm)^{-1} \Delta_* B^* \tag{2.5.11}$$

Therefore, the magnitude of Rm dictates the relative importance of the convection and diffusion terms. In laboratory experiments, the convection condition $Rm \gg 1$

is rarely satisfied, so that the diffusion effects may not be neglected. On the contrary, on the cosmic scale the condition $Rm \gg 1$ is easily fulfilled. Therefore, the induction equation (2.5.8) may be used to a good approximation in such cases.

2.5.3 The Bernoulli integral

For certain hypotheses (to be specified on the way), a first integral resembling the Bernoulli integral for classical fluids can be obtained from the equations of motion [12]. Let us then consider an inviscid, perfectly conducting fluid in isentropic steady motion and let the following equations

$$(\boldsymbol{V}\cdot\nabla)\rho + \rho \operatorname{div} \boldsymbol{V} = 0 \tag{2.5.12}$$

$$(\rho\boldsymbol{V}\cdot\nabla)\boldsymbol{V} = -\operatorname{grad} p + \operatorname{curl} \boldsymbol{H} \times \boldsymbol{B} \tag{2.5.13}$$

$$\operatorname{div} \boldsymbol{B} = 0 \tag{2.5.14}$$

$$\operatorname{curl}(\boldsymbol{V} \times \boldsymbol{B}) = 0 \tag{2.5.15}$$

$$(\boldsymbol{V}\cdot\nabla)S = 0, \ \rho = \mathscr{P}(p)\mathscr{S}(S) \tag{2.5.16}$$

describe its motion.

Multiplying equation (2.5.13) scalarly by $\boldsymbol{V}$ we find

$$(\rho\boldsymbol{V}\cdot\nabla)\frac{1}{2}V^2 = -\boldsymbol{V}\cdot\operatorname{grad} p + \boldsymbol{B}\cdot(\boldsymbol{V}\times\operatorname{curl}\boldsymbol{H}) \tag{2.5.17}$$

By a simple calculation [12], [G.24] based on equations (2.5.16), it follows that

$$\boldsymbol{V}\cdot\operatorname{grad} p = (\rho\boldsymbol{V}\cdot\nabla)\frac{\mathscr{P}(p)\,\mathscr{H}(p)}{\rho}, \ \mathscr{H}(p) \overset{\text{def}}{=} \int_{p_0}^{p} \frac{d\sigma}{\mathscr{P}(\sigma)} \tag{2.5.18}$$

For the barotropic fluid, the following relationship obviously holds:

$$\boldsymbol{V}\cdot\operatorname{grad} p = (\rho\boldsymbol{V}\cdot\nabla)\int_{p_0}^{p}\frac{dp}{\rho} \tag{2.5.18'}$$

To get a different expression for the last term of equation (2.5.17) we use equation (2.5.15) in the form:

$$(\boldsymbol{V}\cdot\nabla)\boldsymbol{B} - (\boldsymbol{B}\cdot\nabla)\boldsymbol{V} + \boldsymbol{B}\operatorname{div}\boldsymbol{V} = 0$$

and eliminate the term $(\boldsymbol{B}.\nabla)\boldsymbol{V}$ by means of equation (A.26). We get

$$\boldsymbol{V}\times\operatorname{curl}\boldsymbol{B} = \operatorname{grad}(\boldsymbol{V}\cdot\boldsymbol{B}) - 2(\boldsymbol{V}\cdot\nabla)\boldsymbol{B} - \boldsymbol{B}\operatorname{div}\boldsymbol{V} - \boldsymbol{B}\times\operatorname{curl}\boldsymbol{V}$$

Or, by scalar multiplication by $\boldsymbol{B}$ and using equation (2.5.12)

$$\boldsymbol{B}\cdot(\boldsymbol{V}\times \operatorname{curl}\boldsymbol{B}) = \boldsymbol{B}\cdot \operatorname{grad}(\boldsymbol{V}\cdot\boldsymbol{B}) - (\rho\boldsymbol{V}\cdot\nabla)(B^2/\rho) \tag{2.5.19}$$

With the results of equations (2.5.18) and (2.5.19), equation (2.5.17) becomes

$$(\rho\boldsymbol{V}\cdot\nabla)\left\{\frac{1}{2}V^2 + \frac{\mathscr{P}(p)\mathscr{H}(p)}{\rho} + \frac{\boldsymbol{H}\cdot\boldsymbol{B}}{\rho}\right\} = (\boldsymbol{H}\cdot\nabla)(\boldsymbol{V}\cdot\boldsymbol{B}) \tag{2.5.20}$$

which, for the case of the barotropic fluid, has the form

$$(\rho\boldsymbol{V}\cdot\nabla)\left\{\frac{1}{2}V^2 + \int_{p_0}^{p}\frac{\mathrm{d}p}{\rho} + \frac{\boldsymbol{H}\cdot\boldsymbol{B}}{\rho}\right\} = (\boldsymbol{H}\cdot\nabla)(\boldsymbol{V}\cdot\boldsymbol{B}) \tag{2.5.20'}$$

It follows that a first integral of motion (constant along the current lines) is obtained in one of the following cases:

$\boldsymbol{V}\perp\boldsymbol{B}$ for any point within the fluid
$\boldsymbol{V}\cdot\boldsymbol{B}$ = constant, along the magnetic field lines
$\boldsymbol{H} = \lambda\rho\,\boldsymbol{V}$ at any point within the fluid.

The first case has been studied, e.g., in [13] and the last one in [14]—[16] and [G.14].

In the last case, we have

$$\boldsymbol{V}\times\boldsymbol{B} = 0$$

so that equation (2.5.15) is automatically satisfied. Since in Ohm's law the fluid velocity does not appear any more, it follows that the magnetic field is determined independently of the motion. Taking into account the continuity equation from equation (2.5.14), it follows that

$$\operatorname{div}(\lambda\rho\boldsymbol{V}) = (\rho\boldsymbol{V}\cdot\nabla)\lambda = 0,$$

this relationship showing that λ is constant along the current lines. Consequently,

$$(\boldsymbol{H}\cdot\nabla)(\boldsymbol{V}\cdot\boldsymbol{B}) = (\lambda\rho\boldsymbol{V}\cdot\nabla)(\boldsymbol{V}\cdot\boldsymbol{B}) = (\rho\boldsymbol{V}\cdot\nabla)(\lambda\boldsymbol{V}\cdot\boldsymbol{B})$$

so that equation (2.5.20) becomes

$$(\rho\boldsymbol{V}\cdot\nabla)\left\{\frac{1}{2}V^2 + \frac{\mathscr{P}(p)\mathscr{H}(p)}{\rho}\right\} = 0 \tag{2.5.21}$$

and thus

$$\frac{1}{2}V^2 + \frac{\mathscr{P}(p)\mathscr{H}(p)}{\rho} = \frac{1}{2}a^2 \tag{2.5.22}$$

where a is constant along the current lines. Particularly, if the motion is uniform at infinity, then a will be constant everywhere within the fluid.

For the case of a barotropic fluid, equation (2.5.22) becomes:

$$\frac{1}{2}V^2 + \int_{p_0}^{p} \frac{\mathrm{d}p}{\rho} = \frac{1}{2}a^2 \tag{2.5.22'}$$

It is to be noticed that in this case the first integral (2.5.22) does not contain the magnetic field. The explanation of this fact is that the magnetic field is independent of the motion. Here, as well as in the other two cases, in which a first integral can be obtained from equation (2.5.20), the phenomenon of magnetofluid dynamic interaction is sacrificed. That is why the Bernoulli integral is only of secondary importance in magnetofluid dynamics.

2.6 Uniqueness and existence theorems

Theorems of uniqueness in magnetofluid dynamics have been established by Nardini [17], Kanwall [18] and Ferrari [19]. The last, developing Graffi's idea [20], has proved uniqueness theorems for the magnetofluid dynamic case for conditions of convergence at infinity weaker than those required by Nardini. However, Ferrari assumes that, at all times, the tangential component of the magnetic field is given on the body surface Σ. Dyer and Edmunds [21] have given the most natural and complete uniqueness theorem. Not only that they preserve the conditions of convergence at infinity but they also assume that the magnetic field and the tangential component of the electric field are continuous across the body surface Σ. In the following the proof of Dyer and Edmunds is presented.

Supposing that the fluid is an incompressible simple medium, one uses the following equations:

$$\operatorname{div} \boldsymbol{V} = 0 \tag{2.6.1}$$

$$\operatorname{div} \boldsymbol{H} = 0 \tag{2.6.2}$$

$$\rho\partial_t \boldsymbol{V} + \rho(\boldsymbol{V}\cdot\nabla)\boldsymbol{V} - (\mu\boldsymbol{H}\cdot\nabla)\boldsymbol{H} = -\nabla P + \tilde{\mu}\Delta\boldsymbol{V} + \rho\boldsymbol{F} \tag{2.6.3}$$

$$\partial_t\boldsymbol{H} + (\boldsymbol{V}\cdot\nabla)\boldsymbol{H} - (\boldsymbol{H}\cdot\nabla)\boldsymbol{V} = \eta\Delta\boldsymbol{H} \tag{2.6.4}$$

$$\operatorname{curl} \boldsymbol{H} = \sigma(\boldsymbol{E} + \boldsymbol{V}\times\mu\boldsymbol{H}) \tag{2.6.5}$$

with the notations of (2.2.8′), (2.2.10′). Let us assume that the fluid motion takes place in the presence of a three-dimensional body $\boldsymbol{B}$ bounded by a piece-wise smooth surface $\partial\boldsymbol{B}$. Let us also assume that $\boldsymbol{B}$ is in motion with the velocity distribution $\boldsymbol{V}_0$ ($\operatorname{div}\boldsymbol{V}_0 = 0$). If σ_0 is the conductivity of body $\boldsymbol{B}$ and η_0 ($= \sigma_0^{-1}\mu^{-1}$) its magnetic viscosity (one assumes that the magnetic permeability has approximately the same value within the fluid and the body) then within the body equations (2.6.2), (2.6.4) and (2.6.5) with V_0 replacing V, σ_0 replacing σ and η_0 replacing η will be satisfied. On ∂B, the following conditions are to be fulfilled

$$[V] = 0,\ [H] = 0,\ [E] \cdot s = 0 \tag{2.6.6}$$

If D is the three-dimensional domain (external to B) occupied by the fluid, and T a closed time interval $[0, t_0]$ ($t_0(> 0)$ being arbitrary but fixed), then one has the following hypotheses:

(i) the velocity components V_i and their first derivatives with respect to x_i and t are *continuous bounded functions* within $D \times T$; the second order space derivatives are continuous within $D \times T$;

(ii) the magnetic field components H_i and their first derivatives with respect to x_i and t are continuous bounded functions in $(D \times T) \bigcup (B \times T)$ (H_i is continuous across ∂B but in this case $H_{i,j}$ may be discontinuous); the second order space derivatives are also continuous in $(D \times T) \bigcup (B \times T)$.

(iii) the pressure p together with its first derivatives with respect to x_i are continuous within $D \times T$; at infinity, p converges to a given value p_0 such that for all $t \in T$, $p = p_0 + 0(r^{-1})$ where r is the distance defined by $r^2 = x_i x_i$.

(iv) the components V_{0i} are given for every $t \geqslant 0$.

In this conditions, there exists at most one solution of the system (2.6.1)—(2.6.5) satisfying the conditions (2.6.6). Indeed, if we had two solutions p, V_i, H_i and p', $V_i + v_i$, $H_i + h_i$, then we would have within the fluid that

$$\operatorname{div} \boldsymbol{v} = 0 \tag{2.6.7}$$

$$\operatorname{div} \boldsymbol{h} = 0 \tag{2.6.8}$$

$$\rho\{\partial_t \boldsymbol{v} + (\boldsymbol{V} \cdot \nabla)\boldsymbol{v} + (\boldsymbol{v} \cdot \nabla)(\boldsymbol{V} + \boldsymbol{v})\} - (\mu \boldsymbol{H} \cdot \nabla)\boldsymbol{h} - (\mu \boldsymbol{h} \cdot \nabla)(\boldsymbol{H} + \boldsymbol{h}) = -\nabla P' + \tilde{\mu}\Delta \boldsymbol{v} \tag{2.6.9}$$

$$\partial_t \boldsymbol{h} + (\boldsymbol{V} \cdot \nabla)\boldsymbol{h} + (\boldsymbol{v} \cdot \nabla)(\boldsymbol{H} + \boldsymbol{h}) - (\boldsymbol{H} \cdot \nabla)\boldsymbol{v} - (\boldsymbol{h} \cdot \nabla)(\boldsymbol{V} + \boldsymbol{v}) = \eta \Delta \boldsymbol{h} \tag{2.6.10}$$

$$P' = p' + (\mu/2)(2h_i H_i + h_i h_i) \tag{2.6.9'}$$

and within the body that

$$\operatorname{div} \boldsymbol{h} = 0 \tag{2.6.11}$$

$$\partial_t \boldsymbol{h} + (\boldsymbol{V}_0 \cdot \nabla)\boldsymbol{h} - (\boldsymbol{h} \cdot \nabla)\boldsymbol{V}_0 = \eta_0 \Delta \boldsymbol{h} \tag{2.6.12}$$

Let now $\Omega(R)$ be a sphere with the centre at the origin and of radius R sufficiently large so that for any $t \in T$, $B \subset \Omega$. Let also $\Omega' = \Omega \bigcap \complement B$ where $\complement B$ denotes the region external to B. The scalar multiplication of equation (2.6.9) by $\boldsymbol{v}$

and the integration on Ω' gives (after Green's theorem and condition $\boldsymbol{v}|_{\partial B} = 0$ have been used)

$$\int_{\Omega'} \left\{ \frac{1}{2} \rho \frac{\partial}{\partial t}(v_i v_i) + \tilde{\mu} \frac{\partial v_i}{\partial x_k} \frac{\partial v_i}{\partial x_k} \right\} \mathrm{d}v$$

$$= - \int_{\Omega'} \left\{ \rho v_i v_k \frac{\partial}{\partial x_k}(V_i + v_i) - \mu v_i \left(H_k \frac{\partial h_i}{\partial x_k} + h_k \frac{\partial}{\partial x_k}(H_i + h_i) \right) \right\} \mathrm{d}v \quad (2.6.13)$$

$$- \int_{\partial\Omega} \left(P' v_i n_i - \tilde{\mu} v_i \frac{\partial v_i}{\partial x_k} n_k \right) \mathrm{d}a - \frac{1}{2} \rho \int_{\partial\Omega} v_i v_i V_k n_k \, \mathrm{d}a$$

$\boldsymbol{n}$ (n_i) being the external normal to Ω' (outwards from Ω'). Similarly, if equation (2.6.10) and (2.6.12) are scalarly multiplied by $\boldsymbol{h}$, the integration over Ω' and B performed, and the resultant expression added to each other, one obtains

$$\int_{\Omega'} \left\{ \frac{1}{2} \frac{\partial}{\partial t}(h_i h_i) + \eta \frac{\partial h_i}{\partial x_k} \frac{\partial h_i}{\partial x_k} \right\} \mathrm{d}v + \int_{B} \left\{ \frac{1}{2} \frac{\partial}{\partial t}(h_i h_i) + \eta_0 \frac{\partial h_i}{\partial x_k} \frac{\partial h_i}{\partial x_k} \right\} \mathrm{d}v$$

$$= \eta \int_{\partial\Omega} h_i \frac{\partial h_i}{\partial x_k} n_k \, \mathrm{d}a + \int_{\partial B} h_i \left[\eta \frac{\partial h_i}{\partial x_k} \right] n_k \, \mathrm{d}a - \int_{B} h_i \left(V_{0k} \frac{\partial h_i}{\partial x_k} - h_k \frac{\partial V_{0i}}{\partial x_k} \right) \mathrm{d}v$$

$$- \int_{\Omega'} h_i \left\{ V_k \frac{\partial h_i}{\partial x_k} + v_k \frac{\partial}{\partial x_k}(H_i + h_i) - H_k \frac{\partial v_i}{\partial x_k} - h_k \frac{\partial}{\partial x_k}(V_i + v_i) \right\} \mathrm{d}v \quad (2.6.14)$$

Now

$$\int_{\partial B} h_i \left[\eta \frac{\partial h_i}{\partial x_k} \right] n_k \, \mathrm{d}a = \int_{\partial B} h_i \left[\eta \frac{\partial h_k}{\partial x_i} \right] n_k \, \mathrm{d}a + \int_{\partial B} \boldsymbol{h} \cdot [\eta \operatorname{curl} \boldsymbol{h} \times \boldsymbol{n}] \, \mathrm{d}a \quad (2.6.15)$$

Denoting by $\boldsymbol{E}$ and $\boldsymbol{E} + \boldsymbol{e}$ the electric field for the two possible motions of the fluid from equation (2.6.5) we have

$$\boldsymbol{E} = -\boldsymbol{V} \times \mu \boldsymbol{H} + \sigma^{-1} \operatorname{curl} \boldsymbol{H}$$

$$\boldsymbol{E} + \boldsymbol{e} = -(\boldsymbol{V} + \boldsymbol{v}) \times \mu(\boldsymbol{H} + \boldsymbol{h}) + \sigma^{-1} \operatorname{curl}(\boldsymbol{H} + \boldsymbol{h})$$

so that

$$\boldsymbol{e} = -\boldsymbol{V} \times \mu \boldsymbol{h} - \boldsymbol{v} \times \mu(\boldsymbol{H} + \boldsymbol{h}) + \sigma^{-1} \operatorname{curl} \boldsymbol{h}$$

A similar expression for $\boldsymbol{e}$ there exists within B but with $\boldsymbol{V}$ replaced by $\boldsymbol{V}_0$ and σ by σ_0. From the condition of continuity of the tangential component of the electric field on ∂B it follows that

$$0 = \boldsymbol{n} \times [\boldsymbol{e}] = \mu \boldsymbol{n} \times [\eta \operatorname{curl} \boldsymbol{h}] \quad (2.6.16)$$

since V and h are continuous across ∂B and $V|_{\partial B} = 0$. To conclude, we have (also applying Green's theorem)

$$\int_{\partial B} h_i\left[\eta \frac{\partial h_i}{\partial x_k}\right] n_k \, da = \int_{\partial B} h_i\left[\eta \frac{\partial h_k}{\partial x_i}\right] n_k \, da$$

$$= -\eta \int_{\partial\Omega} h_i \frac{\partial h_k}{\partial x_i} n_k da + \eta \int_{\Omega'} \left\{ \frac{\partial h_i}{\partial x_k} \frac{\partial h_i}{\partial x_k} - (\mathrm{curl}\, \boldsymbol{h})^2 \right\} dv + \eta_0 \int_B \left\{ \frac{\partial h_i}{\partial x_k} \frac{\partial h_i}{\partial x_k} (\mathrm{curl}\, \boldsymbol{h})^2 \right\} dv$$

Taking into account this result in equation (2.6.14), multiplying this equation by μ and adding it to equation (2.6.13), we obtain

$$\int_{\Omega'} \left\{ \frac{1}{2} \frac{\partial}{\partial t} (\rho v_i v_i + \mu h_i h_i) + \tilde{\mu} \frac{\partial v_i}{\partial x_k} \frac{\partial v_i}{\partial x_k} \right\} dv + \int_B \frac{\mu}{2} \frac{\partial}{\partial t} (h_i h) \, dv$$

$$+ \eta\mu \int_{\Omega'} (\mathrm{curl}\, \boldsymbol{h})^2 \, dv + \tilde{\eta}\mu \int_B (\mathrm{curl}\, \boldsymbol{h})^2 \, dv$$

$$= -\rho \int_{\Omega'} v_i v_k \frac{\partial}{\partial x_k} (V_i + v_i) \, dv - \frac{1}{2} \rho \int_{\partial\Omega} v_i v_i V_k n_k \, da - \int_{\partial\Omega} \left(P' v_i n_i - \tilde{\mu} v_i \frac{\partial v_i}{\partial x_k} n_k\right) da$$

$$+ \mu \int_{\Omega'} \left\{ (v_i h_k - v_k h_i) \frac{\partial}{\partial x_k} (H_i + h_i) + h_i h_k \frac{\partial}{\partial x_k} (V_i + v_i) \right\} dv \qquad (2.6.17)$$

$$+ \mu \int_{\partial\Omega} \left(v_i h_i H_k - \frac{1}{2} h_i h_i V_k\right) n_k \, da + \mu \int_B h_i h_k \frac{\partial V_{0i}}{\partial x_k} \, da + \eta\mu \int_{\partial\Omega} h_i \left(\frac{\partial h_i}{\partial x_k} - \frac{\partial h_k}{\partial x_i} \right) n_k \, da$$

We now first integrate this equation with respect to t from t_0 to t_1 and then with respect to t_1 from 0 to $b(\leqslant t_0)$ where b is a strictly positive constant to be specified later. Let now make the following notations

$$N = \sup_{D\times T} \left| \frac{\partial}{\partial x_k} (V_i + v_i) \right|, \quad N' = \sup_{D\times T} |V_k|, \quad N'' = \sup_{B\times T} \left| \frac{\partial V_{0i}}{\partial x_k} \right|$$

$$Q = \sup_{D\times T} |v_i|, \quad M = \sup_{D\times T} \left| \frac{\partial}{\partial x_k} (H_i + h_i) \right|, \quad M' = \sup_{D\times T} |H_k| \qquad (2.6.18)$$

the suprema being taken over all i and k. Now using repeatedly Cauchy's inequality

$$|\Sigma\, \alpha_i \beta_i| \leqslant (\Sigma\, \alpha_i^2)^{1/2} (\Sigma\, \beta_i^2)^{1/2}$$

we find in $D \times T$

$$\left| v_i v_k \frac{\partial}{\partial x_k}(V_i + v_i) \right| \leqslant 3N v_i v_i, \quad |v_i v_i V_k n_k| \leqslant 3N' v_i v_i$$

$$|h_k h_k v_i n_i| \leqslant 3Q h_i h_i, \quad \left| v_i \frac{\partial v_i}{\partial x_k} n_k \right| \leqslant \frac{3}{2} v_i v_i + \frac{1}{2} \frac{\partial v_i}{\partial x_k} \frac{\partial v_i}{\partial x_k}$$

$$\left| v_i h_k \frac{\partial}{\partial x_k}(H_i + h_i) \right| \leqslant \frac{3}{2} M(v_i v_i + h_i h_i), \quad \left| h_i h_k \frac{\partial}{\partial x_k}(V_i + v_i) \right| \leqslant 3N h_i h_i$$

$$|H_k v_i h_i n_k| \leqslant \frac{3}{2} M'(v_i v_i + h_i h_i), \quad |h_i h_i V_k n_k| \leqslant 3N' h_i h_i$$

$$\left| h_i \left(\frac{\partial h_i}{\partial x_k} - \frac{\partial h_k}{\partial x_i} \right) n_k \right| \leqslant \frac{3}{2} h_i h_i + (\operatorname{curl} \boldsymbol{h})^2$$

and in $B \times T$

$$\left| h_i h_k \frac{\partial V_{0i}}{\partial x_k} \right| \leqslant 3N'' h_i h_i$$

One now uses Schwartz's inequality for the term containing pressure p'

$$\left| \int_{\partial\Omega} p' v_i n_i \, \mathrm{da} \right| \leqslant \left\{ \int_{\partial\Omega} p'^2 \mathrm{da} \times \int_{\partial\Omega} v_i v_i \, \mathrm{da} \right\}^{1/2}$$

and since $p' = 0(r^{-1})$, by condition (iii), it follows that there exists a strictly positive constant c such that

$$\left| \int_{\partial\Omega} p' v_i n_i \, \mathrm{da} \right| \leqslant c \left(\int_{\partial\Omega} v_i v_i \, \mathrm{da} \right)^{1/2}$$

provided R is large enough.

Using all these inequalities in equation (2.6.17) integrated with respect to t and t_1 and with the notation

$$G(R) = \int_0^b \mathrm{d}t_1 \left\{ \int_{\Omega'(R)} \frac{1}{2}(\rho v_i v_i + \mu h_i h_i) \, \mathrm{d}v + \int_B \frac{\mu}{2} h_i h_i \, \mathrm{d}v \right\} \tag{2.6.19}$$

$$+ \int_0^b \mathrm{d}t_1 \int_0^{t_1} \mathrm{d}t \left\{ \int_{\Omega'(R)} \left(\tilde{\mu} \frac{\partial v_i}{\partial x_k} \frac{\partial v_i}{\partial x_k} + \eta\mu(\operatorname{curl} \boldsymbol{h})^2 \right) \mathrm{d}v + \int_B \eta_0 \mu (\operatorname{curl} \boldsymbol{h})^2 \mathrm{d}v \right\}$$

where $G(R)$ is a positive constant, we find

$$0 \leqslant G(R) \leqslant 3b(N\rho + M\mu)\int_0^b \mathrm{d}t_1 \int_{\Omega'} v_i v_i \,\mathrm{d}v + \frac{3}{2} b(N'\rho + \tilde{\mu} + 2M'\mu)\times$$

$$\times \int_0^b \mathrm{d}t_1 \int_{\partial\Omega} v_i v_i \,\mathrm{d}a + \frac{\tilde{\mu}}{2}\int_0^b \mathrm{d}t_1 \int_0^{t_1} \mathrm{d}t \int_{\partial\Omega} \frac{\partial v_i}{\partial x_k}\frac{\partial v_i}{\partial x_k}\,\mathrm{d}a + 3\mu b(M+N)\int_0^b \mathrm{d}t_1 \int_{\Omega'} h_i h_i \,\mathrm{d}v$$

$$+ \frac{3\mu}{2} b(2M' + N' + \eta + Q)\int_0^b \mathrm{d}t_1 \int_{\partial\Omega} h_i h_i \,\mathrm{d}a + 3\mu b N''\int_0^b \mathrm{d}t_1 \int_B h_i h_i \,\mathrm{d}v \tag{2.6.20}$$

$$+ \eta\mu \int_0^b \mathrm{d}t_1 \int_0^{t_1} \mathrm{d}t \int_{\partial\Omega} (\mathrm{curl}\,\boldsymbol{h})^2 \,\mathrm{d}a + c\int_0^b \mathrm{d}t_1 \int_0^{t_1} \mathrm{d}t \left(\int_{\partial\Omega} v_i l_i \,\mathrm{d}a\right)^{1/2}$$

Using Schwartz's inequality in the last term we have

$$\int_0^b \mathrm{d}t_1 \int_0^{t_1} \mathrm{d}t\left(\int_{\partial\Omega} v_i v_i \,\mathrm{d}a\right)^{1/2} \quad \int_0^b \mathrm{d}t_1\left\{\left(\int_0^{t_1} \mathrm{d}t\right)\left(\int_0^{t_1} \mathrm{d}t \int_{\partial\Omega} v_i v_i \,\mathrm{d}a\right)\right\}^{1/2}$$

$$\leqslant b\left\{\left(\int_0^b \mathrm{d}t\right)\left(\int_0^b \mathrm{d}t \int_{\partial\Omega} v_i v_i \,\mathrm{d}a\right)\right\}^{1/2} = b^{3/2}\left(\int_0^b \mathrm{d}t \int_{\partial\Omega} v_i v_i \,\mathrm{d}a\right)^{1/2}$$

Now choosing constant b in a convenient way and with the notation

$$m = 2\max\left\{3b\left(N' + \frac{\tilde{\mu}}{\rho} + 2M'\frac{\mu}{\rho}\right), \frac{1}{2}, 3b(2M' + N' + \eta + Q)\right\}$$

Dyer and Edmunds derive the following inequality

$$G(R) \leqslant mG'(R) + l\{G'(R)\}^{1/2} \tag{2.6.21}$$

where

$$l = 2cb^{3/2}(2/\rho)^{1/2}, \; G'(R) = \mathrm{d}G/\mathrm{d}R$$

In what follows the following argument is used: on account of hypotheses (i) and (ii) one has for large R

$$G(R) = 0(R^3) \tag{2.6.22}$$

Suppose now that there exists an R_0 such that $G(R_0) \neq 0$. Since $G(R)$ is non-negative, it follows that $G(R_0) > 0$. Because $G'(R)$ is also non-negative, $G(R)$ is monotonically increasing and hence

$$0 < G(R_0) \leqslant G(R) \leqslant mG'(R) + l\{G'(R)\}^{1/2} \tag{2.6.23}$$

for all $R > R_0$.

Let now α be a constant so that $\{G'(R)\}^{1/2} \geqslant \alpha$ for all $R > R_0$. From equation (2.6.23) it follows that

$$G(R) \leqslant \left(m + \frac{l}{\alpha}\right)G'(R) \Rightarrow \frac{G'(R)}{G(R)} \geqslant \left(m + \frac{l}{\alpha}\right)^{-1} = \delta > 0$$

so that

$$G(R) \geqslant G(R_0)\exp\{\delta(R - R_0)\} \tag{2.6.24}$$

for all $R > R_0$.
But inequalities (2.6.22) and (2.6.24) contradict each other; whence $G(R) \equiv 0$. To conclude, $v_i \equiv 0$ in $D \times (0, b)$ and $h_i \equiv 0$ in $(D \cup B) \times (0, b)$.

If now one integrates equation (2.6.17) with respect to t from b to $b + t_1$ and then with respect to t_1 from 0 to b and repeates the above arguments one finds out that $v_i \equiv 0$ within $D \times (b, 2b)$ and $h_i \equiv 0$ within $(D \cup B) \times (b, 2b)$. But the interval $T = (0, t_0)$ may be covered by a finite number of intervals of length b so that $v_i \equiv 0$ within $D \times T$ and $h_i \equiv 0$ within $(D \cup B) \times T$.

Using this result from equation (2.6.9) and (iii) it follows that $p' \equiv 0$ within $D \times T$ so that the solution is unique for all $t \in T$. But T is arbitrary and hence the theorem is proved.

In [22], Edmunds gives sufficient conditions for the uniqueness of the solution and proves the existence of such solutions.

Existence theorems are established by Förste [23], Sanchez-Palencia [24], Lassner [25] and Tzinober [26].

2.7 Magnetofluid dynamic equations in orthogonal curvilinear coordinates

2.7.1 Expressions of the differential operators

It is necessary, in some applications, to use the magnetofluid dynamic equations in curvilinear coordinates. In this respect, we shall notice that whenever we may express the position vector $\boldsymbol{x}$ of some point P by means of three curvilinear coordinates q_1, q_2, q_3, i.e., whenever we have

$$\boldsymbol{x} = \boldsymbol{x}(q_1, q_2, q_3) \tag{2.7.1}$$

it follows that

$$\mathrm{d}\boldsymbol{x} = \Sigma \frac{\partial \boldsymbol{x}}{\partial q_i}\mathrm{d}q_i = \Sigma H_i\,\mathrm{d}q_i\boldsymbol{e}_i = \Sigma\,\mathrm{d}s_i\boldsymbol{e}_i$$

$$\boldsymbol{V} = \Sigma H_i\dot{q}_i\boldsymbol{e}_i = \Sigma V_i\boldsymbol{e}_i, \quad V_i \overset{\text{def}}{=} H_i\dot{q}_i \tag{2.7.2}$$

where $\boldsymbol{e}_i$ is the variable unit vector of curve q_i

$$\boldsymbol{e}_i \overset{\text{def}}{=} \frac{1}{H_i}\frac{\partial \boldsymbol{x}}{\partial q_i}, \quad H_i \overset{\text{def}}{=} \left|\frac{\partial \boldsymbol{x}}{\partial q_i}\right| \tag{2.7.2'}$$

and ds_i is the line element on this curve. In this way, the trihedron $(\boldsymbol{e}_1, \boldsymbol{e}_2, \boldsymbol{e}_3)$ is pointed out at any point P defined by the coordinates q_1, q_2, q_3.

In cylindrical coordinates $(r = q_1, \theta = q_2, x_3 = q_3)$

$$x_1 = r\cos\theta, \; x_2 = r\sin\theta, \; x_3 = x_3 \tag{2.7.3}$$

we have

$$ds_1 = dr, \; ds_2 = r\,d\theta, \; ds_3 = dx_3 \tag{2.7.3'}$$

so that

$$H_1 = 1, \; H_2 = r, \; H_3 = 1. \tag{2.7.3''}$$

In spherical coordinates $(r = q_1, \theta = q_2, \varphi = q_3)$

$$x_1 = r\sin\theta\cos\varphi, \; x_2 = r\sin\theta\sin\varphi, \; x_3 = r\cos\theta \tag{2.7.4}$$

we have

$$ds_1 = dr, \; ds_2 = r\,d\theta, \; ds_3 = r\sin\theta\,d\varphi \tag{2.7.4'}$$

so that

$$H_1 = 1, \; H_2 = r, \; H_3 = r\sin\theta \tag{2.7.4''}$$

Let now $F\,(t, \boldsymbol{x})$ be a scalar field which by the change of variables (2.7.1) becomes $F(t, q_i)$. By definition, the projection of operator grad F on the q_i-axis is $\partial F/\partial s_i$. We thus have

$$\operatorname{grad} F = \Sigma\, \boldsymbol{e}_i \frac{\partial F}{\partial s_i} = \Sigma \frac{\boldsymbol{e}_i}{H_i}\frac{\partial F}{\partial q_i} \tag{2.7.5}$$

For a vector field $\boldsymbol{V}(t, \boldsymbol{x}) \to \boldsymbol{V}(t, q_i)$ the definitions of operators *div* and *curl* give [G. 28]:

$$\operatorname{div} \boldsymbol{V} = \frac{1}{H_1H_2H_3}\left\{\frac{\partial}{\partial q_1}(V_1H_2H_3) + \frac{\partial}{\partial q_2}(V_2H_3H_1) + \frac{\partial}{\partial q_3}(V_3H_1H_2)\right\} \tag{2.7.6}$$

$$\operatorname{curl} \boldsymbol{V} = \begin{vmatrix} \dfrac{\boldsymbol{e}_1}{H_2H_3} & \dfrac{\boldsymbol{e}_2}{H_3H_1} & \dfrac{\boldsymbol{e}_3}{H_1H_2} \\ \dfrac{\partial}{\partial q_1} & \dfrac{\partial}{\partial q_2} & \dfrac{\partial}{\partial q_3} \\ V_1H_1 & V_2H_2 & V_3H_3 \end{vmatrix} \tag{2.7.7}$$

Now noticing that $\Delta F = \text{divgrad } F$ it follows that

$$\Delta F = \frac{1}{H_1 H_2 H_3}\left\{\frac{\partial}{\partial q_1}\left(\frac{H_2 H_3}{H_1}\frac{\partial F}{\partial q_1}\right) + \frac{\partial}{\partial q_2}\left(\frac{H_3 H_1}{H_2}\frac{\partial F}{\partial q_2}\right) + \frac{\partial}{\partial q_3}\left(\frac{H_1 H_2}{H_3}\frac{\partial F}{\partial q_3}\right)\right\} \tag{2.7.8}$$

In cylindrical coordinates we have

$$\text{div } \boldsymbol{V} = \frac{1}{r}\frac{\partial}{\partial r}(rV_r) + \frac{1}{r}\frac{\partial}{\partial \theta}V_\theta + \frac{\partial}{\partial z}V_z \tag{2.7.9}$$

$$\text{curl } \boldsymbol{V} = \begin{vmatrix} \dfrac{\boldsymbol{e}_r}{r} & \boldsymbol{e}_\theta & \dfrac{\boldsymbol{e}_z}{r} \\ \dfrac{\partial}{\partial r} & \dfrac{\partial}{\partial \theta} & \dfrac{\partial}{\partial z} \\ V_r & rV_\theta & V_z \end{vmatrix}$$

$$\Delta F = \frac{1}{r}\frac{\partial}{\partial r}\left(r\frac{\partial F}{\partial r}\right) + \frac{1}{r^2}\frac{\partial^2 F}{\partial \theta^2} + \frac{\partial^2 F}{\partial z^2}$$

and in spherical coordinates

$$\text{div } \boldsymbol{V} = \frac{1}{r^2}\frac{\partial}{\partial r}(r^2 V_r) + \frac{1}{r \sin\theta}\left\{\frac{\partial}{\partial \theta}(V_\theta \sin\theta) + \frac{\partial}{\partial \varphi}V_\varphi\right\}$$

$$\text{curl } \boldsymbol{V} = \begin{vmatrix} \dfrac{\boldsymbol{e}_r}{r^2 \sin\theta} & \dfrac{\boldsymbol{e}_\theta}{r \sin\theta} & \dfrac{\boldsymbol{e}_\varphi}{r} \\ \dfrac{\partial}{\partial r} & \dfrac{\partial}{\partial \theta} & \dfrac{\partial}{\partial \varphi} \\ V_r & rV_\theta & r\,V_\varphi \sin\theta \end{vmatrix} \tag{2.7.10}$$

$$\Delta F = \frac{1}{r^2}\frac{\partial}{\partial r}\left(r^2\frac{\partial F}{\partial r}\right) + \frac{1}{r^2 \sin\theta}\left\{\frac{\partial}{\partial \theta}\left(\sin\theta\frac{\partial F}{\partial \theta}\right) + \frac{1}{\sin\theta}\frac{\partial^2 F}{\partial \varphi^2}\right\}$$

2.7.2 The expression for acceleration

In order to project the equation of motion on the trihedron $(\boldsymbol{e}_1, \boldsymbol{e}_2, \boldsymbol{e}_3)$, one should first derive the expression of the acceleration $\boldsymbol{a}$ in curvilinear coordinates.

Using a method due to Lagrange, we have

$$a \cdot e_i H_i = \frac{\mathrm{d}V}{\mathrm{d}t} \cdot \frac{\partial x}{\partial q_i} = \frac{\mathrm{d}}{\mathrm{d}t}\left(V \cdot \frac{\partial x}{\partial q_i}\right) - V \cdot \frac{\mathrm{d}}{\mathrm{d}t}\left(\frac{\partial x}{\partial q_i}\right)$$

$$= \frac{\mathrm{d}}{\mathrm{d}t}\left(V \cdot \frac{\partial V}{\partial \dot{q}_i}\right) - V \cdot \frac{\partial V}{\partial q_i} = \frac{\mathrm{d}}{\mathrm{d}t}\left(\frac{\partial E}{\partial \dot{q}_i}\right) - \frac{\partial E}{\partial q_i}$$

E being the specific kinetic energy (expressed in curvilinear coordinates)

$$E = \frac{1}{2} V^2 = \frac{1}{2} \Sigma H_i^2 \dot{q}_i^2$$

In order to write explicitly the acceleration projection, we shall notice that

$$\frac{\partial E}{\partial q_i} = H_i^2 \dot{q}_i = H_i V_i, \quad \frac{\partial E}{\partial q_i} = \sum_k H_k \frac{\partial H_k}{\partial q_i} \dot{q}_k^2 = \sum_k \frac{\partial H_k}{\partial q_i} \frac{V_k^2}{H_k}$$

But for some field F we have

$$\frac{\mathrm{d}F}{\mathrm{d}t} = \frac{\partial F}{\partial t} + \Sigma \frac{\partial F}{\partial q_k} \dot{q}_k = \frac{\partial F}{\partial t} + \Sigma \frac{\partial F}{\partial q_k} \frac{V_k}{H_k} \tag{2.7.11}$$

so that (noticing that H_i does not explicitly depend on t)

$$\frac{\mathrm{d}}{\mathrm{d}t}(H_i V_i) = H_i \frac{\partial V_i}{\partial t} + \sum_k \frac{V_k}{H_k} \frac{\partial}{\partial q_k}(H_i V_i)$$

To conclude

$$a \cdot e_i = \frac{\partial V_i}{\partial t} + \frac{1}{H_i} \sum_k \frac{V_k}{H_k} \frac{\partial}{\partial q_k}(H_i V_i) - \frac{1}{H_i} \sum_k \frac{V_k^2}{H_k} \frac{\partial H_k}{\partial q_i} \tag{2.7.12}$$

We could also express the acceleration by means of the covariant derivative.

In cylindrical coordinates we have

$$a_r = \frac{\partial V_r}{\partial t} + \left(V_r \frac{\partial}{\partial r} + \frac{V_\theta}{r} \frac{\partial}{\partial \theta} + V_z \frac{\partial}{\partial z}\right) V_r - \frac{V_\theta^2}{r} \tag{2.7.13}$$

. . . .

. . . .

and in spherical coordinates

$$a_r = \frac{\partial V_r}{\partial t} + \left(V_r \frac{\partial}{\partial r} + \frac{V_\theta}{r} \frac{\partial}{\partial \theta} + \frac{V_\varphi}{r \sin \theta} \frac{\partial}{\partial \varphi}\right) V_r - \frac{V_\theta^2 + V_\varphi^2}{r} \tag{2.7.14}$$

. . . .

. . . .

2.7.3 The magnetofluid dynamic equations

We shall use the magnetofluid dynamic equations in the following form

$$\partial_t \rho + \operatorname{div}(\rho \boldsymbol{V}) = 0 \tag{2.7.15}$$

$$\boldsymbol{a} = \boldsymbol{F} - \frac{1}{\rho}\nabla p + \frac{\tilde{\nu}}{3}\nabla \operatorname{div} \boldsymbol{V} + \tilde{\nu}\,\Delta \boldsymbol{V} + \frac{\mu}{\rho}\operatorname{curl} \boldsymbol{H} \times \boldsymbol{H} \tag{2.7.16}$$

$$\frac{\mathrm{d}_t S}{\mathrm{d}t} = 0 \tag{2.7.17}$$

$$\partial_t \boldsymbol{H} = \operatorname{curl}(\boldsymbol{V} \times \boldsymbol{H}) + \eta\,\Delta \boldsymbol{H} \tag{2.7.18}$$

In cylindrical coordinates these equations become

$$r\frac{\partial \rho}{\partial t} + \frac{\partial}{\partial r}(r\rho V_r) + \frac{\partial}{\partial \theta}(\rho V_\theta) + \frac{\partial}{\partial z}(r\rho V_z) = 0$$

$$a_r = \boldsymbol{F}\cdot\frac{\partial \boldsymbol{x}}{\partial r} - \frac{1}{\rho}\frac{\partial p}{\partial r} + \frac{\tilde{\nu}}{3}\frac{\partial}{\partial r}(\operatorname{div} \boldsymbol{V}) + \tilde{\nu}\Delta V_r +$$

$$+ \frac{\mu}{\rho}\left\{H_z\left(\frac{\partial H_r}{\partial z} - \frac{\partial H_z}{\partial r}\right) - \frac{H_\theta}{r}\left(\frac{\partial}{\partial r}(rH_\theta) - \frac{\partial H_r}{\partial \theta}\right)\right\}$$

. . . .

. . . .

$$\frac{\mathrm{d}S}{\mathrm{d}t} + V_r\frac{\partial S}{\partial r} + \frac{V_\theta}{r}\frac{\partial S}{\partial \theta} + H_z\frac{\partial S}{\partial z} = 0$$

$$\frac{\partial H_r}{\partial t} = \frac{1}{r}\frac{\partial}{\partial \theta}(V_r H_\theta - V_\theta H_r) - \frac{\partial}{\partial z}(V_z H_r - V_r H_z) + \eta\Delta H_r$$

. . . .

. . . .

where div $\boldsymbol{V}$ and ΔF are expressed by means of equations (2.7.9).

In a similar way, we have in spherical coordinates

$$r^2 \sin\theta \frac{\partial \rho}{\partial t} + \frac{\partial}{\partial r}\left(\sin\theta\, r^2 \rho V_r\right) + \frac{\partial}{\partial \theta}(r \sin\theta\, \rho V_\theta) + \frac{\partial}{\partial \varphi}(r\rho V_\varphi) =$$

$$a_r = \boldsymbol{F} \cdot \boldsymbol{e}_r - \frac{1}{\rho}\frac{\partial p}{\partial r} + \frac{\tilde{\nu}}{3}\frac{\partial}{\partial r}(\operatorname{div} V) + \tilde{\nu}\Delta V_r$$

$$+ \frac{\mu}{\rho}\left\{\frac{H_\varphi}{r}\left(\frac{1}{\sin\theta}\frac{\partial H_r}{\partial \varphi} - \frac{\partial}{\partial r}(rH_\varphi)\right) - \frac{H_\theta}{r}\left(\frac{\partial}{\partial r}(rH_\theta) - \frac{\partial H_r}{\partial \theta}\right)\right\}$$

. . . .

. . . .

$$\frac{dS}{dt} + V_r \frac{\partial S}{\partial r} + \frac{V_\theta}{r}\frac{\partial S}{\partial \theta} + \frac{V_\varphi}{r \sin\theta}\frac{\partial S}{\partial \varphi} = 0$$

$$\frac{\partial H_r}{\partial t} = \frac{1}{r}\left\{\frac{1}{\sin\theta}\frac{\partial}{\partial \theta}(H_r V_\varphi - H_\varphi V_r) - \frac{\partial}{\partial \varphi}(V_r H_\theta - V_\theta H_r)\right\} + \eta \Delta H_r$$

. . . .

. . . .

where div V and ΔF are defined by equations (2.7.10).

3

Laminar flow between parallel plates and through ducts

3.1 Introduction

As already mentioned in the introduction, the first magnetofluid dynamic problem was that of the flow of a viscous fluid through a parallel-wall duct. The theoretical and experimental (using mercury) study of this problem was performed by Hartmann and Lazarus (1937) [1] [2]. Although good agreement between theoretical and experimental results was found, the investigations were not continued due to the uncertainty of their usefulness. Only during 1950 when the possibility of studying laboratory plasmas arised and magnetofluid dynamics developed rapidly, the flow of conducting fluids through ducts began to be studied. Two circumstances stimulated these studies: the fewer mathematical difficulties as compared with other problems and a number of immediate technological applications (flowmeters, pumps, MHD generators).

Besides the studies mentioned above, some other important studies should be named. These are due to the following research workers: Lehnert [3] considered Couette flow; Shercliff [4] solved the problem of viscous fluid flow through cylindrical ducts of rectangular cross-section and insulating walls; Resler and Sears [5] gave a thorough analysis of the motion between parallel-planes; Blewiss [6] solved the Couette problem for very general conditions with the dissipation coefficients as functions of the thermodynamic variables p and I (i.e., $\sigma = \sigma(p, I)$, $\mu = \mu(p, I)$) and Chang and Lundgren [7] stated the general problem of flow through ducts and solved it for the rectangular duct with perfectly conducting walls. During the same period, Greenberg constructed the Green function for the case of a rectangular cross-section duct with two insulating and two perfectly conducting walls. This chapter is mainly based on the works of Resler and Sears, Shercliff, Chang and Lundgren and Greenberg. The important results of Sato [9] on the Hall effect are also included.

A great number of works on this subject have appeared since 1960 and their review would be possible only in a monograph like that of Vatajin, Lyubimov and Reghirer (G. 27). Concerning the problem of flow through rectangular ducts, we should like to mention the work of Hunt and his coworkers (1965—1969) in which very simple solutions based on the theory of the boundary layer were given. Hunt's method has also been used by other workers, e.g., Sloan. A solution also based on the boundary layer theory was given by Chang and Lundgren (1967).

3.2 The motion between parallel plates

3.2.1 The Poiseuille — Hartmann problem

In this section, we shall deal with the motion of a viscous fluid between two parallel insulating walls in the presence of a homogeneous magnetic field of induction $\boldsymbol{B}_0$ perpendicular to the walls. The motion is determined by a constant pressure gradient P_0 in the Poiseuille-Hartmann problem and by the motion of one of the walls in the Couette problem.

Assuming that the motion is due to a pressure gradient, we shall choose the reference system such that the xOz plane is coincident with the median plane between the two walls, the Ox-axis being along the direction of the pressure gradient ($P_0 = \partial p/\partial x$) and the Oy-axis being perpendicular to the walls. We shall use the dimensionless variables defined in Section 2.4 with B_0 being the external magnetic induction, L_0 half the distance between the two walls, and V_0 a quantity defined by means of the pressure gradient by the formula $V_0 = -P_0 L_0^2/\mu$. Then it follows that

$$\partial p^*/\partial x^* = -Re^{-1}$$

(the asterisks will be dropped in the following).

Due to the physical and geometrical conditions in which the motion takes place, we may assume that this is steady and plane and even that the velocity and magnetic induction depend on the variable y only. Then, from the equations

$$\operatorname{div} \rho \boldsymbol{V} = 0, \ \operatorname{div} \boldsymbol{B} = 0, \ \operatorname{div} \boldsymbol{J} = 0 \tag{3.2.1}$$

it follows that $\rho V_y \equiv C_1$; $B_y = C_2$; $J_y = C_3$ (C_1, C_2, C_3 being determined by the boundary conditions). From the condition of adherence of the fluid to the walls it follows that $C_1 = 0$ (and thus $V_y \equiv 0$); from the condition of continuity of the normal component of the induction $B_y \equiv 1$ and from the insulating-wall hypothesis $J_y = 0$. In this manner, we have

$$\boldsymbol{J} = \operatorname{curl} \boldsymbol{B} = \left(\frac{\mathrm{d}B_z}{\mathrm{d}y}, \ 0, \ -\frac{\mathrm{d}B_x}{\mathrm{d}y} \right) \tag{3.2.2}$$

so that from the equation of motion it follows

$$0 = 1 + \frac{\mathrm{d}^2 V_x}{\mathrm{d}y^2} + ReRh \frac{\mathrm{d}B_x}{\mathrm{d}y} \tag{3.2.3}$$

$$0 = -\frac{\mathrm{d}p}{\mathrm{d}y} - Rh\left(B_x \frac{\mathrm{d}B_x}{\mathrm{d}y} + B_z \frac{\mathrm{d}B_z}{\mathrm{d}y} \right) \tag{3.2.4}$$

$$0 = Rh \frac{\mathrm{d}B_z}{\mathrm{d}y} \tag{3.2.5}$$

Equation (3.2.5) together with the condition of continuity of the tangential component of the magnetic field (1.5.16) gives $B_z \equiv 0$.

In these conditions the magnetic induction equation is reduced to

$$\frac{d^2 B_x}{dy^2} + Rm \frac{dV_x}{dy} = 0 \tag{3.2.6}$$

Equations (3.2.3) and (3.2.6) determine the unknowns V_x and B_x provided that we take into account the boundary conditions

$$V_x(\pm 1) = 0, \ B_x(\pm 1) = 0 \tag{3.2.7}$$

which follow from the adherence condition and from the continuity condition (1.5.16).

One obtains

$$\frac{d^3 V_x}{dy^3} - R^2 \frac{dV_x}{dy} = 0 \tag{3.2.8}$$

where

$$R = \sqrt{ReRmRh} = B_0 L_0 \sqrt{\sigma/\tilde{\mu}}$$

Consequently

$$V_x = \frac{\operatorname{ch} R - \operatorname{ch} Ry}{R \operatorname{sh} R} = V_x(0) \frac{\operatorname{ch} R - \operatorname{ch} Ry}{\operatorname{ch} R - 1} \tag{3.2.9}$$

$$B_x = \frac{1}{ReRh}\left(-y + \frac{\operatorname{sh} Ry}{\operatorname{sh} R}\right) \tag{3.2.10}$$

$V_x(0)$ being the maximum velocity (the velocity in the plane $y = 0$). The flow rate is proportional to the mean velocity $\langle V_x \rangle$:

$$\langle V_x \rangle \overset{\text{def}}{=} \frac{1}{2}\int_{-1}^{+1} V_x \, dy = \frac{R - \operatorname{th} R}{R^2 \operatorname{th} R} \tag{3.2.11}$$

From Ohm's law

$$\mathbf{J} = Rm\,(\mathbf{E} + \mathbf{V} \times \mathbf{B}) \tag{3.2.12}$$

and from the above results it follows that

$$E_x \equiv E_y \equiv 0, \ E_z = -\langle V_x \rangle \tag{3.2.13}$$

The same result could be obtained noticing that the total electric current flowing along the Oz axis vanishes:

$$\int_{-1}^{+1} J_z \, dy = -\int_{-1}^{+1} \frac{dB_x}{dy} \, dy = B_x(-1) - B_x(+1) = 0$$

Indeed, from curl $\boldsymbol{E} = 0$ it follows that $E_z = E_0 =$ constant, so that, integrating the projection on the Oz-axis of equation (3.2.12) it follows that

$$0 = Rm(2E_0 + \int_{-1}^{+1} V_x \, dy)$$

The pressure is determined from equation (3.2.4) in the form

$$p = \frac{1}{2} SB_x^2 - P_0 x + p(0, 1) \tag{3.2.14}$$

For the non-conducting fluids velocity is found from equation (3.2.9) making $R \to 0$ ($B_0 = 0$). One obtains

$$\lim_{R \to 0} V_x(0) = \frac{1}{2}, \quad \lim_{R \to 0} V_x = \frac{1}{2}(1 - y^2) \tag{3.2.15}$$

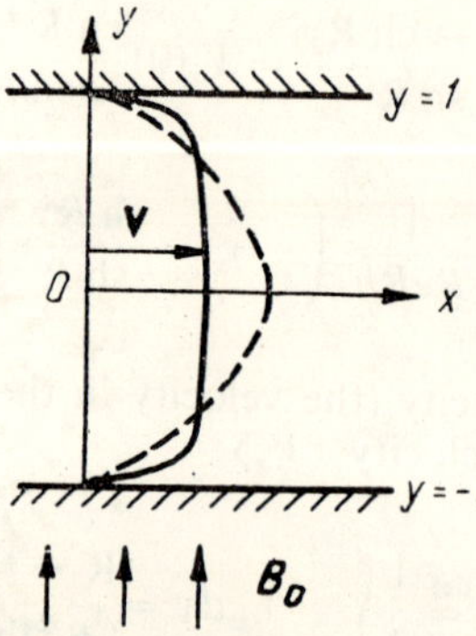

Fig. 3

The parabolic velocity profile (Fig. 3) is a result of the viscosity effect. If we now increase R (starting from zero), the magnetic field action is superposed on the viscosity effect so that for large R the latter predominates. Since

$$\lim_{R \to \infty} V_x(0) = 0, \quad \lim_{R \to \infty} \frac{V_x(y)}{V_x(0)} = 1$$

(excepting $y = \pm 1$ when the latter vanishes) it follows that velocity becomes uniform in the central region of the channel and tends rapidly to zero at the walls (the boundary layer narrows with increasing magnetic field intensity). Since the mean velocity also vanishes for $R \to \infty$, it follows that the flow rate decreases as well. To conclude, *the magnetic field slows down the motion.* This process may be easily explained by taking the projection of the magnetic field along the velocity direction.

$$Rh(\boldsymbol{J} \times \boldsymbol{B}) \cdot \boldsymbol{V} = RhRm\{(\boldsymbol{E} + \boldsymbol{V} \times \boldsymbol{B}) \times \boldsymbol{B}\} \cdot \mathrm{V} = RhRm\, V_x(\langle V_x \rangle - V_x)$$

It can be seen that in the region where $V_x > \langle V_x \rangle$ the magnetic force slows down the motion and it accelerates it in the region where $V_x < \langle V_x \rangle$; hence the uniformization effect.

Some simple solutions for the case of unsteady motion are given in [10] and [11].

3.2.2 The Couette problem

Let us now assume that the distance between the plates is L_0, that the Ox-axis is coincident with the fixed plate and that the other plate (of equation $y = 1$) is moving along the Ox-axis in the positive direction with a constant velocity V_0. The values of L_0, B_0 and V_0 will be taken as measurement units defining in this manner the dimensionless variables of Section 2.4.

This problem is of great practical importance since it represents a model, although much simplified, of the boundary layer (the fixed plate representing the fixed wall and the moving plate representing the separation surface between the boundary layer and the fluid whose viscosity is neglected). For hypersonic velocities the gas within the laminar boundary layer is ionized (heat is produced due to the viscous friction) and thus becomes electrically conducting. Therefore, an exact study of the hypersonic boundary layer should take into account both the high temperatures and the electrical conductivity of the gas. Of course, all these will influence not only the fluid motion but also the composition of the walls. An exact study of the magnetofluid dynamic processes in the boundary layer at high velocities remains still to be done. Some of the difficulties of such a study are presented in [12] where the boundary layer adherent to a plane plate in the presence of an orthogonal magnetic field is considered.

Thus, considering the Couette plane problem, we shall assume the fluid to be compressible, (the motion of a viscous fluid between two coaxial cylinders, one fixed and the other in axial uniform motion may be reduced to a problem similar to the plane Couette one). From the problem conditions we have

$$\frac{\partial}{\partial t} = \frac{\partial}{\partial x} = \frac{\partial}{\partial z} = 0$$

so that, using the above argument, we find $V_y \equiv 0$ and $B_y \equiv 1$. Equation (3.2.8) is derived for V_x so that, using the boundary conditions

$$V_x(0) = 0,\ V_x(\pm 1) = \pm 1 \tag{3.2.16}$$

it follows

$$V_x = \frac{\operatorname{sh} Ry}{\operatorname{sh} R},\ B_x = \sqrt{\frac{Rm}{ReRh}}\frac{1 - \operatorname{ch} Ry}{\operatorname{sh} R} \tag{3.2.17}$$

The first two conditions of (3.2.16) express the adherence of the fluid to the two plates. The third comes from the observation that the fluid motion is the same if we exclude the wall of equation $y = 0$ and consider another wall of equation $y = -1$ moving with velocity V_0 in the direction opposite to the Ox-axis. As a matter of fact, this condition is equivalent to the hypothesis that the mean velocity in the channel of opening $2L_0$ vanishes. Consequently, the electric field is zero. A more detailed study in which the wall conditions for which no electric field appears in the fluid are determined is done in [10].

For the non-conducting fluid we have

$$\lim_{R\to 0} V_x = y \tag{3.2.17'}$$

except for $y = \pm 1$. To conclude, the magnetic field presence slows down the motion ($V_x < y$). While for the non-conducting fluid the velocity distribution within the boundary layer may be assumed linear, for the conducting fluid this becomes parabolic and the motion is the more slowed down the higher the Hartmann number R (Fig. 4).

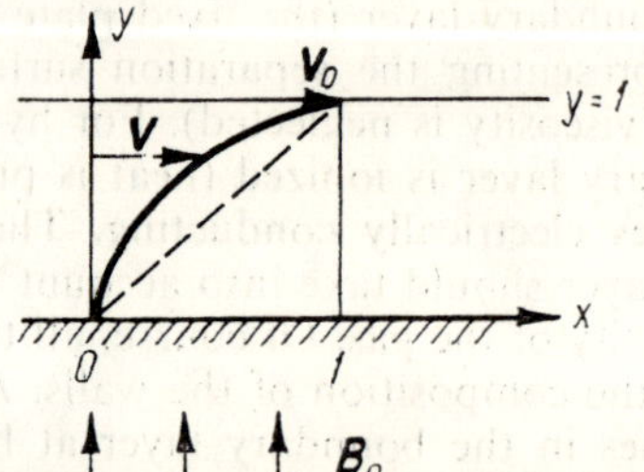

Fig. 4

It is of interest to calculate the unit stress on the plates. Since

$$\tau_{ji} = \begin{pmatrix} 0 & \dfrac{dV_x}{dy} & 0 \\ \dfrac{dV_x}{dy} & 0 & 0 \end{pmatrix} \tag{3.2.18}$$

the stress on the wall $y = 1$ is reduced to the vector

$$\left(\frac{dV_x}{dy}, 0, 0\right)$$

so that

$$|\tau| = \frac{R \operatorname{ch} Ry}{\operatorname{sh} R}, \quad |\tau|_{y=1} = R \operatorname{cth} R$$

$$\lim_{R\to 0} R \operatorname{cth} R = 0, \quad \lim_{R\to\infty} R \operatorname{cth} R = 0 \tag{3.2.19}$$

The stress increases with increasing Hartmann number. On the other hand, on the fixed wall

$$|\tau|_{y=0} = R/\operatorname{sh} R, \quad \lim_{R\to\infty} |\tau| = 0 \tag{3.2.20}$$

the stress decreases. Definite conclusions can only be drawn by introducing the pressure and electromagnetic stress tensor into the study.

In the boundary layer theory, it is important to determine the fluid temperature. Assuming that the fluid obeys the perfect gas law, we shall use equation (2.4.10). Noticing that

$$\tau_{ij}\frac{\partial V_j}{\partial x_i} = \left(\frac{\partial V_x}{dy}\right)^2, \; J'^2 = J^2 = \left(\frac{dB_x}{dy}\right)^2 = Rm^2 V_x^2$$

we find

$$\Delta T = (1-\gamma)M^2 Pr\left\{\left(\frac{dV_x}{dy}\right)^2 + R^2 V_x^2\right\}$$

Now, using (3.2.17) it follows that

$$\frac{d^2T}{dy^2} = (1-\gamma)M^2R^2Pr\frac{\operatorname{ch} 2Ry}{\operatorname{sh}^2 R}$$

and then

$$T = \frac{(1-\gamma)M^2Pr}{4\operatorname{sh}^2 R}\operatorname{ch} 2Ry + Cy + C' \tag{3.2.21}$$

The two constants of equation (3.2.21) are determined by the temperature boundary conditions. If, for instance, we assume that the temperature of the moving plate is T_0 (constant) and if we take this value as the temperature unit, then we have $T(1) = 1$. Moreover, if we assume that the heat flow through the $y = 0$ wall is q_0, then from the Fourier hypothesis (2.3.38) we have $(\partial T/\partial y) = -q_0$, so that

$$T = \frac{(1-\gamma)M^2Pr}{4\,\mathrm{sh}^2 R}(\mathrm{ch}\,2Ry - \mathrm{ch}\,2R) + q_0(1-y) + 1 \tag{3.2.22}$$

Then, the fixed-wall temperature is

$$T = \frac{\gamma - 1}{2}M^2Pr + q_0 + 1 \tag{3.2.23}$$

and does not depend on the magnetic field.

If, for instance, the fixed wall temperature is given as T_0, then $T(0) = 1$ so that

$$T = \frac{(1-\gamma)M^2Pr}{4\,\mathrm{sh}^2 R}(\mathrm{ch}\,2Ry - 1) - Cy + 1$$

constant C being determined from data on the moving wall. Estimating the heat flow through the fixed wall

$$q = -\left.\frac{\mathrm{d}T}{\mathrm{d}y}\right|_{y=0}$$

we find that this quantity does not depend on the applied magnetic field intensity

The Couette problem for unsteady motions $V_0 = V_0(t)$ is solved in [13].

3.3 The Hall effect in the Poiseuille-Hartmann flow

3.3.1 The conductivity tensor

Hartmann's solution is valid only for the case of isotropic conductivity and ohmic fluids. While these conditions are quite well satisfied for liquid metals (e.g. mercury) this is not the case for ionized gases. For the latter, the current carriers are the electrons which interact with the magnetic field so that the current is no longer proportional to the electric field intensity. Ohm's law gains a complex form derived in the last chapter. Assuming a completely ionized gas and steady motion, Ohm's law has the form

$$\boldsymbol{J} = \sigma(\boldsymbol{E} + \boldsymbol{E}_e + \boldsymbol{V} \times \boldsymbol{B}) + \frac{\sigma}{nq}\boldsymbol{B} \times \boldsymbol{J} \tag{3.3.1}$$

where

$$\sigma = \frac{nq^2\tau}{m_e}, \quad \boldsymbol{E}_e = \frac{1}{nq} \operatorname{grad} p_e \tag{3.3.1'}$$

q being the electron charge, m_e its mass, n the particle density and τ the mean free path between electron collisions with other particles. The electron pressure being denoted by p_e we have $p_e = sp$, s being a parameter characterizing the state of ionization ($s = 1/2$ for the completely ionized plasma and approximately zero for a weakly ionized gas). In what follows, the ionization state is assumed constant during the motion.

In dimensionless variables, equation (3.3.1) is written

$$\boldsymbol{J} = Rm(\boldsymbol{E} + \boldsymbol{E}_e + \boldsymbol{V} \times \boldsymbol{B}) + v\boldsymbol{B} \times \boldsymbol{J} \tag{3.3.2}$$

where

$$v = \frac{\sigma B_0}{nq}, \quad \boldsymbol{E}_e = \operatorname{grad} \frac{vs}{RmRh} p \tag{3.3.2'}$$

As usually, the asterisks have been dropped. By projecting equation (3.3.2) on the reference system axes we find

$$\begin{aligned} J_x &= \mathscr{E}_x + \mathscr{B}_y J_z - \mathscr{B}_z J_y \\ J_y &= \mathscr{E}_y + \mathscr{B}_z J_x - \mathscr{B}_x J_z \\ J_z &= \mathscr{E}_z + \mathscr{B}_x J_y - \mathscr{B}_y J_x \end{aligned} \tag{3.3.3}$$

where

$$\boldsymbol{\mathscr{E}} = Rm(\boldsymbol{E} + \boldsymbol{E}_e + \boldsymbol{V} \times \boldsymbol{B}), \quad \boldsymbol{\mathscr{B}} = v\boldsymbol{B}$$

From system (3.3.3) the quantities J_x, J_y, J_z can be found. With the notation

$$\underline{\boldsymbol{Rm}} = \frac{Rm}{1 + \mathscr{B}^2} \begin{pmatrix} 1 + \mathscr{B}_x^2 & \mathscr{B}_x\mathscr{B}_y - \mathscr{B}_z & \mathscr{B}_x\mathscr{B}_z + \mathscr{B}_y \\ \mathscr{B}_x\mathscr{B}_y + \mathscr{B}_z & 1 + \mathscr{B}_y^2 & \mathscr{B}_y\mathscr{B}_z - \mathscr{B}_x \\ \mathscr{B}_x\mathscr{B}_z - \mathscr{B}_y & \mathscr{B}_y\mathscr{B}_z + \mathscr{B}_x & 1 + \mathscr{B}_z^2 \end{pmatrix} \tag{3.3.4}$$

we find

$$\boldsymbol{J} = \underline{\boldsymbol{Rm}} \cdot (\boldsymbol{E} + \boldsymbol{E}_e + \boldsymbol{V} \times \boldsymbol{B}), \tag{3.3.5}$$

$\underline{\boldsymbol{Rm}}$ being called the conductivity tensor. Equation (3.3.5) gives Ohm's law for a completely ionized gas. This is complicated not only because of the tensor form of the conductivity, but also because the conductivity tensor depends on the *a priori* unknown magnetic induction. The conductivity tensor becomes constant and known only for those problems in which the induced magnetic field may be neglected (in this case conductivity only depends on the given applied magnetic field). The induced magnetic field may be neglected only if the scalar Rm is sufficiently small (which is so when σ is sufficiently small). Then, from Ohm's law it follows that curl $\boldsymbol{B} = \boldsymbol{J} \cong 0$ so that together with div $\boldsymbol{B} = 0$ we have $\boldsymbol{B} \cong \boldsymbol{B}_{\text{ext}}$.

3.3.2 The Poiseuille-Hartmann flow

We consider the Poiseuille-Hartmann problem of the previous section and preserve its geometrical hypotheses. Concerning the fluid behaviour we assume that: it is viscous and incompressible, its ionization state is constant, it obeys Ohm's law in the form of equation (3.3.2), and it has a low conductivity, so that the induced magnetic field may be neglected. We shall thus have

$$\frac{\partial}{\partial t} = \frac{\partial}{\partial x} = \frac{\partial}{\partial z} = 0$$

except for the pressure variable. With the above notations we have $\boldsymbol{B} = \nu(0, 1, 0)$, and

$$\underline{Rm} = Rm\begin{pmatrix} \sigma_1 & 0 & \sigma_2 \\ 0 & 1 & 0 \\ -\sigma_2 & 0 & \sigma_1 \end{pmatrix} \tag{3.3.6}$$

where

$$\sigma_1 = \frac{1}{1 + \nu^2}, \; \sigma_2 = \frac{\nu}{1 + \nu^2} \tag{3.3.6'}$$

From the continuity equation it follows that $V_y \equiv 0$, so that,

$$\boldsymbol{V} = (V_x(y),\ 0,\ V_z(y)),\ \boldsymbol{B} = (0, 1, 0) \tag{3.3.7}$$

From (3.3.3) we derive

$$\begin{aligned} J_x &= \sigma_1 Rm\left(E_x - \frac{\nu s}{R^2} - V_z\right) + \sigma_2 Rm(E_z + V_x) \\ J_y &= Rm\left(E_y + \frac{\nu s}{Rm\, Rh}\frac{dp}{dy}\right) \\ J_z &= -\sigma_2 Rm\left(E_x - \frac{\nu s}{R^2} - V_z\right) + \sigma_1 Rm(E_z + V_x) \end{aligned} \tag{3.3.8}$$

so that the equation of motion

$$0 = -\operatorname{grad} p + Re^{-1}\Delta V + RhJ \times B$$

gives ($\partial p/\partial y = 0$).

$$t_1 + \frac{d^2 V_x}{dy^2} - \sigma_1 R^2(E_z + V_x) + \sigma_2 R^2(E_x - V_z) = 0 \tag{3.3.9}$$

$$t_2 + \frac{d^2 V_z}{dy^2} + \sigma_1 R^2(E_x - V_z) + \sigma_2 R^2(E_z + V_x) = 0 \tag{3.3.10}$$

where

$$t_1 = 1 - s(1 - \sigma_1), \; t_2 = - s\sigma_2$$

using the same definition of the characteristic velocity as in Section 3.2.

Eliminating V_z from equations (3.3.9) and (3.3.10), we get a fourth order linear differential equation. Using the boundary conditions we find that

$$V_x = A_1 \operatorname{ch} \alpha y \cos \beta y + A_2 \operatorname{sh} \alpha y \sin \beta y + C_1 \tag{3.3.11}$$

and then

$$V_z = A_2 \operatorname{ch} \alpha y \cos \beta y - A_1 \operatorname{sh} \alpha y \sin \beta y + C_2 \tag{3.3.12}$$

where

$$\sqrt{2}\,\alpha = R(\sigma_1 + \sqrt{\sigma_1^2 + \sigma_2^2})^{1/2}, \; \sqrt{2}\,\beta = R(-\sigma_1 + \sqrt{\sigma_1^2 + \sigma_2^2})^{1/2}$$

$$C_1 = -E_{z0} + \frac{t_1\sigma_1 - t_2\sigma_2}{R^2(\sigma_1^2 + \sigma_2^2)}, \; C_2 = E_{x0} + \frac{t_1\sigma_2 + t_2\sigma_1}{R^2(\sigma_1^2 + \sigma_2^2)}$$

$$A_1 = \frac{-C_1 \operatorname{ch} \alpha \cos \beta + C_2 \operatorname{sh} \alpha \sin \beta}{\operatorname{sh}^2 \alpha + \cos^2 \beta}, \; A_2 = \frac{-C_1 \operatorname{sh} \alpha \sin \beta + C_2 \operatorname{ch} \alpha \cos \beta}{\operatorname{sh}^2 \alpha + \cos^2 \beta}$$

We have used the fact that from equation curl $\mathbf{E} = 0$ it follows that $E_x = E_{x0}$, $E_{z0} = E_{z0}$, E_{x0} and E_z being constants to be determined from the boundary conditions. One also notices that from the equation div $\boldsymbol{J} = 0$ it follows that $J_y = J_{y0}$, and from (3.3.8) that $J_y = RmE_y$. Therefore, for insulating walls we have $E_y \equiv 0$.

Before we go over to the determination of constants E_{x0} and E_{z0}, (which certainly will involve boundary conditions on side walls) we notice that due to the Hall effect the V_z component is not vanishing as for the ohmic fluid. Consequently, the side walls (of equations $z = \pm L_1/L_0$) should be such that the fluid motion is not hampered. Since such walls do not actually exist, the above solution may only be used to describe the motion in the central region of a very large channel ($L_1 \to \infty$).

Let us now specify the boundary conditions for the following ideal cases: insulating side walls and ideally conducting side walls. For the insulating walls, we have:

$$\int_{-1}^{+1} J_z \, dy = 0 \Rightarrow \int_0^1 J_z \, dy = 0 \tag{3.3.13}$$

since, like for V_x and V_z, we have $J_z(y) = J_z(-y)$. Assuming that the walls ($x = \pm L_2/L_0$) bounding the flow along the Ox-axis are insulators, we also have

$$\int_0^1 J_x \, dy = 0 \tag{3.3.14}$$

so that we are able to determine the two constants E_{x0} and E_{z0}. Such conditions are physically possible if, for instance, the flowing ionized gas is sidewise limited by a cool non-conducting gas.

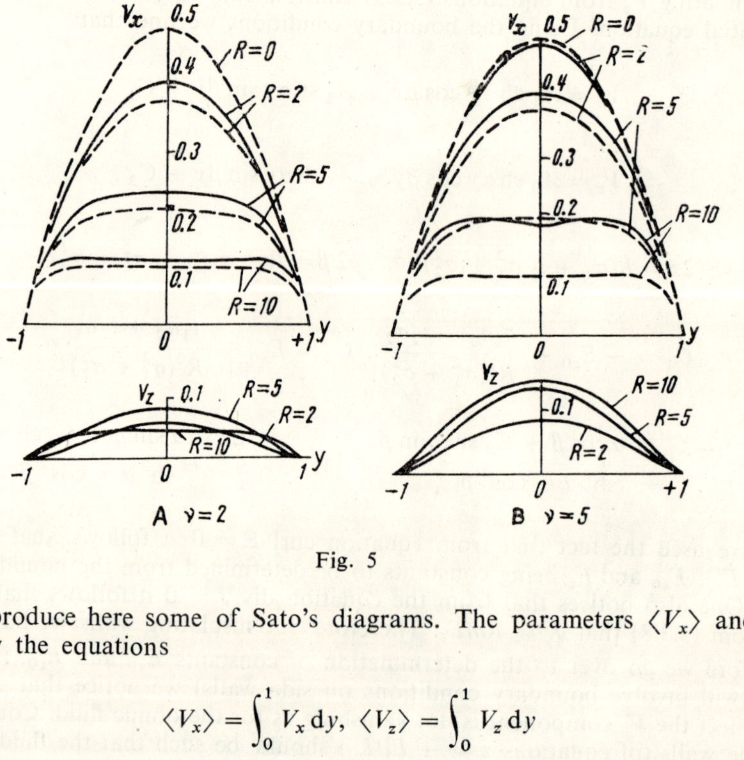

Fig. 5

We reproduce here some of Sato's diagrams. The parameters $\langle V_x \rangle$ and $\langle V_z \rangle$ defined by the equations

$$\langle V_x \rangle = \int_0^1 V_x \, dy, \; \langle V_z \rangle = \int_0^1 V_z \, dy$$

are functions of R and ν only. From Figures 5 and 6 we see that the magnetic field influence on the fluid flow decreases with increasing ν (the dotted lines represent the Hartmann flow). Since the velocity is defined relative to V_0 (given in Section 3.2) the velocity on the channel axis for the Hartmann motion ($R=0$) is 0.5. We see from Figure 5 that the velocity gradient V_x at the walls ($y = \pm 1$) has a finite constant value, while the velocity gradient V_z vanishes for any value of ν and R. This fact presents some physical interest since the tangential flow with zero velocity gradient at the walls is very unstable, in other words, it easily tends to become a turbulent motion. If V_z is unstable, the overall motion may be turbulent even if V_x is stable against external perturbations. To conclude, we cannot expect *the Hall effect to destabilize the laminar motion.*

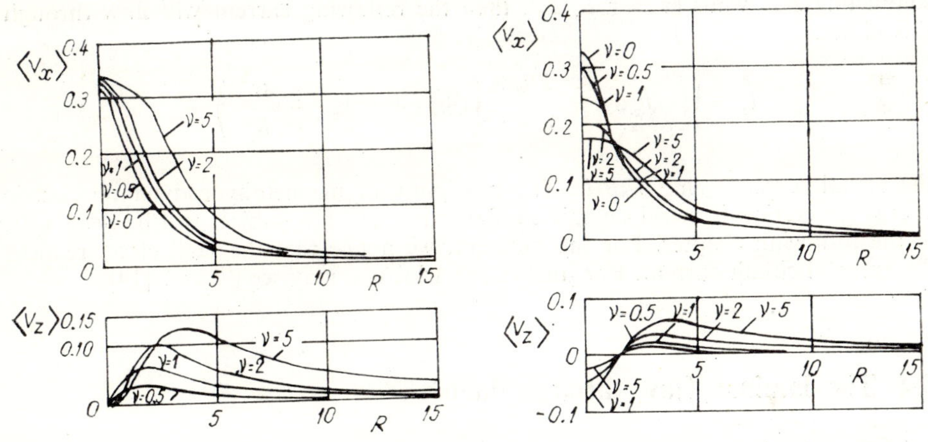

Fig. 6

If the side walls are of conducting material and externally short-circuited then the induced electric currents will flow in a circuit external to the channel. In this case, the potential difference between the lateral walls is zero and thus $E_z = 0$. Supposing also that $E_x = 0$, one determines $\langle V_x \rangle$ and $\langle V_z \rangle$. These quantities are shown in Figure 6a for a weakly ionized gas ($s = 0$) and in Figure 6b for a fully ionized gas ($s = 1/2$). The magnetic field effect (and the decrease in the flow rate) is more pronounced in this case than in that of insulating walls.

One more remark about the fully ionized gas can be made: the limiting value for $R = 0$ of the parameter $\langle V_x \rangle$ depends on the value of ν. Concerning $\langle V_z \rangle$, one notices that this changes from negative values, passes through zero, and becomes positive with increasing R. The negative values of $\langle V_z \rangle$ are explained by the electron pressure action when the magnetic field tends to zero. Indeed, from (3.3.8) we have

$$J_e = \frac{\nu s}{ReRh}(-\sigma_1, 0, \sigma_2)$$

so that the force due to this gradient

$$RhJ_e \times i_2 = -\frac{vs}{Re}(\sigma_2, 0, \sigma_1)$$

slows down the motion even when R is very small. The effect of this force disappears only when B_0 is rigourously zero ($R = 0$).

For conducting side walls the gradient of both V_x and V_z components is independent of v and R and the gradient of V_z vanishes at the walls (contrary to the insulating wall case).

If a resistance is introduced in the external circuit connecting the perfectly conducting side walls ($z = \pm L_1/L_0$), then the following current will flow through this circuit

$$I_z = \int_{-1}^{+1} J_z \, dy = \frac{2Rm}{1 + v^2}\left(\langle V_x \rangle + v\langle V_z \rangle + \frac{sv^2}{R^2}\right)$$

the actual current being $V_0 B_0 I_z$. The possibility of magnetohydrodynamic electric energy generation is based on this configuration and this subject will be dealt with in the following chapter. For the moment, we notice that the Hall effect reduces the external circuit current. For further remarks, one may see [9] and [14].

3.4 The laminar flow through ducts

3.4.1 Introduction

The motion of a non-conducting fluid through a duct can only be maintained by some cause of a mechanical nature: a pressure gradient along the duct, a given upstream motion, the gravitational field, etc. These causes transfer a certain energy to the fluid and the motion is performed through the consumption of this energy. Conducting fluids having an additional property give us the possibility of slowing down or accelerating the motion through the external action of an electromagnetic field. The wellknown phenomena of classical fluid dynamics can in this manner be influenced by the application of an external electromagnetic field of characteristics and structures adequate to the proposed purpose.

Let us, for instance, consider the motion of a conducting fluid through a duct and let us externally apply a magnetic field of induction $\boldsymbol{B}$ normal to the channel axis and an electric field $\boldsymbol{E}$ perpendicular to the plane determined by $\boldsymbol{B}$ and the axis. According to Ohm's law, $\boldsymbol{J}$ will have the direction of $\boldsymbol{E}$ except when the trihedron ($\boldsymbol{V}, \boldsymbol{B}, \boldsymbol{E}$) is negatively orientated and $|\boldsymbol{V} \times \boldsymbol{B}| > |\boldsymbol{E}|$ (in which case $\boldsymbol{J}$ is opposite to $\boldsymbol{E}$). Then the magnetic force $\boldsymbol{J} \times \boldsymbol{B}$ will slow down the motion in the first case and will accelerate it in the second. This is the basic idea of many technical applications like plasma jet engines, magnetohydrodynamic brakes, etc.

Moreover, the external magnetic field determines the appearance within the fluid of an induced current which can be made to flow in an external circuit. In this manner, some of the internal energy of the fluid is given up to the exterior as utilizable electrical energy (magnetohydrodynamic generators). These are some of the reasons for the motion through ducts to be of considerable theoretical and practical importance.

3.4.2 The general problem

Let us consider a straight cylindrical duct*, with constant thickness walls of sufficient length, so that the end effect may be neglected. We assume that the fluid flowing through is viscous, incompressible and has electrical and magnetic constants close to those of the external free space ($\varepsilon \approx \varepsilon_0$, $\mu \approx \mu_0$). The external magnetic field of induction $\boldsymbol{B}_0$ is supposed constant along the duct and perpendicular to the generatrice. A cross-section of the duct is shown in Figure 7, the Ox-axis being coincident with the duct axis and the Oy-axis coincident with the induction axis. We denote by $\mathscr{R}_1$ the region occupied by the fluid, by $\mathscr{R}_2$ the wall region and by $\mathscr{R}_3$ the outside region.

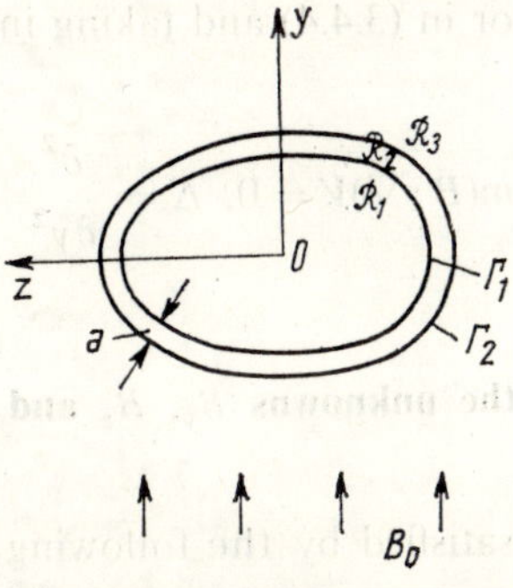

Fig. 7

We assume that the fluid motion is due to a pressure gradient constant along the channel and use the same dimensionless variables as in Section 3.2 (the length L_0 being unspecified for the moment). From the conditions of motion it follows that

$$\frac{\partial}{\partial t} = \frac{\partial}{\partial x} = 0, \quad \frac{\partial p}{\partial x} = -\frac{1}{Re} \tag{3.4.1}$$

* In most of the practical problems the cross-section of the duct changes slowly along the axis (compared with other characteristics of the problem) so that, by considering it constant, the generality of the problem is not very much restricted.

Moreover, we will use the assumption $V = (V_x, 0, 0)$ consistent with the continuity equation.

Using the identity (A.26′), the equation of motion is written as

$$0 = -\operatorname{grad} P + Re^{-1}\Delta V + Rh(B \cdot \nabla)B \tag{3.4.2}$$

where

$$P = p + RhB^2/2 \tag{3.4.2′}$$

is the total pressure. For the solution of the problem we also have the Maxwell equations

$$\operatorname{curl} B = J,\ \operatorname{div} B = 0,\ \operatorname{curl} E = 0 \tag{3.4.3}$$

as well as Ohm's law

$$J = Rm(E + V \times B) \tag{3.4.4}$$

Since we assumed the constants ε and μ to be the same all over, the space equations (3.4.3) should be satisfied in $\mathscr{R}_1 \cup \mathscr{R}_2 \cup \mathscr{R}_3$. Equation (3.4.4) takes different forms in the three regions according to the conductivity and velocity of the respective media (in $\mathscr{R}_2$ and $\mathscr{R}_3$ $V \equiv 0$). Finally equation (3.4.2) is valid only in $\mathscr{R}_1$.

Applying the *curl* operator in (3.4.4) and taking into account (3.4.3) we get the induction equation

$$\Delta B + Rm(B \cdot \nabla)V = 0,\ \Delta = \frac{\partial^2}{\partial y^2} + \frac{\partial^2}{\partial z^2} \tag{3.4.5}$$

3.4.3 The determination of the unknowns B_y, B_z and P

Equation div $B = 0$ will be satisfied by the following

$$B_y = \frac{\partial A}{\partial z},\ B_z = -\frac{\partial A}{\partial y},\ A = A(y, z) \tag{3.4.6}$$

Using (3.4.5) for B_y and B_z it follows that

$$\frac{\partial}{\partial y}\Delta A = \frac{\partial}{\partial z}\Delta A = 0 \Rightarrow \Delta A = A_0 = \text{const.} \tag{3.4.7}$$

But

$$A_0 = \Delta A = -\left(\frac{\partial B_z}{\partial y} - \frac{\partial B_y}{\partial z}\right) = -\,i_1 \cdot \operatorname{curl} B = -\,i_1 \cdot J \tag{3.4.8}$$

so that we have the possibility of determining constant A_0 from the condition at infinity on the current density. Particularly, if the projection on the axis of the current density is zero at infinity, it will be so everywhere inside the duct and A becomes a harmonic function.

Let now

$$y = y(s), \; z = z(s) \tag{3.4.9}$$

be the equation of curve Γ_1 separating the cross-section of $\mathscr{R}_1$ and $\mathscr{R}_2$ and $\boldsymbol{n}$ its external normal:

$$\boldsymbol{n} = \left(0, \frac{dz}{ds}, -\frac{dy}{ds}\right) \tag{3.4.10}$$

From the condition of continuity of the normal component of the induction, as well as from (3.4.6) and (3.4.10), we find

$$A|_{\Gamma 1} = z|_{\Gamma 1} \tag{3.4.11}$$

Consequently, the determination of the components B_y and B_z comes to the solution of the Dirichlet problem (3.4.11) for the Poisson equation (3.4.7) within the cross-section bounded by Γ_1. Particularly, if $A_0 = 0$, then according to the maximum principle of harmonic functions, it follows that $A \equiv z$ and thus $B_y \equiv 1$ and $B_z \equiv 0$.

Projecting now the equation of motion (2) on the $Oy-$ and $Oz-$ axis, we obtain

$$\frac{\partial P}{\partial y} = Rh\left\{\frac{1}{2}\frac{\partial}{\partial y}(B_y^2 + B_z^2) - B_z\left(\frac{\partial B_z}{\partial y} - \frac{\partial B_y}{\partial z}\right)\right\}$$

$$\frac{\partial P}{\partial z} = Rh\left\{\frac{1}{2}\frac{\partial}{\partial z}(B_y^2 + B_z^2) + B_y\left(\frac{\partial B_z}{\partial y} - \frac{\partial B_y}{\partial z}\right)\right\}$$

so that with the notation

$$P^* = p + RhB_x^2/2$$

it follows that

$$\frac{\partial P^*}{\partial y} = RhA_0B_z = -RhA_0\frac{\partial A}{\partial y}$$

$$\frac{\partial P^*}{\partial z} = -RhA_0B_y = -Rh\,A_0\frac{\partial A}{\partial z}$$

and then

$$P^* = -RhA_0A(y,z) + f(x) \tag{3.4.12}$$

Finally, from (3.4.1) it follows that $f(x) = -Re^{-1}x + p_0$, where p_0 is a constant.
In this manner, the problem of determining the unknowns B_y, B_z and P is separated from that of determining the unknowns V_x and B_x.

3.4.4 The determination of the unknowns V_x and B_x

For the determination of V_x and B_x we have the projections of equations (3.4.2) and (3.4.5) on the Ox-axis:

$$\begin{aligned} \Delta V_x + ReRh\left(B_y\frac{\partial B_x}{\partial y} + B_z\frac{\partial B_x}{\partial z}\right) &= -1 \\ \Delta B_x + Rm\left(B_y\frac{\partial V_x}{\partial y} + B_z\frac{\partial V_x}{\partial z}\right) &= 0 \end{aligned} \tag{3.4.13}$$

where B_y and B_z may be considered to be known (previously determined).
The system (3.4.13) is put into a symmetrical form with the notation

$$V_x = V, \; ReRhB_x = RB \tag{3.4.14}$$

and then it becomes

$$\begin{aligned} \Delta V + R\left(B_y\frac{\partial B}{\partial y} + B_z\frac{\partial B}{\partial z}\right) &= -1 \\ \Delta B + R\left(B_y\frac{\partial V}{\partial y} + B_z\frac{\partial V}{\partial z}\right) &= 0 \end{aligned} \tag{3.4.15}$$

In the problems with homogeneous boundary conditions on V and B the following is a suitable change of variables

$$\begin{aligned} 2V &= U_1 + U_2 \\ 2B &= U_1 - U_2 \end{aligned} \tag{3.4.16}$$

This separates the system

$$\begin{aligned} \Delta U_1 + R\left(B_y\frac{\partial U_1}{\partial y} + B_z\frac{\partial U_1}{\partial z}\right) &= -1 \\ \Delta U_2 - R\left(B_y\frac{\partial U_2}{\partial y} + B_z\frac{\partial U_2}{\partial z}\right) &= -1 \end{aligned} \tag{3.4.17}$$

Finally, when B_y and B_z are constant, equations (3.4.17) have the form

$$\begin{aligned} \Delta u_1 - k^2 u_1 &= -\exp\{(R/2)(B_y y + B_z z)\} \\ \Delta u_2 - k^2 u_2 &= -\exp\{-(R/2)(B_y y + B_z z)\} \end{aligned} \tag{3.4.18}$$

With the notations

$$\begin{aligned} u_1 &= U_1 \exp\{(R/2)(B_y y + B_z z)\} \\ u_2 &= U_2 \exp\{-(R/2)(B_y y + B_z z)\} \\ 4k^2 &= R^2(B_y^2 + B_z^2) \end{aligned} \tag{3.4.18'}$$

Let us now establish the boundary conditions for B_x. In this respect, we assume that the medium outside the duct is an insulating one (curl $\boldsymbol{B} = 0$). It follows that $B_x^{(3)} = 0$ (Indices will be used for the interesting quantities in $\mathscr{R}_2$ and $\mathscr{R}_3$; in $\mathscr{R}_1$ these indices will be dropped). In $\mathscr{R}_2$ equation (3.4.5) with $V \equiv 0$ is satisfied, so that we have

$$\Delta B_x^{(2)} = 0 \tag{3.4.19}$$

Since the tangential component of the induction is continuous across Γ_2 and Γ_1, it follows that

$$B_x^{(2)}\big|_{\Gamma_2} = 0 \tag{3.4.20}$$

$$B_x\big|_{\Gamma_1} = B_x^{(2)}\big|_{\Gamma_1} \tag{3.4.21}$$

Let us now also use the continuity condition of the tangential component of the electric field (1.5.12). From Ohm's law we have

$$\left[\frac{1}{Rm}\boldsymbol{J}\cdot\boldsymbol{s}\right]_{\Gamma_1} = [\boldsymbol{E}\cdot\boldsymbol{s}]_{\Gamma_1} + [(\boldsymbol{V}\times\boldsymbol{B})\cdot\boldsymbol{s}]_{\Gamma_1} \tag{3.4.22}$$

But $\boldsymbol{V}$ vanishes on Γ_1, within the fluid (the adherence condition) and wall, so that the last term is vanishing. Then we notice that

$$\boldsymbol{s} = \left(0, \frac{dy}{ds}, \frac{dz}{ds}\right)$$

and

$$J_y = \frac{\partial B_x}{\partial z}, \quad J_z = -\frac{\partial B_x}{\partial y}$$

The continuity condition of the electric field entails

$$\left[\frac{1}{Rm}\frac{dB_x}{dn}\right] = 0 \tag{3.4.23}$$

For V_x we have the adherence condition

$$V_x|_{\Gamma_1} = 0 \tag{3.4.24}$$

To conclude, the problem of determining the unknowns V_x and B_x is reduced to finding that solution of system (3.4.13) which satisfies condition (3.4.24) and through conditions (3.4.21) and (3.4.23) joins itself up with the solution of equation (3.4.19) determined from condition (3.4.20). This is a very difficult mathematical problem.

However, two ideal cases exist in which the solution of (3.4.13) closes itself on Γ_1 without joining up with the solution in $\mathscr{R}_2$. This is the case for insulating and perfectly conducting walls. Indeed, for insulating walls ($\sigma^{(2)}=0$) from equation (3.4.23) it follows that

$$(dB_x^{(2)}/dn)_{\Gamma_1} = 0$$

so that $B_x^{(2)} \equiv 0$. From (3.4.21) one has

$$B_x|_{\Gamma_1} = 0 \tag{3.4.25}$$

System (3.4.17) together with the boundary conditions (3.4.24) and (3.4.25) determine the solution of the problem. In the other limiting case (perfectly conducting walls), condition (3.4.23) reduces to

$$\left.\frac{dB_x}{dn}\right|_{\Gamma_1} = 0 \tag{3.4.26}$$

so that system (3.4.13) together with conditions (3.4.24) and (3.4.26) determine the solution.

Finally, Shercliff [15] [G. 15] has also pointed out the case of the thin wall duct for which one may assume to a good approximation that the harmonic function $B^{(2)}$ is linear. Taking into account (3.4.20) it follows that

$$dB_x^{(2)}/dn = -\,L_0 B_x^{(2)}/a$$

on Γ_1 (a being the wall thickness). Then equation (3.4.23) becomes

$$\left(\frac{dB_x}{dn} + \frac{B_x}{\varphi}\right)_{\Gamma_1} = 0,\ \varphi = \frac{a\sigma^{(2)}}{L_0\sigma} \tag{3.4.27}$$

Conditions (3.4.24) and (3.4.27) are sufficient for the solution of the problem. We notice that if $\sigma^{(2)} = 0$ from (3.4.27) we get condition (3.4.25) and if $\sigma^{(2)} = \infty$ we get condition (3.4.26).

3.5 The rectangular duct

3.5.1 Motion between parallel planes

This represents a case in which the approximate boundary conditions (3.4.27) become exact. Indeed, supposing the channel to be of rectangular cross-section with walls perpendicular to Oz at infinity, we may conclude that function $B_x^{(2)}$ does not depend on the z-variable. Since this is also a harmonic function, it will become linear in y, i.e., $B_x^{(2)} = C_1 y + C_2$. If a is the actual thickness of the wall, from the conditions of continuity of $B_x^{(2)}$ across the straight lines $y = 1$ and $y = 1 + a/L$ we obtain

$$C_1 + C_2 = B_x, \quad C_1\left(1 + \frac{a}{L}\right) + C_2 = 0 \Rightarrow C_1 = -\frac{L}{a} B_x$$

so that conditions (3.4.23) and (3.4.24) become

$$\left(\frac{dB}{dy} \pm \frac{B}{\varphi}\right)_{y=\pm 1} = 0, \quad V(\pm 1) = 0 \tag{3.5.1}$$

Assuming also that there is no current in the Ox-axis direction, system (3.4.15) becomes

$$\frac{d^2V}{dy^2} + R\frac{dB}{dy} = -1, \quad \frac{d^2B}{dy^2} + R\frac{dV}{dy} = 0 \tag{3.5.2}$$

and has the solution

$$V = \frac{1}{R}\frac{\varphi + 1}{R\varphi + \operatorname{th} R}\left(1 - \frac{\operatorname{ch} Ry}{\operatorname{ch} R}\right)$$

$$B = -\frac{y}{R} + \frac{1}{R}\frac{\varphi + 1}{R\varphi + \operatorname{th} R}\frac{\operatorname{sh} Ry}{\operatorname{ch} R} \tag{3.5.3}$$

which for insulating walls ($\varphi = 0$) coincides with the solution of Section 3.2.

For a given φ the velocity has a parabolic profile if $R = 0$. With increasing R, velocity tends to become uniform except for the boundary layer of thickness of the order of R^{-1}. Asymptotically,

$$V = \frac{1}{R}\frac{\varphi + 1}{R\varphi + 1}$$

which points out that for insulating walls $V \approx R^{-1}$ while for perfectly conducting ones $V \approx R^{-2}$. We find again the following result: *the velocity is inversely proportional to the wall conductivity.*

The ratio of the Lorentz force to the pressure gradient

$$Rh \frac{dB_x}{dy} \Big/ \left(- \frac{\partial p}{\partial x} \right) = ReRh \frac{dB_x}{dy} = R \frac{dB}{dy}$$

as well as the current density

$$J_x \equiv J_y \equiv 0, \; J_z = - \frac{R}{ReRh} \frac{dB}{dy}$$

are proportional to the quantity

$$R \frac{dB}{dy} = -1 + R \frac{\varphi + 1}{R\varphi + \text{th}\, R} \frac{\text{ch}\, Ry}{\text{ch}\, R} \tag{3.5.4}$$

The graph of this function for $R = 3$, $\varphi = 1$ and $a = 0.5\, L_0$ is given in Figure 8 (after Chang and Lundgren). One notices that near the central part of the duct the current flows to the right which fact indicates that the Lorentz force is opposed to the motion (the Lorentz force adds itself to the viscosity to balance the pressure

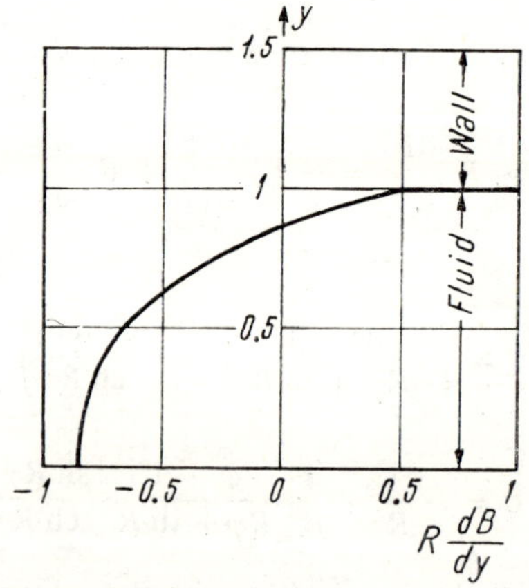

Fig. 8

gradient). Near the walls the current flows to the left (looking along the direction of fluid flow) so that a Lorentz force opposed to the fluid viscosity appears. For large R we have $R\,(dB/dy) \simeq -1$ except for the boundary layer of thickness R^{-1}.

One also finds that the fluid velocity decreases with increasing wall conductivity.

We have the following formula for the flow rate

$$Q = \int_{-1}^{+1} V\,dy = 2\,\frac{\varphi + 1}{R^2}\,\frac{R - \operatorname{th} R}{R\varphi + \operatorname{th} R}$$

so that

$$\Lambda = \frac{Q(R = 0)}{Q(R)} = \frac{R^2(R\varphi + \operatorname{th} R)}{3(\varphi + 1)(R - \operatorname{th} R)} \tag{3.5.5}$$

this quantity being plotted in Figure 9 [7]. From the meaning of Q (inversely proportional to the pressure gradient), we deduce that Λ represents the ratio of the pressure gradient needed to transfer a certain quantity of conducting fluid to the

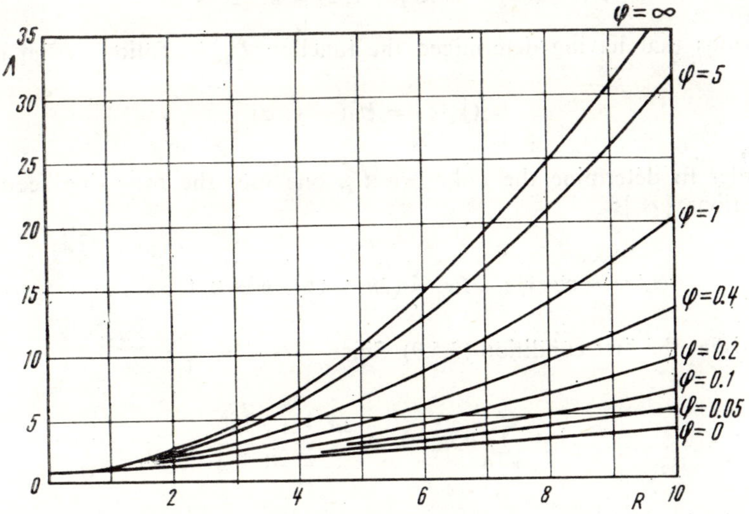

Fig. 9

pressure gradient needed to transfer the same quantity of non-conducting fluid. One thus finds that for a given mass current a higher pressure gradient is needed for the perfectly conducting than for the insulating walls duct.

3.5.2 The rectangular duct with insulating walls

For this case, the boundary conditions are homogeneous in V and B.

$$V(y, \pm L) = 0,\ B(y, \pm L) = 0,\ |y| \leqslant 1 \tag{3.5.6}$$

$$V(\pm 1, z) = 0,\ B(\pm 1, z) = 0,\ |z| \leqslant L \tag{3.5.7}$$

($L = L_1/L_0$, $2L_1$ being the channel width along the Oz-axis), so that we can successfully use the transformation (3.4.16). If, moreover, we assume that there is no electric current along the Ox-axis, then we have to integrate the following system:

$$\Delta U_1 + R\frac{\partial U_1}{\partial y} = -1, \ \Delta U_2 - R\frac{\partial U_2}{\partial y} = -1 \tag{3.5.8}$$

with the boundary conditions

$$U_1(y, \pm L) = 0, \ U_2(y, \pm L) = 0, \ |y| \leqslant 1 \tag{3.5.9}$$

$$U_1(\pm 1, z) = 0, \ U_2(\pm 1, z) = 0, \ |z| \leqslant L$$

It is obvious that having determined the function U_1, U_2 follows from the relationship

$$U_2(y, z) = U_1(-y, z) \tag{3.5.10}$$

In order to determine the unknown U_1, one uses the following sequence of eigenfunctions $\{\varphi_n\}$:

$$\varphi_n(z) = \cos \nu_n z, \ 2L\nu_n \overset{\text{def}}{=} (2n+1)\pi, \ n = 0, 1, \ldots \tag{3.5.11}$$

which satisfies the first condition (3.5.9). Since

$$1 = \sum_{n=0}^{\infty} a_n \varphi_n(z), \ a_n \overset{\text{def}}{=} \frac{4}{\pi}\frac{(-1)^n}{2n+1}, \tag{3.5.12}$$

being an absolutely convergent series for $|z| < L$, we shall look for solutions of the form

$$U_1 = \sum_{n=0}^{\infty} A_n(y) a_n \varphi_n(z) \tag{3.5.13}$$

and will determine the functions $A_n(y)$ such that the boundary condition relative to y should be also satisfied. From (3.5.8) one has

$$A_n'' + RA_n' - \nu_n^2 A_n = -1$$

so that, if we denote by r_{1n} and r_{2n} the roots of equation

$$r^2 + Rr - \nu_n^2 = 0, \tag{3.5.14}$$

it follows that

$$\nu_n^2 A_n(y) = 1 + \frac{\exp(r_{1n}y)\,\mathrm{sh}\,r_{2n} - \exp(r_{2n}y)\,\mathrm{sh}\,r_{1n}}{\mathrm{sh}(r_{1n} - r_{2n})}$$

A simple calculation gives

$$\nu_n^2\{A_n(y) + A_n(-y)\} = 2\left\{1 + \frac{\mathrm{sh}\,r_{2n}\,\mathrm{ch}(r_{1n}y) - \mathrm{sh}\,r_{1n}\,\mathrm{ch}(r_{2n}y)}{\mathrm{sh}(r_{1n} - r_{2n})}\right\}$$

$$\nu_n^2\{A_n(y) - A_n(-y)\} = 2\left\{\frac{\mathrm{sh}\,r_{2n}\,\mathrm{sh}(r_{1n}y) - \mathrm{sh}\,r_{1n}\,\mathrm{sh}(r_{2n}y)}{\mathrm{sh}(r_{1n} - r_{2n})}\right\}$$

so that

$$2\begin{Bmatrix} V \\ B \end{Bmatrix} = U_1(y, z) \pm U_1(-y, z) = \sum_{n=0}^{\infty} \{A_n(y) \pm A_n(-y)\}a_n\varphi_n(z)$$

and finally [4]

$$\begin{Bmatrix} V \\ B \end{Bmatrix} = \frac{16L^2}{\pi^3}\sum_{n=0}^{\infty}\frac{(-1)^n}{(2n+1)^3}\left\{\begin{matrix} 1 + \dfrac{\mathrm{sh}\,r_{2n}\,\mathrm{ch}(r_{1n}y) - \mathrm{sh}\,r_{1n}\,\mathrm{ch}(r_{2n}y)}{\mathrm{sh}(r_{1n} - r_{2n})} \\[2ex] \dfrac{\mathrm{sh}\,r_{2n}\,\mathrm{sh}(r_{1n}y) - \mathrm{sh}\,r_{1n}\,\mathrm{sh}(r_{2n}y)}{\mathrm{sh}(r_{1n} - r_{2n})} \end{matrix}\right\}\cos\nu_n z \qquad (3.5.15)$$

For the asymptotic behaviour of the solution one may refer to [16].

3.5.3 The rectangular duct with perfectly conducting walls

In this case [7], we have the following boundary conditions:

$$V(y, \pm L) = 0, \quad \left.\frac{\mathrm{d}B}{\mathrm{d}z}\right|_{z=\pm L} = 0, \quad |y| \leqslant 1 \qquad (3.5.16)$$

$$V(\pm 1, z) = 0, \quad \left.\frac{\mathrm{d}B}{\mathrm{d}y}\right|_{y=\pm 1} = 0, \quad |z| \leqslant L \qquad (3.5.17)$$

Looking for a solution of the form

$$V = \sum_{n=0}^{\infty} a_n V_n(z)\cos\beta_n y, \quad B = \sum_{n=0}^{\infty} a_n B_n(z)\sin\beta_n y \qquad (3.5.18)$$

the boundary conditions (3.5.17) will be satisfied if

$$\beta_n = (2n + 1)\pi/2 \tag{3.5.19}$$

With the above meaning of a_n we have

$$1 = \Sigma\, a_n \cos \beta_n y$$

so that, from equations (3.4.15), we have

$$V_n'' - \beta_n^2 V_n + R\beta_n B_n = -1$$

$$B_n'' - \beta_n^2 B_n - R\beta_n V_n = 0$$

The solution of this system obeying the boundary conditions (3.5.16) is

$$\begin{Bmatrix} V_n \\ B_n \end{Bmatrix} = \frac{1}{\beta_n^2 + R^2} \begin{Bmatrix} 1 - \dfrac{r_{2n}\,\mathrm{sh}(r_{2n}L)\,\mathrm{ch}(r_{1n}z) + r_{1n}\,\mathrm{sh}(r_{1n}L)\,\mathrm{ch}(r_{2n}z)}{r_{2n}\,\mathrm{ch}(r_{1n}L)\,\mathrm{sh}(r_{2n}L) + r_{1n}\,\mathrm{ch}(r_{2n}L)\,\mathrm{sh}(r_{1n}L)} \\ -\dfrac{R}{\beta_n} + \mathrm{i}\,\dfrac{r_{2n}\,\mathrm{sh}(r_{2n}L)\,\mathrm{ch}(r_{1n}z) - r_{1n}\,\mathrm{sh}(r_{1n}L)\,\mathrm{ch}(r_{2n}z)}{r_{2n}\,\mathrm{ch}(r_{1n}L)\,\mathrm{sh}(r_{2n}L) + r_{1n}\,\mathrm{ch}(r_{2n}L)\,\mathrm{sh}(r_{1n}L)} \end{Bmatrix} \tag{3.5.20}$$

where the complex quantities

$$r_{1n} = \sqrt{\beta_n^2 + \mathrm{i}R\beta_n}, \; r_{2n} = \sqrt{\beta_n^2 - \mathrm{i}R\beta_n} \tag{3.5.21}$$

may be separated into their real and imaginary parts:

$$r_{1n} = \alpha_n + \mathrm{i}\gamma_n, \; r_{2n} = \alpha_n - \mathrm{i}\gamma_n$$

$$\sqrt{2}\,\alpha_n = \sqrt{\beta_n(\beta_n + \sqrt{\beta_n^2 + R^2})}, \; \sqrt{2}\,\gamma_n = \sqrt{\beta_n(-\beta_n + \sqrt{\beta_n^2 + R^2})}$$

After some simple calculations one obtains the final solution

$$\begin{Bmatrix} V \\ B \end{Bmatrix} = \sum_{n=0}^{\infty} \frac{2(-1)^n}{\beta_n(\beta_n^2 + R^2)} \begin{Bmatrix} \left(1 - \dfrac{\alpha_n E_n(z) - \gamma_n F_n(z)}{\alpha_n\,\mathrm{sh}(2\alpha_n L) - \gamma_n\,\mathrm{sh}(2\gamma_n L)}\right) \cos \beta_n y \\ \left(-\dfrac{R}{\beta_n} + \dfrac{\alpha_n F_n(z) + \gamma_n E_n(z)}{\alpha_n\,\mathrm{sh}(2\alpha_n L) - \gamma_n\,\mathrm{sh}(2\gamma_n L)}\right) \sin \beta_n y \end{Bmatrix} \tag{3.5.22}$$

where

$$4E_n(z) = \cos\gamma_n(L-z)\,\mathrm{sh}\,\alpha_n(L+z) + \cos\gamma_n(L+z)\,\mathrm{sh}\,\alpha_n(L-z)$$
$$4F_n(z) = \sin\gamma_n(L-z)\,\mathrm{ch}\,\alpha_n(L+z) + \sin\gamma_n(L+z)\,\mathrm{ch}\,\alpha_n(L-z) \qquad (3.5.22')$$

In order to calculate the mass flow per unit pressure gradient (i.e. Q), we should use the complex form of the velocity. Finally we have

$$Q = \sum_{n=0}^{\infty} \frac{8}{\beta_n^2(\beta_n^2+R^2)} \left\{ L - \frac{\beta_n}{\sqrt{\beta_n^2+R^2}} \; \frac{\mathrm{ch}(2\alpha_n L) - \cos(2\alpha_n L)}{\alpha_n \mathrm{sh}(2\alpha_n L) - \gamma_n \sin(2\gamma_n L)} \right\} \qquad (3.5.23)$$

the function Λ (defined in (3.5.5)) for a rectangular duct being plotted in Figure 10. In this figure, the dependence of Λ for insulating walls is also included (after Shercliff).

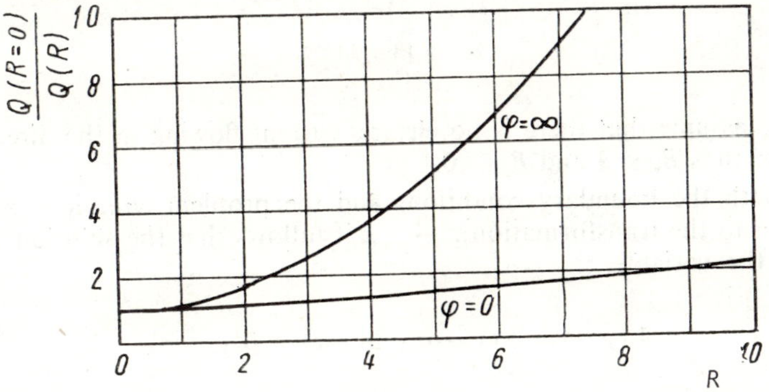

Fig. 10

This problem has also been studied by Pan Chang Lu [17] by means of a variational principle.

3. 6 The rectangular duct of two perfectly conducting and two insulating walls

3.6.1 The integral equation

Let us assume that the side walls parallel to the external induction are perfectly conducting, and the horizontal walls, perpendicular to the magnetic induction, are insulators (Fig. 11).

In such conditions, we have

$$V(y, \pm L) = 0, \left.\frac{dB}{\partial z}\right|_{z=\pm L} = 0, \ |y| \leqslant 1 \tag{3.6.1}$$

$$V(\pm 1, z) = 0, \ B(\pm 1, z) = 0, \ |z| \leqslant L \tag{3.6.2}$$

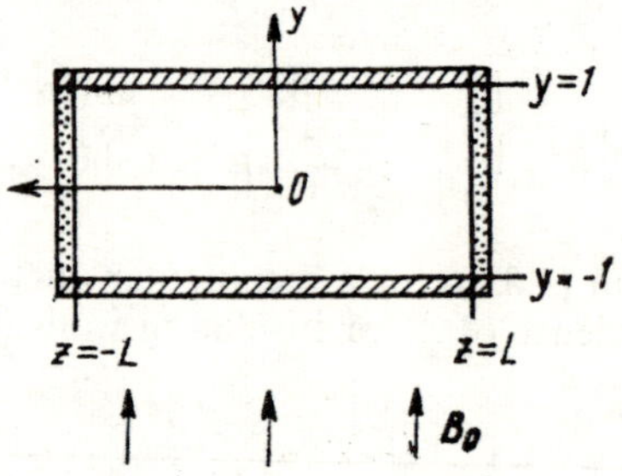

Fig. 11

Let us also assume that there is no electric current flowing in the direction of the Ox-axis and thus $B_y = 1$ and $B_z = 0$.

Since both the boundary conditions and the problem equations are invariant with respect to the transformation $z \to -z$ it follows that the solution is even with respect to the variable z:

$$V(y, z) = V(y, -z), \ B(y, z) = B(y, -z) \tag{3.6.3}$$

and thus

$$\frac{\partial V}{\partial z}(y, z) = -\frac{\partial V}{\partial z}(y, -z), \ \frac{\partial B}{\partial z}(y, z) = -\frac{\partial B}{\partial z}(y, -z) \tag{3.6.3'}$$

Since variables are continuous it follows that

$$\frac{\partial V}{\partial z}(y, 0) = 0, \ \frac{\partial B}{\partial z}(y, 0) = 0 \tag{3.6.4}$$

so that it is sufficient to solve the problem in the semi-duct $0 \leqslant z \leqslant L$.

With the change of variables (3.4.16), it follows that

$$\Delta U_1 + R\frac{\partial U_1}{\partial y} = -1, \ \Delta U_2 - R\frac{\partial U_2}{\partial y} = -1 \tag{3.6.5}$$

and also

$$\begin{cases} \left.\dfrac{\partial U_1}{\partial z}\right|_{z=0} = 0, \quad \left.\dfrac{\partial U_1}{\partial z}\right|_{z=L} = f(y), \quad |y| \leqslant 1 \\ U_1(\pm 1, z) = 0, \quad 0 \leqslant z \leqslant L \end{cases} \tag{3.6.6}$$

$$\begin{cases} \left.\dfrac{\partial U_2}{\partial z}\right|_{z=0} = 0, \quad \left.\dfrac{\partial U_2}{\partial z}\right|_{z=L} = f(y), \quad |y| \leqslant 1 \\ U_2(\pm 1, z) = 0, \quad 0 \leqslant z \leqslant L \end{cases} \tag{3.6.7}$$

where $f(y)$ is an unknown function. The condition (3.6.1) for V (not used so far) becomes

$$U_1(y, L) + U_2(y, L) = 0 \tag{3.6.8}$$

which equation will determine the function f (by means of an integral equation).

The problem is further simplified by the following change of variables:

$$RU_1 = RW_1 - y, \quad RU_2 = RW_2 + y \tag{3.6.9}$$

and then by the transformation (3.4.18′):

$$W_1 = u_1 \exp(-ky), \quad W_2 = u_2 \exp(ky), \quad 2k = R \tag{3.6.10}$$

Finally one obtains

$$\Delta u_1 - k^2 u_1 = 0, \quad \Delta u_2 - k^2 u_2 = 0 \tag{3.6.11}$$

$$\left.\frac{\partial u_1}{\partial z}\right|_{z=0} = 0, \quad \left.\frac{\partial u_1}{\partial z}\right|_{z=L} = f(y)\,\mathrm{e}^{ky}, \quad 2ku_1(\pm 1, z) = \pm\,\mathrm{e}^{\pm k} \tag{3.6.12}$$

$$\left.\frac{\partial u_2}{\partial z}\right|_{z=0} = 0, \quad \left.\frac{\partial u_2}{\partial z}\right|_{z=L} = f(y)\,\mathrm{e}^{-ky}, \quad 2ku_2(\pm 1, z) = \mp\,\mathrm{e}^{\mp k} \tag{3.6.13}$$

In order to homogenize the conditions on the boundaries $y = \pm 1$, one should also make the following change

$$\begin{aligned} u_1(y, z) &= a(y) + v_1(y, z) \\ u_2(y, z) &= a(-y) + v_2(y, z) \end{aligned} \tag{3.6.14}$$

where $a(y)$ is a particular solution of equation (3.6.11) satisfying the last condition of (3.6.12)

$$2ka(y) = \text{th } k \text{ ch } ky + \text{cth } k \text{ sh } ky \tag{3.6.15}$$

With this we get

$$\Delta v_1 - k^2 v_1 = 0, \ \Delta v_2 - k^2 v_2 = 0 \tag{3.6.16}$$

$$\left.\frac{\partial v_1}{\partial z}\right|_{z=0} = 0, \ \left.\frac{\partial v_1}{\partial z}\right|_{z=L} = f(y)\, e^{ky}, \quad v_1(\pm 1, z) = 0 \tag{3.6.17}$$

$$\left.\frac{\partial v_2}{\partial z}\right|_{z=0} = 0, \ \left.\frac{\partial v_2}{\partial z}\right|_{z=L} = f(y)\, e^{-ky}, \ v_2(\pm 1, z) = 0 \tag{3.6.18}$$

while condition (3.6.8) becomes

$$v_1(y, L) \exp(-ky) + v_2(y, L) \exp(ky) = 2(\text{ch}^2 ky - \text{ch}^2 k)/k \text{ sh } 2k \tag{3.6.19}$$

In this manner, it is sufficient to find the general expression of function v_1.

If $P(y, z)$ is an arbitrary point within the domain D: $(|y| < 1, 0 < z < L)$ and $Q(\eta, \zeta)$ an arbitrary point on the boundary ∂D and if we denote by $G(P, Q)$ the Green function associated with operator (3.6.16), then the general solution has the form

$$v_1(P) = \frac{1}{2\pi} \oint_{\partial D} \left\{ v_1(Q) \frac{\partial}{\partial n_Q} G(P, Q) - G(P, Q) \frac{\partial v_1}{\partial n_Q} \right\} ds_Q \tag{3.6.20}$$

If moreover the Green function satisfies the conditions

$$\left.\frac{\partial G}{\partial \zeta}\right|_{\zeta=0} = 0, \ \left.\frac{\partial G}{\partial \zeta}\right|_{\zeta=L} = 0, \ G(\pm 1, \zeta, y, z) = 0 \tag{3.6.21}$$

then

$$v_1(y, z) = \frac{1}{2\pi} \int_{-1}^{+1} G(\eta, L, y, z) f(\eta)\, e^{k\eta}\, d\eta \tag{3.6.22}$$

and in a similar way

$$v_2(y, z) = \frac{1}{2\pi} \int_{-1}^{+1} G(\eta, L, y, z) f(\eta)\, e^{-k\eta}\, d\eta \tag{3.6.23}$$

so that condition (3.6.19) becomes

$$\frac{1}{2\pi} \int_{-1}^{+1} G(\eta, L, y, L) \text{ ch } k(\eta - y) f(\eta)\, d\eta = \frac{\text{ch}^2 ky - \text{ch}^2 k}{k \text{ sh } 2k} \tag{3.6.24}$$

This is the integral equation which is going to determine the function f and thus solve the problem. If this is done, then

$$2V = a(y)\,e^{-ky} + a(-y)\,e^{ky} + v_1\,e^{-ky} + v_2\,e^{ky}$$

$$2B = a(y)\,e^{-ky} - a(-y)\,e^{ky} + v_1\,e^{-ky} - v_2\,e^{ky} - 2y/R$$

so that

$$V(y, z) = a^*(y) + \frac{1}{2\pi}\int_{-1}^{+1} G(\eta, L, y, z) f(\eta)\,\mathrm{ch}\,k(\eta - y)\,d\eta$$

$$B(y, z) = a^{**}(y) + \frac{1}{2\pi}\int_{-1}^{+1} G(\eta, L, y, z) f(\eta)\,\mathrm{sh}\,k(\eta - y)\,d\eta \tag{3.6.25}$$

where

$$2ka^*(y) = \mathrm{th}\,k\,\mathrm{ch}^2\,ky - \mathrm{cth}\,k\,\mathrm{sh}^2\,ky$$

$$2k\,\mathrm{th}k a^{**}(y) = (1 - \mathrm{th}^2\,k)\,\mathrm{ch}\,ky\,\mathrm{sh}\,ky$$

The Green function associated to operator (3.6.16) within the domain D with the boundary condition (3.6.21) is determined by the reflection method. Indeed, the fundamental solution of operator (3.6.16) having a logarithmic singularity at $r = 0$ is $K_0(kr)$;

$$2K_0(kr) = \pi i H_0^1(ikr), \quad r = \sqrt{(y - \eta)^2 + (z - \zeta)^2}$$

H_0^1 being the Henkel function [G.37]. In the vicinity of point $r = 0$ we have

$$K_0(kr) = \ln\frac{1}{r} + \ldots .$$

with the bounded terms being dropped. If now starting from $P(y, z)$ we place negative sources at points $P_1(2 - y, z)$ and P_{-1} $(-2 - y, z)$, symmetrical about the straight lines $y = \pm 1$, and positive sources at the points $P_2(4 + y, z)$ and P_{-2} $(-4 + y, z)$ and so on, the last condition of (3.6.21) will be fulfilled. Consequently, the function

$$G(P, Q) = \sum_{n=-\infty}^{+\infty} (-1)^n K_0\left(k\sqrt{(y_n - \eta)^2 + (z - \zeta)^2}\right)$$

$$y_n = 2n + (-1)^n y, \quad n = 0, \pm 1, \pm 2, \ldots$$

satisfies the last condition (3.6.21). The first two conditions of (3.6.21) are satisfied if one places positive sources at points symmetrical about $z = 0$ and $z = 1$ and then sum up. Then the function

$$G(P, Q) = \sum_{m=-\infty}^{+\infty} K_0\left(k\sqrt{(y-\eta)^2 + (z_m - \zeta)^2}\right)$$

$$2z_m - L = 2mL + (-1)^m(2z - L)$$

satisfies the first two boundary conditions (3.6.21). This is easily checked by grouping together two by two the functions under Σ and noticing that each group is an even function in ζ and $\zeta - L$ respectively, so that its derivative with respect to ζ vanishes (for $\zeta = 0$ and $\zeta - L = 0$). To conclude, the Green function satisfying all the conditions is

$$G(P, Q) = \sum_{m=-\infty}^{+\infty} \sum_{n=-\infty}^{+\infty} (-1)^n K_0\left(k\sqrt{(y_n - \eta)^2 + (z_m - \zeta)^2}\right) \qquad (3.6.26)$$

This has been established for the first time by Greenberg [8].

The above solution represents a simplified form of Greenberg's solution due to N. Marcov. Although this is a rigorous solution, from the mathematical point of view, its practical usefulness is limited since the integral equation cannot be solved easily. Hence other solutions—based on the boundary layer theory have been put forward (Hunt [18], Chang and Lundgren [19]). The reader may also find asymptotic solutions valid for large and small R in [20] and [21].

4

The theory of magnetohydrodynamics generators

4.1 Introduction

One of the most important applications of magnetofluid dynamics (magnetohydrodynamics) is undoubtedly the direct conversion of thermal energy into electrical energy — the magnetohydrodynamic (MHD) conversion. This can be produced by means of MHD generators.

A MHD generator is essentially made up of a cylindrical tube (Figure 12) through which an ionized gas flows, the tube having insulating walls with two perfectly conducting regions which represent the electrodes. A magnetic field having the role of braking the gas motion and orienting the charged particles towards the electrodes is applied externally in the electrode region. The two electrodes — one positive and the other negative — collect the charges and make them flow in an external load circuit. In this way one obtains an electric current directly from the thermal energy of the ionized gas without the use of an intermediate mechanical

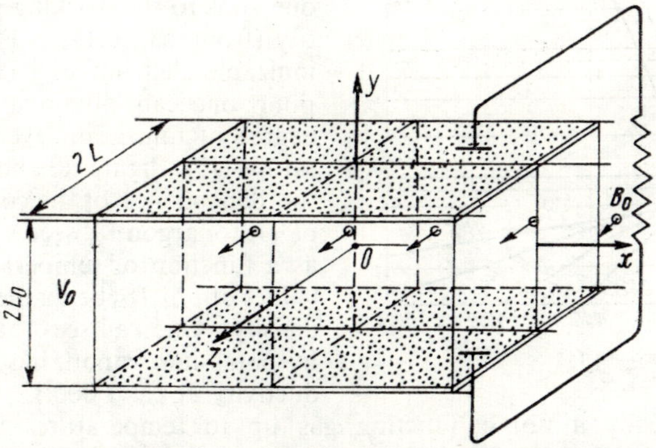

Fig. 12

conversion as in classical electrical power generators. It is well known that it is this successive conversion from thermal to mechanical energy and then from mechanical to electrical energy that causes the low efficiency (maximum 40%) of classical generators. By eliminating the mechanical stage in conversion, the MHD generators can work with a 70—80% efficiency. From the fuel point of view, the MHD generators again present some advantages. While in classical generators one uses raw materials (e.g. coal) which exist in limited quantities in nature, for the MHD generators fuel is practically unlimited (ionized gas can be obtained by heating water to a sufficiently high temperature).

There are at the present time some problems that hinder the development of MHD generators. These problems concern in the first place the method of obtaining the ionized gas and secondly the production of suitable refractory materials for the generator walls. Such materials should withstand temperatures of 2 000 —5 000°C without changing their mechanical and electrical properties. There are two classical methods of obtaining the ionized gas; the *thermal* and *superthermal* procedures. The thermal procedure consists of heating the gas up to temperatures of the order of 5 000—10 000°C. By transferring thermal energy to the gas, the kinetic energy per particle increases up to the point where ionization by collisions becomes possible. Positively (ions) and negatively (electrons) charged particles appear within the gas. The basic theory of thermal ionization of gases was developed by Saha in 1920 [1] [2]. The superthermal (non-equilibrium) procedure resembles the processes taking place in a low pressure gas discharge. At the present time the superthermal ionization procedure is more widespread although its mechanism is more complicated. Thermal ionization is however more realistic and more promising.

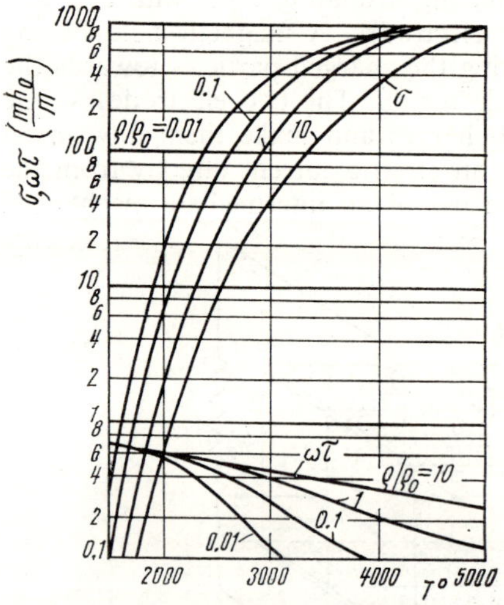

Fig. 13

The majority of common gases like air, CO, CO_2, etc. have a quite high ionization potential and therefore cannot be thermally ionized as long as one has not sufficiently high temperatures. However, if one adds to the working gas some quantity (from ca. 0.1 to 1%) of an easily ionizable element, e. g. alkali metal vapour, one can obtain a sufficiently high degree of ionization even at low temperatures [3]. Figure 13 shows after Rosa [3] the dependence of the conductivity σ and of $\omega\tau$ for argon 1% seeded with potassium as a function of temperature for various values of ρ/ρ_0 (ρ_0 being the normal pressure density). One can see that argon seeded by potassium vapour has a detectable conductivity at ca. 2 000°C while pure argon remains practically a non-conducting gas up to temperatures around 4 000°C. It is obvious that the magnetic field intensity and the generator dimensions will be defined in terms of gas conductivity.

Due to the way in which the conductivity depends on the mean free path it is obvious that there exists an optimum concentration of the seeding material which depends on the nature of the two materials. If one uses a formula put forward by Lin, Resler and Kantrowitz [4],

$$\sigma = 0 \cdot 532 \frac{q^2}{\sqrt{m_e kT}} \frac{n_i}{n_i Q_i + n_a Q_a + n_0 Q_0} \tag{4.1.1}$$

one could show that in some conditions σ has a maximum value for $n_a Q_a = n_0 Q_0$. In equation (4.1.1) Q_i, Q_a and Q_0 represent the collisional cross-sections of the ions, neutral atoms and seed atoms, and n_i, n_a and n_0 their concentrations; q and m_e are the electron charge and mass, respectively, k Boltzmann's constant and T the temperature in °K. According to the above formula for argon as the working gas (the cross-section averaged over a Maxwellian distribution of velocities being $Q_a \simeq 6 \times 10^{-17}$ cm²) and potassium vapour as the ionizing seed ($Q_0 = 3 \times 10^{-14}$ cm²), it follows that σ has a maximum value for a concentration of the potassium seed of 0.2%.

As regards the wall refractory materials it seems that considerable progress has been recently made in the field of chemical synthesis of inorganic materials with high melting temperatures (2 500—3 500°C), like magnesium, cerium, zirconium and thorium oxides.

All these results make us hope that the technological problems of the MHD generators will soon be solved. In any case, MHD generator theory should be developed.

4.1.1 The mathematical model

In the following we shall consider a MHD generator of rectangular cross-section (Figure 12) this configuration being more widespread. We shall consider the motion with respect to a reference system having the Ox-axis along the channel axis, the Oy-axis perpendicular to the electrodes and the Oz-axis parallel to the external magnetic field (in order that the magnetic field brake the gas motion and collect the charges on the electrodes, it should be parallel to the latter). The yOz-plane cuts the electrodes along their median line.

We shall consider the dimensionless variables introduced in Section 2.4 where L_0 is the half-distance between the electrodes, V_0 is the magnitude of the injection velocity and B_0 the magnitude of the external induction. Then the external induction has the form:

$$\boldsymbol{B}_0 = \varepsilon(x)\boldsymbol{i}_3$$

where:

$$\varepsilon(x) = \begin{cases} 1 \text{ if } |x| < a \\ 0 \text{ if } |x| > a \end{cases}$$

$\boldsymbol{i}_3$ being the Oz-axis unit-vector and $2a$ the dimensionless length of the electrodes. The fluid occupies the domain $\mathscr{D}: \{-\infty < x < \infty,\ -1 < y < 1,\ -l < z < l\}$, $2l$ being the dimensionless width of the electrodes.

We shall assume that the fluid is a simple, ionized, incompressible medium. We shall also assume the existence of ionization equilibrium such that the presence of the electromagnetic field cannot change the fluid ionization state. To a first approximation we shall assume the working gas to be an Ohmic fluid and shall consider Ohm's law in the form (2.4.8). Afterwards, we shall also consider an ionized gas obeying Ohm's law in the form (3.3.2).

Let us notice that the uniform motion of the injected fluid will be modified by the presence of the magnetic field and by the electrode conductivity. Therefore a discontinuity in the motion is produced at the entrance to the electrode region. The phenomena related to this discontinuity are referred to as *end effects.* It is obvious that the fluid motion through the generator and hence its characteristic parameters will be influenced by these (inlet and outlet) discontinuity phenomena. These discontinuities appear as a result of the MHD interaction.

The first paper in which the end effects and hence the MHD interaction are taken into account is due to Sutton and Carlson [5]. These authors consider a set of semi-infinite electrodes (and thus take into consideration only the inlet end effects) and an Ohmic fluid. Expanding the solution with respect to the interaction parameter $N = Rh\, Rm$ they find the zero-order approximation of the electrical potential and the first order approximation of the fluid velocity. Using the electrical analogy method, Nguen-Ngoc-Tran [6] further determines the first-order approximation of the electrical potential. At the same time, L. Dragoş [7] elaborates an iterative procedure for finding the solution.

However, the practical generators have finite length electrodes. The problem of MHD interaction in a finite electrode generator was studied by L. Dragoş in [8]. Using the Green-function technique and some elements of the theory of generalized functions, the first approximation is determined and an iterative procedure for obtaining the higher order approximations (the solution is expanded in terms of the parameter N) is given. Due to the discontinuity introduced into the motion by the electrode ends, the problem can be formulated in a natural way if the theory of distributions (generalized functions) is used. Such a formulation is for the first time given in [8]. Section 3.3 is based on this paper.

Due to the Hall effect, conductivity becomes anisotropic for the ionized gas. Leaving aside the works in which the MHD interaction is neglected* (the motion is considered to be uniform everywhere or one-dimensional), the first paper in this field are those of Nguen-Ngoc-Tran [9] and L. Dragoş [10]. Using again the electrical analogy method, Tran determines numerically the zero order approximation of the electrical potential. L. Dragoş independently determines analytically the first approximation of the electrical potential and of the current function and elaborates an iterative procedure (using the Green-function technique and the theory of distributions) for the determination of the general solution. Moreover, the solution given by L. Dragoş is valid for the case of finite electrodes. Section 4.4 is based on the paper [10].

* These papers are mentioned in the bibliography in chronological order.

In his PhD thesis [11], Nguen-Ngoc-Tran presents his main results (some of them unpublished) on the MHD generator theory. The semi-infinite electrode case is considered therein both in the hypothesis of isotropic and anisotropic conductivity. The case of finite electrodes is also considered in the isotropic conductivity hypothesis. The solution is given piecewise using series expansions in terms of proper functions and the matching conditions. By truncating the series expansions the author gives numerical results which seem to be remarkable.

A study of the end effects with the consideration of the induced magnetic field can be found in [12] [G. 24]. For other results we should mention the review work presented in [G. 27].

4.2 The general problem

4.2.1 The first approximation system

For Figure 12 and with the dimensionless variables defined above, the following equations should be satisfied within the fluid (domain $\mathscr{D}$):

$$\operatorname{div} \boldsymbol{V} = 0 \tag{4.2.1}$$

$$(\boldsymbol{V}\cdot\nabla)\boldsymbol{V} = -\operatorname{grad} p + Rh\boldsymbol{J} \times \boldsymbol{B} \tag{4.2.2}$$

$$\operatorname{curl} \boldsymbol{E} = 0 \tag{4.2.3}$$

$$\operatorname{div} \boldsymbol{B} = 0 \tag{4.2.4}$$

$$\operatorname{curl} \boldsymbol{B} = \boldsymbol{J}\ (\operatorname{div} \boldsymbol{J} = 0) \tag{4.2.5}$$

$$\boldsymbol{J} = \underline{\boldsymbol{Rm}}(v)\cdot(\boldsymbol{E} + \boldsymbol{V}\times\boldsymbol{B}) \tag{4.2.6}$$

where $\underline{\boldsymbol{Rm}}(v)$ is the conductivity tensor (3.3.4).

Since the conditions of the problem do not change in time, the motion is assumed to be steady.

Generally in Ohm's law (4.2.6) the electric field $\boldsymbol{E}$ should be $\boldsymbol{E} + \boldsymbol{E}_e$ where $\boldsymbol{E}_e$ is the electric field due to the electron pressure gradient. But since $\boldsymbol{E}_e$ also obeys an equation of the form (4.2.3), one may assume that in (4.2.3) and (4.2.6) $\boldsymbol{E}$ represents the total field. As a matter of fact, for weakly ionized gases (as it is the case for MHD generators) $\boldsymbol{E}_e$ is negligible.

Rm is a small parameter since generally the generator fluid conductivity is small. Consequently, we shall expand the solution of equations (4.2.1)–(4.2.6) in a series of integral powers of *Rm*:

$$(p, \boldsymbol{V}, \boldsymbol{B}, \boldsymbol{J}, \boldsymbol{E}) = \sum_{n=0}^{\infty} (Rm)^n (p^{(n)}, \boldsymbol{V}^{(n)}, \boldsymbol{B}^{(n)}, \boldsymbol{J}^{(n)}, \boldsymbol{E}^{(n)}) \tag{4.2.7}$$

Substituting into (4.2.1) — (4.2.6) we notice that, although *Rm* is a small parameter due to the applied magnetic field, the product *RmRh* is not small. Indeed we have:

$$N \overset{\text{def}}{=} RmRh = \sigma B_0^2 L_0/\rho_0 V_0 \tag{4.2.8}$$

The high magnetic field B_0 externally applied in the MHD generators is due just to the small conductivity σ. Consequently, the solution will depend on two parameters *Rm* and *N*.

Using the expansion (4.2.7) from (4.2.6) we get $\boldsymbol{J}^{(0)} = 0$, so that for the first approximation it follows that:

$$\operatorname{div} \boldsymbol{B}^{(0)} = 0, \ \operatorname{curl} \boldsymbol{B}^{(0)} = 0 \tag{4.2.9}$$

$$\left.\begin{aligned} &\operatorname{div} \boldsymbol{V}^{(0)} = 0 \\ &(\boldsymbol{V}^{(0)} \cdot \nabla)\boldsymbol{V}^{(0)} = -\operatorname{grad} p^{(0)} + N\boldsymbol{J}^{(1)} \times \boldsymbol{B}^{(0)} \\ &\operatorname{curl} \boldsymbol{E}^{(0)} = 0, \quad \operatorname{div} \boldsymbol{J}^{(1)} = 0 \\ &\boldsymbol{J}^{(1)} = (\nu)^{(0)} \cdot (\boldsymbol{E}^{(0)} + \boldsymbol{V}^{(0)} \times \boldsymbol{B}^{(0)}) \end{aligned}\right\} \tag{4.2.10}$$

Taking into account the continuity conditions of the magnetic field across the separation planes between the fluid and the external medium, from (4.2.9) it follows that $\boldsymbol{B}^{(0)}$ has the expression (4.1.2). The $x = \pm a$ planes are discontinuity planes on which we should assume the existence of some surface currents determined by the discontinuities in the tangential components of $B_z^{(0)}$.

Using this result, we replace $\boldsymbol{B}^{(0)}$ in (4.2.10) by (4.1.2) so that we have (3.3.6):

$$(\nu)^{(0)} = \begin{pmatrix} \sigma_1 & -\sigma_2 & 0 \\ \sigma_2 & \sigma_1 & 0 \\ 0 & 0 & 1 \end{pmatrix}$$

$$\sigma_1 = \frac{1}{1 + \varepsilon \nu^2}, \quad \sigma_2 = \frac{\varepsilon \nu}{1 + \varepsilon \nu^2} \tag{4.2.10'}$$

Once the first approximation determined we can find $\boldsymbol{B}^{(1)}$ since from (4.2.4) and (4.2.5) we have:

$$\operatorname{div} \boldsymbol{B}^{(1)} = 0, \ \operatorname{curl} \boldsymbol{B}^{(1)} = \boldsymbol{J}^{(1)} \tag{4.2.11}$$

After this determination is performed, from (4.2.1) — (4.2.6) we have:

$$\begin{aligned} &\operatorname{div} \boldsymbol{V}^{(1)} = 0, \ (\boldsymbol{V}^{(0)} \cdot \nabla)\boldsymbol{V}^{(1)} + (\boldsymbol{V}^{(1)} \cdot \nabla)\boldsymbol{V}^{(0)} \\ &= -\operatorname{grad} p^{(1)} + N(\boldsymbol{J}^{(2)} \times \varepsilon \boldsymbol{i}_3) + N(\boldsymbol{J}^{(1)} \times \boldsymbol{B}^{(1)}) \\ &\operatorname{curl} \boldsymbol{E}^{(1)} = 0, \quad \operatorname{div} \boldsymbol{J}^{(2)} = 0 \\ &\boldsymbol{J}^{(2)} = (\nu)^{(1)} \cdot (\boldsymbol{E}^{(1)} + \boldsymbol{V}^{(1)} \times \varepsilon \boldsymbol{i}_3) + (\nu)^{(1)} \cdot (\boldsymbol{V}^{(0)} \times \boldsymbol{B}^{(1)}) \end{aligned} \tag{4.2.12}$$

which system determines the next approximation ($\boldsymbol{V}^{(0)}$, $\boldsymbol{J}^{(1)}$ and $\boldsymbol{B}^{(1)}$ being known). The next approximations are obtained in a similar way. The systems of equations which determine the succesive approximations are essentially identical to one another. We shall restrict ourselves in what follows to finding the first approximation. This represents the first hypothesis.

4.2.2 Plane motion

The second hypothesis is to consider the motion to be plane:

$$(\boldsymbol{V}^{(0)}, \boldsymbol{E}^{(0)}, \boldsymbol{J}^{(1)}) \cdot \boldsymbol{i}_3 \equiv 0$$
$$\frac{\partial}{\partial z}(\boldsymbol{V}^{(0)}, \boldsymbol{E}^{(0)}, \boldsymbol{J}^{(1)}, p) \equiv 0 \tag{4.2.13}$$

This hypothesis is compatible with the boundary conditions:

$$\boldsymbol{V}^{(0)} \cdot \boldsymbol{i}_3 = 0, \ \boldsymbol{E}^{(0)} \cdot \boldsymbol{i}_3 = 0, \ \boldsymbol{J}^{(1)} \cdot \boldsymbol{i}_3 = 0$$

to be satisfied on the $z = \pm l$ wall. The first condition expresses the fact that the fluid is slipping along the walls, the third that these are insulators and the second follows from equation (4.2.10 e).

If the motion is a plane one from (4.2.10 e) it follows that:

$$V_x^{(0)} = \frac{\partial \Psi}{\partial y}, \ V_y^{(0)} = -\frac{\partial \Psi}{\partial x}, \ \Psi = \Psi(x, y) \tag{4.2.14}$$

and from (4.2.10 c)

$$\boldsymbol{E}^{(0)} = -\operatorname{grad} \varphi, \quad \varphi = \varphi(x, y) \tag{4.2.15}$$

As a matter of fact, in the following we shall point out the basic (uniform) and the perturbed motions by writing $\boldsymbol{V}^{(0)} = \boldsymbol{i}_1 + \boldsymbol{v}^{(0)}$ where:

$$v_x^{(0)} = \frac{\partial \psi}{\partial y}, \quad v_y^{(0)} = -\frac{\partial \psi}{\partial x}, \quad \psi = \psi(x, y) \tag{4.2.16}$$

From (4.2.14) and (4.2.16) it follows that:

$$\Psi = y + \psi \tag{4.2.17}$$

ψ being the current function for the perturbed motion. The functions ψ and φ will represent the unknowns of the problem.

By eliminating the pressure from the equation of motion (4.2.10b) (differentiating with respect to y the projection on the Ox-axis, with respect to x the projection on the Oy-axis and subtracting one result from the other) and taking into account (4.2.10d) we get (outside the segments $x = \pm a$):

$$\left(\frac{\partial \Psi}{\partial y}\frac{\partial}{\partial x} - \frac{\partial \Psi}{\partial x}\frac{\partial}{\partial y}\right)\Delta \Psi = N\varepsilon' J_x^{(1)} \tag{4.2.18}$$

From (4.2.10e) it follows:

$$\begin{aligned} J_x^{(1)} &= -\sigma_1\left(\frac{\partial \varphi}{\partial x} + \varepsilon\frac{\partial \Psi}{\partial x}\right) + \sigma_2\left(\frac{\partial \varphi}{\partial y} + \varepsilon\frac{\partial \Psi}{\partial y}\right) \\ J_y^{(1)} &= -\sigma_2\left(\frac{\partial \varphi}{\partial x} + \varepsilon\frac{\partial \Psi}{\partial x}\right) - \sigma_1\left(\frac{\partial \varphi}{\partial y} + \varepsilon\frac{\partial \Psi}{\partial y}\right) \end{aligned} \tag{4.2.19}$$

so that using (4.2.10d) we find:

$$\sigma_1(\Delta\varphi + \varepsilon\Delta\Psi) + \varepsilon'\left(\sigma_1\frac{\partial \Psi}{\partial x} - \frac{\partial \Psi}{\partial y}\right) = 0 \tag{4.2.20}$$

with the hypothesis that the second-order derivatives of the functions φ and Ψ may be permuted with one another. If φ and Ψ are classical functions, then equations (4.2.18) and (4.2.20) will be valid within the domain D: $\{-\infty < x < \infty, -1 < y < 1\}$ except the segments $x = \pm a$ (with this hypothesis $\varepsilon' \equiv 0$). If φ and Ψ are generalized functions, then equations (4.2.18) and (4.2.20) may be considered to be valid everywhere within D. For the latter case we shall introduce:

$$\varepsilon(x) = \theta(x + a) - \theta(x - a)$$

so that [10] [G.33]:

$$\varepsilon' = \delta(x + a) - \delta(x - a)$$

where θ is the Heaviside function and δ the Dirac distribution.

4.2.3 Boundary conditions

At infinity, the upstream fluid motion will be uniform so that:

$$\lim_{x \to -\infty} \Psi = y \quad (\lim_{x \to -\infty} \psi = 0) \tag{4.2.21}$$

The fluid slips over the walls so that $V_y^{(0)}(x, \pm 1) = 0$. Taking into account the velocity representation we find:

$$\Psi(x, \pm 1) = \pm 1 \quad (\psi(x, \pm 1) = 0) \tag{4.2.22}$$

We shall also use the condition:

$$\lim_{|x|\to\infty} \frac{\partial \Psi}{\partial x} = 0 \quad \left(\lim_{|x|\to\infty} \frac{\partial \psi}{\partial x} = 0 \right) \tag{4.2.23}$$

In order to derive the conditions to be satisfied by φ, we shall use the fact that outside the electrodes the walls are insulators:

$$J_y^{(1)}(x, \pm 1) = 0, \quad |x| > a \tag{4.2.24}$$

Taking into account (4.2.19) it follows (outside the walls $\sigma_2 = 0$):

$$\left.\frac{\partial \varphi}{\partial y}\right|_{y=\pm 1} = 0, \quad |x| > a \tag{4.2.24'}$$

The electrode potential being $\mp \varphi_w (= \varphi_w^{\text{dim}}/L_0 V_0 B_0)$, it follows that we also have the condition:

$$\varphi(x, \pm 1) = \mp \varphi_w, \quad |x| < a \tag{4.2.25}$$

Finally, we can also use the condition:

$$\lim_{|x|\to\infty} |\text{grad}\, \varphi| = 0 \tag{4.2.26}$$

One can show that if the magnetic field is vanishing for $|x| \to \infty$ then $\boldsymbol{J}$ (and thus $\boldsymbol{J}^{(1)}$) will also be vanishing for $|x| \to \infty$ [G.27]. Equation (4.2.26) expresses this condition.

4.2.4 The classical formulation of the problem

If φ and ψ are classical functions from (4.2.18) and (4.2.20). we get the following equations:

$$\left(\frac{\partial}{\partial x} + \frac{\partial \psi}{\partial y} \frac{\partial}{\partial x} - \frac{\partial \psi}{\partial x} \frac{\partial}{\partial y} \right) \Delta\psi = \begin{cases} 0, & \text{for } |x| > a \\ 0, & \text{for } |x| < a \end{cases} \tag{4.2.27}$$

$$\begin{aligned} \Delta\varphi &= 0, && \text{for } |x| > a \\ \Delta(\varphi + \psi) &= 0, && \text{for } |x| < a \end{aligned} \tag{4.2.28}$$

The segments $x = \mp a$ are discontinuity lines. Consequently, we should use across them the field jump equations (1.5.12), (1.5.16) and (1.5.17), and the jump equations of motion (11.1.5) and (11.1.6). The first set of equations yields:

$$[E_y] = 0, \quad [B_y] = 0, \quad [B_x] = 0 \tag{4.2.29}$$

and the second (taking into account the fact that the fluid is incompressible, the discontinuity lines are fixed and the magnetic induction is continuous) also yields:

$$[V_x] = 0, \quad [V_y] = 0, \quad [p] = 0 \tag{4.2.30}$$

By [ ] we mean the difference between the right- and left-hand side boundary values.

Taking into account the representation of the solution from (4.2.29a), (4.2.30a) and (4.2.30b) it follows that:

$$[\varphi] = 0, \quad [\psi] = 0, \quad \left[\frac{\partial\psi}{\partial x}\right] = 0 \tag{4.2.31}$$

Conditions (4.2.29b), (4.2.29c) yield no information on the first approximation On the other hand, we should use the condition:

$$[J_x^{(1)}] = 0 \tag{4.2.32}$$

which follows from (1.5.19). Using (4.2.19), it follows that:

$$-\frac{1}{1+\nu^2}\left(\frac{\partial\varphi}{\partial x} + \frac{\partial\psi}{\partial x}\right)_+ + \frac{\nu}{1+\nu^2}\left(\frac{\partial\varphi}{\partial y} + \frac{\partial\psi}{\partial y} + 1\right)_+ + \left.\frac{\partial\varphi}{\partial x}\right|_- = 0, \quad \text{if } x = -a$$

$$-\frac{1}{1+\nu^2}\left(\frac{\partial\varphi}{\partial x} + \frac{\partial\psi}{\partial x}\right)_- + \frac{\nu}{1+\nu^2}\left(\frac{\partial\varphi}{\partial y} + \frac{\partial\psi}{\partial y} + 1\right)_- + \left.\frac{\partial\varphi}{\partial x}\right|_+ = 0, \quad \text{if } x = a \tag{4.2.33}$$

where we denoted by (+) and (−) the right-hand and left-hand side boundary values.

In order to utilize the condition (30 c) as well, we shall write the projection of the equation of motion (10 b) on the Oy-axis to the right and to the left of the segments $x = \mp a$. Since $V_x^{(0)}$, $V_y^{(0)}$ and $J_x^{(1)}$ are continuous across these segments, condition (4.2.30 c) becomes:

$$V_x^{(0)}\left[\frac{\partial V_y^{(0)}}{\partial x}\right] = \begin{cases} -NJ_x^{(1)}, & \text{for } x = -a \\ NJ_x^{(1)}, & \text{for } x = a \end{cases}$$

and then:

$$\left(1+\frac{\partial\psi}{\partial y}\right)\left[\frac{\partial^2\psi}{\partial x^2}\right]=\begin{cases}-N\left.\dfrac{\partial\varphi}{\partial x}\right|_-, & \text{for } x=-a\\[2ex] N\left.\dfrac{\partial\varphi}{\partial x}\right|_+, & \text{for } x=a\end{cases} \tag{4.2.34}$$

Therefore we have to solve equations (4.2.27) and (4.2.28) with the boundary conditions (4.2.21)–(4.2.26) and with the matching conditions (4.2.31), (4.2.33) and (4.2.34).

By formulating the problem by means of the generalized functions (distributions) the matching conditions are included in the equations of motion. The theory of distributions provides the means of treating both the continuous and discontinuous motions in a unified manner.

4.2.5 The linearized problem

Neglecting the products of perturbations, equation (4.2.27) is reduced to the form:

$$\Delta\psi = N\begin{cases} f_0(y), & \text{for } x<-a\\ f_0+f_1(y), & \text{for } |x|<a\\ f_0+f_1+f_2(y), & \text{for } x>a\end{cases} \tag{4.2.35}$$

and the condition (4.2.34) to the form:

$$\left[\frac{\partial^2\psi}{\partial x^2}\right]=\begin{cases}-N\left.\dfrac{\partial\varphi}{\partial x}\right|_-, & \text{for } x=-a\\[2ex] N\left.\dfrac{\partial\varphi}{\partial x}\right|_+, & \text{for } x=+a\end{cases} \tag{4.2.36}$$

Since at infinity the upstream motion is irrotational ($\Delta\psi=0$) it follows that $f_0(y)\equiv\equiv 0$. Using the matching conditions (4.2.31), (4.2.36) and (4.2.33) from (4.2.35) we have:

$$\begin{aligned} f_1(y) &= -\frac{1}{1+v^2}\left(\frac{\partial\varphi}{\partial x}+\frac{\partial\psi}{\partial x}\right)_+ + \frac{v}{1+v^2}\left(\frac{\partial\varphi}{\partial y}+\frac{\partial\psi}{\partial y}+1\right)_+, \quad x=-a\\ f_2(y) &= \frac{1}{1+v^2}\left(\frac{\partial\varphi}{\partial x}+\frac{\partial\psi}{\partial x}\right)_- - \frac{v}{1+v^2}\left(\frac{\partial\varphi}{\partial y}+\frac{\partial\psi}{\partial y}+1\right)_-, \quad x=a \end{aligned} \tag{4.2.37}$$

For a negligible Hall effect and semi-infinite electrodes $(x > 0)$ we have to solve the equations:

$$\Delta\psi = 0, \quad \Delta\varphi = 0 \quad \text{in} \quad D^{0}_{-\infty}(x < 0) \tag{4.2.38}$$

$$\Delta\psi = -N\left(\frac{\partial\varphi}{\partial x} + \frac{\partial\psi}{\partial x}\right)_{x=+0} \quad \text{and}$$

$$\Delta(\psi + \varphi) = 0 \quad \text{in} \quad D^{\infty}_{0}\,(x > 0)$$

with the matching conditions (4.2.31) and with the boundary conditions defined above. This problem was considered by Sutton and Carlson in [5] and then by Dragoş [7] and Tran [6], [11]. The exact solution can be obtained from an integral equation obtained in the following manner: using the Green-function technique we get the representation of the solution (functions ψ and φ) by means of some integrals depending on $f_1(y)$; having determined ψ and φ the expression of $f_1(y)$ is formed. This will be an integral whose kernel will be the product of $f_1(\eta)$ and the sum of the two Green functions. The integral equation (depending on the parameter N) found in this manner will solve the problem, i.e. will determine the function $f_1(y)$.

When the Hall effect is no longer neglected we have the equations:

$$\Delta\psi = 0, \quad \Delta\varphi = 0 \quad \text{in} \quad D^{0}_{-\infty}$$

$$\Delta\psi = -\frac{N}{1+v^2}\left(\frac{\partial\varphi}{\partial x} + \frac{\partial\psi}{\partial x}\right)_{+0} + \frac{Nv}{1+v^2}\left(\frac{\partial\varphi}{\partial y} + \frac{\partial\psi}{\partial y} + 1\right)_{+0} \quad \text{and} \tag{4.2.39}$$

$$\Delta(\psi + \varphi) = 0 \quad \text{in} \quad D^{\infty}_{0}$$

to which the boundary and matching conditions are added. For this case the exact solution is obtained by means of a system of integral equations. A quite thorough study of this problem can be found in [11].

For finite electrodes without the Hall effect we have to solve the following equations:

$$\Delta\psi = 0, \quad \Delta\varphi = 0 \quad \text{in} \quad D^{-a}_{-\infty}$$

$$\Delta\psi = -N\left(\frac{\partial\varphi}{\partial x} + \frac{\partial\psi}{\partial x}\right)_{x=-a+} \quad \text{and}$$

$$\Delta(\psi + \varphi) = 0 \quad \text{in} \quad D^{+a}_{-a} \tag{4.2.40}$$

$$\Delta\psi = -N\left(\frac{\partial\varphi}{\partial x} + \frac{\partial\psi}{\partial x}\right)_{-a+} + N\left(\frac{\partial\varphi}{\partial x} + \frac{\partial\psi}{\partial x}\right)_{a-} \quad \text{and}$$

$$\Delta\varphi = 0 \quad \text{in} \quad D^{\infty}_{+a}$$

which can be reduced to a system of two integral equations with two unknowns.

Finally, when the Hall effect is also considered (the electrodes being of finite dimensions) we have the following system of equations:

$$\Delta\psi = 0, \quad \Delta\varphi = 0 \quad \text{in} \quad D_{-\infty}^{-a}$$

$$\Delta\psi = -\frac{N}{1+\nu^2}\left(\frac{\partial\varphi}{\partial x} + \frac{\partial\psi}{\partial x}\right)_{-a+} + \frac{N\nu}{1+\nu^2}\left(\frac{\partial\varphi}{\partial y} + \frac{\partial\psi}{\partial y} + 1\right)_{-a+} \quad \text{and}$$

$$\Delta(\psi + \varphi) = 0 \quad \text{in} \quad D_{-a}^{+a} \tag{4.2.41}$$

$$\Delta\psi = -\frac{N}{1+\nu^2}\left\{\left(\frac{\partial\varphi}{\partial x} + \frac{\partial\psi}{\partial x}\right)_{-a+} - \nu\left(\frac{\partial\varphi}{\partial y} + \frac{\partial\psi}{\partial y}\right)_{-a+} - \left(\frac{\partial\varphi}{\partial x} + \frac{\partial\psi}{\partial x}\right)_{a-} + \nu\left(\frac{\partial\varphi}{\partial y} + \frac{\partial\psi}{\partial y}\right)_{a-}\right\} \quad \text{and}$$

$$\Delta\varphi = 0 \quad \text{in} \quad D_{+a}^{\infty}$$

In this latter case the determination of the exact solution could be reduced to solving a system of four integral equations depending on the parameter N (and ν) with four unknowns.

Since the only method of solution of integral equations depending on a small parameter is the successive approximations method it is simpler to apply this method directly to the system of differential equations. Therefore we shall look for the solutions of the form:

$$\psi = \sum_{j=0}^{\infty} N^j \psi_j, \quad \varphi = \sum_{j=0}^{\infty} N^j \varphi_j \tag{4.2.42}$$

and shall deal only with the systems (4.2.40) and (4.2.41) these being the most general.

4.2.6 The generator characteristics

Let us start from the expression:

$$A = -\int_D \boldsymbol{V}\cdot(\boldsymbol{J}\times\boldsymbol{B})\,\mathrm{d}v = \int_D \boldsymbol{J}\cdot(\boldsymbol{V}\times\boldsymbol{B})\,\mathrm{d}v \tag{4.2.43}$$

which represents the power (in dimensionless variables) developed by the medium in the motion against the electromagnetic field (with changed sign this represents the power dissipated by the Lorentz force). Replacing the value of $\boldsymbol{V}\times\boldsymbol{B}$ obtained from Ohm's law (3.3.1) we obtain:

$$A = Q + W \tag{4.2.44}$$

where:

$$Q = \int_D (J^2/\sigma)\,\mathrm{d}v, \quad W = -\int_D \boldsymbol{E}\cdot\boldsymbol{J}\,\mathrm{d}v \tag{4.2.44'}$$

The interpretation of equation (44) is straightforward: the total work is used in Joule dissipation (Q) on the one hand and as useful output power of the generator (W) on the other hand.

In dimensionless variables, equation (4.2.44) becomes:

$$A^* = Q^* + W^* \tag{4.2.45}$$

where:

$$Q^* = \int_{D^*} Rm^{-1} J^{*2}\, dv^*, \quad W^* = -\int_{D^*} J^* \cdot E^*\, dv^*$$

$$(Q, W) = \mu H_0^2 V_0 l^2 (Q^*, W^*) \tag{4.2.45'}$$

Since the dimensionless values of the dissipation and power should be multiplied by H_0^2 in order to get the dimensional values we shall retain in their calculation the terms of the order of Rm (like for the product $RmRh$). Therefore using the expansion (4.2.7), we get for the unit of length:

$$Q^* = Rm \int_{-\infty}^{+\infty} \int_{-1}^{+1} J^{(1)2}\, dx\, dy$$

$$W^* = -Rm \int_{-\infty}^{+\infty} \int_{-1}^{+1} J^{(1)} \cdot E^{(0)} dx\, dy = Rm \int_{-\infty}^{+\infty} \int_{-1}^{+1} \operatorname{div}(\varphi J^{(1)})\, dx\, dy$$

$$= Rm \oint \varphi J^{(1)} \cdot n\, ds = -\varphi_w Rm \int_{-a}^{+a} \{J_y^{(1)}(x, -1) + J_y^{(1)}(x, 1)\}\, dx$$

where we have taken into account the boundary conditions (4.2.24) and (4.2.25) and the fact that $J_x^{(1)}$ vanishes on the segments $x = \pm$ d for d $\to \infty$.

4.3 Scalar conductivity fluids

4.3.1 The determination of the first approximation

When the Hall effect is negligible we have to solve the system (4.2.40). Using the expansion (4.3.42) for the first approximation we find:

$$\Delta\psi_0 = 0 \quad \text{in} \quad D_{-\infty}^{-a} \cup D_{-a}^{+a} \cup D_a^{\infty} \tag{4.3.1}$$

$$\Delta\varphi_0 = 0 \quad \text{,,} \tag{4.3.2}$$

where by D_a^b we denote the domain: $a < x < b, -1 < y < 1$.

Using the Green formula for the domain D_{-d}^{+d} $(d \to \infty)$ and the boundary conditions satisfied by ψ_0 (the same as for ψ) we have:

$$\int_{D_{-d}^{+d}} \{\psi_0 \Delta\psi_0 + (\text{grad}\,\psi_0)^2\}\, \mathrm{d}x\, \mathrm{d}y = \oint \psi_0 \frac{\mathrm{d}\psi_0}{\mathrm{d}n}\, \mathrm{d}s = 0 \tag{4.3.3}$$

whence $\psi_0 \equiv 0$. It should be mentioned that due to the jump conditions:

$$[\psi_0] = 0, \quad \left[\frac{\partial\psi_0}{\partial x}\right] = 0, \quad \left[\frac{\partial^2\psi_0}{\partial x^2}\right] = 0 \tag{4.3.4}$$

which follow from (4.2.31) and (4.2.36) the functions appearing in (4.3.3) are continuous within the fluid so that the Green formula is valid. The result obtained $\psi_0 \equiv 0$ is a natural one from the physical point of view since when the interaction parameter is zero the motion remains uniform.

The function φ_0 is determined from the equation (4.3.2), from the boundary conditions (4.2.24′) — (4.2.26) and from the matching conditions of the form (4.3.4) (which follow from (4.2.31), (4.2.33) and (4.2.2)). One first proves that the solution of this problem is unique. Let us denote by φ_0^* the difference between two solutions; φ_0^* will be a harmonic function within the fluid, will satisfy the homogeneous boundary conditions and the conditions of applicability of the Green formula. Using the Green formula it follows that $\varphi_0^* \equiv 0$.

Due to the uniqueness of the solution it follows that:

$$\varphi_0(x, y) = -\varphi_0(x, -y) \tag{4.3.5}$$

Indeed, one can check directly that if $\varphi_0(x, y)$ satisfies the problem conditions then the function $-\varphi_0(x, -y)$ also satisfies these conditions. From the uniqueness theorem equation (4.3.5) follows.

In order to determine the function $\varphi_0(x, y)$ we shall introduce the function $\chi_0(x, y)$ the harmonic conjugate of φ_0:

$$\frac{\partial\varphi_0}{\partial x} = \frac{\partial\chi_0}{\partial y}, \qquad \frac{\partial\varphi_0}{\partial y} = -\frac{\partial\chi_0}{\partial x} \tag{4.3.6}$$

It follows that the function:

$$f_0(z) = \varphi_0 + \mathrm{i}\chi_0 \tag{4.3.7}$$

is holomorphic within the fluid (due to the conditions $[\varphi_0] = 0$, $[\partial\varphi_0/\partial x] = 0$ the derivative of this function is also defined on the discontinuity lines $x = \mp a$). From (4.3.5) and (4.3.6) it follows that:

$$\chi_0(x, y) = \chi_0(x, -y) \tag{4.3.8}$$

Due to the equations (4.3.6) and (4.3.8) the boundary conditions (4.2.24) can be transposed on the function χ_0 in the form:

$$\begin{aligned} \chi_0(x, \pm 1) &= b, \quad \text{for} \quad x < -a \\ \chi_0(x, \pm 1) &= 0, \quad \text{for} \quad x > a \end{aligned} \tag{4.3.9}$$

b being an undetermined constant (in the last condition the integration constant may be considered to be zero since the function χ_0 is determined up to an additive constant).

With the conformal mapping:

$$z' = \mathrm{i} \exp \frac{\pi}{2}(z + a), \quad z' = x' + \mathrm{i}y' \tag{4.3.10}$$

the domain occupied by the fluid is transformed into the upper half-plane of the (z') — plane so that we have the following point-to-point corespondence:

$$\begin{array}{lcccccc} (z): & (-\infty, \pm 1), & (-a, -1), & (a, -1), & (\infty, \mp 1), & (a, 1), & (-a, 1) \\ & \downarrow & \downarrow & \downarrow & \downarrow & \downarrow & \downarrow \\ (z'): & (0, 0), & (1, 0), & (\exp \pi a, 0), & (\pm\infty, 0), & (-\exp \pi a, 0), & (-1, 0) \end{array}$$

In this manner the determination of function $f_0(z)$ within the fluid is reduced to the problem of finding the function $f_0(z')$ holomorphic in the half-plane $y' > 0$ which should satisfy the following conditions on the real axis:

$$\begin{array}{llll} \text{on the segment} & (-\infty, -\exp \pi a), & \chi_0 = 0 & \\ \text{,, — ,,} & (-\exp \pi a, -1), & \varphi_0 = -\varphi_w & \\ \text{,, — ,,} & (-1, 1), & \chi_0 = b & \\ \text{,, — ,,} & (1, \exp \pi a), & \varphi_0 = \varphi_w & \\ \text{,, — ,,} & (\exp \pi a, \infty), & \chi_0 = 0 & \end{array} \tag{4.3.11}$$

This is a Volterra-Signorini-type problem [G. 28] [8] whose solution is:

$$f(z') = \frac{1}{\pi} \sqrt{P(z')} \int_{-\exp \pi a}^{+\exp \pi a} \frac{v(\xi)}{\sqrt{|P(\xi)|}} \frac{\mathrm{d}\xi}{z' - \xi} \tag{4.3.12}$$

where:

$$P(z') = (z'^2 - e^{2\pi a})(z'^2 - 1)$$

$$v(\xi) = \begin{cases} \varphi_w, & \text{for } \xi \in (-\exp \pi a, -1) \cup (1, \exp \pi a) \\ b, & \text{for } \xi \in (-1, +1) \end{cases} \tag{4.3.12'}$$

We have chosen that argument under the square-root that is positive for $z' \in (\exp \pi a, \infty)$.

The constant b is determined from the condition (2.26). By a direct calculation one can show that for any conformal mapping one has:

$$|\mathrm{grad}_{x',y'}\varphi_0|^2 = |\mathrm{grad}_{x,y}\varphi_0|^2 \left|\frac{\mathrm{d}z}{\mathrm{d}z'}\right|^2 \tag{4.3.13}$$

Since in our case:

$$\left|\frac{\mathrm{d}z}{\mathrm{d}z'}\right|^2 = \left(\frac{2}{\pi}\right)^2 \mathrm{e}^{-\pi(x+a)}$$

it follows that the boundedness condition (4.2.26) entails:

$$\lim_{|x'|\to\infty} |\mathrm{grad}_{x',y'}\,\varphi_0| = 0 \tag{4.3.14}$$

But one can easily check that:

$$|\mathrm{grad}_{x',y'}\varphi_0|^2 = \left|\frac{\mathrm{d}f_0}{\mathrm{d}z'}\right| \tag{4.3.15}$$

Therefore the derivative of the function $f_0(z')$ should vanish at infinity. Taking into account (4.3.12) this condition becomes:

$$\int_{-\exp \pi a}^{\exp \pi a} \frac{v(\xi)}{\sqrt{|P(\xi)|}}\,\mathrm{d}\xi = 0$$

whence:

$$b\int_{-1}^{+1} \frac{\mathrm{d}\xi}{\sqrt{P(\xi)}} = -2\varphi_w \int_{1}^{\exp \pi a} \frac{\mathrm{d}\xi}{\sqrt{|P(\xi)|}} \tag{4.3.16}$$

This is a final solution of the problem [8].

In order to determine the generator characteristics, we need the boundary values of the function $f_0(z)$ at the ends of the electrodes (that is, the values of $f_0(z')$ at the points ± 1, $\pm \exp \pi a$ on the real axis). From a theorem whose proof can be found in [G. 34] it follows that the function $f_0(z')$ determined by the equation (4.3.12) takes at the points on the real axis at which $P(x')$ vanishes the values (4.3.11) which it has on the adjacent segments. Consequently:

$$\chi_0(a, \pm 1) = 0, \quad \varphi_0(\pm a, 1) = -\varphi_w \tag{4.3.17}$$

$$\chi_0(-a, \pm 1) = b, \; \varphi_0(\pm a, -1) = \varphi_w$$

Noticing that in this case (in the electrode region) we have:

$$J_y^{(1)} = -\left(\frac{\partial \varphi_0}{\partial y} + 1\right) = \frac{\partial \chi_0}{\partial x} - 1$$

$$J_y^{(1)}(x, -1) = J_y^{(1)}(x, 1)$$

from (4.2.46) and (4.3.17) it follows that:

$$W^* = -2\varphi_w Rm \int_{-a}^{+a} \left(\left.\frac{\partial \chi_0}{\partial x}\right|_{y=1} - 1 \right) dx \tag{4.3.18}$$

$$= 2\varphi_w Rm (2a + b)$$

This is the expression for the generator power (b being determined by (4.3.16)).

4.3.2 The determination of higher order approximations

4.3.2.1 All the approximations are odd functions

For determining the $n (n \geqslant 1)$-order approximations from the system (4.2.40) we obtain the following equations:

$$\Delta\psi_n = \mathscr{F}_{n-1}(x, y) = \begin{cases} 0, & \text{for } x < -a \\ f_{n-1}^{(1)}(y), & \text{for } |x| < a \\ f_{n-1}^{(1)}(y) + f_{n-1}^{(2)}(y), & \text{for } x > a \end{cases}$$

$$\Delta\varphi_n = \mathscr{H}_{n-1}(x, y) = \begin{cases} 0 & \text{for } x < -a \\ -f_{n-1}^{1}(y), & \text{for } |x| < a \\ 0, & \text{for } x > a \end{cases} \tag{4.3.19}$$

where

$$f_{n-1}^{(1)}(y) = -\left(\frac{\partial \varphi_{n-1}}{\partial x} + \frac{\partial \psi_{n-1}}{\partial x}\right)_{x=-a+}$$

$$f_{n-1}^{(2)}(y) = \left(\frac{\partial \varphi_{n-1}}{\partial x} + \frac{\partial \psi_{n-1}}{\partial x}\right)_{x=a-} \tag{4.3.19'}$$

which should be integrated together with the homogeneous conditions ($\varphi_w = 0$). The jump conditions indicate that ψ_n together with its first order derivatives (and the second-order derivative with respect to y) are continuous across $x = \mp a$. The function φ_n and thus its derivatives with respect to the variable y are continuous as well.

Concerning system (4.3.19) one first notices that $\mathscr{F}_{n-1}$ and $\mathscr{H}_{n-1}$ are known functions if the previous approximations are known.

Secondly one notices that if:

$$\begin{aligned} \mathscr{F}_{n-1}(x, y) &= -\mathscr{F}_{n-1}(x, -y) \\ \mathscr{H}_{n-1}(x, y) &= -\mathscr{H}_{n-1}(x, -y) \end{aligned} \tag{4.3.20}$$

then:

$$\begin{aligned} \psi_n(x, y) &= -\psi_n(x, -y) \\ \varphi_n(x, y) &= -\varphi_n(x, -y) \end{aligned} \tag{4.3.21}$$

The proof is similar to that for equation (4.3.5). One first proves the uniqueness theorem by means of the Green formula (the difference of two solutions is again a harmonic function everywhere within the fluid).

Since the first approximation (determined above) is antisymmetrical in the y variable from the definitions of the functions $\mathscr{F}_{n-1}$ and $\mathscr{H}_{n-1}$ it follows that $\mathscr{F}_0$ and $\mathscr{H}_0$ have the same property. Applying theorem (4.3.21), it follows that the functions ψ_1 and φ_1 are anti-symmetrical, and so on. To conclude, all these approximations are anti-symmetrical functions in y.

4.3.2.2 Determination of function ψ_n

We shall now prove that ψ_n is given by the equation [10]:

$$\psi_n(x, y) = \int_{-a}^{\infty} \int_{0}^{1} g(x, y, \xi, \eta)\, \mathscr{F}_{n-1}(\xi, \eta)\, d\xi\, d\eta \tag{4.3.22}$$

where:

$$g(x, y, \xi, \eta) = -\frac{1}{4\pi} \ln \left\{ \frac{\sin^2 \frac{\pi}{2}(y + \eta) + \operatorname{sh}^2 \frac{\pi}{2}(x - \xi)}{\sin^2 \frac{\pi}{2}(y - \eta) + \operatorname{sh}^2 \frac{\pi}{2}(x - \xi)} \right\} \tag{4.3.23}$$

Indeed, one can write the solution of the first equation (4.3.19) and of the boundary conditions (4.2.22), (4.2.23) in the form:

$$\psi_n(x, y) = \int_{-\infty}^{+\infty} G(x, y, \xi)\, d\xi \tag{4.3.24}$$

where G is the Green function determined by the equations:

$$\Delta_{x,y}G = \mathscr{F}_{n-1}(\xi, y)\,\delta(\xi - x)$$

$$G(x, \pm 1, \xi) = 0, \quad \forall x \tag{4.3.25}$$

$$\lim_{|x|\to\infty} \frac{\partial G}{\partial x} = 0$$

and the condition:

$$G(x, y, \xi) = -G(x, -y, \xi) \tag{4.3.26}$$

It can easily be checked that except for the point $x = \xi$ the proper function system for the problem (4.3.25) is:

$$\{\sin(\alpha_j y)\,e^{-\alpha_j|x-\xi|}\}, \quad \alpha_j = j\pi, \; j = 1,2,\cdots \tag{4.3.27}$$

$$\{\cos(\beta_j y)\,e^{-\beta_j|x-\xi|}\}, \quad \beta_j = (2j+1)\frac{\pi}{2}, \quad j = 0,1,\cdots \tag{4.3.28}$$

Taking into account (4.3.26) only the sequence (4.3.27) is retained, so that:

$$G(x, y, \xi) = \sum_{j=1}^{\infty} a_j \sin(\alpha_j y)\,e^{-\alpha_j|x-\xi|} \tag{4.3.29}$$

Since there is a singularity at the point $x = \xi$, we shall impose the condition that the first equation (4.3.25) be also satisfied:

$$\int_{\xi-\varepsilon}^{\xi+\varepsilon} \left(\frac{\partial^2 G}{\partial x^2} + \frac{\partial^2 G}{\partial y^2}\right) dx = \mathscr{F}_{n-1}(\xi, y)$$

Taking the limit for $\varepsilon \to 0$, we get:

$$\lim_{\varepsilon\to 0} \frac{\partial G}{\partial x}\bigg|_{\xi-\varepsilon}^{\xi+\varepsilon} = \mathscr{F}_{n-1}(\xi, y) \tag{4.3.30}$$

whence, taking into account (4.3.29), it follows that:

$$2\sum_{j=1}^{\infty} \alpha_j a_j \sin(\alpha_j y) = -\mathscr{F}_{n-1}(\xi, y) \tag{4.3.31}$$

This relationship indicates that $-2\alpha_j a_j$ are coefficients in the Fourier series expansion of the function $\mathscr{F}_{n-1}$. Consequently:

$$\alpha_j a_j = -\int_0^1 \mathscr{F}_{n-1}(\xi, \eta)\sin(\alpha_j \eta)\,d\eta \tag{4.3.32}$$

so that (4.3.29) becomes:

$$G(x, y, \xi) = -\sum_{j=1}^{\infty}\int_0^1 \frac{\sin(\alpha_j y)\sin(\alpha_j \eta)}{\alpha_j} e^{-\alpha_j|x-\xi|}\mathscr{F}_{n-1}(\xi, \eta)\,d\eta \qquad (4.3.33)$$

Now, noticing that the series under the integral is absolutely convergent, we may introduce the summation sign Σ under the integral. Since [13] we have:

$$-\sum_{j=1}^{\infty} \frac{\sin(\alpha_j y)\sin(\alpha_j \eta)}{\alpha_j} e^{-\alpha_j|x-\xi|} = g(x, y, \xi, \eta) \qquad (4.3.34)$$

from (4.3.24) and (4.3.33), (4.3.22) follows.

For numerical calculations it is useful to write the solution (4.3.22) explicitly. Taking into account the expression of the function $\mathscr{F}_{n-1}$(4.3.19) and with the notations:

$$A(\alpha_j, x) = \int_{-a}^{+a} e^{-\alpha_j|x-\xi|}d\xi = \frac{2}{\alpha_j}\begin{cases} \exp(\alpha_j x)\,\mathrm{sh}(\alpha_j a) & , \text{ for } x < -a \\ 1 - \exp(-\alpha_j a)\,\mathrm{ch}(\alpha_j x) & , \text{ for } |x| < a \\ \exp(-\alpha_j x)\,\mathrm{sh}(\alpha_j a) & , \text{ for } x > a \end{cases}$$

$$B(\alpha_j, x) = \int_a^{\infty} e^{-\alpha_j|x-\xi|}d\xi = \frac{1}{\alpha_j}\begin{cases} e^{\alpha_j(x-a)} & , \text{ for } x < a \\ 2 - e^{-\alpha_j(x-a)} & , \text{ for } x > a \end{cases}$$

$$S_{n-1}^{(k)}(\alpha_j) = \int_0^1 f_{n-1}^{(k)}(\eta)\sin(\alpha_j \eta)\,d\eta, \quad k = 1, 2 \qquad (4.3.35)$$

we have:

$$\psi_n(x, y) = -\sum_{j=1}^{\infty} \frac{\sin(\alpha_j y)}{\alpha_j}\{A(\alpha_j, x)\,S_{n-1}^{(1)}(\alpha_j) + B(\alpha_j, x)(S_{n-1}^{(1)} + S_{n-1}^{(2)})\} \qquad (4.3.36)$$

which is the final form of the solution. It can be easily checked that it satisfies the jump conditions (4.2.31).

The behaviour of the velocity at the channel outlet is given by the following formulae:

$$\lim_{x\to\infty} \frac{\partial \psi_n}{\partial y} = -2\sum_{j=1}^{\infty} \cos(\alpha_j y)\,(S_{n-1}^{(1)} + S_{n-1}^{(2)}), \quad \lim_{x\to\infty} \frac{\partial \psi_n}{\partial x} = 0 \qquad (4.3.37)$$

and at the central part of the channel by:

$$\begin{aligned} \lim_{x\to 0} \frac{\partial \psi_n}{\partial y} &= -\sum_{j=1}^{\infty} \frac{\cos(\alpha_j y)}{\alpha_j}\{(2 - e^{-\alpha_j a})\,S_{n-1}^{(1)} + e^{-\alpha_j a}\,S_{n-1}^{(2)}\} \\ \lim_{x\to 0} \frac{\partial \psi_n}{\partial x} &= -\sum_j \sin(\alpha_j y)\,e^{-\alpha_j a}\,(S_{n-1}^{(1)} + S_{n-1}^{(2)}) \end{aligned} \qquad (4.3.38)$$

Consequently, everything depends on the numbers $S_{n-1}^{(1)}$ and $S_{n-1}^{(2)}$.

For the first approximation ($n = 1$) we have:

$$f_0^{(1)}(y) = -\left.\frac{\partial \varphi_0}{\partial x}\right|_{x=-a}, \quad f_0^{(2)}(y) = \left.\frac{\partial \varphi_0}{\partial x}\right|_{x=a}$$

The boundary values of $\partial\varphi_0/\partial x$ at the points $\mp a$ can be calculated from (4.3.12) noticing that we have:

$$\frac{\mathrm{d}f_0}{\mathrm{d}z} = \frac{\partial \varphi_0}{\partial x} - \mathrm{i}\,\frac{\partial \varphi_0}{\partial y} = \left(\frac{\mathrm{d}f_0}{\partial z'}\right)_{(4.3.10)} \frac{\mathrm{d}z'}{\mathrm{d}z} \tag{4.3.39}$$

One obtains:

$$\left.\frac{\partial \varphi_0}{\partial x}\right|_{x=-a} = \left\{\sin\frac{\pi}{4}(5y-\theta) + \mathrm{e}^{\pi a}\mathrm{ch}\,(\pi a)\sin\frac{\pi}{4}(y-\theta)\right.$$

$$\left. + R\cos\left(\frac{\pi}{2}y\right)\sin\frac{\pi}{4}(3y+\theta)\right\} I_1R_1^{-1} - \left\{\cos\frac{\pi}{4}(7y-\theta)\right.$$

$$\left. + \mathrm{e}^{\pi a}\mathrm{ch}(\pi a)\cos\frac{\pi}{4}(3y-\theta)\right\} I_2R_1^{-1}$$

$$- \left\{I_3\sin\frac{\pi}{4}(y+\theta) - I_4\cos\frac{\pi}{4}(3y+\theta)\right\}\cos\left(\frac{\pi}{2}y\right)R_1$$

$$\left.\frac{\partial \varphi_0}{\partial x}\right|_{x=a} = \left\{\sin\frac{\pi}{4}(5y-\theta) + \mathrm{e}^{-2\pi a}\mathrm{ch}(\pi a)\sin\frac{\pi}{4}(y-\theta)\right.$$

$$\left. + R\cos\left(\frac{\pi}{2}y\right)\sin\frac{\pi}{4}(3y+\theta)\right\} \times$$

$$\times\, \mathrm{e}^{3\pi a}I_1'R_1^{-1} - \left\{\cos\frac{\pi}{4}(7y-\theta) + \mathrm{e}^{-\pi a}\mathrm{ch}(\pi a)\cos\frac{\pi}{4}(3y-\theta)\right\}\mathrm{e}^{2\pi a}I_2'R_1^{-1}$$

$$+ \mathrm{e}^{5\pi a}\cos\left(\frac{\pi}{2}y\right)\sin\frac{\pi}{4}(y+\theta)\,I_3'R_1$$

$$+ \mathrm{e}^{3\pi a}\cos\left(\frac{\pi}{2}y\right)\cos\frac{\pi}{4}(3y+\theta)\,I_4'R_1$$

where we used the notations:

$$\operatorname{tg}\frac{\pi\theta}{2} = \frac{\sin(\pi y)}{\cos(\pi y) + \exp(2\pi a)}, \quad R_1 = \sqrt{2R\cos\left(\frac{\pi}{2}y\right)},$$

$$R = \sqrt{1 + 2e^{2\pi a}\cos(\pi y) + e^{4\pi a}}$$

$$(I_1, I_2) = \int_{-\exp(\pi a)}^{+\exp(\pi a)} \frac{\nu}{\sqrt{|P(\xi)|}} \frac{(1, \xi)\,d\xi}{1 + 2\xi\sin\left(\frac{\pi}{2}y\right) + \xi^2}$$

$$(I_3, I_4) = \int_{-\exp(\pi a)}^{+\exp(\pi a)} \frac{\nu}{\sqrt{|P(\xi)|}} \frac{(1, \xi)\,d\xi}{\left\{1 + 2\xi\sin\left(\frac{\pi}{2}y\right) + \xi^2\right\}^2}$$

$$(I'_1, I'_2) = \int_{-\exp(\pi a)}^{+\exp(\pi a)} \frac{\nu}{\sqrt{|P(\xi)|}} \frac{(1, \xi)\,d\xi}{1 + 2\xi e^{\pi a}\sin\left(\frac{\pi}{2}y\right) + \xi^2}$$

$$(I'_3, I'_4) = \int_{-\exp(\pi a)}^{+\exp(\pi a)} \frac{\nu}{\sqrt{|P(\xi)|}} \frac{(1, \xi)\,d\xi}{\left\{1 + 2\xi e^{\pi a}\sin\left(\frac{\pi}{2}y\right) + \xi^2\right\}^2}$$

4.3.2.3 The determination of function φ_n

Denoting by φ_{n0} the solution of equation (4.3.19*b*) with the boundary condition

$$\left.\frac{\partial \varphi_{n0}}{\partial y}\right|_{y=\pm 1} = 0, \quad \forall x; \quad \lim_{|x|\to\infty} |\operatorname{grad}\varphi_{n0}| = 0 \tag{4.3.40}$$

and by φ_{ng} the general solution of the homogeneous equation $\Delta\varphi_{ng} = 0$ with boundary conditions:

$$\varphi_{ng}(x, \pm 1) = -\varphi_{n0}(x, \pm 1), \text{ for } |x| < a$$

$$\left.\frac{\partial \varphi_{ng}}{\partial y}\right|_{y=\pm 1} = 0, \text{ for } |x| > a; \quad \lim_{|x|\to\infty} |\operatorname{grad}\varphi_{ng}| = 0 \tag{4.3.41}$$

the solution of equation (4.3.19*b*) is written in the form:

$$\varphi_n = \varphi_{ng} + \varphi_{n0} \tag{4.3.42}$$

The solution of problem (4.3.41) is given by equation (4.3.12) in which φ_w is substituted by $\varphi_{n0}(x, +1)$ (4.3.48).

The solution of equation (4.3.40) is written in the form:

$$\varphi_{n0}(x, y) = \int_{-\infty}^{+\infty} \int_0^1 g_0(x, y, \xi, \eta) \mathscr{H}_{n-1}(\xi, \eta) \, d\xi \, d\eta \tag{4.3.43}$$

where:

$$g_0(x, y, \xi, \eta) = -\sum_{j=0}^{\infty} \frac{\sin(\beta_j y) \sin(\beta_j \eta)}{\beta_j} e^{-\beta_j |x-\xi|} \tag{4.3.43'}$$

Indeed, looking for a solution of equation (4.3.19b) of the form:

$$\varphi_{n0}(x, y) = \int_{-\infty}^{+\infty} G_0(x, y, \xi) \, d\xi \tag{4.3.44}$$

it follows that:

$$\begin{gathered} \Delta_{x,y} G_0 = \mathscr{H}_{n-1}(\xi, y) \delta(\xi - x) \\ \left.\frac{\partial G_0}{\partial y}\right|_{y=\pm 1} = 0, \ \forall x; \ \lim_{|x| \to \infty} |\text{grad } G_0| = 0 \\ G_0(x, y, \xi) = -G_0(x, -y, \xi) \end{gathered} \tag{4.3.44'}$$

The function G_0 satisfying the conditions (4.3.44'), except for the point $x = \xi$, has the form:

$$G_0 = \sum_{j=0}^{\infty} a_j \sin(\beta_j y) e^{-\beta_j |x-\xi|}, \quad \beta_j = (2j+1)\frac{\pi}{2} \tag{4.3.45}$$

By imposing the condition that G_0 should overall satisfy the equation (4.3.44′ a) it follows:

$$2 \sum_{j=0}^{\infty} \beta_j a_j \sin(\beta_j y) = \mathscr{H}_{n-1}(\xi, y) \tag{4.3.46}$$

whence:

$$\beta_j a_j = -\int_0^1 \mathscr{H}_{n-1}(\xi, \eta) \sin(\beta_j \eta) \, d\eta \tag{4.3.46'}$$

Substituting (4.3.42) into (4.3.40) and (4.3.40) into (4.3.39), (4.3.38) is obtained.

Using the notations (4.3.35), φ_{n0} becomes:

$$\varphi_{n0}(x, y) = \sum_{j=0}^{\infty} \frac{\sin(\beta_j y)}{\beta_j} A(\beta_j, x) S_{n-1}^{(1)}(\beta_j) \tag{4.3.47}$$

and

$$\varphi_{n0}(x, 1) = 2 \sum_{j=0}^{\infty} \frac{(-1)^j}{\beta_j^2} (1 - e^{-\beta_j a} \operatorname{ch}(\beta_j x)) S_{n-1}^{(1)}(\beta_j) \qquad (4.3.48)$$

This method can be easily extended to the multiple electrode pair generator [8].

4.4 Tensor conductivity fluids

4.4.1 The determination of functions ψ_n

Whenever we cannot neglect the Hall effect (ionized gases) we have to solve the system (4.2.41) with the boundary conditions (4.2.21) – (4.2.26). It can be easily noticed that to a first approximation the solution determined in Section (4.3.1) is valid. This solution is anti-symmetrical with respect to the variable y, but due to the Hall effect the higher order approximations will no longer have the same property. We shall therefore write:

$$\psi_n = \psi_n^{(a)} + \psi_n^{(s)}, \quad \varphi_n = \varphi_n^{(a)} + \varphi_n^{(s)} \qquad (4.4.1)$$

Introducing now the notations:

$$\mathscr{F}_{n-1}^{(a)}(x, y) = \begin{cases} 0, & \text{for } x < -a \\ F_{n-1}^{1a}(y), & \text{for } |x| < a \\ F_{n-1}^{1a}(y) + F_{n-1}^{2a}(y), & \text{for } x > a \end{cases}$$

$$\mathscr{H}_{n-1}^{(a)}(x, y) = \begin{cases} 0, & \text{for } x < -a \\ -F_{n-1}^{1a}(y), & \text{for } |x| < a \\ 0, & \text{for } x > a \end{cases}$$

$$(4.4.2)$$

$$\mathscr{F}_{n-1}^{(s)}(x, y) = \begin{cases} 0, & \text{for } x < -a \\ F_{n-1}^{1s}(y), & \text{for } |x| < a \\ F_{n-1}^{1s}(y) + F_{n-1}^{2s}(y), & \text{for } x > a \end{cases}$$

$$\mathscr{H}_{n-1}^{(s)}(x, y) = \begin{cases} 0, & \text{for } x < -a \\ -F_{n-1}^{1s}(y), & \text{for } |x| < a \\ 0, & \text{for } x > a \end{cases}$$

where:

$$F^{1a}_{n-1}(y) = -\frac{1}{1+v^2}\left(\frac{\partial\varphi^{(a)}_{n-1}}{\partial x} + \frac{\partial\psi^{(a)}_{n-1}}{\partial x}\right)_{x=-a+} + \frac{v}{1+v^2}\left(\frac{\partial\varphi^{(s)}_{n-1}}{\partial y} + \frac{\partial\psi^{(s)}_{n-1}}{\partial y}\right)_{x=-a+}$$

$$F^{2a}_{n-1}(y) = \frac{1}{1+v^2}\left(\frac{\partial\varphi^{(a)}_{n-1}}{\partial x} + \frac{\partial\psi^{(a)}_{n-1}}{\partial x}\right)_{x=a-} - \frac{v}{1+v^2}\left(\frac{\partial\varphi^{(s)}_{n-1}}{\partial y} + \frac{\partial\psi^{(s)}_{n-1}}{\partial y}\right)_{x=a-} \tag{4.4.2'}$$

$$F^{1s}_{n-1}(y) = -\frac{1}{1+v^2}\left(\frac{\partial\varphi^{(s)}_{n-1}}{\partial x} + \frac{\partial\psi^{(s)}_{n-1}}{\partial x}\right)_{x=-a+} + \frac{v}{1+v^2}\left(\frac{\partial\varphi^{(a)}_{n-1}}{\partial y} + \frac{\partial\psi^{(a)}_{n-1}}{\partial y} + 1\right)_{x=-a+}$$

$$F^{2s}_{n-1}(y) = \frac{1}{1+v^2}\left(\frac{\partial\varphi^{(s)}_{n-1}}{\partial x} + \frac{\partial\psi^{(s)}_{n-1}}{\partial x}\right)_{x=a-} - \frac{v}{1+v^2}\left(\frac{\partial\varphi^{(a)}_{n-1}}{\partial y} + \frac{\partial\psi^{(a)}_{n-1}}{\partial y} + 1\right)_{x=a-}$$

We shall have to solve the following equations ($n \geqslant 1$):

$$\Delta\psi^{(a)}_n = \mathscr{F}^{(a)}_{n-1}, \quad \Delta\varphi^{(a)}_n = \mathscr{H}^{(a)}_{n-1} \tag{4.4.3}$$

$$\Delta\psi^{(s)}_n = \mathscr{F}^{(s)}_{n-1}, \quad \Delta\varphi^{(s)}_n = \mathscr{H}^{(s)}_{n-1} \tag{4.4.4}$$

with the homogeneous boundary conditions.

The sequence of anti-symmetrical approximations $\{\psi^{(a)}_n, \varphi^{(a)}_n\}$ was determined in the previous section. One has now to determine the symmetrical approximation sequence $\{\psi^{(s)}_n, \varphi^{(s)}_n\}$. As for the first case, one can prove that all these latter approximations are symmetrical functions.

The solution of the equation:

$$\Delta\psi^{(s)}_n = \mathscr{F}^{(s)}_{n-1}(x, y) \tag{4.4.5}$$

with the boundary conditions:

$$\psi^{(s)}_n(x, \pm 1) = 0, \ \forall x$$

$$\lim_{|x|\to\infty} \frac{\partial\psi^{(s)}_n}{\partial x} = 0 \tag{4.4.5'}$$

where:

$$\mathscr{F}^{(s)}_{n-1}(x, y) = \mathscr{F}^{(s)}_{n-1}(x, -y) \tag{4.4.5''}$$

is given by the formula:

$$\psi^{(s)}_n(x, y) = \int_{-\infty}^{+\infty}\int_0^1 g^{(s)}(x, y, \xi, \eta)\mathscr{F}^{(s)}_{n-1}(\xi, \eta)\, \mathrm{d}\xi\, \mathrm{d}\eta \tag{4.4.6}$$

where:

$$g^{(s)}(x, y, \xi, \eta) = -\sum_{j=0}^{\infty} \frac{\cos(\beta_j y)\cos(\beta_j \eta)}{\beta_j} e^{-\beta_j|x-\xi|} \tag{4.4.6'}$$

the series being absolutely convergent. This statement is proved as in the previous section noticing however that, due to the property (4.4.5''), formula (4.3.29) is replaced by:

$$G^{(s)}(x, y, \xi) = \sum_{j=0}^{\infty} a_j \cos(\beta_j y) \exp(-\beta_j|x-\xi|)$$

and the formulae (4.3.31) and (4.3.32) by:

$$-2\sum_{j=0}^{\infty} \beta_j a_j \cos(\beta_j y) = \mathscr{F}^{(s)}_{n-1}(\xi, y)$$

$$\beta_j a_j = -\int_0^1 \mathscr{F}^{(s)}_{n-1}(\xi, \eta)\cos(\beta_j \eta)\, d\eta$$

Taking into account the definition of function $\mathscr{F}^{(s)}_{n-1}$ the solution (4.4.6) can be also written:

$$\psi_n^{(s)}(x, y) = -\sum_{j=0}^{\infty} \frac{\cos(\beta_j y)}{\beta_j}\{A(\beta_j, x)C^{1s}_{n-1}(\beta_j) + B(\beta_j, x)(C^{1s}_{n-1} + C^{2s}_{n-1})\} \tag{4.4.7}$$

A and B being defined by (4.3.35) and $C^{ks}_{n-1}(\beta_j)$ by the equations:

$$C^{ks}_{n-1}(\beta_j) = \int_0^1 F^{ks}_{n-1}(\eta)\cos(\beta_j\eta)\, d\eta, \quad k = 1, 2 \tag{4.4.8}$$

4.4.2 Determination of function φ_n

Taking into account the solution of the anti-symmetrical problem determined in the previous section, here one only needs to determine the solution of the equation:

$$\Delta\varphi_n^{(s)} = \mathscr{H}^{(s)}_{n-1}$$

$$\mathscr{H}^{(s)}_{n-1}(x, y) = \mathscr{H}^{(s)}_{n-1}(x, -y) \tag{4.4.9}$$

with the following boundary conditions:

$$\varphi_n^{(s)}(x, \pm 1) = 0, \text{ for } |x| < a; \quad \left.\frac{\partial \varphi_n^{(s)}}{\partial y}\right|_{y=\pm 1} = 0, \text{ for } |x| > a$$

$$\lim_{|x|\to\infty} |\text{grad } \varphi_n^{(s)}| = 0 \tag{4.4.9'}$$

A particular solution $\varphi_{n0}^{(s)}$ which satisfies the boundary conditions:

$$\left.\frac{\partial \varphi_{n0}^{(s)}}{\partial y}\right|_{y=\pm 1} = 0, \ \forall x; \ \lim_{|x|\to\infty} |\text{grad } \varphi_{n0}^{(s)}| = 0 \tag{4.4.10}$$

for equation (4.4.9) is

$$\varphi_{n0}^{(s)}(x, y) = \int_{-\infty}^{+\infty} \int_0^1 g_0^{(s)}(x, y, \xi, \eta) \, \mathscr{H}_{n-1}^{(s)}(\xi, \eta) \, d\xi \, d\eta \tag{4.4.11}$$

where:

$$g_0^{(s)}(x, y, \xi, \eta) = -\sum_{j=1}^{\infty} \frac{\cos(\alpha_j y)\cos(\alpha_j \eta)}{\alpha_j} \, e^{-\alpha_j |x-\xi|} \tag{4.4.11'}$$

Indeed looking for a solution of the form

$$\varphi_{n0}^{(s)}(x, y) = \int_{-\infty}^{+\infty} G_0^{(s)}(x, y, \xi) \, d\xi$$

and taking into account (4.4.10) we have:

$$G_0^{(s)}(x, y, \xi) = \sum_{j=1}^{\infty} a_j \cos(\alpha_j y) \, e^{-\alpha_j |x-\xi|}$$

The condition that $G_0^{(s)}$ overall satisfies the equation:

$$\Delta_{x,y} G_0^{(s)} = \mathscr{H}_{n-1}^{(s)}(\xi, y) \, \delta(x - \xi)$$

as in (4.3.30) reduces itself to:

$$-2 \sum_{j=1}^{\infty} \alpha_j a_j \cos(\alpha_j y) = \mathscr{H}_{n-1}^{(s)}(\xi, y)$$

whence the coefficients a_j and then $G_0^{(s)}$ are determined.

Taking into account the above notations, $\varphi_{n0}^{(s)}$ becomes:

$$\varphi_{n0}^{(s)}(x, y) = \sum_{j=1}^{\infty} \frac{\cos(\alpha_j y)}{\alpha_j} A(\alpha_j, x) C_{n-1}^{1s}(\alpha_j) \tag{4.4.12}$$

whence:

$$\varphi_{n0}^{(s)}(x, 1) = 2 \sum_{j=1}^{\infty} \frac{(-1)^j}{\alpha_j^2} [1 - e^{-\alpha_j a} \operatorname{ch}(\alpha_j x)] C_{n-1}^{1s}(\alpha_j) \tag{4.4.12'}$$

Once having determined a particular solution of equation (4.4.9), the general solution $\varphi_{ng}^{(s)}(\varphi_n^{(s)} = \varphi_{ng}^{(s)} + \varphi_{n0}^{(s)})$ of the homogeneous equation:

$$\Delta\varphi_{ng}^{(s)} = 0, \ \varphi_{ng}^{(s)}(x, y) = \varphi_{ng}^{(s)}(x, -y) \tag{4.4.13}$$

with the boundary conditions:

$$\begin{gathered} \varphi_{ng}^{(s)}(x, \pm 1) = -\varphi_{n0}^{(s)}(x, \pm 1), \text{ for } |x| < a \\ \left.\frac{\partial \varphi_{ng}^{(s)}}{\partial y}\right|_{y=\pm 1} = 0, \text{ for } |x| > a; \ \lim_{|x| \to \infty} |\operatorname{grad} \varphi_{ng}^{(s)}| = 0 \end{gathered} \tag{4.4.13'}$$

remains to be determined. The solution of this problem is obtained as in Section 4.3.1 by introducing the function χ, the harmonic conjugate of $\varphi_{ng}^{(s)}$ and considering the function:

$$f_{ng}^{(s)} = \varphi_{ng}^{(s)}(x, y) + i\chi(x, y) \tag{4.4.14}$$

holomorphic within the fluid. For this case:

$$\chi(x, y) = \chi(x, -y)$$

In a manner similar to Section 4.3.1, [10], one obtains:

$$f_{ng}^{(s)}(z') = \frac{1}{\pi} \sqrt{P(z')} \int_{-\exp \pi a}^{+\exp \pi a} \frac{v(\xi')}{\sqrt{|P(\xi')|}} \frac{d\xi'}{z' - \xi'} \tag{4.4.15}$$

where z' and $P(z')$ are defined by the equations (4.3.10) and (4.3.12′) and:

$$v(x') = \begin{cases} \bar{\varphi}_{n0}^{(s)}(x'), & \text{for } x \in (-\exp \pi a, -1) \\ 0, & \text{for } x \in (-1, 0) \cup (0, 1) \\ -\bar{\varphi}_{n0}^{(s)}(x'), & \text{for } x \in (1, \exp \pi a) \end{cases}$$

$\bar{\varphi}_{n0}^{(s)}(x')$ being the function corresponding by the conformal mapping (4.3.10) to the function $\bar{\varphi}_{n0}^{(s)}(x, \pm 1)$. In (4.4.14) we considered that argument of the square root which is positive for $z' \in (-\infty, -\exp \pi a)$.

4.4.3 The second order approximation

The determination of the second order approximation $\{\psi_1, \varphi_1\}$ can be performed numerically if we calculate the coefficients $C_0(\alpha_j)$ and $C_0(\beta_j)$. In this respect we have:

$$F_0^{1a} = -\frac{1}{1+v^2}\frac{\partial\varphi_0}{\partial x}\bigg|_{x=-a}, \quad F_0^{2a} = \frac{1}{1+v^2}\frac{\partial\varphi_0}{\partial x}\bigg|_{x=a}$$

$$F_0^{1s} = \frac{v}{1+v^2}\left(\frac{\partial\varphi_0}{\partial y}\bigg|_{x=-a} + 1\right), \quad F_0^{2s} = -\frac{v}{1+v^2}\left(\frac{\partial\varphi_0}{\partial y}\bigg|_{x=a} + 1\right)$$

the functions $\partial\varphi_0/\partial x$ and $\partial\varphi_0/\partial y$ being determined by the formula (3.39).

5

Theory of thin airfoils in perfectly conducting fluids

5.1 Introduction

As in the classical aerodynamics, the motion of conducting fluids past thin airfoils can be examined by means of linearized equations. This fact facilitates substantially the solution of the problem and explains to a great extent the reason why in this field, as in that of the flow through ducts, important results have been obtained.

Practically, this problem arises mainly in hypersonic aerodynamics where, due to the friction between the body and the air, the latter is ionized and becomes a conductor. The problem also arises in cosmic flights where the surrounding medium is a rarefied and ionized medium.

We will assume in this chapter that the fluid is of infinite electrical conductivity. Such an assumption does not correspond exactly to the actual phenomenon but simplifies to a great extent the equations of motion. Accordingly, the solution is relatively simple and constitutes a transition from the solution given by classical aerodynamics to that from the aerodynamics of fluids with finite electrical conductivity which will be discussed in the next chapter. In addition, such a solution is of importance in laboratory work where fluid metal is used, as well as in the case of high temperature gases.

Historically, the examination of the flow of conducting fluids past thin airfoils was carried out by W. R. Sears and E. L. Resler jr. in 1959 [1]. They considered the fluid as being incompressible, inviscid and of infinite electrical conductivity. In free motion, a uniform magnetic field orientated perpendicular to, or aligned with, the fluid velocity is present. In [2], K. Stewartson extended this study to the case of an arbitrarily orientated magnetic field. These solutions considered again by L. Dragoş in [3] are presented in Section 5.2.

The motion of compressible fluids with infinite electrical conductivity was discussed by W. R. Sears [4] in the case of aligned fields and by J. E. McCunne and E. L. Resler jr. [5] for aligned and orthogonal fields. The case of orthogonal fields was further examined by P. Greenberg [6] and L. Dragoş [7]. A review of these results is presented in Section 5.3 which also includes D. Homentcovschi's solution [8] concerning the hyperbolic flow in the presence of thick airfoils.

The general case of oblique fields was studied by E. Cumberbatch, L. Sarason and H. Weitzner [9], P. Greenberg [10], E. P. Salathé and L. Sirowich [11] and I. Imai [12]. In [9], the solution is represented using Riemann's invariants and in [11] using generalized functions. Although of a particular importance, these works can not be discussed here.

The following section deals with the motion of conducting fluids in the presence of non-conductivity walls. This problem can be also handled by means of linearized equations. The problem of incompressible fluids was studied by L. Dragoş and D. Homentcovschi [13]. To the author's knowledge, the first results concerning the motion of compressible fluids were given by T. Taniuti [14] and M. N. Kogan [15]. The problem was further discussed by C. K. Chy and Y. K. Lynn [16], Y. Mimura [17], C. K. Chu [18] and G. W. Swan [19]. The results given in [13] and [18] are included in Section 5.4.

Unsteady motion was studied by K. Stewartson [20], L. Ring [21] and D. Homentcovschi [22]. Section 5.5 is based on Homentcovschi's paper.

The motion of (perfectly conducting) incompressible fluids in aligned fields is not included in the general theory presented in Section 5.2. This has been the subject of a dispute between Sears and Resler [1] who treated the problem using the linearized theory and Stewartson [20] who proved that such a theory was not applicable. This problem has been studied by means of the non-linear equations in a series of papers by G. S. Ludford and co-workers [23]—[26]. Finally, Homentcovschi [27] has shown that the problem can be solved using linearized equations and non-linear boundary conditions. This last solution is presented in Section 5.6.

The last section (5.7) deals with three-dimensional motion. The potential representation of the solution of the magnetofluid dynamic equations due to S. Ando [28] is presented here as well as the representation given by L. Dragoş [29] for aligned and orthogonal fields (for perfectly conducting and compressible fluids). Some contributions to the three-dimensional problem are also due to K. Kusukawa [30].

5.1.1 Formulation of the problem

Let us consider uniform motion with velocity V_0 of an ideal and perfectly conducting fluid in the presence of a homogeneous magnetic field of induction $\boldsymbol{B}_0$. The motion is assumed to be perturbed by the presence of a cylindrical body whose cross-section represents a thin airfoil in the plane of the vectors V_0 and $\boldsymbol{B}_0$. The airfoil chord is assumed to be parallel to V_0. The problem is to determine the perturbed motion.

For this purpose we choose a Cartesian system of coordinates having the origin at the middle of the airfoil chord, the Ox-axis parallel to V_0 and the Oy-axis in the plane of the airfoil (Figure 14). In order to obtain the equations of motion in dimensionless variables, the coordinates will be referred to L (as in Section 2.4), $2L$ being the airfoil chord length. We hence assume that the airfoil equations are:

$$y = h_{\pm}(x), \ |x| \leqslant 1 \tag{5.1.1}$$

where $h_+(x)$ indicates the upper and $h_-(x)$ the lower surface of the body. The functions $h_{\pm}$ are assumed to be smooth and the functions $h'_{\pm}$ to be Hölderian.

The fluid is assumed to be a simple medium (1.2.20).

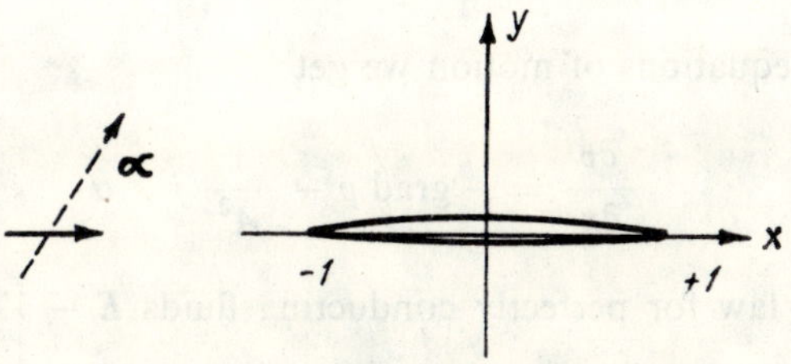

Fig. 14

Denoting by $\boldsymbol{V}$, $\boldsymbol{B}$, p, $\boldsymbol{E}$, $\boldsymbol{J}$, ρ^* the dimensionless variables (Section 2.4) we have

$$\boldsymbol{V} = \boldsymbol{i}_1 + \boldsymbol{v}, \ \boldsymbol{B} = \boldsymbol{\alpha} + \boldsymbol{b}, \ p = p$$
$$\boldsymbol{E} = \boldsymbol{e}_0 + \boldsymbol{e}, \ \boldsymbol{J} = \boldsymbol{j}_0 + \boldsymbol{j}, \ \rho^* = 1 + \rho \tag{5.1.2}$$

where $\boldsymbol{v}$, $\boldsymbol{b}$, p, $\boldsymbol{e}$, $\boldsymbol{j}$ and ρ designate the perturbations and $\boldsymbol{i}_1$ and $\boldsymbol{\alpha}$ are the unit vectors of the Ox-axis and of the $\boldsymbol{B}_0$ direction, respectively.

It is assumed that at large distances the perturbations vanish:

$$\lim_{x^2+y^2\to\infty} (\boldsymbol{v}, \boldsymbol{b}, \boldsymbol{e}, p, \rho) = 0 \tag{5.1.3}$$

For perfectly conducting fluids, assumption (5.1.3) is only partially valid. It will be however completely valid in the case of fluids with finite electrical conductivity (Chapter 6). Its validity for perfectly conducting fluids is justified by the fact that any fluid possesses some electrical resistivity.

Since the problem data are the same at any instant and in any plane parallel to the xOy-plane, it may be assumed that the perturbed motion is a steady and plane one. Accordingly:

$$\boldsymbol{v} = \boldsymbol{v}(x, y), \ \boldsymbol{b} = \boldsymbol{b}(x, y), \ p = p(x, y)$$
$$\boldsymbol{v} = (v_x, v_y, 0), \ \boldsymbol{b} = (b_x, b_y, 0) \tag{5.1.4}$$

As the profile is thin, the equations of motion can be linearized.

5. 2 Incompressible fluids

5.2.1 The equations of motion

Since the motion is steady, we have:

$$\operatorname{curl} \boldsymbol{b} = \boldsymbol{j} \ (\boldsymbol{j}_0 = 0), \quad \operatorname{curl} \boldsymbol{e} = 0 \tag{5.2.1}$$

On linearizing the equations of motion we get:

$$\frac{\partial \boldsymbol{v}}{\partial x} = - \operatorname{grad} p + \frac{1}{A^2} \boldsymbol{j} \times \boldsymbol{\alpha} \tag{5.2.2}$$

Finally, Ohm's law for perfectly conducting fluids $\boldsymbol{E} + \boldsymbol{V} \times \boldsymbol{B} = 0$ by linearization gives:

$$\boldsymbol{e}_0 = \boldsymbol{\alpha} \times \boldsymbol{i}_1 \tag{5.2.3}$$

$$\boldsymbol{e} = \boldsymbol{\alpha} \times \boldsymbol{v} + \boldsymbol{b} \times \boldsymbol{i}_1 \tag{5.2.4}$$

From (5.2.3) it follows that in free motion an electric field is induced only if $\boldsymbol{B}_0$ is not parallel to $\boldsymbol{V}_0$. From (5.2.4), we get:

$$\boldsymbol{e} = \boldsymbol{e}(x, y), \quad \boldsymbol{e} = (0, 0, e)$$

If we also consider the second equation of (5.2.1) and condition (5.2.13) we get $e \equiv 0$. This conclusion ($\boldsymbol{e} \equiv 0$) as well as equation (5.2.3) are valid for fluids with finite electrical resistivity as well.

On projecting equations (5.2.2) and (5.2.4) on the coordinate axes, we find:

$$\frac{\partial v_x}{\partial x} = - \frac{\partial p}{\partial x} - \frac{\alpha_y}{A^2} \left(\frac{\partial b_y}{\partial x} - \frac{\partial b_x}{\partial y} \right) \tag{5.2.5}$$

$$\frac{\partial v_y}{\partial x} = - \frac{\partial p}{\partial y} + \frac{\alpha_x}{A^2} \left(\frac{\partial b_y}{\partial x} - \frac{\partial b_x}{\partial y} \right) \tag{5.2.6}$$

$$b_y + \alpha_y v_x - \alpha_x v_y = 0 \tag{5.2.7}$$

such that, by adding equations:

$$\frac{\partial v_x}{\partial x} + \frac{\partial v_y}{\partial y} = 0 \quad (5.2.8), \qquad \frac{\partial b_x}{\partial x} + \frac{\partial b_y}{\partial y} = 0, \quad (5.2.9)$$

the system of equations for the determination of the perturbations is complete.

The system of equations (5.2.5) — (5.2.9) may be reduced to a simple equation by the following procedure. Firstly, we eliminate the pressure from (5.2.5) and (5.2.6) and from the resulting equation we eliminate v_x and b_x with the aid of equations (5.2.8) and (5.2.9). We obtain in this manner the equation:

$$\frac{\partial}{\partial x}\,\Delta v_y = \frac{1}{A^2}\,\frac{\mathrm{d}}{\mathrm{d}\alpha}\,\Delta b_y, \quad \Delta \overset{\text{def}}{=} \frac{\partial^2}{\partial x^2} + \frac{\partial^2}{\partial y^2} \tag{5.2.10}$$

From (5.2.7) and (5.2.8) we also have:

$$\frac{\partial b_y}{\partial x} = \frac{\mathrm{d}v_y}{\mathrm{d}\alpha}, \quad \frac{\mathrm{d}}{\mathrm{d}\alpha} \overset{\text{def}}{=} \alpha_x \frac{\partial}{\partial x} + \alpha_y \frac{\partial}{\partial y} \tag{5.2.11}$$

Equations (5.2.10) and (5.2.11) yield:

$$T\Delta(v_y, b_y) = 0, \quad T \overset{\text{def}}{=} \frac{\partial^2}{\partial x^2} - \frac{1}{A^2}\,\frac{\mathrm{d}^2}{\mathrm{d}\alpha^2} \tag{5.2.12}$$

This equation is also verified by v_x, b_x and p.

The operator T is of hyperbolic type except for the case $\alpha_y = 0$ (the characteristic equation has the discriminant equal to α_y^2/A^2). The case $\alpha_y = 0$ will be considered in Section 5.6.

5.2.2 General solution

Since the linear operators T and Δ are independent of each other, the solution of equation (5.2.12) can be represented by a sum of two terms, the first term being the solution of operator T and the second the solution of operator Δ. The general condition for the representation of the solution of a multiple operator may be found in [31] and [32].

The hyperbolic operator T has solutions of the form $F(x - cy)$, where:

$$c = c_\pm, \quad c_\pm = \frac{\alpha_x \pm A}{\alpha_y} \tag{5.2.13}$$

Without restricting the generality we assume $\alpha_y > 0$. Since the operator T contains only derivatives of the same order, one may conclude that the perturbations propagate without dispersion. Consequently, any function F depending on the plane wave $x - cy$ is a solution of the operator T.

The perturbations induced by the airfoil propagate along the direction of the characteristic lines. Obviously, in the upper half-plane the characteristic lines will have the slope c_{+}^{-1} and in the lower half-plane the slope c_{-}^{-1}. Accordingly, the perturbation in the upper half-plane are described by $F_{+}(x - c_{+}y)$ and those in the lower half-plane by $F_{-}(x - c_{-}y)$. If $c_{-}^{-1} > 0$, then the perturbations in the lower half-plane propagate in the opposite direction such that the wave will be propagating upstream. For orthogonal fields, the characteristic lines are placed symmetrically with respect to the Ox-axis.

In this manner the general solution of equation (5.2.12) is of the following form:

$$v_y^{\pm}(x, y) = F_{\pm}(x - c_{\pm} y) + \frac{\partial \varphi}{\partial y}, \quad \Delta\varphi = 0$$
$$b_y^{\pm}(x, y) = G_{\pm}(x - c_{\pm} y) + \frac{\partial \psi}{\partial y}, \quad \Delta\psi = 0 \tag{5.2.14}$$

where the (+) sign indicates the solution valid in the upper half-plane ($y > 0$) and the (—) sign the solution valid in the lower half-plane. Functions F_{+}, F_{-}, G_{+} and G_{-} are for the moment arbitrary. For the solution of operator Δ we have chosen the derivatives of the harmonic functions φ and ψ.

From equations (5.2.8) and (5.2.9) we have:

$$v_x^{\pm}(x, y) = c_{\pm} F_{\pm}(x - c_{\pm} y) + \frac{\partial \varphi}{\partial x}$$
$$b_x^{\pm}(x, y) = c_{\pm} G_{\pm}(x - c_{\pm} y) + \frac{\partial \psi}{\partial x} \tag{5.2.15}$$

By using the damping condition (5.1.3) and noticing that outside the strip of characteristic lines the hyperbolic perturbation does not exist, we get:

$$\lim_{x^2 + y^2 \to \infty} \left(\frac{\partial \varphi}{\partial x}, \frac{\partial \varphi}{\partial y}, \frac{\partial \psi}{\partial x}, \frac{\partial \psi}{\partial y} \right) = 0 \tag{5.2.16}$$

By using the solutions (5.2.14) and (5.2.15) from (5.2.7) we obtain the following relationships:

$$G_{\pm} = \mp A F_{\pm} \tag{5.2.17}$$

$$\frac{\partial \psi}{\partial y} = \alpha_x \frac{\partial \varphi}{\partial y} - \alpha_y \frac{\partial \varphi}{\partial x} \tag{5.2.18}$$

By differentiating (5.2.18) with respect to y and taking into account the fact that φ and ψ are harmonic functions, we have:

$$\frac{\partial \psi}{\partial x} = \alpha_x \frac{\partial \varphi}{\partial x} + \alpha_y \frac{\partial \varphi}{\partial y} \tag{5.2.19}$$

This relation could be also derived from the equations of motion (5.2.5) and (5.2.6).

Using equations (5.2.5) and (5.2.6) the pressure may be obtained as:

$$p^{\pm}(x, y) = \frac{\alpha_x \pm A^{-1}}{\alpha_y} F_{\pm}(x - c_{\pm} y) - \frac{\partial \varphi}{\partial y} \tag{5.2.20}$$

In this manner, the solution obtained verifies the system of equations of motion. The general solution depends on three arbitrary functions F_+, F_- and φ, the last one being harmonic outside the airfoil. These functions are to be determined from the boundary conditions.

5.2.3 Boundary conditions

As the fluid is assumed ideal we have:

$$\boldsymbol{V} \cdot \boldsymbol{n} = 0, \text{ if } y = h_{\pm}(x) \tag{5.2.21}$$

whence we get:

$$v_y^{\pm}(x, 0) = h'_{\pm}(x), \ |x| \leqslant 1 \tag{5.2.22}$$

because the parameters of the normal to the profile are $(h'_{\pm}, -1)$. To obtain the condition (5.2.22) we have disregarded products $v_x h'_{\pm}$ and imposed the boundary conditions on the segment $y = 0$ and not on the curves $y = h_{\pm}$. This assumption is justified by the continuity of the functions $v_y^{\pm}$ as well as by the fact that the airfoil is thin. This is a classical assumption in aerodynamics.

To solve the problem, it would be necessary to write Maxwell's equations inside the body (taking into account its electromagnetic characteristics) and to determine the solution of the system (5.2.5)—(5.2.9) which may be related by the jump equations (1.5.16) (1.5.17) to the internal field.

Let us notice that these conditions entail

$$[B_x] = 0, \ [B_y] = 0 \tag{5.2.23}$$

Indeed, if $x = x(s)$ and $y = y(s)$ are the equations of the surface of separation we have

$$s_x = \frac{\mathrm{d}x}{\mathrm{d}s} = -\frac{\mathrm{d}y}{\mathrm{d}n} = -n_y, \quad s_y = \frac{\mathrm{d}y}{\mathrm{d}s} = \frac{\mathrm{d}x}{\mathrm{d}n} = n_x$$

so that equations (1.5.16) and (1.5.17) become

$$[B_x]s_x + [B_y]s_y = 0$$

$$[B_x]s_y - [B_y]s_x = 0$$

Since $s_x^2 + s_y^2 = 1$ it follows that (5.2.23)

If the body is an insulator, the variation of the internal field (determined by the equations div $\boldsymbol{B} = 0$, curl $\boldsymbol{B} = 0$) is small [2] such that we may impose the conditions (5.2.23) for connecting the field on the upper surface to that on the lower surface. Taking into account (5.2.2), these conditions reduce to:

$$\begin{aligned} b_x^+(x,0) &= b_x^-(x,0) \\ b_y^+(x,0) &= b_y^-(x,0) \end{aligned} \tag{5.2.24}$$

Using the general solution, we find:

$$F_+(x) + \left.\frac{\partial \varphi}{\partial y}\right|_{y=+0} = h'_+(x), \quad |x| \leqslant 1 \tag{5.2.25}$$

$$[F] + \left[\frac{\partial \varphi}{\partial y}\right] = \eta(x)[h'], \quad \forall x \tag{5.2.26}$$

where

$$[\Phi] = \Phi^+(x,0) - \Phi^-(x,0)$$

$$\eta(x) = \begin{cases} 1 \text{ if } |x| < 1 \\ 0 \text{ if } |x| > 1 \end{cases} \tag{5.2.26'}$$

Condition (5.2.26) was obtained by subtracting equations (5.2.22) from one another and using the fact that outside the airfoil on the Ox-axis the normal velocity component is continuous (this theorem follows from the continuity equation exactly in the manner the continuity of the normal induction component follows from div $\boldsymbol{B} = 0$)

Conditions (5.2.24) yield:

$$c_+F_+ + c_-F_- - \frac{\alpha_x}{A}\left[\frac{\partial \varphi}{\partial x}\right] - \frac{\alpha_y}{A}\left[\frac{\partial \varphi}{\partial y}\right] = 0, \quad |x| \leqslant 1 \tag{5.2.27}$$

$$F_+ + F_- - \frac{\alpha_x}{A}\left[\frac{\partial \varphi}{\partial y}\right] + \frac{\alpha_y}{A}\left[\frac{\partial \varphi}{\partial x}\right] = 0, \quad \forall x \tag{5.2.28}$$

Since the derivatives of the harmonic function are continuous outside the airfoil from (5.2.26) and (5.2.28) we have:

$$F_+(x) \equiv F_-(x) \equiv 0, \quad |x| > 1 \tag{5.2.29}$$

With the notation:

$$f(x) = \left[\frac{\partial \varphi}{\partial x}\right] \tag{5.2.30}$$

from (5.2.26)–(5.2.28) we find ($f(x) \equiv 0$ if $|x| > 1$):

$$2kF_+ = \alpha_y A^{-1}(2A\alpha_x - 1 - A^2)f + (A\alpha_x - \alpha_x^2 + \alpha_y^2)[h']$$

$$2kF_- = -\alpha_y A^{-1}(2A\alpha_x + 1 + A^2)f + (A\alpha_x + \alpha_x^2 - \alpha_y^2)[h'] \quad (5.2.31)$$

$$k\left[\frac{\partial\varphi}{\partial y}\right] = -2\alpha_x\alpha_y f + A^2[h']$$

where:

$$k = A^2 - 1 + 2\alpha_y^2 \quad (5.2.32)$$

From (5.2.20) and (5.2.31) we get the following equation to be used for calculating the lift:

$$k[p] = -(k - 2\alpha_x^2 + 1 + A^{-2})f + 2\alpha_x\alpha_y\eta[h'] \quad (5.2.33)$$

We notice that the solution is determined only if $k \neq 0$. D. Homentcovschi has shown (5.2.27) that this singularity is due to the linearization of the boundary conditions.

5.2.4 Determination of function f

Equations (5.2.31) determine the hyperbolic part of the solution by means of the function f. Let us now show how the elliptical part of the solution can be determined with the aid of f and then derive the equation that determines the function f.

To this purpose we consider the complex plane $z = x + iy$ and the complex variable function:

$$\Phi(z) = \frac{\partial\varphi}{\partial x} - i\frac{\partial\varphi}{\partial y} \quad (5.2.34)$$

This function is holomorphic outside the segment $(-1, +1)$ and vanishes at infinity. The discontinuity of Φ across the Ox-axis can be determined from (5.2.30) and (5.2.31) as follows:

$$[\Phi] = \eta\left(f + \frac{2i\,\alpha_x\alpha_y}{k}f - \frac{iA^2}{k}[h']\right) \quad (5.2.35)$$

The holomorphic function whose discontinuity on the real axis is given by (5.2.35) has the expression:

$$\Phi(z) = \frac{1}{2\pi i}\int_{-1}^{+1}\left\{f + \frac{2i\,\alpha_x\alpha_y}{k}f - \frac{iA^2}{k}[h']\right\}\frac{dt}{t - z}. \quad (5.2.36)$$

This relation determines the elliptical part of the solution.

In order to determine the function f we use the condition (5.2.25). For this purpose in (5.2.36) we impose $z \to x + i(+0)$, $x \in (-1, +1)$. For the Cauchy-type integral we use Plemelj's formulas [G. 34]. By separating the imaginary part in the result we find:

$$\left.\frac{\partial \varphi}{\partial y}\right|_{y=+0} = -\frac{\alpha_x \alpha_y}{k} f + \frac{A^2}{2k}[h'] + \frac{1}{2\pi}\int_{-1}^{+1} \frac{f(t)}{t-x}\, dt$$

the integral being taken in the Cauchy principal value. By using this result condition (5.2.24) becomes:

$$-af(x) + \frac{k}{\pi}\int_{-1}^{+1} \frac{f(t)}{t-x}\, dt = R(x), \quad |x| < 1 \tag{5.2.37}$$

where:

$$a = \alpha_y(1 + A^2)A^{-1}, \quad R(x) = k(h'_+ + h'_-) - A\alpha_x[h'] \tag{5.2.37'}$$

The singular integral equation (5.2.37) determines function f. For $A \to \infty$ (5.2.37) reduces to:

$$\frac{1}{\pi}\int_{-1}^{+1} \frac{f(t)}{t-x}\, dt = h'_+ + h'_-, \quad |x| < 1 \tag{5.2.38}$$

Equation (5.2.37) is characteristic for perfectly conducting fluids exactly as equation (5.2.38) is characteristic for non-conducting fluids.

Equation (5.2.37) may be solved by reducing it to a Hilbert-type problem [G.34] [10] [11]. In this respect the papers of Rott and Cheng [33], Williams [34] and D. Homentcovschi [35] gave substantial contributions.

The solution of equation (5.2.37) is:

$$f(x) = -\frac{a}{a^2 + k^2} R(x) - \frac{k}{\pi(a^2 + k^2)} \left(\frac{1-x}{1+x}\right)^{\theta} \int_{-1}^{+1} \left(\frac{1+t}{1-t}\right)^{\theta} \frac{R(t)}{t-x}\, dt$$

$$+ \frac{2C \sin \theta\pi}{(1-x)^{1-\theta}(1+x)^{\theta}}, \quad \operatorname{tg} \theta\pi = \frac{k}{a}, \quad 0 \leqslant \theta < 1 \tag{5.2.39}$$

C being a constant to be determined by imposing Kutta-Joukovski's condition. If this condition is imposed at the trailing edge, then $C = 0$. Stewartson notices in [2] that this determination entails a quick variation of the solution when k changes its sign. He suggests the application of the Kutta-Joukovski (K.J.) condition at the trailing edge for $k > 0$ and at the leading edge for $k < 0$. It should be also mentioned that Kutta-Joukovski's condition in magnetofluid dynamics has constituted the object of papers [36]—[38].

5.2.5 Determination of lift

We consider the mechanical and electromagnetical stress tensor under the assumption of an ideal fluid and in dimensionless variables:

$$T_{ji} = -\left(p + \frac{1}{2A^2} B^2\right)\delta_{ji} + \frac{1}{A^2} B_j B_i \tag{5.2.40}$$

This expression is derived from (2.4.17) by neglecting the terms of the order of the relativistic approximation. The constants in the expression of T_{ji} will be of no importance in the following treatment. Since $\boldsymbol{B} = \alpha + \boldsymbol{b}$, by linearizing (5.2.40) we find:

$$\begin{aligned} T_{11} &= -p + A^{-2}(\alpha_x b_x - \alpha_y b_y) \\ T_{22} &= -p + A^{-2}(-\alpha_x b_x + \alpha_y b_y) \\ T_{12} &= T_{21} = A^{-2}(\alpha_x b_y + \alpha_y b_x) \end{aligned} \tag{5.2.40'}$$

The lift may be calculated from the formula:

$$L = \oint_\Gamma (T_{12}n_1 + T_{22}n_2)\,\mathrm{d}s = \oint_\Gamma T_{12}\,\mathrm{d}y - T_{22}\,\mathrm{d}x \tag{5.2.41}$$

where Γ is the curve that bounds the airfoil in positive sense (the sense that leaves the airfoil on the left hand side) and $\boldsymbol{n}$ is the external normal. If we take into account the fact that on the airfoil $\mathrm{d}y = h'\mathrm{d}x$, then T_{12} $\mathrm{d}y$ is of the order of the products $b_x h'$, $b_y h'$, and it is thus negligible.

In the expression $T_{22}\mathrm{d}x$ we used the fact that the airfoil reduces to the segment $(-1, +1)$ and the components $\boldsymbol{b}_x$ and $\boldsymbol{b}_y$ are continuous across this segment. In this manner the lift formula becomes:

$$L = -\int_{-1}^{+1} [p]\,\mathrm{d}x \tag{5.2.42}$$

The drag:

$$D = \oint_\Gamma T_{11}\,\mathrm{d}y - T_{21}\,\mathrm{d}x$$

is of the order of the perturbation products.

The lift in dimensional variables has the expression $\mathscr{L} = \rho_0 V_0^2 L$.

For the case under consideration we have:

$$L = \left(1 - \frac{2\alpha_x^2}{k} + \frac{1 + A^{-2}}{k}\right)I - \frac{2\alpha_x\alpha_y}{k}\int_{-1}^{+1} [h']\,\mathrm{d}x \tag{5.2.43}$$

where*:

$$I = \int_{-1}^{+1} f(x)\,\mathrm{d}x = \frac{k\cos\theta\pi - \mathrm{a}\sin\theta\pi}{(a^2 + k^2)\sin\theta\pi}\int_{-1}^{+1} R(x)\,\mathrm{d}x$$

$$- \frac{k}{(a^2 + k^2)\sin\theta\pi}\int_{-1}^{+1}\left(\frac{1 - x}{1 - x}\right)^{\theta} R(x)\,\mathrm{d}x \tag{5.2.44}$$

If the *K.J.* condition is applied at the trailing edge the integral (44) becomes:

$$I = \frac{1}{\sqrt{a^2 + k^2}}\int_{-1}^{+1}\left(\frac{1 + x}{1 - x}\right)^{\theta} R(x)\,\mathrm{d}x, \text{ if } k < 0 \quad \left(\theta > \frac{1}{2}\right) \tag{5.2.45}$$

---

* In the above calculations we use the following results:

$$I_1 = \int_{-1}^{+1}\left(\frac{1 + t}{1 - t}\right)^{\theta}\frac{\mathrm{d}t}{t - x} = \frac{\pi}{\sin\theta\pi}\left\{1 - \left(\frac{1 + x}{1 - x}\right)^{\theta}\cos\theta\pi\right\}$$

$$I_2 = \int_{-1}^{+1}\left(\frac{1 - t}{1 + t}\right)^{\theta}\frac{\mathrm{d}t}{t - x} = -\frac{\pi}{\sin\theta\pi}\left\{1 - \left(\frac{1 - x}{1 + x}\right)^{\theta}\cos\theta\pi\right\}$$

One may notice that the two equations may be derived from one another by changing θ into $-\theta$. Accordingly, only the first equation will be proven.

For this purpose we consider the holomorphic function:

$$K(z) = \frac{1}{2\pi\mathrm{i}}\int_{-1}^{+1}\left(\frac{1 + t}{1 - t}\right)^{\theta}\frac{\mathrm{d}t}{t - z}$$

and observe that according to Plemelj's formulae we have:

$$K_+(x) - K_-(x) = \left(\frac{1 + x}{1 - x}\right)^{\theta} \tag{*}$$

$$K_+(x) + K_-(x) = \frac{1}{\pi\mathrm{i}}\int_{-1}^{+1}\left(\frac{1 + t}{1 - t}\right)^{\theta}\frac{\mathrm{d}t}{t - x} \tag{**}$$

Hilbert's problem (*) has the following obvious solution:

$$K(z) = \gamma\left\{\left(\frac{z + 1}{z - 1}\right)^{\theta} - 1\right\}, \qquad \gamma = \text{const} \tag{***}$$

From (***) we get:

$$K_{\pm}(x) = \gamma\left\{\left(\frac{1 + x}{1 - x}\right)^{\theta}\mathrm{e}^{\mp i\theta\pi} - 1\right\} \tag{****}$$

such that using (*) we get $2\gamma\sin\theta\pi = \mathrm{i}$. In this manner from (**) and (****) we obtain I_1. We also have:

$$\lim_{\theta\to 0} I_1 = \ln\frac{1 - x}{1 + x}, \qquad \lim_{\theta\to 1/2} I_1 = \pi$$

and

$$I = -\frac{1}{\sqrt{a^2+k^2}}\int_{-1}^{+1}\left(\frac{1+x}{1-x}\right)^{\theta} R(x)\,dx, \text{ if } k > 0 \left(\theta < \frac{1}{2}\right) \quad (5.2.46)$$

In the particular case of orthogonal fields, the most important case in aerodynamics, the solution simplifies considerably. For instance the integral equation (5.2.37) reduces to:

$$-\frac{1}{A}f(x) + \frac{1}{\pi}\int_{-1}^{\prime +1}\frac{f(t)}{t-x}\,dt = h'_{+} + h'_{-}, \quad |x| < 1 \quad (5.2.37')$$

and the lift reduces itself to:

$$L = -\sqrt{1+\frac{1}{A^2}}\int_{-1}^{+1}\left(\frac{1+x}{1-x}\right)^{\theta}(h'_{+} + h'_{-})\,dx, \quad \mathrm{tg}\,\theta\pi = A \quad (5.2.47)$$

the *K.J.* condition being applied at the trailing edge.

In practice it is sometimes required to calculate integrals of the form of those appearing in (5.2.45)—(5.2.47). Such computations can be found in [35]. Recently, Sheriazdanov [39] has considered again the problem of perpendicular fields in view of giving some calculation formulas.

5.2.6 The flat plate

If the airfoil reduces to a flat plate of incidence ε, using the approximation $\mathrm{tg}\,\varepsilon \simeq \varepsilon$ we obtain $h'_{+} = h'_{-} = -\varepsilon$, $R = -2\varepsilon k$,

$$f(x) = \frac{2\varepsilon|k|}{\sqrt{a^2+k^2}}\left(\frac{1-x}{1+x}\right)^{\theta}, \quad 0 \leqslant \theta < 1 \quad (5.2.48)$$

By using:

$$\int_{-1}^{+1}\left(\frac{1+x}{1-x}\right)^{\theta} dx = \frac{2\pi\theta}{\sin\theta\pi} \quad (5.2.49)$$

we get:

$$L = \frac{(A^2-1)^2 + 4A^2\alpha_y^2}{A^2(A^2-1+2\alpha_y^2)}\,4\varepsilon\theta\pi \quad (5.2.50)$$

This expression is valid for all cases. Accordingly, the lift does not depend on the manner in which the *K.J.* condition is imposed.

In the case of orthogonal fields we have:

$$L = \left(1 + \frac{1}{A^2}\right) 4\varepsilon\theta\pi \tag{5.2.51}$$

For the non-conducting fluid ($A \to \infty \Leftrightarrow k \to \infty$) from (5.2.50) or (5.2.51) we obtain the known formula $L = 2\pi\varepsilon$.

Figure 15 [3] is a plot of the expression:

$$l(A, \alpha_y) = \frac{L}{2\pi\varepsilon} = \frac{2}{\pi} \frac{(A^2 - 1)^2 + 4A^2\alpha_y^2}{A^2(A^2 - 1 + 2\alpha_y^2)} \operatorname{arc\,tg} \frac{A(A^2 - 1 + 2\alpha_y^2)}{\alpha_y(A^2 + 1)}$$

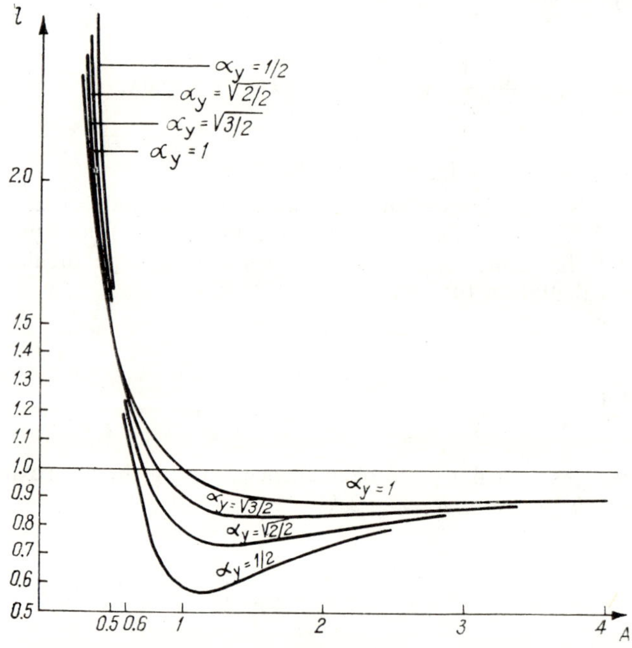

Fig. 15

for the following α-angles determined by the magnetic induction and the free stream velocity:

$$\frac{\pi}{6}\left(\alpha_y = \frac{1}{2}\right), \frac{\pi}{4}\left(\alpha_y = \frac{\sqrt{2}}{2}\right), \frac{\pi}{3}\left(\alpha_y = \frac{\sqrt{3}}{2}\right), \frac{\pi}{2}\ (\alpha_y = 1)$$

The following conclusions may be drawn:

(i) The lift decreases with the increase of α.

(ii) For a given angle α the lift decreases with the increase of the parameter A. The lift reaches a minimum value, smaller than the value of the classical lift, after which it begins to increase, tending asymptotically towards the classical value. When α increases, the minimum lift value goes towards higher value of A.

Numerical calculations show that the minimum lift values are:

$$l(\simeq 1.13, 1/2) = 0.5706, \qquad l(\simeq 1.33, \sqrt{2}/2) = 0.7365$$

$$l(\simeq 1.67, \sqrt{3}/2) = 0.8287, \quad l(\simeq 2.33, 1) = 0.8786$$

Other results on this matter can be found in paper [3].

5.3 Compressible fluids

For compressible fluids one needs in addition the unknown ρ such that account should be taken of the energy equation. However, the energy equation introduces as an unknown temperature T such that the fluid equation of state should be also considered. We have already noted in Section 2.3 that if the fluid is perfectly conducting the electromagnetic dissipating effects do not appear in the energy equation. By neglecting the heat exchange from the inviscid fluid energy equation it follows that the motion is isentropic. We assume that the fluid satisfies the equation of state of the ideal gas. Accordingly:

$$\frac{\mathrm{d}}{\mathrm{d}t}\left(\frac{p_{\mathrm{dim}}}{\rho_{\mathrm{dim}}^{\gamma}}\right) = 0 \tag{5.3.1}$$

Taking into account the results of Section 2.4 and linearizing this equation we have:

$$\frac{\partial \rho}{\partial x} = M^2 \frac{\partial p}{\partial x}, \quad M^2 \overset{\mathrm{def}}{=} \frac{V_0^2}{a_0^2} \tag{5.3.1$'$}$$

Also by using the damping condition (5.1.3) from (5.3.1′) we have $\rho = M^2 p$.

Consequently the equations of motion of the compressible fluid are (5.2.5), (5.2.6), (5.2.7) and (5.2.9) to which we add the continuity equation under the form:

$$M^2 \frac{\partial p}{\partial x} + \frac{\partial v_x}{\partial x} + \frac{\partial v_y}{\partial y} = 0 \tag{5.3.2}$$

In this case too the equations of motion can be reduced to a single equation. Indeed, by eliminating the pressure from (5.2.5) and (5.3.2) we find:

$$\beta^2 \frac{\partial v_x}{\partial x} + \frac{\partial v_y}{\partial y} = M^2 \frac{\alpha_y}{A^2}\left(\frac{\partial b_y}{\partial x} - \frac{\partial b_x}{\partial y}\right), \quad \beta^2 \overset{\mathrm{def}}{=} 1 - M^2 \tag{5.3.3}$$

Then, eliminating the pressure from equations (5.2.5) and (5.2.6), we get:

$$\frac{\partial}{\partial x}\left(\frac{\partial v_y}{\partial x} - \frac{\partial v_x}{\partial y}\right) = \frac{1}{A^2}\frac{d}{d\alpha}\left(\frac{\partial b_y}{\partial x} - \frac{\partial b_x}{\partial y}\right) \tag{5.3.4}$$

From (5.3.3), (5.3.4) and (5.2.9) we have:

$$A^2 H \frac{\partial v_y}{\partial x} = \left(\beta^2 \alpha_x \frac{\partial}{\partial x} + \alpha_y \frac{\partial}{\partial y}\right) \Delta b_y \tag{5.3.5}$$

where:

$$H = \beta^2 \frac{\partial^2}{\partial x^2} + \frac{\partial^2}{\partial y^2}, \quad \Delta = \frac{\partial^2}{\partial x^2} + \frac{\partial^2}{\partial y^2} \tag{5.3.5'}$$

Equation (5.3.5) is also valid under the assumption that the fluid has finite electrical conductivity.

An additional equation can be also obtained if v_x is eliminated from (5.3.3) using equation (5.2.7) and from the equation thus found b_x is eliminated using equation (5.2.9). In this manner, we find:

$$A^2\left(\beta^2 \alpha_x \frac{\partial}{\partial x} + \alpha_y \frac{\partial}{\partial y}\right)\frac{\partial v_y}{\partial x} = \left(A^2 \beta^2 \frac{\partial^2}{\partial x^2} + M^2 \alpha_y^2 \Delta\right) b_y \tag{5.3.6}$$

Equations (5.3.5) and (5.3.6) yield:

$$T(v_y, b_y) = 0 \tag{5.3.7}$$

where

$$T = \left(\beta^2 \alpha_x \frac{\partial}{\partial x} + \alpha_y \frac{\partial}{\partial y}\right)^2 \Delta - \left(A^2 \beta^2 \frac{\partial^2}{\partial x^2} + M^2 \alpha_y^2 \Delta\right) H \tag{5.3.7}$$

It can be shown this time too that v_x, b_x and p satisfy the same equation (5.3.7).

In the case of aligned fields ($\alpha_x = 1$, $\alpha_y = 0$) the operator T reduces to:

$$T_1 = (1 - M^2)(1 - A^2)\frac{\partial^2}{\partial x^2} - (A^2 + M^2 - 1)\frac{\partial^2}{\partial y^2} \tag{5.3.8}$$

and in the case of orthogonal fields it reduces to:

$$T_2 = \left(\frac{\partial^2}{\partial x^2} - \frac{1}{s^2}\frac{\partial^2}{\partial y^2}\right)\left(\frac{\partial^2}{\partial x^2} - \frac{s^2}{B}\frac{\partial^2}{\partial y^2}\right) \tag{5.3.9}$$

where:

$$B = A^2M^2 - A^2 - M^2$$

$$2s^2 = A^2 + M^2 - 1 + \sqrt{1 + 2(A^2 + M^2) + (A^2 - M^2)^2} > 0 \quad (5.3.9')$$

s^2 being the solution of the equation:

$$s^4 - (A^2 + M^2 - 1)s^2 + B = 0 \quad (5.3.9'')$$

In the domains D_2 and D_4 of Figure 16 the operator T_1 is of hyperbolic type while in the domains D_1, D_3 and D_5 it is of elliptical type. The operator T_2 has the form of a product of two operators, the first being of hyperbolic type under any conditions and the second of hyperbolic type in the domain $\mathscr{D}_1$ in Figure 17* and of elliptical type in the domain $\mathscr{D}_2$. Accordingly in $\mathscr{D}_1$ there exist at any point two pairs of characteristic real straight lines (this flow will be referred to as doubly hyperbolic flow) while in $\mathscr{D}_2$ there exists at any point a single pair of characteristic real straight lines (this flow will be referred to as hyperelliptic flow).

The curves $A = 1$, $M = 1$ and $A^2 + M^2 = 1$ from the case of aligned fields and the curve $A^2 + M^2 = A^2M^2$ from the case of orthogonal fields are singular lines. Across these lines the motion cannot be studied using linearized equations. Such singular motions are the object of various studies ([40]—[45]).

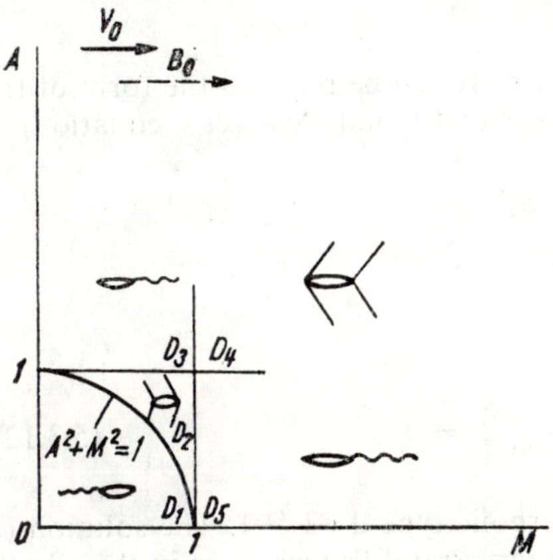

Fig. 16

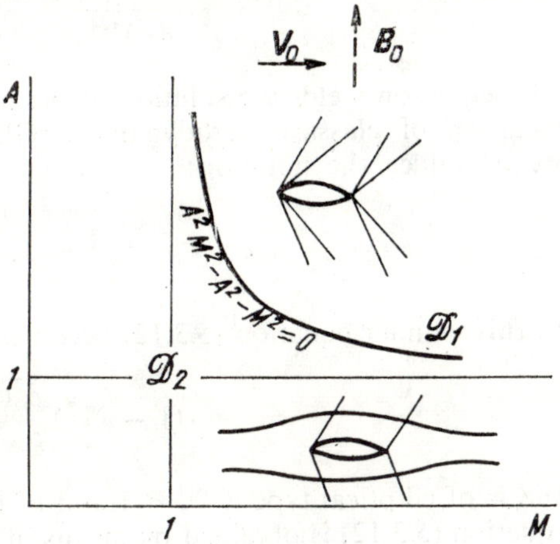

Fig. 17

In the two figures the classical aerodynamics results as a limit case ($A \to \infty$). The same is true for the theory of incompressible fluids ($M \to 0$). The graph given in Figure 16 was obtained by Taniuti [14] using the non-linear theory of characteiistics. The graphs of Figures 3 and 4 are also valid in the three dimensional problem [29].

In the following considerations we shall deal in detail with the cases of aligned and orthogonal fields.

* In the plane $(1/A, 1/M)$ the curves of Figure 17 have the same shape as those given in Figure 16 for the plane (A, M).

5.3.1 Aligned fields

For aligned fields we obtain a solution similar to that supplied by classical aerodynamics. Indeed, from (5.2.5) we have $p = -v_x$ and from (5.3.2) we get:

$$(1 - M^2)\frac{\partial v_x}{\partial x} + \frac{\partial v_y}{\partial y} = 0 \tag{5.3.10}$$

Equation (5.3.10) shows (except for $M = 1$) that there exists a function $\psi(x, y)$ such that:

$$v_x = \frac{1}{1 - M^2}\frac{\partial \psi}{\partial y}, \quad v_y = -\frac{\partial \psi}{\partial x} \tag{5.3.11}$$

On the other hand from (5.2.7) we get $b_y = v_y$, and from (5.2.9) and (5.3.10) we have $b_x = (1 - M^2)\, v_x$. Consequently, all the unknowns may be expressed by means of the functions v_x and v_y. Using this result as well as equations (5.3.11) equation (5.2.6) becomes:

$$\frac{(A^2 - 1)(1 - M^2)}{A^2 + M^2 - 1}\frac{\partial^2 \psi}{\partial x^2} + \frac{\partial^2 \psi}{\partial y^2} = 0 \tag{5.3.12}$$

This equation yields the solution of the problem. It can be put into the form of the equation of classical aerodynamics (Glauert-Prandtl and Ackeret's equation) if we introduce the notation:

$$\mathfrak{M}^2 = \frac{A^2 M^2}{A^2 + M^2 - 1}$$

In this manner equation (5.3.12) becomes:

$$(1 - \mathfrak{M}^2)\frac{\partial^2 \psi}{\partial x^2} + \frac{\partial^2 \psi}{\partial y^2} = 0 \tag{5.3.12'}$$

and is of elliptical type if $\mathfrak{M} < 1$ and of hyperbolic type if $\mathfrak{M} > 1$. The solution of equation (5.3.12) is obtained by means of Glauert-Prandtl's technique in the ellipticity domain and with the aid of Ackeret's technique in the hyperbolicity domain. The velocity boundary condition determines the general solution and the lift.

We notice that (5.2.40) yields:

$$[T_{22}] = -\left[p + \frac{1}{A^2}\, b_x\right] = \left(1 - \frac{\beta^2}{A^2}\right)[v_x] = \left(\frac{1}{\beta^2} - \frac{1}{A^2}\right)[b_x]$$

$$L = -\int_{-1}^{+1} [T_{22}]\, dx \tag{5.3.13}$$

As the lift value differs from zero (being determined by $[v_x]$) one concludes that the first boundary condition (5.2.23) cannot be used. This condition must be replaced by $[b_x] = i(x)$, $i(x)$ being the surface current density along the Oz-axis. This density is determined by the lift value through equation (3.3.13).

Another remark that should be made is that while in D_4 the characteristic lines are orientated downstream, in D_2 they are oriented upstream. Indeed, from the equation of the characteristic lines

$$(1 - \mathfrak{M}^2)\,dy^2 + dx^2 = 0$$

we find:

$$\frac{dy}{dx} = \pm \frac{1}{\sqrt{\mathfrak{M}^2 - 1}}$$

and then:

$$\sin\theta = \pm\sqrt{\frac{1}{A^2} + \frac{1}{M^2} - \frac{1}{A^2 M^2}}$$

θ being the angle made by the characteristic lines with the Ox-axis ($\operatorname{tg}\theta = dy/dx$). Taking into account the relationships $A = V_0 V_A^{-1}$, $M = V_0 a_0^{-1}$ we get:

$$\frac{d}{dV_0}\sin\theta = \mp \frac{a_0^2(1 - A^{-2}) + V_A^2(1 - M^{-2})}{V_0^2\sqrt{a_0^2(1 - A^{-2}) + V_A^2}}$$

We notice that the denominator of this expression is negative in D_2 and positive in D_4. If we restrict ourselves only to the interval $(0, \pi)$ we deduce that in D_4 the angle θ decreases with the increase of V_0 (the characteristic lines are downstream as in classical aerodynamics), while in D_2 the angle θ increases when V_0 increases (the characteristic lines are sloping upstream).

5.3.2 Orthogonal fields

5.3.2.1 Hyperelliptic flow ($B < 0$)

If we use the affine transformation:

$$x = X, \quad y = \alpha Y, \quad \alpha \stackrel{\text{def}}{=} s/\sqrt{-B} > 0 \tag{5.3.14}$$

the operator T_2 becomes:

$$T_2 = \left(\frac{\partial^2}{\partial X^2} - \frac{1}{\alpha^2 s^2}\frac{\partial^2}{\partial Y^2}\right)\left(\frac{\partial^2}{\partial X^2} + \frac{\partial^2}{\partial Y^2}\right)$$

and has the general solution:

$$v_y^{\pm} = F_{\pm}(X \mp \alpha s Y) + \frac{\partial \varphi}{\partial Y}$$
$$b_y^{\pm} = G_{\pm}(X \mp \alpha s Y) + \frac{\partial \psi}{\partial Y} \tag{5.3.15}$$

where:

$$\left(\frac{\partial^2}{\partial X^2} + \frac{\partial^2}{\partial Y^2}\right)(\varphi, \psi) = 0 \tag{5.3.15'}$$

From (5.2.7) one finds $v_x = -b_y$. The transformation of equation (5.2.9) in the XOY — plane and the use of (5.3.15) yields:

$$b_x^{\pm} = \pm\, s G_{\pm}(X \mp \alpha s Y) + \frac{1}{\alpha}\frac{\partial \psi}{\partial X} \tag{5.3.16}$$

Finally, equation (5.2.6) gives the pressure:

$$p^{\pm} = \pm \frac{1}{s} F_{\pm}(X \mp \alpha s Y) - \alpha \frac{\partial \varphi}{\partial X} \tag{5.3.17}$$

the upper sign again indicating the solution in the upper half-plane and the lower sign the solution in the lower half-plane.

By equalizing the elliptical parts and the hyperbolic parts from the continuity equation we find:

$$G_{\pm} = \mp s(1 - M^2 s^{-2}) F_{\pm} \tag{5.3.18}$$

$$\frac{\partial \psi}{\partial Y} = -\left(\alpha M^2 + \frac{1}{\alpha}\right)\frac{\partial \varphi}{\partial X}, \quad \frac{\partial \psi}{\partial X} = \left(\alpha M^2 + \frac{1}{\alpha}\right)\frac{\partial \varphi}{\partial Y} \tag{5.3.19}$$

By using the same boundary conditions as in Section 5.2, noticing that the transformation (5.3.14) along the Ox-axis reduces itself to an identity we get:

$$F_+(X) + \left.\frac{\partial \varphi}{\partial Y}\right|_{+0} = h' + (X), \quad |X| \leqslant 1 \tag{5.3.20}$$

$$[F] + \left[\frac{\partial \varphi}{\partial Y}\right] = [h'], \qquad |X| \leqslant 1 \tag{5.3.21}$$

$$(M^2 - s^2)[F] + \left(M^2 + \frac{1}{\alpha^2}\right)\left[\frac{\partial \varphi}{\partial Y}\right] = 0, \qquad \forall X \tag{5.3.22}$$

$$\left(\frac{M^2}{s} - s\right)(F_+ + F_-) - \left(\alpha M^2 + \frac{1}{\alpha}\right)\left[\frac{\partial \varphi}{\partial X}\right] = 0, \quad \forall X \tag{5.3.23}$$

For compressible fluids, the continuity condition of the normal component v_x across the Ox-axis is no longer valid. But we may require that b_x be continuous across this axis (this is the condition (5.3.22)). Since the derivatives of the harmonic functions φ are continuous outside the airfoil from (5.3.22) and (5.3.23) we have:

$$F_+ \equiv F_- \equiv 0 \quad , \qquad |x| > 1 \tag{5.3.24}$$

With the notation:

$$f = \left[\frac{\partial \varphi}{\partial X}\right] \tag{5.3.25}$$

from equations (5.3.23), (5.3.21) and (5.3.22) we find:

$$F_+ + F_- = k_1 f,\ [F] = \frac{a}{k_2 s}[h'],\ \left[\frac{\partial \varphi}{\partial Y}\right] = k_3[h'] \tag{5.3.26}$$

where:

$$k_1 = \frac{1}{\sqrt{-B}}\frac{M^2s^2 - B}{M^2 - s^2} < 0,\ k_2 = \frac{1}{s\sqrt{-B}}\frac{s^4 - B}{M^2 - s^2},\ k_3 = \frac{s^2(s^2 - M^2)}{s^4 - B} \tag{5.3.26'}$$

Finally, using the expression for pressure we get:

$$[p] = k_2 f,\ (f \equiv 0 \text{ if } |X| > 1) \tag{5.3.27}$$

In this manner the hyperbolic part of the solution is expressed by means of function f. In order to express the elliptical part using the same function f we consider the complex plane $Z = X + iY$ and the function:

$$\Phi(Z) = \frac{\partial \varphi}{\partial X} - i\frac{\partial \varphi}{\partial Y}$$

which is holomorphic outside the segment $(-1, +1)$. Since:

$$[\Phi] = \begin{cases} 0, & \text{if } |X| > 1 \\ f - ik_3[h'], & \text{if } |X| < 1 \end{cases}$$

$$\lim_{Z \to \infty} \Phi(Z) = 0$$

we have:

$$\Phi(Z) = \frac{1}{2\pi i}\int_{-1}^{+1} \{f(t) - ik_3[h']_t\}\frac{dt}{t - Z} \tag{5.3.28}$$

This formula determines the elliptical part of the solution.

The integral equation of the problem may be obtained from (5.3.20) in the same manner as in Section 5.2. The solution of this equation is:

$$f = \frac{k_1}{1 + k_1^2}(h'_+ + h'_-) - \frac{1}{\pi(1 + k_1^2)}\left(\frac{1 - X}{1 + X}\right)^\theta \int_{-1}^{+1} \left(\frac{1+t}{1-t}\right)^\theta \frac{h'_+ + h'_-}{t - X}\,dt$$

$$\operatorname{tg} \theta\pi = -1/k_1 > 0, \quad 0 \leqslant \theta \leqslant 1/2 \tag{5.3.29}$$

the *K.J.* condition being imposed at the trailing edge as for the incompressible fluid.

The lift value may be determined from equations (5.2.42), (5.3.27) and (5.3.29). We get:

$$L = \frac{k_2}{\sqrt{1 + k_1^2}} \int_{-1}^{+1} \left(\frac{1+t}{1-t}\right)^\theta (h'_+ + h'_-)\,dt \tag{5.3.30}$$

For the plate of incidence ε we find:

$$L = -4\varepsilon k_2 \theta\pi \tag{5.3.31}$$

Taking into account the expressions of k_2 and k_1, from (5.3.31) we have:

$$l \overset{\text{def}}{=} \frac{L}{2\varepsilon\pi} = \frac{2}{\pi} \frac{s^4 - B}{s\sqrt{-B(M^2 - s^2)}} \operatorname{arc\,tg} \frac{\sqrt{-B(M^2 - s^2)}}{M^2 s^2 - B}$$

This expression is plotted [7] in Figure 18. One may notice that for given M the lift value decreases with the increase of the number A, reaches a minimum and then increases. If $0.2 \leqslant M < 1$ the lift value exceeds the value for the compressible fluid. If $1.6 \leqslant M \leqslant 2.1$ then the minimum lift value is obtained for $A = 0.2$. Another effect of the magnetic field is that it produces a uniform decrease of the lift value in the region where M exceeds the value 1.

From the above solution one may obtain as a particular case the solutions both for the incompressible and for the non-conducting fluids. Indeed, by setting $M = 0$ (incompressible fluid) we get $s = A$ and $\alpha = 1$, such that transformation (5.3.14) reduces itself to an identity. In these conditions the solution of system (5.3.15)—(5.3.19) coincides with that given in Section 5.2. The lift value also reduces to that given in Section 5.2 since:

$$\lim_{M\to 0} k_1 = -\frac{1}{A}, \quad \lim_{M\to 0} k_2 = -\left(1 + \frac{1}{A^2}\right)$$

For the non-conducting fluid ($A \to \infty$) we find:

$$\lim \frac{s^2}{A^2} = 1, \ \lim \frac{-B}{A^2} = 1 - M^2, \ \lim k_1 A = \lim k_2 = -\frac{1}{\sqrt{1 - M^2}}$$

so that the lift value reduces to that of the conventional aerodynamics:

$$L = -\frac{1}{\sqrt{1 - M^2}} \int_{-1}^{+1} \left(\frac{1 + t}{1 - t} \right)^{1/2} (h'_+ + h'_-)\, dt$$

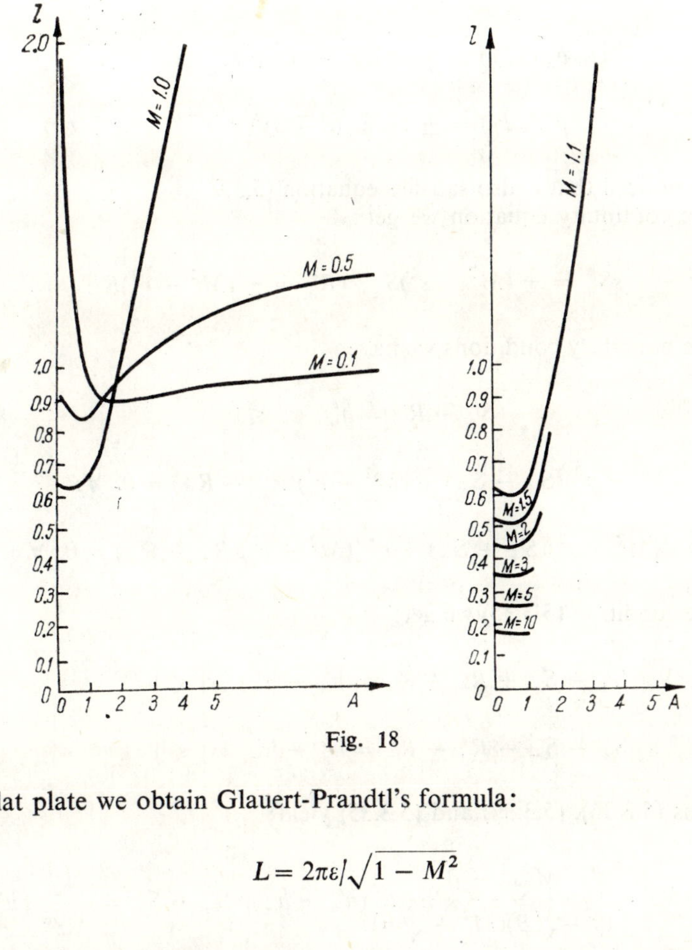

Fig. 18

For the flat plate we obtain Glauert-Prandtl's formula:

$$L = 2\pi\varepsilon / \sqrt{1 - M^2}$$

5.3.2.2 Doubly hyperbolic regime ($B > 0$)

In this case two pairs of characteristic lines emerge from each point. With the Ox-axis they make the angles θ_1 and θ_2 determined by the relationships ($s > 0$, $r > 0$):

$$\operatorname{tg} \theta_1 = \pm \frac{1}{s}, \quad \operatorname{tg} \theta_2 = \pm \frac{s}{\sqrt{-B}} \stackrel{\text{def}}{=} \pm \frac{1}{r} \tag{5.3.32}$$

so that the solution is a superposition of two waves:

$$v_y^{\pm}(x, y) = S_{\pm}(x \mp sy) + R_{\pm}(x \mp ry)$$

$$b_y^{\pm}(x, y) = S_{\pm}^{*}(x \mp sy) + R_{\pm}^{*}(x \mp ry) \tag{5.3.33}$$

$$b_x^{\pm}(x, y) = \pm\, sS_{\pm}^{*}(x \mp sy) \pm rR_{\pm}^{*}(x \mp ry)$$

$$p^{\pm}(x, y) = \pm\, s^{-1}S_{\pm}(x \mp sy) \pm r^{-1}R_{\pm}(x \mp ry)$$

It should be noticed that r also satisfies equation (5.3.9″).

From the continuity equation we get:

$$sS_{\pm}^{*} = \pm\,(M^2 - s^2)S_{\pm},\ rR_{\pm}^{*} = \pm\,(M^2 - r^2)R_{\pm} \tag{5.3.34}$$

and from the boundary conditions we have:

$$S_{\pm} + R_{\pm} = h'_{\pm},\ |x| \leqslant 1 \tag{5.3.35}$$

$$(M^2 - s^2)(S_+ - S_-) + (M^2 - r^2)(R_+ - R_-) = 0,\ \forall x \tag{5.3.36}$$

$$s^{-1}(M^2 - s^2)(S_+ + S_-) + r^{-1}(M^2 - r^2)(R_+ + R_-) = 0,\ \forall x \tag{5.3.37}$$

From the condition (5.3.35) we get:

$$S_+ + S_- + R_+ + R_- = h'_+ + h'_-,\ |x| \leqslant 1$$

$$S_+ - S_- + R_+ - R_- = h'_+ - h'_-,\ |x| \leqslant 1 \tag{5.3.35'}$$

Conditions (5.3.36), (5.3.37) and (5.3.35) yield:

$$S_+ + S_- = \frac{M^2 s^2 - B}{(s^2 - \sqrt{B})(M^2 + \sqrt{B})}(h'_+ + h'_-),\ S_+ - S_- = \frac{k_1}{k_2 s}[h']$$

$$R_+ + R_- = \frac{(s^2 - M^2)\sqrt{B}}{(s^2 - \sqrt{B})(M^2 + \sqrt{B})}(h'_+ + h'_-),\ R_+ - R_- = b[h'] \tag{5.3.38}$$

For $|x| > 1$ we take $S_+ = S_- = R_+ = R_- = 0$, as the airfoil is the only source of perturbations. Equations (5.3.38) give the solution of the problem.

For the lift value we obtain the following expression:

$$L = -\frac{s^2 + \sqrt{B}}{s(M^2 + \sqrt{B})} \int_{-1}^{+1} (h'_+ + h'_-)\,dx \tag{5.3.39}$$

which for the plate of incidence ε reduces to:

$$L = \frac{4\varepsilon(s^2 + \sqrt{B})}{s(M^2 + \sqrt{B})} \tag{5.3.40}$$

For the non-conducting fluid from (5.3.40) we obtain Ackeret's formula:

$$L = \frac{4\varepsilon}{\sqrt{M^2 - 1}} \tag{5.3.40'}$$

In this case S_+ does not appear in the solution (since $s \to \infty$) and in the wave $R_\pm$ one substitutes $r = \sqrt{M^2 - 1}$.

Figure 19 is a plot [7] of the following expression:

$$l \overset{\text{def}}{=} L \Big/ \frac{4\varepsilon}{\sqrt{M^2 - 1}} = \frac{(s^2 + \sqrt{B})\sqrt{M^2 - 1}}{s(M^2 + \sqrt{B})}$$

It may be noticed that the lift value lies all the time below the classical value (th magnetic field presence decreases the lift). For given M the lift value decreases with

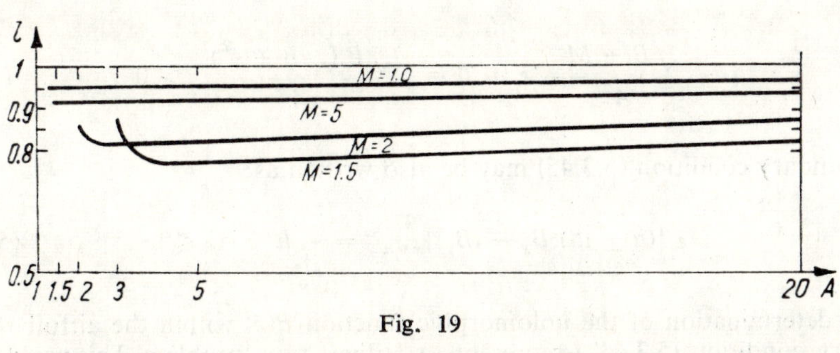

Fig. 19

increasing A, reaches a minimum value and then increases tending asymptotically towards the classical value. The minima are obtained for smaller A the higher M. The absolute value of these minima increases with increasing M.

5.3.3 Hyperbolic flow in the presence of a thick airfoil

Assuming a hyperbolic flow regime the mathematical problem can also be solved if the internal magnetic field is included. In this sense the boundary conditions may be written as:

$$v_y(x, h_\pm) = h'_\pm, \; |x| \leqslant 1 \tag{5.3.41}$$

$$b_x(x, h_\pm) = B_x(x, h_\pm), \; |x| \leqslant 1 \tag{5.3.42}$$

$$b_y(x, h_\pm) = B_y(x, h_\pm), \; |x| \leqslant 1 \tag{5.3.43}$$

where B_x and B_y stand for the components of the dimensionless internal induction. If the body is an insulator, then:

$$\frac{\partial B_x}{\partial x} + \frac{\partial B_y}{\partial y} = 0, \; \frac{\partial B_y}{\partial x} - \frac{\partial B_x}{\partial y} = 0 \tag{5.3.44}$$

These equations prove that the function $B(z) = B_x - iB_y$ is holomorphic within the body.

Taking into account (5.3.33) and (5.3.34) we eliminate $S_\pm\,(x \mp sh_\pm)$ and $R_\pm\,(x \mp rh_\pm)$ from (5.3.42), (5.3.43) and (5.3.41). For this purpose we use the identity $(M^2 - s^2)(M^2 - r^2) = -A^2$ which results from equation (5.3.9″). Thus we find the condition:

$$-\alpha B_x(x, h_\pm) \pm \beta B_y(x, h_\pm) = h'_\pm, \; |x| \leqslant 1 \tag{5.3.45}$$

where:

$$\alpha = \frac{\sqrt{B} + M^2}{A^2} > 0, \; \beta = \frac{\sqrt{B}\,(\sqrt{B} + s^2)}{sA^2} > 0 \tag{5.3.45'}$$

The boundary condition (5.3.45) may be also written as:

$$\mathcal{R}e\,\{(\alpha \pm i\beta)(B_x - iB_y)\}_{y=h_\pm} = -\,h'_\pm, \; |x| \leqslant 1 \tag{5.3.45''}$$

The determination of the holomorphic function $B(z)$ within the airfoil with the boundary condition (5.3.45″) represents a Hilbert-type problem. This problem was solved by D. Homentcovschi [8]. Its solution is found in the following way:

Let $z = z(\zeta)$ be the function that conformally maps the airfoil interior in the z plane into the upper half-plane of the complex plane $\zeta = \xi + i\eta$, such that the points ± 1 remain unchanged and the point c having the property $h'_+(c) = 0$ goes into $\zeta = \infty$. By this mapping condition (5.3.45″) becomes (for $\eta = 0$):

$$\mathscr{Re}\{(\alpha + i\beta)(B_x - iB_y)\} = -h'_-(x(\xi)) \stackrel{\text{def}}{=} g(\xi), \ |\xi| < 1$$

$$\mathscr{Re}\{(\alpha - i\beta)(B_x - iB_y)\} = -h'_+(x(\xi)) \stackrel{\text{def}}{=} g(\xi), \ |\xi| > 1 \tag{5.3.46}$$

Using the notation:

$$\gamma = \begin{cases} \alpha + i\beta = r\,e^{-i\delta\pi}, \ |\xi| < 1 \\ \alpha - i\beta = r\,e^{i\delta\pi}, \ |\xi| > 1 \end{cases} \quad \text{and } \delta \in \left(0, \frac{1}{2}\right) \tag{5.3.47}$$

the solution of the problem becomes:

$$B(\zeta) = \frac{i}{\pi}(\zeta - 1)^{2\delta}(\zeta + 1)^{1-2\delta} \int_{-\infty}^{+\infty} \frac{g(\xi)}{\gamma} \frac{1}{(\xi - 1)^{2\delta}(\xi + 1)^{1-2\delta}} \frac{d\xi}{\zeta - \xi} \tag{5.3.48}$$

This solution vanishes both at the leading and the trailing edge. Unfortunately the use of this solution even in the simplest cases involves great difficulties.

5.4 Electromagnetic flow past non-conducting walls

5.4.1 Formulation of the problem

We shall examine in this section the flow of perfectly conducting fluids past an insulating wall of equation

$$y = h(x), \ x \in (-\infty, +\infty) \tag{5.4.1}$$

More exactly we assume that the fluid motion takes place in the upper half-plane $y > h(x)$ the lower half-plane $y < h(x)$ being a non-conducting medium. The functions $h(x)$ and $h'(x)$ are supposed sufficiently smooth so that the linearization of the equations is made possible and the boundary conditions on the $y = 0$ axis are satisfied. In order to ensure the convergence of the solution we should further assume that $h'(x)$ is Hölderian and that, in the vicinity of the point at infinity, it satisfies the condition:

$$|h'(x)| < \frac{k}{|x|^{\nu}}, \quad 0 < \nu < 1, \; k = \text{const} \tag{5.4.1'}$$

In the insulating half-plane the magnetic induction is characterized by the equations:

$$\frac{\partial B_x}{\partial x} + \frac{\partial B_y}{\partial y} = 0, \quad \frac{\partial B_y}{\partial x} - \frac{\partial B_x}{\partial y} = 0 \tag{5.4.2}$$

which prove that the complex variable function

$$B(z) = B_x - \mathrm{i}B_y, \; z = x + \mathrm{i}y \tag{5.4.2'}$$

is holomorphic for $(y < 0)$.

In the half-plane $y > 0$ the linearized equations of motion are valid.

5.4.2 Incompressible fluids

In the upper half-plane the equations of motion have the following solution:

$$\begin{aligned}
v_y &= F(x - cy) + \frac{\partial \varphi}{\partial y}, \qquad c = c_+ \\
b_y &= -AF(x - cy) + \alpha_x \frac{\partial \varphi}{\partial y} - \alpha_y \frac{\partial \varphi}{\partial x} \\
v_x &= cF(x - cy) + \frac{\partial \varphi}{\partial x} \\
b_x &= -cAF(x - cy) + \alpha_x \frac{\partial \varphi}{\partial x} + \alpha_y \frac{\partial \varphi}{\partial y} \\
p &= A^{-1}F(x - cy) - \frac{\partial \varphi}{\partial x}
\end{aligned} \tag{5.4.3}$$

derived from (5.2.13) and (5.2.19).

The boundary condition (5.2.21) gives:

$$F(x) = -\left.\frac{\partial \varphi}{\partial y}\right|_{y=0} + h'(x),\ \forall x \tag{5.4.4}$$

Using this equation the conditions (5.2.23) become:

$$\begin{aligned} &\alpha_x \frac{\partial \varphi}{\partial x}(x, 0) + (\alpha_y + cA)\frac{\partial \varphi}{\partial y}(x, 0) - B_x(x, 0) = cAh'(x),\ \forall x \\ &-\alpha_y \frac{\partial \varphi}{\partial x}(x, 0) + (\alpha_y + A)\frac{\partial \varphi}{\partial y}(x, 0) - B_y(x, 0) = Ah'(x),\ \forall x \end{aligned} \tag{5.4.5}$$

Thus we have to determine the holomorphic function:

$$\Phi(z) = \begin{cases} \dfrac{\partial \varphi}{\partial x} - \mathrm{i}\dfrac{\partial \varphi}{\partial y},\ y > 0,\ z = x + \mathrm{i}y \\ B_x - \mathrm{i}B_y,\ y < 0 \end{cases} \tag{5.4.6}$$

which satisfies conditions (5.4.5) on the boundary $y = 0$. To this end one looks for a solution having the form of a Cauchy-type integral.

$$\Phi(z) = \frac{1}{\pi \mathrm{i}} \int_{-\infty}^{+\infty} \frac{m(t) + \mathrm{i}n(t)}{t - z}\, \mathrm{d}t \tag{5.4.7}$$

where $m(t)$ and $n(t)$ are real functions of real variables to be derived from (5.4.5). Using Plemelj's equations from (5.4.7) it follows that

$$\left(\frac{\partial \varphi}{\partial x} - \mathrm{i}\frac{\partial \varphi}{\partial y}\right)_{y=0} = m(x) + \mathrm{i}n(x) + \frac{1}{\pi \mathrm{i}} \int_{-\infty}^{\prime +\infty} \frac{m(t) + \mathrm{i}n(t)}{t - x}\, \mathrm{d}t$$

$$(B_x - \mathrm{i}B_y)_{y=0} = -\,m(x) - \mathrm{i}n(x) + \frac{1}{\pi \mathrm{i}} \int_{-\infty}^{\prime +\infty} \frac{m(t) + \mathrm{i}n(t)}{t - x}\, \mathrm{d}t$$

whence

$$\frac{\partial \varphi}{\partial x}(x, 0) = m(x) + \frac{1}{\pi}\int_{-\infty}^{'+\infty} \frac{n(t)}{t - x}\,dt$$

$$\frac{\partial \varphi}{\partial y}(x, 0) = -n(x) + \frac{1}{\pi}\int_{-\infty}^{'+\infty} \frac{m(t)}{t - x}\,dt$$

$$B_x(x, 0) = -m(x) + \frac{1}{\pi}\int_{-\infty}^{'+\infty} \frac{n(t)}{t - x}\,dt \qquad (5.4.8)$$

$$B_y(x, 0) = n(x) + \frac{1}{\pi}\int_{-\infty}^{'+\infty} \frac{m(t)}{t - x}\,dt$$

Taking into account (5.4.8) conditions (5.4.5) become:

$$(1 + \alpha_x)\, m(x) - (\alpha_y + cA)\, n(x)$$

$$+ \frac{\alpha_y + cA}{\pi}\int_{-\infty}^{'+\infty} \frac{m(t)}{t - x}\,dt + \frac{\alpha_x - 1}{\pi}\int_{-\infty}^{'+\infty} \frac{n(t)}{t - x}\,dt = cAh'(x), \ \forall x$$

$$\alpha_y m(x) + (A + \alpha_x + 1)\, n(x) \qquad (5.4.9)$$

$$+ \frac{1 + \alpha_x - A}{\pi}\int_{-\infty}^{'+\infty} \frac{m(t)}{t - x}\,dt + \frac{\alpha_y}{\pi}\int_{-\infty}^{'+\infty} \frac{n(t)}{t - x}\,dt = -Ah'(x), \forall x$$

Multiplying by $d\xi/\pi(x - \xi)$ each equation (5.4.9), transforming to the real axis and using the Poincaré-Bertrand equation [G. 34], we get the following equations:

$$-(\alpha_y + cA)\, m(x) + (1 - \alpha_x)\, n(x)$$

$$+ \frac{1 + \alpha_x}{\pi}\int_{-\infty}^{'+\infty} \frac{m(t)}{t - x}\,dt - \frac{\alpha_y + cA}{\pi}\int_{-\infty}^{'+\infty} \frac{n(t)}{t - x}\,dt = \frac{cA}{\pi}\int_{-\infty}^{'+\infty} \frac{h'(t}{t - x}\,dt$$

$$(1 - \alpha_x - A)\, m(x) + \alpha_y n(x) \qquad (5.4.9')$$

$$- \frac{\alpha_y}{\pi}\int_{-\infty}^{'+\infty} \frac{m(t)}{t - x}\,dt - \frac{1 + \alpha_x + A}{\pi}\int_{-\infty}^{'+\infty} \frac{n(t)}{t - x}\,dt = \frac{A}{\pi}\int_{-\infty}^{'+\infty} \frac{h'(t)}{t - x}\,dt$$

From (5.4.9) and (5.4.9′) it follows that:

$$m(x) = a_1 h'(x) + \frac{b_1}{\pi} \int_{-\infty}^{'+\infty} \frac{h'(t)}{t - x} \mathrm{d}t$$

$$n(x) = a_2 h'(x) + \frac{b_2}{\pi} \int_{-\infty}^{'+\infty} \frac{h'(t)}{t - x} \mathrm{d}t \tag{5.4.10}$$

$$\frac{1}{\pi} \int_{-\infty}^{'+\infty} \frac{m(t)}{t - x} \mathrm{d}t = -b_1 h'(x) + \frac{a_1}{\pi} \int_{-\infty}^{'+\infty} \frac{h'(t)}{t - x} \mathrm{d}t$$

$$\frac{1}{\pi} \int_{-\infty}^{'+\infty} \frac{n(t)}{t - x} \mathrm{d}t = -b_2 h'(x) + \frac{a_2}{\pi} \int_{-\infty}^{'+\infty} \frac{h'(t)}{t - x} \mathrm{d}t \tag{5.4.11}$$

where a_1, a_2, b_1 and b_2 are some constants which can be obtained easily.

The equations (5.4.10) and (5.4.11) uniquely determine the solution. Indeed, the elliptical part

$$\frac{\partial \varphi}{\partial x} - \mathrm{i} \frac{\partial \varphi}{\partial y} = \left(\frac{a_1 + \mathrm{i} a_2}{\pi \mathrm{i}} + \frac{b_1 + \mathrm{i} b_2}{\pi} \right) \int_{-\infty}^{+\infty} \frac{h'(t)\,\mathrm{d}t}{t - z} \tag{5.4.12}$$

is obtained from (5.4.7) and (5.4.10) using the equation:

$$\int_{-\infty}^{+\infty} \frac{\mathrm{d}t}{(\xi - t)(t - z)} = \frac{1}{\xi - z} \int_{-\infty}^{+\infty} \frac{\mathrm{d}t}{t - z} = \frac{\pi}{\xi - z} \tag{5.4.12′}$$

and the hyperbolic part

$$F(x) = (1 + a_2 + b_1)\, h'(x) + \frac{b_2 - a_1}{\pi} \int_{-\infty}^{'+\infty} \frac{h'(t)}{t - x} \mathrm{d}t \tag{5.4.13}$$

from (5.4.4), (5.4.8) and (5.4.11). One can easily see that for $h' \equiv 0$ the solution is vanishing, this proving the uniqueness of the solution.

If the function $h'(x)$ had a first kind discontinuity at x_0 then a logarithmic discontinuity propagating along the characteristic line passing through x_0 would appear in the solution. Indeed, in the neighbourhood of x_0 the Cauchy-type integral has the following behaviour [G. 34]:

$$\frac{1}{\pi i} \int_{-\infty}^{+\infty} \frac{h'(t)}{t - z} \mathrm{d}t = \frac{1}{\pi i} \ln \frac{h'(x_0 - 0)}{h'(x_0 + 0)} \ln (z - x_0) + \Gamma(z) \tag{5.4.14}$$

where $\Gamma(z) = \Gamma_1 + i\Gamma_2$ represents an analytical bounded function and $h'(x_0 - 0)$ and $h'(x_0 + 0)$ represent the limits of function h' to the left and the right, respectively. The integral (5.4.11) defined in the upper half-plane yields at the limit

$$h'(x) = \frac{1}{\pi} \ln \frac{h'(x_0 - 0)}{h'(x_0 + 0)} \arg(x - x_0) + \Gamma_1$$

$$\frac{1}{\pi} \int_{-\infty}^{\prime +\infty} \frac{h'(t)}{t - x} \, dt = \frac{1}{\pi} \ln \frac{h'(x_0 - 0)}{h'(x_0 + 0)} \ln |x - x_0| - \Gamma_2$$

so that (5.4.13) becomes

$$F(x) = \frac{b_2 - a_1}{\pi} \ln \frac{h'(x_0 - 0)}{h'(x_0 + 0)} \ln |x - x_0| + \Gamma_3(x) \qquad (5.4.13')$$

Γ_3 being bounded at $x = x_0$. Equation (5.4.13′) proves the above statement.

This solution is derived in [13] where the singular case of aligned fields is also studied in detail.

5.4.3 Compressible fluids

For compressible fluids one obtains a simple solution if the hypothesis of orthogonal fields in the hyperbolic flow regime is used. Indeed, in this case, the solution on the upper half plane has the following form:

$$v_y = S(x - sy) + R(x - ry)$$

$$b_y = s^{-1}(M^2 - s^2)\, S(x - sy) + r^{-1}(M^2 - r^2)\, R(x - ry) \qquad (5.4.15)$$

$$b_x = (M^2 - s^2)\, S(x - sy) + (M^2 - r^2)\, R(x - ry)$$

$$p = s^{-1} S(x - sy) + r^{-1} R(x - ry), \; v_x = -b_y$$

derived from (5.3.33) and (5.3.34).

Let us now suppose the following boundary conditions:

$$v_y(x, 0) = h'(x), \forall x \qquad (5.4.16)$$

$$b_x(x, 0) = B_x(x, 0), \forall x \qquad (5.4.17)$$

$$b_y(x, 0) = B_y(x, 0), \forall x \qquad (5.4.18)$$

From (5.4.15), (5.4.17) and (5.4.18) we obtain:

$$(M^2 - s^2)S = \frac{s(b_x - rb_y)}{s - r} = \frac{s(B_x - rB_y)}{s - r}, \; y = 0$$

$$(M^2 - r^2)R = \frac{r(-b_x + sb_y)}{s - r} = \frac{r(-B_x + sB_y)}{s - r}, \; y = 0 \tag{5.4.19}$$

So that substituting into (5.4.16) it follows that

$$-\alpha B_x(x, 0) + \beta B_y(x, 0) = h'(x), \; \forall x \tag{5.4.20}$$

the expressions of α and β being those of (5.3.45′). The problem is therefore reduced to determining the holomorphic function $B(z)$ in the half-plane $y < 0$ with the boundary condition (5.4.20) (a Hilbert type problem).

Chu [18] solved this problem noticing that the function:

$$\Phi(x, y) = -\alpha B_x(x, y) + \beta B_y(x, y), \; y < 0 \tag{5.4.21}$$

satisfies the following conditions:

$$\Delta\Phi = 0, \quad y < 0$$

$$\Phi(x, 0) = h'(x), \quad \forall x \tag{5.4.22}$$

$$|\Phi| < \infty, \qquad y < 0$$

The solution of problem (5.4.22) is

$$\Phi = -\frac{1}{\pi}\int_{-\infty}^{+\infty} \frac{yh'(\xi)\,d\xi}{(x - \xi)^2 + y^2}, \; y < 0 \tag{5.4.22′}$$

Then determining the function Ψ whose harmonic conjugate is Φ, from the equations $\Psi = \beta B_x - \alpha B_y$ and (5.4.21) it follows that:

$$B_x = \frac{\beta\Psi - \alpha\Phi}{\alpha^2 + \beta^2}, \; B_y = \frac{\alpha\Psi + \beta\Phi}{\alpha^2 + \beta^2} \tag{5.4.23}$$

However, in actual problems the integrals of the form (5.4.22′) are rather difficult to handle. We thus present another solution of the problem (5.4.20) which has the advantage of presenting the results in a concise form. In this respect we shall determine the solution of problem (5.4.20) in the form of a Cauchy integral:

$$B(z) = \frac{1}{\pi i}\int_{-\infty}^{+\infty} \frac{v(t)}{t - z}\,dt, \quad y < 0 \tag{5.4.24}$$

Applying Plemelj's equations from (5.4.24) it follows that:

$$B_x(x, 0) = -v(x)$$

$$B_y(x, 0) = \frac{1}{\pi}{\int_{-\infty}^{+\infty}}' \frac{v(t)}{t - x}\,dt \tag{5.4.25}$$

so that the condition (5.4.20) reduces to the following singular integral equation

$$\alpha v(x) + \frac{B}{\pi}{\int_{-\infty}^{+\infty}}' \frac{v(t)}{t - x}\,dt = h'(x), \ \forall x \tag{5.4.26}$$

The solution of this equation is [G. 34], [19]:

$$v(x) = \frac{\alpha}{\alpha^2 + \beta^2}\,h'(x) - \frac{\beta}{\pi(\alpha^2 + \beta^2)}{\int_{-\infty}^{+\infty}}' \frac{h'(\xi)}{\xi - x}\,d\xi \tag{2.4.27}$$

This solution directly determines $B_x(x, 0)$. In order to determine $B_y(x, 0)$ we shall use equations (5.4.25) and (5.4.27). A simple calculation in which the Poincaré-Bertrand equation and the relationship

$$\int_{-\infty}^{+\infty} \frac{dt}{(t - x)(\xi - t)} = \frac{1}{\xi - x}\left(\int_{-\infty}^{+\infty} \frac{dt}{t - x} - \int_{-\infty}^{+\infty} \frac{dt}{t - \xi}\right) = 0 \tag{5.4.28}$$

are used yields the following result:

$$B_y(x, 0) = \frac{\beta}{\alpha^2 + \beta^2}\,h'(x) + \frac{\alpha}{\pi(\alpha^2 + \beta^2)}{\int_{-\infty}^{+\infty}}' \frac{h'(t)}{t - x}\,dt \tag{5.4.29}$$

Having determined the functions $B_x(x, 0)$ and $B_y(x, 0)$ the solution within the fluid follows from (5.4.19) and (5.4.15).

For the particular case of a wall with a finite perturbation region

$$h'(x) = \eta(x)\,k, \quad k = \text{const}$$

$$\eta(x) = \begin{cases} 0, & -\infty < x < 0, 1 < x < \infty \\ 1, & 0 < x < 1 \end{cases} \tag{5.4.30}$$

the equations (5.4.27) and (5.4.29) become

$$\begin{aligned} B_x(x,0) &= -\frac{\alpha k}{\alpha^2+\beta^2}\eta + \frac{\beta k}{\alpha^2+\beta^2}\frac{1}{\pi}\ln\left|\frac{x-1}{x}\right| \\ B_y(x,0) &= \frac{\beta k}{\alpha^2+\beta^2}\eta + \frac{\alpha k}{\alpha^2+\beta^2}\frac{1}{\pi}\ln\left|\frac{x-1}{x}\right| \end{aligned} \tag{5.4.31}$$

The solution (5.4.31) coincides with that given by Chy.

We notice again that the discontinuities of function $h'(x)$ produce logarithmic singularities of the magnetic induction on the frontier. These singularities propagate along the two characteristic lines C_s^+ and C_r^+ passing through the singular points (for our case $x = 0$, $x = 1$).

5.5 Unsteady motion (incompressible fluids)

5.5.1 General solution

Let us now consider an insulating arc airfoil AB moving with velocity $V_0(t)$ in a constant direction through a perfectly conducting fluid at rest. For this case one

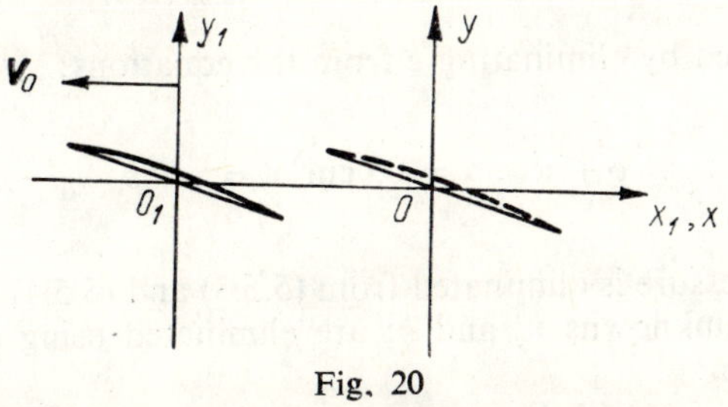

Fig. 20

chooses a reference frame $x_1 0_1 y_1$ bound to the airfoil and another reference frame $x0y$ fixed at the initial position of $x_1 0_1 y_1$. The axes of the two systems are supposed to be parallel the 0_1x_1 — and $0x$-axes having the direction opposite to V_0 (Figure 20).

The transformation from one system to another is given by:

$$x_1 = x + \int_0^t V_0(t)\,\mathrm{d}t = x + s(t) \tag{5.5.1}$$

$$y_1 = y, \; z_1 = z$$

Only plane motion is considered.

Denoting by:

$$\boldsymbol{v}(t, x, y), \; \boldsymbol{\alpha} + \boldsymbol{b}(t, x, y), \; p(t, x, y), \; e(t, x, y)$$

the dimensionless variables characterizing the fluid motion, we have the following equations:

$$\frac{\partial v_x}{\partial x} + \frac{\partial v_y}{\partial y} = 0, \; \frac{\partial b_x}{\partial x} + \frac{\partial b_y}{\partial y} = 0 \tag{5.5.2}$$

$$\frac{\partial v_x}{\partial t} = -\frac{\partial p}{\partial x} - \frac{\alpha_y}{A^2}\left(\frac{\partial b_y}{\partial x} - \frac{\partial b_x}{\partial y}\right) \tag{5.5.3}$$

$$\frac{\partial v_y}{\partial t} = -\frac{\partial p}{\partial y} + \frac{\alpha_x}{A^2}\left(\frac{\partial b_y}{\partial x} - \frac{\partial b_x}{\partial y}\right) \tag{5.5.4}$$

$$\frac{\partial \boldsymbol{b}}{\partial t} = \frac{\mathrm{d}\boldsymbol{v}}{\mathrm{d}\alpha} \tag{5.5.5}$$

the last being derived by eliminating $\boldsymbol{e}$ from the equations:

$$\boldsymbol{e} + \boldsymbol{v} \times \boldsymbol{\alpha} = 0, \; \mathrm{curl}\, \boldsymbol{e} = -\partial_t \boldsymbol{b} \tag{5.5.5'}$$

As usual the pressure is eliminated from (5.5.3) and (5.5.4) and from the remaining equation the unknowns v_y and b_y are eliminated using equation (5.5.2). One obtains the equation:

$$\frac{\partial}{\partial t}\Delta v_x = \frac{1}{A^2}\frac{\mathrm{d}}{\mathrm{d}\alpha}\Delta b_x, \; \Delta = \frac{\partial^2}{\partial x^2} + \frac{\partial^2}{\partial y^2}$$

Eliminating the unknowns v_x and b_x a similar equation for v_y and b_y would be obtained. Now using (5.5.3) it follows that:

$$T\Delta(v_x, v_y, b_x, b_y) = 0,$$

$$T \overset{\text{def}}{=} \frac{\partial^2}{\partial t^2} - \frac{1}{A^2}\frac{d^2}{d\alpha^2} \tag{5.5.6}$$

The pressure also obeys equation (5.5.6). Indeed, if one differentiates equation (5.5.3) with respect to x, equation (5.5.4) with respect to y, adds up the equations obtained in this way and uses (5.5.2) one obtains:

$$0 = \Delta p + \frac{\alpha_y}{A^2}\Delta b_y + \frac{\alpha_x}{A^2}\Delta b_x$$

Applying here the operator T and taking into account (5.5.6) the above statement is proved.

From equations (5.5.3), (5.5.4) and (5.5.5) it also follows that:

$$T(j, \omega) = 0,\ j \overset{\text{def}}{=} \frac{\partial b_y}{\partial x} - \frac{\partial b_x}{\partial y},\ \omega \overset{\text{def}}{=} \frac{\partial v_y}{\partial x} - \frac{\partial v_x}{\partial y} \tag{5.5.6'}$$

If the applied induction is parallel to $V_0(t)$ ($\alpha_x = 1$, $\alpha_y = 0$) equation (5.5.6) has the following general solution:

$$b_x = F_1(x - \frac{t}{A}, y) + G_1(x + \frac{t}{A}, y) + \frac{\partial \psi}{\partial x}(t, x, y)$$

$$b_y = F_2(x - \frac{t}{A}, y) + G_2(x + \frac{t}{A}, y) + \psi_1(t, x, y)$$

$$v_x = F_3(x - \frac{t}{A}, y) + G_3(x + \frac{t}{A}, y) + \frac{\partial \varphi}{\partial x}(t, x, y) \tag{5.5.7}$$

$$v_y = F_4(x - \frac{t}{A}, y) + G_4(x + \frac{t}{A}, y) + \varphi_1(t, x, y)$$

$$p = F_5(x - \frac{t}{A}, y) + G_5(x + \frac{t}{A}, y) + \chi(t, x, y)$$

$$\Delta(\varphi, \varphi_1, \psi, \psi_1, \chi) = 0 \tag{5.5.8}$$

Substituting this solution into equations (5.5.2), (5.5.5) and (5.5.3) we find that:

$$\frac{\partial}{\partial x}(F_3 + G_3) + \frac{\partial}{\partial y}(F_4 + G_4) = 0, \ \varphi_1 = \frac{\partial \varphi}{\partial y} \tag{5.5.9}$$

$$\frac{\partial}{\partial x}(F_1 + G_1) + \frac{\partial}{\partial y}(F_2 + G_2) = 0, \ \psi_1 = \frac{\partial \psi}{\partial y} \tag{5.5.10}$$

$$F_3 = -\frac{1}{A}F_1, \ G_3 = \frac{1}{A}G_1, \ \frac{\partial \psi}{\partial t} = \frac{\partial \varphi}{\partial x} \tag{5.5.11}$$

$$F_4 = -\frac{1}{A}F_2, \ G_4 = \frac{1}{A}G_2, \tag{5.5.12}$$

$$F_5 = -\frac{1}{A^2}F_1, \ G_5 = -\frac{1}{A^2}G_1, \ \chi = -\frac{\partial \varphi}{\partial t} \tag{5.5.13}$$

Equation (5.5.4) is satisfied identically.

Thus, to conclude, we have the following general solution:

$$b_x = F_1\left(x - \frac{t}{A}, y\right) + G_1\left(x + \frac{t}{A}, y\right) + \frac{\partial \psi}{\partial x}$$

$$b_y = F_2\left(x - \frac{t}{A}, y\right) + G_2\left(x + \frac{t}{A}, y\right) + \frac{\partial \psi}{\partial y}$$

$$v_x = -\frac{1}{A}F_1 + \frac{1}{A}G_1 + \frac{\partial \varphi}{\partial x} \tag{5.5.14}$$

$$v_y = -\frac{1}{A}F_2 + \frac{1}{A}G_2 + \frac{\partial \varphi}{\partial y}$$

$$p = -\frac{1}{A^2}F_1 - \frac{1}{A^2}G_1 - \frac{\partial \varphi}{\partial t}$$

with the conditions

$$\frac{\partial F_1}{\partial x} + \frac{\partial F_2}{\partial y} = 0, \ \frac{\partial G_1}{\partial x} + \frac{\partial G_2}{\partial y} = 0 \tag{5.5.15}$$

which follow from (5.5.9), (5.5.10), as well as the condition:

$$\frac{\partial \psi}{\partial t} = \frac{\partial \varphi}{\partial x} \tag{5.5.16}$$

One notices that equations (5.5.15) represent the necessary and sufficient conditions for the existence of the functions $H(x - t/A, y)$ and $K(x + t/A, y)$ and for having

$$F_1 = \frac{\partial H}{\partial y},\ F_2 = -\frac{\partial H}{\partial x},\ G_1 = \frac{\partial K}{\partial y},\ G_2 = -\frac{\partial K}{\partial x}$$

By expressing the solution (5.5.14) by means of these functions the solution given by Homentcovschi [22] is obtained.

In exactly the same manner as above one can prove that when the external magnetic induction $\boldsymbol{B}_0$ is orthogonal to the velocity $\boldsymbol{V}_0\,(t)$ the general solution has the form:

$$b_x = F_1\left(x, y - \frac{t}{A}\right) + G_1\left(x, y + \frac{t}{A}\right) + \frac{\partial \psi}{\partial x}$$

$$b_y = F_2\left(x, y - \frac{t}{A}\right) + G_2\left(x, y + \frac{t}{A}\right) + \frac{\partial \psi}{\partial y}$$

$$v_x = -\frac{1}{A} F_1 + \frac{1}{A} G_1 + \frac{\partial \varphi}{\partial x}$$

$$v_y = -\frac{1}{A} F_2 + \frac{1}{A} G_2 + \frac{\partial \varphi}{\partial y}$$

$$p = -\frac{1}{A^2} F_2 - \frac{1}{A^2} G_2 - \frac{\partial \varphi}{\partial t}$$

where

$$\frac{\partial F_1}{\partial x} + \frac{\partial F_2}{\partial y} = 0,\ \frac{\partial G_1}{\partial x} + \frac{\partial G_2}{\partial y} = 0,\ \frac{\partial \psi}{\partial t} = \frac{\partial \varphi}{\partial y}$$

$$\Delta \varphi = \Delta \psi = 0$$

5.5.2 Aligned fields

One assumes that the airfoil is initially in rest. Under these conditions, taking into account the hyperbolic characteristics of the current density (5.5.6′), this will be non-vanishing at any time only within a finite region R of the plane. According to equation (2.5.3) the velocity circulation along any material curve external to R should be vanishing. In order to satisfy this condition, Homentcovschi [22] (as well as Birnbaum for the classical case [G. 28]) supposes that the trailing edge leaves behind (along the Ox-axis) a stream of vortices. However, contrary to the classical situation in this case, the vortices will not remain fixed but will be carried downstream by the hyperbolic current so that at time t the stream of vortices will occupy the segment $(1 - s(t), 1 + t/A)$ on the Ox-axis. This stream is associated to a surface current sheet.

Since the body is the only source of perturbations all the flow characteristics will be determined by the airfoil. The functions F_1 and F_2 of (5.5.14) describe waves propagating downstream and G_1 and G_2 waves propagating upstream (for $A < 1$). For $A > 1$ one has only waves propagating downstream.

One also assumes that the propagations vanish at infinity for any t.

On the airfoil we have the jump conditions (1.5.22) and (1.5.17) which, like for the stationary case yield:

$$[b_x] = 0, \; [b_y] = 0 \tag{5.5.17}$$

Using Ohm's law $\boldsymbol{e}$ is eliminated from (1.5.21) and one obtains (for $\boldsymbol{s}$ = the unit vector of the Oz-axis):

$$[v_y] = 0 \tag{5.5.18}$$

Finally we have the sliding condition for the velocity on the airfoil:

$$\boldsymbol{v} \cdot \boldsymbol{n} = \boldsymbol{V}_0 \cdot \boldsymbol{n} = d \tag{5.5.19}$$

which should be satisfied both on the upper and the lower side of the airfoil. Denoting by n_x and n_y the coordinates in the fixed system of the normal to the airfoil from (5.5.19) it follows that:

$$[v_x]n_x + [v_y]n_y = 0 \tag{5.5.20}$$

Since $n_x \neq 0$ on the profile, from (5.5.18) and (5.5.20) it follows that:

$$[v_x] = 0 \tag{5.5.21}$$

v_x, v_y, b_x, and b_y being the coordinates with respect to the fixed system, i.e., the coordinates determined by equations (5.5.14).

If we denote by:

$$y = h(t, x), \; -1 - s < x < 1 - s \tag{5.5.22}$$

the airfoil equation with respect to the fixed system, equations (5.5.17), (5.5.18), (5.5.19) and (5.5.21) should be satisfied under conditions (5.5.22). One also assumes that $h_x' < 0$ which fact makes sure that each characteristic intersects the airfoil at one point only.

The vortex stream $(1 - s < x < 1 + t/A)$ is a discontinuity line for which $d = 0$ and $n_x = 0$. From (1.5.10), (1.5.14) and (1.5.17) we find:

$$[v_y] = 0, \; [b_x] = -i(t, x), \; [b_y] = 0 \tag{5.5.23}$$

$i(t, x)$ being the dimensionless surface current density along the Oz-axis. Since the stream is a discontinuity line the jump equations following from the conservation of mass and momentum (11.1.5) and (11.1.5′) and (11.1.6) should also be satisfied. The first condition coincides with the first equation of (5.5.23). The second reduces itself to

$$[p + A^{-2} b_x] = 0 \tag{5.5.24}$$

after linearization.

Let us now write that the general solution (5.5.14) satisfies conditions (5.5.17), (5.5.18) and (5.5.21). For $A < 1$ we have:

$$F_1\left(x - \frac{t}{A}, h\right) + \frac{\partial \psi}{\partial x}(t, x, h_+) = G_1\left(x + \frac{t}{A}, h\right) + \frac{\partial \psi}{\partial x}(t, x, h_-)$$

$$F_2\left(x - \frac{t}{A}, h\right) + \frac{\partial \psi}{\partial y}(t, x, h_+) = G_2\left(x + \frac{t}{A}, h\right) + \frac{\partial \psi}{\partial y}(t, x, h_-)$$

$$-\frac{1}{A} F_2\left(x - \frac{t}{A}, h\right) + \frac{\partial \varphi}{\partial y}(t, x, h_+) = \frac{1}{A} G_2\left(x + \frac{t}{A}, h\right) + \frac{\partial \varphi}{\partial y}(t, x, h_-)$$

$$-\frac{1}{A} F_1\left(x - \frac{t}{A}, h\right) + \frac{\partial \varphi}{\partial x}(t, x, h_+) = \frac{1}{A} G_1\left(x + \frac{t}{A}, h\right) + \frac{\partial \varphi}{\partial x}(t, x, h_-)$$

whence

$$2F_1\left(x - \frac{t}{A}, h\right) = A\left[\frac{\partial\varphi}{\partial x}\right] - \left[\frac{\partial\psi}{\partial x}\right] \stackrel{\text{def}}{=} 2\tilde{F}_1(t, x)$$

$$2G_1\left(x + \frac{t}{A}, h\right) = A\left[\frac{\partial\varphi}{\partial x}\right] + \left[\frac{\partial\psi}{\partial x}\right] \stackrel{\text{def}}{=} 2\tilde{G}_1(t, x)$$

$$2F_2\left(x - \frac{t}{A}, h\right) = A\left[\frac{\partial\varphi}{\partial y}\right] - \left[\frac{\partial\psi}{\partial y}\right] \stackrel{\text{def}}{=} 2\tilde{F}_2(t, x) \tag{5.5.25}$$

$$2G_2\left(x + \frac{t}{A}, h\right) = A\left[\frac{\partial\varphi}{\partial y}\right] + \left[\frac{\partial\psi}{\partial y}\right] \stackrel{\text{def}}{=} 2\tilde{G}_2(t, x)$$

These equations determine the rotational part of the solution as a function of its irrotational part. It is obvious that the rotational perturbation starts on the airfoil and propagates along the characteristic lines.

The functions on the right hand side of the equations (5.5.25) are not independent since conditions (5.5.15) are still to be satisfied. In order to derive their form we notice that:

$$\frac{\partial\tilde{F}_1}{\partial x} = \frac{\partial F_1}{\partial x} + \frac{\partial F_1}{\partial y} h', \quad \frac{\partial\tilde{F}_1}{\partial t} = -\frac{1}{A}\frac{\partial F_1}{\partial x} + \frac{\partial F_1}{\partial y}\dot{h}$$

whence

$$\frac{\partial F_1}{\partial x}\left(\dot{h} + \frac{1}{A}h'\right) = \frac{\partial\tilde{F}_1}{\partial t}\dot{h} - \frac{\partial\tilde{F}_1}{\partial t}h'$$

In a similar way we find

$$\frac{\partial F_2}{\partial y}\left(\dot{h} + \frac{1}{A}h'\right) = \frac{1}{A}\frac{\partial\tilde{F}_2}{\partial x} + \frac{\partial\tilde{F}_2}{\partial t}$$

Consequently the first equation (5.5.15) becomes:

$$\frac{\partial\tilde{F}_1}{\partial x}\dot{h} - \frac{\partial\tilde{F}_1}{\partial t}h' + \frac{1}{A}\frac{\partial\tilde{F}_2}{\partial x} + \frac{\partial\tilde{F}_2}{\partial t} = 0 \tag{5.5.26}$$

In a similar way the second equation (5.5.15) yields:

$$\frac{\partial\tilde{G}_1}{\partial x}\dot{h} - \frac{\partial\tilde{F}_1}{\partial t}h' - \frac{1}{A}\frac{\partial\tilde{G}_2}{\partial x} + \frac{\partial\tilde{F}_2}{\partial t} = 0 \tag{5.5.27}$$

5.5.3 Determination of the potential part

Using the general solution and equations (5.5.25) condition (5.5.19) becomes:

$$\left[\frac{\partial\psi}{\partial y}\right] + A\left(\left.\frac{\partial\varphi}{\partial y}\right|_{+} + \left.\frac{\partial\varphi}{\partial y}\right|_{-}\right) - h'\left[\frac{\partial\psi}{\partial x}\right] - Ah'\left(\left.\frac{\partial\varphi}{\partial x}\right|_{+} + \left.\frac{\partial\varphi}{\partial x}\right|_{-}\right)$$

$$= 2A\dot{h}, \quad -1 - s < x < 1 - s \tag{5.5.28}$$

Then, employing a customary procedure in the thin airfoil theory, we shall replace the boundary conditions on the airfoil by the conditions on the Ox-axis and shall linearize these conditions. In this way, conditions (5.5.28) become:

$$\left[\frac{\partial\psi}{\partial y}\right](t, x, 0) + A\frac{\partial\varphi}{\partial y}(t, x, 0+) + A\frac{\partial\varphi}{\partial y}(t, x, 0-) = 2A\dot{Y}(t, x)$$

$$-1 - s < x < 1 - s \tag{5.5.29}$$

and (5.5.26) and (5.5.27):

$$\frac{1}{A}\frac{\partial\widetilde{F}_2}{\partial x} + \frac{\partial\widetilde{F}_2}{\partial t} = 0, \quad -\frac{1}{A}\frac{\partial\widetilde{G}_2}{\partial x} + \frac{\partial\widetilde{G}_2}{\partial t} = 0 \tag{5.5.30}$$

Taking into account (5.5.25), equations (5.5.30) yield:

$$\left[\frac{\partial^2\varphi}{\partial x\partial y}\right] - \frac{1}{A}\left[\frac{\partial^2\psi}{\partial x\partial y}\right] + A\left[\frac{\partial^2\varphi}{\partial t\partial y}\right] - \left[\frac{\partial^2\psi}{\partial t\partial y}\right] = 0, \ y = 0$$

$$-\left[\frac{\partial^2\varphi}{\partial x\partial y}\right] - \frac{1}{A}\left[\frac{\partial^2\psi}{\partial x\partial y}\right] + A\left[\frac{\partial^2\varphi}{\partial t\partial y}\right] + \left[\frac{\partial^2\psi}{\partial t\partial y}\right] = 0, \ y = 0$$

whence adding and subtracting we get:

$$\left[\frac{\partial^2\varphi}{\partial t\,\partial y}\right] = \frac{1}{A^2}\left[\frac{\partial^2\psi}{\partial x\,\partial y}\right], \left[\frac{\partial^2\varphi}{\partial x\,\partial y}\right] = \left[\frac{\partial^2\psi}{\partial t\,\partial y}\right], \quad y = 0 \tag{5.5.30'}$$

Taking into account (5.5.16), equation (5.5.30′b) reduces to an identity while (5.5.30′a) yields:

$$\frac{\partial^2}{\partial t^2}\left[\frac{\partial\varphi}{\partial y}\right] = \frac{1}{A^2}\frac{\partial^2}{\partial x^2}\left[\frac{\partial\varphi}{\partial y}\right]$$

The solution of this equation is vanishing since outside the region swept by the airfoil (the hatched region in Figure 21) we have $[\partial\varphi/\partial y] = 0$. To conclude:

$$\left[\frac{\partial\varphi}{\partial y}\right] = 0,\ \left[\frac{\partial\psi}{\partial y}\right] = 0,\ x\in(-\infty, 1-s)\cup\left(1+\frac{t}{A}, +\infty\right) \quad (5.5.31)$$

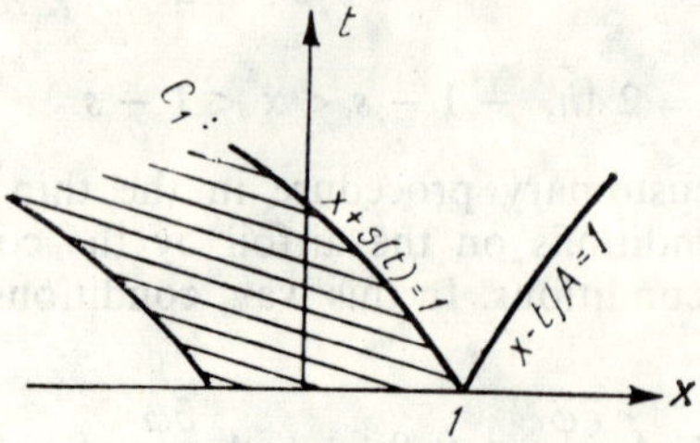

Fig. 21

On the discontinuity line from (5.5.23) it follows that:

$$\left[\frac{\partial\varphi}{\partial y}\right] = 0,\ \left[\frac{\partial\psi}{\partial y}\right] = 0,\ x\in\left(1-s,\ 1+\frac{t}{A}\right) \quad (5.5.32)$$

so that

$$\begin{aligned}\varphi(t, x, y) &= -\varphi(t, x, -y)\\ \psi(t, x, y) &= -\psi(t, x, -y)\end{aligned} \quad (5.5.33)$$

On the same line, from (5.5.24) it follows that:

$$\left[\frac{\partial\varphi}{\partial t}\right] - \frac{1}{A^2}\left[\frac{\partial\psi}{\partial x}\right] = 0$$

so that, taking into account (5.5.16) and (5.5.33) we find:

$$\left(\frac{\partial^2}{\partial t^2} - \frac{1}{A^2}\frac{\partial^2}{\partial x^2}\right)\varphi(t, x, 0+) = 0,\ x\in\left(1-s, 1+\frac{t}{A}\right) \quad (5.5.34)$$

whence

$$\varphi(t, x, 0+) = \mathscr{F}_-\left(x-\frac{t}{A}\right) + \mathscr{F}_+\left(x+\frac{t}{A}\right) \quad (5.5.34')$$

It can be easily checked that the curve C_1 described by the trailing edge is intersected only by the characteristics $(x - t/A) = \text{const}$.

Indeed, the slope of curve C_1 is

$$m_1 = \frac{dt}{dx} = -\frac{1}{V_0(t)} = \frac{\widetilde{V}_0}{V_0^{\text{dim}}(t)}$$

while the slope of the second family curve is $m_2 = -A = -\widetilde{V}_0/V_A$. It follows that:

$$m_1 - m_2 = \widetilde{V}_0 \frac{V_0^{\text{dim}}(t) - V_A}{V_A V_0^{\text{dim}}(t)} < 0$$

since we assumed $A < 1$.

Consequently, we shall write condition (5.5.34′) in the form

$$\frac{\partial \varphi}{\partial x}(t, x, 0+) = f\left(x - \frac{t}{A}\right) \tag{5.5.35}$$

f being a function to be determined later on. For $A \to \infty$ equation (5.5.35) coincides with the corresponding classical aerodynamics equation.

Now, taking into account equations (5.5.33), (5.5.29) and (5.5.35) it follows that the holomorphic function:

$$\frac{df}{dz} = \frac{\partial \varphi}{\partial x} - i\frac{\partial \varphi}{\partial y}$$

will be determined by the following boundary conditions:

$$\frac{\partial \varphi}{\partial x} = 0, \quad -\infty < x < -1 - s(t), \quad 1 + \frac{t}{A} < x < +\infty$$

$$\frac{\partial \varphi}{\partial y} = \dot{h}(t, x), \quad -1 - s < x < 1 - s \tag{5.5.36}$$

$$\frac{\partial \varphi}{\partial x}(t, x, 0+) = f\left(x - \frac{t}{A}\right), \quad 1 - s < x < 1 + \frac{t}{A}$$

This is a Volterra problem whose solution is:

$$\frac{df}{dz} = \frac{i}{\pi}\sqrt{\frac{z+1+s}{z-1+s}}\left\{-\int_{-1}^{+1} \frac{\dot{h}(t, \xi - s)}{z + s - \xi}\sqrt{\frac{1-\xi}{1+\xi}}\, d\xi \right.$$
$$\left. + \int_{1}^{1+s+t/A} \frac{f(\xi - s - t/A)}{z + s - \xi}\sqrt{\frac{\xi - 1}{\xi + 1}}\, d\xi\right\} \tag{5.5.37}$$

The function f is determined from the condition that the velocity circulation on a material curve enclosing the perturbed region at any time should vanish. The following integral equation is obtained [22]:

$$\int_{1}^{1+s+t/A} f\left(\xi - s - \frac{t}{A}\right)\sqrt{\frac{\xi-1}{\xi+1}}\,\mathrm{d}\xi = \int_{-1}^{+1} \dot{h}(t, \xi - s)\sqrt{\frac{1-\xi}{1+\xi}}\,\mathrm{d}\xi \quad (5.5.38)$$

which for $A \to \infty$ reduces itself (leaving aside the K. J. condition) to the (Wagner) equation of classical aerodynamics.
This equation has a unique solution which can be determined by using the Laplace transformation.

5.5.4 The limiting case of steady motion

Let us now suppose that the airfoil is at rest and at time $t = 0$ begins to move with constant velocity $V_0(t)$ of magnitude 1. Assuming that the airfoil does not undergo vibrations we are interested in the solution for $t \to \infty$ (ultimate steady motion). Let

$$y_1 = h(x_1 - t, t) \stackrel{\text{def}}{=} \tilde{h}(x_1)$$

be the equation of the profile with respect to the reference system bound to the airfoil. Let also:

$$\varphi(x_1 - t, y, t) \stackrel{\text{def}}{=} \tilde{\varphi}(x_1, y, t), \cdots$$

Then we have:

$$\dot{h}(x, t) = \frac{\partial \tilde{h}}{\partial x_1} + \frac{\partial \tilde{h}}{\partial t} = \frac{\partial \tilde{h}}{\partial x_1}, \quad \frac{\partial \varphi}{\partial t} = \frac{\partial \tilde{\varphi}}{\partial x_1} + \frac{\partial \tilde{\varphi}}{\partial t}$$

As in the hydrodynamic case from equation (5.5.38) it follows that $f(x_1 - t - t/A) \to \to 0$ for $t \to \infty$ so that (it follows from (5.5.36)) the function $\tilde{\varphi}(x_1, y, t)$ harmonic in the upper half-plane will satisfy the following boundary conditions on the half-plane boundary:

$$\frac{\partial \tilde{\varphi}}{\partial x_1}(x_1, 0, t) = 0, \; |x_1| > 1$$

$$(5.5.39)$$

$$\frac{\partial \tilde{\varphi}}{\partial y}(x_1, 0+, t) = \tilde{h}'(x_1), \; (x_1| < 1$$

The solution of this Volterra-type problem is independent of time. For a plate $(h(x_1) \equiv -\varepsilon x_1)$ we have

$$\frac{\partial\tilde{\varphi}}{\partial x_1} - \mathrm{i}\frac{\partial\tilde{\varphi}}{\partial y} = \mathrm{i}\varepsilon\left\{1 - \sqrt{\frac{z_1+1}{z_1-1}}\right\}, \quad z_1 = x_1 + \mathrm{i}y_1 \tag{5.5.40}$$

Equation (16) becomes:

$$\frac{\partial\tilde{\psi}}{\partial t}(x_1, y, t) + \frac{\partial\tilde{\psi}}{\partial x_1}(x_1, y, t) = \frac{\partial\tilde{\varphi}}{\partial x_1}(x_1, y) \tag{5.5.41}$$

whence

$$\begin{aligned}&\tilde{\psi}(x_1, y, t) = \tilde{\varphi}(x_1, y) + g(x_1 - t, y) \Rightarrow \\ &\psi(x, y, t) = \varphi(x, y, t) + g(x, y)\end{aligned} \tag{5.5.42}$$

Since at $t = 0$ we may suppose that $\psi(x, y, 0) \equiv \varphi(x, y, 0) \equiv 0$ it follows that:

$$\tilde{\psi}(x_1, y, t) \equiv \tilde{\varphi}(x_1, y) \tag{5.5.43}$$

Then equations (5.5.25) will determine the non-potential part of the motion in the form:

$$\begin{aligned}&F_1(-\infty, \tilde{h}(x_1)) = (A-1)\frac{\partial\tilde{\varphi}}{\partial x_1}(x_1, 0) \\ &G_1(-\infty, \tilde{h}(x_1)) = (A+1)\frac{\partial\tilde{\varphi}}{\partial x_1}(x_1, 0) \\ &F_2(-\infty, \tilde{h}(x_1)) = G_2(-\infty, \tilde{h}(x_1)) = 0\end{aligned} \tag{5.5.44}$$

Particularly, for the case of a plate the rotational part of component v_x downstream the airfoil will be:

$$-\frac{1}{A}F_1 = \varepsilon\left(1 - \frac{1}{A}\right)\sqrt{\frac{1+x_1}{1-x_1}} \tag{5.5.45}$$

5.6 Steady motion (of incompressible fluids) in aligned fields

In this section we shall present Homentcovschi's solution [27] of the incompressible fluid motion for aligned fields. As mentioned already this solution is derived by using the linearized equations and the non-linearized boundary conditions. This solution is justified by the fact that it is obtained as a limiting case (for $t \to \infty$) from the unsteady solution given above.

Thus for aligned fields ($\alpha_x = 1, \alpha_y = 0$) equations (5.2.5)–(5.2.9) have (for $A < 1$) the following solution [13]:

$$v_y = \frac{\partial \varphi}{\partial y}, \quad v_x = \frac{\partial \varphi}{\partial x} + \begin{cases} F_-(y) \text{ in } D_- \\ F_+(y), \text{ in } D_+ \end{cases}$$

$$b_y = \frac{\partial \varphi}{\partial y}, \quad b_x = \frac{\partial \varphi}{\partial x} + \begin{cases} AF_-(x) \text{ in } D_- \\ -AF_+(y) \text{ in } D_+ \end{cases} \tag{5.6.1}$$

$$p = -\frac{\partial \varphi}{\partial x} + \begin{cases} -A^{-1} F_-(y) \text{ in } D_- \\ A^{-1} F_+(y) \text{ in } D_+ \end{cases}$$

$\varphi(x, y)$ being a harmonic function within the fluid. $F_\mp$ are arbitrary functions. The representation (5.6.1) is also valid outside the domains D_-, D_+ where the functions $F_\mp$ vanish.

If we now denote by:

$$y = \varepsilon h(x), \ |x| \leqslant 1 \tag{5.6.2}$$

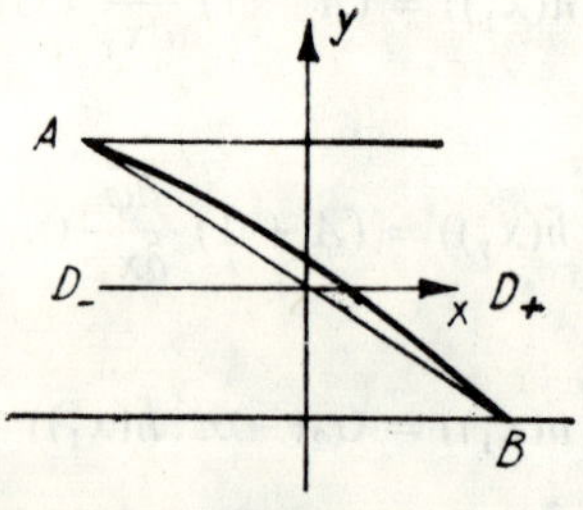

Fig. 22

the equation of the airfoil arc AB (Figure 22) the boundary condition (2.21) becomes:

$$\left(1 + \left.\frac{\partial \varphi}{\partial x}\right|_\pm + F_\pm\right) \varepsilon h' - \left.\frac{\partial \varphi}{\partial y}\right|_\pm = 0, \text{ for } y = \varepsilon h, |x| \leqslant 1 \tag{5.6.3}$$

and the conditions of magnetic field continuity become:

$$\left.\frac{\partial \varphi}{\partial x}\right|_{+} - \left.\frac{\partial \varphi}{\partial x}\right|_{-} - A(F_{+} + F_{-}) = 0 \tag{5.6.4}$$

$$\left.\frac{\partial \varphi}{\partial y}\right|_{-} - \left.\frac{\partial \varphi}{\partial y}\right|_{-} = 0, \text{ for } y = \varepsilon h, |x| \leqslant 1 \tag{5.6.5}$$

where as usual by the (+) and (−) sign we denote the limit values for $y = \varepsilon h$ in D_{+} and D_{-}, respectively.

From equation (5.6.3) one has:

$$F_{\pm} = \frac{1}{\varepsilon h'} \left.\frac{\partial \varphi}{\partial y}\right|_{\pm} - i - \left.\frac{\partial \varphi}{\partial x}\right|_{\pm} \tag{5.6.6}$$

and from (5.6.4) and (5.6.5):

$$(1 + A)\,\varepsilon h' \left.\frac{\partial \varphi}{\partial x}\right|_{+} - (1 - A)\,\varepsilon h' \left.\frac{\partial \varphi}{\partial x}\right|_{-} - 2A \left.\frac{\partial \varphi}{\partial y}\right|_{+} = -\, 2A\,\varepsilon h', \tag{5.6.7}$$

$$|x| \leqslant 1$$

It follows that we have to determine a function φ harmonic in the fluid which satisfies the condition (5.6.7) on the arc AB.

In order to solve this problem one substitutes the arc AB by its chord and assumes that the latter makes an angle $-\alpha$ with the Ox-axis. One now considers another reference frame XOY whose OX axis is coincident with the chord AB. The transformation from the reference frame xOy to XOY is performed through the equation:

$$Z = ze^{i\alpha},\ Z = X + iY,\ z = x + iy \tag{5.6.8}$$

By this transformation the holomorphic function:

$$f(z) = \frac{\partial \varphi}{\partial x} - i\frac{\partial \varphi}{\partial y} \tag{5.6.9}$$

becomes:

$$F(Z) = U + iV \tag{5.6.9'}$$

a holomorphic function in the plane (Z) apart from the segment (−1, +1). We have:

$$U(X, \pm 0) = \left.\frac{\partial \varphi}{\partial x}\right|_{\pm},\quad V(X, \pm 0) = -\left.\frac{\partial \varphi}{\partial y}\right|_{\pm} \tag{5.6.9''}$$

Due to condition (5.6.5) function $V(X, Y)$ is continuous across the segment $(-1, +1)$ so that we have:

$$U(X, Y) = -U(X, -Y) \tag{5.6.10}$$

From this equation and from the continuity condition of function U on the real axis outside the segment $(-1, +1)$ we have:

$$U(X, 0) = 0, \ |X| > 1 \tag{5.6.11}$$

Taking into account (5.6.10), the boundary condition (5.6.7) becomes:

$$\varepsilon h'(X)\, U(X, +0) + AV(X, +0) = -A\varepsilon h'(X), \ |X| \leqslant 1 \tag{5.6.12}$$

Therefore we have to determine a function $F(Z)$ holomorphic in the (Z)-plane apart from the segment $(-1, +1)$, which vanishes at infinity and satisfies conditions (5.6.11) and (5.6.12) on the real axis. This is a Hilbert type problem with discontinuous coefficients.

The homogeneous problem has two solutions:

$$F_1(Z) = \mathrm{i}\Gamma \exp\left\{\frac{1}{\pi}\int_{-1}^{+1} \operatorname{arc\,tg}\left(\frac{A}{\varepsilon h'(\xi)}\right)\frac{\mathrm{d}\xi}{\xi - Z}\right\}$$

$$F_2(Z) = \mathrm{i}\Gamma \exp\left\{\frac{1}{\pi}\int_{-1}^{+1} \left\{\operatorname{arc\,tg}\left(\frac{A}{\varepsilon h'(\xi)}\right) + \pi\right\}\frac{\mathrm{d}\xi}{\xi - Z}\right\} \tag{5.6.13}$$

$$= F_1(Z)\left(\frac{Z-1}{Z+1}\right)$$

The inhomogeneous problem has the following general solution:

$$F(Z) = AF_0(Z)\frac{\mathrm{i}}{\pi}\int_{-1}^{+1} \frac{\varepsilon h'(\xi)}{\{\varepsilon h'(\xi) - \mathrm{i}A\}\, F_0(\xi)}\frac{\mathrm{d}\xi}{\xi - Z} \tag{5.6.14}$$

where F_0 is either F_1 or F_2. For $F_0 = F_1$ the K.J. condition will be fulfilled on the leading edge while for $F_0 = F_2$ the K.J. condition will be fulfilled on the trailing edge. Choosing $F_0 = F_1$ on the trailing edge we have an (integrable) singularity weaker than the one which would be obtained on the leading edge if $F_0 = F_2$ were chosen. This is the reason for choosing in (5.6.14) $F_0 = F_1$.

For the flat plate ($y = -\varepsilon x$) we have:

$$F_1(Z) = \mathrm{i}\Gamma\left(\frac{Z+1}{Z-1}\right)^{\theta}, \quad F(Z) = \mathrm{i}\varepsilon\left\{1 - \left(\frac{Z+1}{Z-1}\right)^{\theta}\right\}$$

$$f(z) = \mathrm{i}\varepsilon\left\{1 - \left(\frac{z+\mathrm{e}^{-\mathrm{i}\alpha}}{z-\mathrm{e}^{-\mathrm{i}\alpha}}\right)^{\theta}\right\}$$

where

$$\theta\pi = \operatorname{arc\,tg}(A/\varepsilon), \quad 0 < \theta < 1/2$$

We also have:

$$[p] = 2\left(1 - \frac{1}{A^2}\right)\frac{A\varepsilon}{\sqrt{\varepsilon^2 + A^2}}\left(\frac{X+1}{X-1}\right)^{\theta}$$

$$L = -\left(1 - \frac{1}{A^2}\right)4\varepsilon\theta\pi$$

Expanding the general solution in a series with respect to ε we get

$$F_0 = \mathrm{i}\Gamma\left(\frac{Z+1}{Z-1}\right)^{1/2} + O(\varepsilon)$$

$$F(Z) = \varepsilon\left(\frac{Z+1}{Z-1}\right)^{1/2}\frac{1}{\pi}\int_{-1}^{+1}\left(\frac{1-\xi}{1+\xi}\right)^{1/2}\frac{h'(\xi)\,\mathrm{d}\xi}{Z-\xi} + O(\varepsilon^2)$$

$$V(X, +0) = \varepsilon h'(X) + O(\varepsilon^2)$$

$$\left.\frac{\partial\varphi}{\partial y}\right|_+ = \varepsilon h'(x) + O(\varepsilon^2), \quad \left.\frac{\partial\varphi}{\partial x}\right|_+ = O(\varepsilon)$$

$$F_+ = \frac{O(\varepsilon^2)}{\varepsilon h'(x)} - O(\varepsilon) = O(\varepsilon).$$

It follows that the rotational part of the motion is also of the order of ε this fact justifying the use of the linearized theory. One also notices that in the first-order approximation (with respect to ε) of the non-potential part of the solution, the second-order approximation of the potential part appears. This means that using the linearized boundary conditions theory we can determine the potential part of the solution (and hence the lift) but not the non-potential part. For the plate the rotational part following from (5.6.6) will be:

$$F_+ = \varepsilon\left(1 - \frac{1}{A}\right)\sqrt{\frac{1+x}{1-x}} + O(\varepsilon^2)$$

One should notice that it coincides with (5.5.45).

For $A > 1$, Alfvén waves will exist only behind the airfoil (within D_+) so that the solution of the equations of motion has the form [13]:

$$v_y = \frac{\partial\varphi}{\partial y},\quad v_x = \frac{\partial\varphi}{\partial x} + \begin{cases} F_-(y) + F_+(y) & \text{in } D_+ \\ 0 & \text{in } D_- \end{cases}$$

$$b_y = \frac{\partial\varphi}{\partial y},\quad b_x = \frac{\partial\varphi}{\partial x} + \begin{cases} AF_-(y) - AF_+(y) & \text{in } D_+ \\ 0 & \text{in } D_- \end{cases} \tag{5.6.15}$$

$$p = -\frac{\partial\varphi}{\partial x} + \begin{cases} -A^{-1}F_-(y) + A^{-1}F_+(y) & \text{in } D_+ \\ 0 & \text{in } D_- \end{cases}$$

$F_{\mp}(y)$ vanishing within $D \cup D_-$ and φ being harmonic within the fluid.

The boundary conditions yield the following equations:

$$\left(1 + F_- + F_+ + \left.\frac{\partial\varphi}{\partial x}\right|_+\right)\varepsilon h' - \left.\frac{\partial\varphi}{\partial y}\right|_+ = 0 \tag{5.6.16}$$

$$\left(1 + \left.\frac{\partial\varphi}{\partial x}\right|_-\right)\varepsilon h' - \left.\frac{\partial\varphi}{\partial y}\right|_- = 0 \tag{5.6.17}$$

$$A(F_- - F_+) + \left.\frac{\partial\varphi}{\partial x}\right|_+ = \left.\frac{\partial\varphi}{\partial x}\right|_- \tag{5.6.18}$$

$$\left.\frac{\partial\varphi}{\partial y}\right|_+ = \left.\frac{\partial\varphi}{\partial y}\right|_- \tag{5.6.19}$$

From (5.6.16) and (5.6.18) we determine the non-potential part of the solution as a function of the potential one and from the (5.6.17) and (5.6.19) we determine, as shown above, the potential part. With the definitions of (5.6.9) and (5.6.9′) in this case we have:

$$F(Z) = F_0(Z)\frac{i}{\pi}\int_{-1}^{+1}\frac{\varepsilon h'(\xi)}{\{\varepsilon h'(\xi)+i\}\,F_0(\xi)}\frac{d\xi}{Z-\xi}$$

$$F_0 = i\Gamma \exp\left\{\frac{1}{\pi}\int_{-1}^{+1}\operatorname{arc\,tg}\left(\frac{1}{\varepsilon h'(\xi)}\right)\frac{d\xi}{Z-\xi}\right. \tag{5.6.20}$$

The K.J. condition being fulfilled on the trailing edge.

For lift we have:

$$[p] = \left(1-\frac{1}{A^2}\right)\left(\frac{\partial\varphi}{\partial x}\bigg|_+ - \frac{\partial\varphi}{\partial x}\bigg|_-\right)$$

For the flat plate we obtain:

$$F(Z) = i\varepsilon\left\{1-\left(\frac{Z-1}{Z+1}\right)^{\theta}\right\},\ \operatorname{tg}\theta\pi = \varepsilon^{-1},\ 0<\theta<\frac{1}{2}$$

$$L = \left(1-\frac{1}{A^2}\right)4\varepsilon\theta\pi$$

5. 7 Potential representation of the solution. Three-dimensional flow

5.7.1 Potential representation of the solution

In three-dimensional problems it is rather difficult to reduce the system of equations of motion to a single equation. However we can find a single equation equivalent to the basic system if we represent the general solution by potentials. The first representation of this kind is due to S. Ando [28]. We present here with some changes this representation.

Since the representation is also valid for finite electrical conductivity fluids we shall consider the equations of motion for such a fluid (compressible fluid in unsteady motion).

$$M^2\dot{p} + \operatorname{div}\boldsymbol{v} = 0 \quad (\rho = M^2 p) \tag{5.7.1}$$

$$\dot{\boldsymbol{v}} = -\operatorname{grad} p + A^{-2}\operatorname{curl}\boldsymbol{b}\times\boldsymbol{\alpha} \tag{5.7.2}$$

$$\operatorname{div}\boldsymbol{b} = 0,\ \operatorname{curl}\boldsymbol{e} = -\partial_t\boldsymbol{b} \tag{5.7.3}$$

$$\operatorname{curl}\boldsymbol{b} = Rm(\boldsymbol{e} + \boldsymbol{i}_1\times\boldsymbol{b} + \boldsymbol{v}\times\boldsymbol{\alpha}) \tag{5.7.4}$$

where due to the linearization

$$\cdot = \frac{\mathrm{d}}{\mathrm{d}t} = \frac{\partial}{\partial t} + \frac{\partial}{\partial x} \tag{5.7.5}$$

Equation (5.7.4) is Ohm's law for finite electrical conductivity fluids (Chapter 7).
Applying the *curl* operator in (5.7.2) we have:

$$\frac{\mathrm{d}}{\mathrm{d}t}\,\mathrm{curl}\,\boldsymbol{v} = \frac{1}{A^2}\frac{\mathrm{d}}{\mathrm{d}\alpha}\,\mathrm{curl}\,\boldsymbol{b}, \quad \frac{\mathrm{d}}{\mathrm{d}\alpha} = \boldsymbol{\alpha}\cdot\nabla\cdot \tag{5.7.6}$$

The pressure can be eliminated between (5.7.1) and (5.7.2). We thus obtain:

$$M^2\frac{\mathrm{d}^2\boldsymbol{v}}{\mathrm{d}t^2} = \mathrm{grad}\,\mathrm{div}\,\boldsymbol{v} + \frac{M^2}{A^2}\frac{\mathrm{d}}{\mathrm{d}t}(\mathrm{curl}\,\boldsymbol{b})\times\boldsymbol{\alpha}$$

whence applying the *div* operator it follows that:

$$\left(\Delta - M^2\frac{\mathrm{d}^2}{\mathrm{d}t^2}\right)\mathrm{div}\,\boldsymbol{v} = -\frac{M^2}{A^2}\boldsymbol{\alpha}\cdot\mathrm{curl}\frac{\mathrm{d}\boldsymbol{b}}{\mathrm{d}t} = \frac{M^2}{A^2}\frac{\mathrm{d}}{\mathrm{d}t}\Delta(\boldsymbol{\alpha}\cdot\boldsymbol{b}) \tag{5.7.7}$$

Finally, applying the *curl* operator to (5.7.4) and taking into account (5.7.3) we have:

$$\frac{\mathrm{d}\boldsymbol{b}}{\mathrm{d}t} = \frac{\mathrm{d}\boldsymbol{v}}{\mathrm{d}\alpha} - \boldsymbol{\alpha}\,\mathrm{div}\,\boldsymbol{v} + \frac{1}{Rm}\Delta\boldsymbol{b}, \quad \Delta = \frac{\partial^2}{\partial x^2} + \frac{\partial^2}{\partial y^2} + \frac{\partial^2}{\partial z^2} \tag{5.7.8}$$

which is the linearized magnetic induction equation.

Equations (5.7.6) and (5.7.7) together with the equation obtained by applying the *curl* operator to (5.7.8) determine once again [G. 24] the equations satisfied by the unknowns *div* $\boldsymbol{v}$, *curl* $\boldsymbol{v}$ and *curl* $\boldsymbol{b}$. The equation for *curl* $\boldsymbol{b}$ was derived for the first time by E. L. Resler jr., J.E. Mc Cune [6.20]. For plane motion the quantities *div* $\boldsymbol{v}$, *curl* $\boldsymbol{v}$ and *curl* $\boldsymbol{b}$ satisfy the same equation.

In obtaining Ando's representation one starts from equations (5.7.6), (5.7.7) and (5.7.8). In this respect one notices that equation (5.7.6) determines the existence of a potential φ so that:

$$\frac{\mathrm{d}\boldsymbol{v}}{\mathrm{d}t} = \frac{1}{A^2}\frac{\mathrm{d}\boldsymbol{b}}{\mathrm{d}\alpha} + \frac{\mathrm{d}}{\mathrm{d}t}\,\mathrm{grad}\,\varphi \tag{5.7.9}$$

Using this relationship we shall eliminate $\boldsymbol{v}$ from (5.7.8). One thus obtains:

$$\left(\frac{\mathrm{d}^2}{\mathrm{d}t^2} - \frac{1}{A^2}\frac{\mathrm{d}^2}{\mathrm{d}\alpha^2} - \frac{1}{Rm}\Delta\frac{\mathrm{d}}{\mathrm{d}t}\right)\boldsymbol{b} = \left(\frac{\mathrm{d}}{\mathrm{d}\alpha}\,\mathrm{grad} - \boldsymbol{\alpha}\Delta\right)\frac{\partial\varphi}{\mathrm{d}t}$$

which equation determines the existence of a generalized potential $\Phi(t, \boldsymbol{x})$ such that:

$$\boldsymbol{b} = \left(\frac{\mathrm{d}}{\mathrm{d}\alpha}\,\mathrm{grad} - \boldsymbol{\alpha}\Delta\right)\frac{\partial \Phi}{\mathrm{d}t}$$

$$\varphi = \left(\frac{\mathrm{d}^2}{\mathrm{d}t^2} - \frac{1}{A^2}\frac{\mathrm{d}^2}{\mathrm{d}\alpha^2} - \frac{1}{Rm}\Delta\frac{\mathrm{d}}{\mathrm{d}t}\right)\Phi \tag{5.7.10}$$

The equation satisfied by the potential Φ is obtained from (5.7.7). In this respect from (5.7.9) and (5.7.10) we obtain:

$$\frac{\mathrm{d}}{\mathrm{d}t}\,\mathrm{div}\,\boldsymbol{v} = \frac{\mathrm{d}}{\mathrm{d}t}\Delta\varphi,\ \boldsymbol{\alpha}\cdot\boldsymbol{b} = \left(\frac{\mathrm{d}^2}{\mathrm{d}\alpha^2} - \Delta\right)\frac{\mathrm{d}\Phi}{\mathrm{d}t}$$

so that (5.7.7) becomes:

$$T\Phi = 0 \tag{5.7.11}$$

where

$$T = \left(\Delta - M^2\frac{\mathrm{d}^2}{\mathrm{d}t^2}\right)\left(\frac{\mathrm{d}^2}{\mathrm{d}t^2} - \frac{1}{Rm}\Delta\frac{\mathrm{d}}{\mathrm{d}t}\right) + \frac{1}{A^2}\left(M^2\frac{\mathrm{d}^2}{\mathrm{d}t^2} - \frac{\mathrm{d}^2}{\mathrm{d}\alpha^2}\right)\Delta$$

Equation (5.7.11) will determine the potential Φ.

The operator T is of the fifth order for the finite electrical conductivity fluid and of the fourth order for the infinite conductivity fluid. For the latter case ($Rm \to \infty$) the operator reduces to the operators of Sections 5.2 and 5.3 for steady motions and to the operator of Section 5.5 for the unsteady motion. For the finite conductivity fluid it reduces to some operators to be pointed out in the next chapter (Sections 6.2 and 6.3).

Using (5.7.9) we can also represent the velocity field by means of the potential Φ. One has:

$$\boldsymbol{v} = \left\{\left(\frac{\mathrm{d}^2}{\mathrm{d}t^2} - \frac{1}{Rm}\Delta\frac{\mathrm{d}}{\mathrm{d}t}\right)\mathrm{grad} - \frac{1}{A^2}\boldsymbol{\alpha}\frac{\mathrm{d}}{\mathrm{d}\alpha}\Delta\right\}\Phi \tag{5.7.12}$$

Finally the expression for pressure is obtained from equation (5.7.2) written in the form

$$\mathrm{grad}\left\{p + \frac{\mathrm{d}\varphi}{\mathrm{d}t} + \frac{1}{A^2}(\boldsymbol{\alpha}\cdot\boldsymbol{b})\right\} = 0$$

whence

$$p = -\frac{\mathrm{d}\varphi}{\mathrm{d}t} - \frac{1}{A^2}(\boldsymbol{\alpha}\cdot\boldsymbol{b}) = \left(\frac{1}{Rm}\Delta\frac{\mathrm{d}}{\mathrm{d}t} + \frac{1}{A^2}\Delta - \frac{\mathrm{d}^2}{\mathrm{d}t^2}\right)\frac{\mathrm{d}\varphi}{\mathrm{d}t} \tag{5.7.13}$$

In the papers quoted above Ando considers various particular cases of the above solution.

5.7.2 Three-dimensional steady motion of perfectly conducting fluids

(a) *Aligned fields.* For this case the solution follows easily from the equations of motion like in Section 5.3. So, from the projection of equation (5.7.2) on the Ox-axis it follows that $p = -v_x$, and from the projection of (5.7.8) on the Oy and Oz axes:

$$b_y = v_y, \quad b_z = v_z \tag{5.7.14}$$

Projecting equation (5.7.8) on the Ox-axis and taking into account (5.7.1) it follows that $b_x = (1 - M^2)\, v_x$. Therefore, once again the magnetic induction is expressed by means of the velocity field.

Projecting equation (5.7.4) (with the hypothesis $Rm = \infty$) on the reference system axes and taking into account (5.7.14) it follows that $e_x = e_y = e_z = 0$.

Projecting equation (5.7.6) on the Ox-axis and taking into account (5.7.14) it follows (except for $A = 1$):

$$\frac{\partial v_z}{\partial y} - \frac{\partial v_y}{\partial z} = 0 \tag{5.7.15}$$

whence

$$v_y = \frac{\partial \varphi}{\partial y}, \quad v_z = \frac{\partial \varphi}{\partial z}, \quad \varphi = \varphi(x, y, z) \tag{5.7.16}$$

Finally, from the projections of equation (5.7.6) on the Oy and Oz axes it follows that

$$v_x = \frac{A^2 - 1}{A^2 + M^2 - 1} \frac{\partial \varphi}{\partial x} \tag{5.7.17}$$

so that equation (5.7.1) reduces itself to

$$\frac{(1 - M^2)(A^2 - 1)}{A^2 + M^2 - 1} \frac{\partial^2 \varphi}{\partial x^2} + \frac{\partial^2 \varphi}{\partial y^2} + \frac{\partial^2 \varphi}{\partial z^2} = 0 \tag{5.7.18}$$

This equation was pointed out in [29].

The analogy between this equation and the Glauert-Prandtl and Ackeret equation of the non-conducting fluids is obvious. We should however point out that this time the fluid motion is not irrotational like it is for the non-conducting fluids. For

$A \to \infty$ we obtain from (5.7.16) – (5.7.18) the classical solution (in this case the motion is irrotational).

The above proof once more demonstrates that the motion of a non-conducting fluid is to the first approximation necessarily irrotational. In the classical aerodynamics this is usually introduced as a hypothesis.

(b) *Orthogonal fields.* We shall use in this case the representation (5.7.10) – (5.7.13). With the usual simplifications:

$$Rm = \infty, \ \boldsymbol{\alpha} = \boldsymbol{i}_2, \ \frac{\mathrm{d}}{\mathrm{d}\alpha} = \frac{\partial}{\partial y}, \ \frac{\mathrm{d}}{\mathrm{d}t} = \frac{\partial}{\partial x}$$

the operator T reduces itself to:

$$(A^2M^2 - A^2 - M^2)\frac{\partial^4}{\partial x^4} - (A^2 + M^2 - 1)\frac{\partial^4}{\partial x^2 \partial y^2} + \frac{\partial^4}{\partial y^4}$$

$$- (A^2 + M^2)\frac{\partial^4}{\partial x^2 \partial z^2} + \frac{\partial^4}{\partial y^2 \partial z^2} \tag{5.7.19}$$

and the solution takes the form:

$$v_x = \frac{\partial^3 \Phi}{\partial x^3}, \ v_y = \left(\frac{\partial^2}{\partial x^2} - \frac{1}{A^2}\Delta\right)\frac{\partial \Phi}{\partial y}, \ v_z = \frac{\partial^3 \Phi}{\partial x^2 \partial z}$$

$$b_x = \frac{\partial^3 \Phi}{\partial x^2 \partial y}, \ b_y = \left(\frac{\partial^2}{\partial y^2} - \Delta\right)\frac{\partial \Phi}{\partial x}, \ b_z = \frac{\partial^3 \Phi}{\partial x \partial y \partial z} \tag{5.7.20}$$

$$p = -\left(\frac{\partial^2}{\partial x^2} - \frac{1}{A^2}\Delta\right)\frac{\partial \Phi}{\partial x}$$

It is of interest to specify the type of the operator T. In this respect we shall notice that:

$$a = A^2M^2 - A^2 - M^2, \ b = A^2 + M^2 - 1, \ c = A^2 + M^2 \tag{5.7.21}$$

We also denote by $G(x, y, z)$ the equation of a characteristic surface and by:

$$s = \frac{\partial G}{\partial x}, \ \beta = \frac{\partial G}{\partial y}, \ \gamma = \frac{\partial G}{\partial z} \tag{5.7.22}$$

the components of its normal. With these specifications the equation of the characteristic equation associated to operator T reduces itself to:

$$E(s) = as^4 - (b\beta^2 + c\gamma^2)s^2 + \beta^2(\beta^2 + \gamma^2) = 0 \tag{5.7.23}$$

The general theory of the partial derivative equations (Section 10.1) shows that if (for any real β and γ) s is real then operator T has real characteristics and thus is completely hyperbolic.

Or, the discriminant of the biquadratic equation (5.7.23) is positive since we have:

$$(b\beta^2 + c\gamma^2)^2 - 4a\beta^2(\beta^2 + \gamma^2) = (b^2 - 4a)\beta^4 + 2(bc - 2a)\beta^2\gamma^2 + c^2\gamma^4$$

$$b^2 - 4a = 1 + 2(A^2 + M^2) + (A^2 - M^2)^2 > 0$$

$$bc - 2a = A^2 + M^2 + A^4 + M^4 > 0$$

Consequently, from (5.7.23) it follows that s^2 is real for any real β and γ. From (5.7.23) it follows as well:

$$as_1^2 s_2^2 = \beta^2(\beta^2 + \gamma^2)$$

This relationship shows that within the domain $\mathcal{D}_1$ of Figure 17 ($a > 0$) both roots s_1^2 and s_2^2 are either positive or negative while within $\mathcal{D}_2$ ($a < 0$) one root is positive and the other is negative. In order to prove that both roots have positive signs in $\mathcal{D}_1$ it is sufficient to take some values of β and γ. For instance for $\beta = 0$ equation (5.7.23) reduces itself to $as^4 - c\gamma^2 s^2 = 0$ with the roots $s_1^2 = 0$, $s_2^2 > 0$.

Consequently, for this case as well as for the plane motion there exist at any point in $\mathcal{D}_1$ two pairs of real characteristic surfaces (T is a double-hyperbolic operator) while in $\mathcal{D}_2$ at any point there exists only one pair of real characteristic surfaces (T is an ellipso-hyperbolic operator).

Finally, we have noticed in [29] that if in free flow the relationship:

$$V_0 = V_A^2 \sqrt{\frac{1}{a_0^2} + \frac{1}{V_A^2}} \tag{5.7.24}$$

satisfied then operator T may be expanded in a product of two operators

$$T = T_1 T_2 \tag{5.7.25}$$

$$T_1 = \frac{\partial^2}{\partial x^2} - \frac{1}{s^2}\frac{\partial^2}{\partial y^2}, \quad T_2 = \frac{a}{s^2}\frac{\partial^2}{\partial x^2} - \frac{\partial^2}{\partial y^2} - \frac{\partial^2}{\partial z^2}, \quad s^2 = A^2 + \frac{M^2}{A^2 + 1}$$

which fact makes easier the determination of the solution.

6

Theory of thin airfoils in fluids of finite electrical conductivity

6.1 Introduction

The study of fluids with finite electrical conductivity is nowadays of particular importance since, as already noticed in the previous chapter, the fluid electrical resistivity cannot be disregarded in actual problems of aerodynamics.

The first paper which takes into account the electromagnetic dissipating effects is due to J. McCune [1]. However his solution is valid only for high Reynolds magnetic numbers (incompressible fluid under the assumption of orthogonal fields). The general problem of finite electrical conductivity fluids (arbitrary Reynolds magnetic number) was dealt with independently by E. C. Lary [2] and L. Dragoş [3] for aligned fields and by K. Stewarston [4] and L. Dragoş [5] for orthogonal fields. A general investigation of the problem in oblique fields was carried out by L. Dragoş [6]. This is presented in section 6.2.

The motion of finite electrical conductivity compressible fluids past thin airfoils was dealt with by L. Dragoş for aligned fields [7] and for orthogonal fields [8]. In these papers the solution was expressed by generalized functions. Proceeding from the observations of W. R. Sears and E. L. Resler jr. [9], J. Tang and R. Seebass [10] found the solution of the problem of symmetrical airfoils in aligned fields. D. Fan and G.S.S. Ludford [11] reconsider the general problem studied by L. Dragoş [7], [8] and make two important remarks which simplify and complete the solution.

Finally, L. Dragoş reanalyses the problem for the case of aligned fields [12] and for the orthogonal fields [13] and expresses the solutions by means of classical functions. Using concepts of the generalized function theory, the principal part of the solution (coincident with that of classical aerodynamics) is pointed out [12], [13] and the boundary problem is reduced to an integral equation of the Fredholm type. Sections 6.3 and 6.4 are based on papers [12] and [13], respectively.

In connection with the general theory of aligned fields, the theory of quasi-aligned fields has also been developed (the projection of the free magnetic field on the plane of motion is aligned with the free stream velocity). The first works in this field are due to R. Thibault [14], [15]. The problem is reconsidered by L. Dragoş and D. Homentcovschi [16] and D. Ivănescu in [17].

6.2 Incompressible fluids in oblique fields

Since the problem is the same as in Chapter 5, equations (5.2.5), (5.2.6), (5.2.8) and (5.2.9) are still valid. However, equation (5.2.7) is replaced by the following equation:

$$\operatorname{curl} \boldsymbol{b} = R(\boldsymbol{i}_1 \times \boldsymbol{b} + \boldsymbol{v} \times \boldsymbol{\alpha}), \; R \overset{\text{def}}{=} Rm \tag{6.2.1}$$

which expresses Ohm's law under the assumption of a fluid with finite electrical conductivity fluid. On applying the *curl* operator from (6.2.1) we get:

$$\Delta \boldsymbol{b} = R\left(\frac{\partial \boldsymbol{b}}{\partial x} - \frac{\mathrm{d}\boldsymbol{v}}{\mathrm{d}\alpha}\right) \tag{6.2.1'}$$

From this equation and from (5.2.10) we have:

$$T\Delta(v_y, b_y) = 0 \tag{6.2.2}$$

where the operator T has this time the expression:

$$T = \left(\Delta - R\frac{\partial}{\partial x}\right)\frac{\partial}{\partial x} + RS\frac{\mathrm{d}^2}{\mathrm{d}\alpha^2}, \; S \overset{\text{def}}{=} \frac{1}{A^2} \tag{6.2.2'}$$

T is a third-order operator also including second-order derivatives. This form is a result of the fact that the perturbation waves propagate with dispersion. Mathematically, this means that not any plane wave function is a solution of operator (6.2.2'). Only the exponential functions have this property. Accordingly, we shall look for the solution of the operator T of the form $\exp(-i\lambda x + sy)$, $s = -i\lambda r$, λ being a real parameter. By using the expression of the operator T to determine r we get the following dispersion equation:

$$i\lambda(r^2 + 1) + R - RS(\alpha_x^2 + 2\alpha_x\alpha_y r + \alpha_y^2 r^2) = 0 \tag{6.2.3}$$

whence:

$$r_{\pm} = \frac{\alpha_x\alpha_y RS \pm |\lambda| a}{i\lambda - \alpha_y^2 RS} \tag{6.2.3'}$$

with the notation:

$$a = \left\{1 + \frac{R}{i\lambda}(1 - S) + \frac{R^2 S}{\lambda^2}\alpha_y^2\right\}^{1/2} = a_1 + ia_2 \operatorname{sign} \lambda$$

$$\sqrt{2}\, a_1 = \left\{\sqrt{\left(1 + \frac{R^2 S}{\lambda^2}\alpha_y^2\right)^2 + \frac{R^2}{\lambda^2}(1 - S)^2} + 1 + \frac{R^2 S}{\lambda^2}\alpha_y^2\right\}^{1/2}$$

$$\sqrt{2}\, a_2 = \left\{\sqrt{\left(1 + \frac{R^2 S}{\lambda^2}\alpha_y^2\right)^2 + \frac{R^2}{\lambda^2}(1 - S)^2} - 1 - \frac{R^2 S}{\lambda^2}\alpha_y^2\right\}^{1/2}$$

In order to construct the general solutions some considerations on the roots r_+ and r_- are necessary. Firstly, we observe that $\mathscr{R}e\, s_+$ is negative while $\mathscr{R}e\, s_-$ is positive for any real value of the parameter λ (for any R, S, α_x and α_y). Indeed, we have:

$$s_+ = -\lambda^2 \frac{\alpha_x\alpha_y RS + |\lambda| a_1 + \alpha_y^2 RS a_2}{\lambda^2 + \alpha_y^4 R^2 S^2} + \mathrm{i}\lambda \frac{\cdots}{\cdots}$$

$$s_- = -\lambda^2 \frac{\alpha_x\alpha_y RS - |\lambda| a_1 - \alpha_y^2 RS a_2}{\lambda^2 + \alpha_y^4 R^2 S^2} + \mathrm{i}\lambda \frac{\cdots}{\cdots}$$

and

$$(\alpha_x\alpha_y RS + |\lambda| a_1 + \alpha_y^2 RS a_2)(\alpha_x\alpha_y RS - |\lambda| a_1 - \alpha_y^2 RS a_2) < 0$$

(for aligned and orthogonal fields these formulae are simple).

From this observation we may conclude that the solutions corresponding to the root s_+ are damping waves for $y \to +\infty$ while those corresponding to the root s_- for $y \to -\infty$. A second remark refers to the behaviour of the roots r_+, r_- for large λ. We have:

$$r_\pm = -\mathrm{i}\,\mathrm{sign}\,\lambda + \frac{\alpha_x\alpha_y RS}{\mathrm{i}\lambda} \pm \frac{Rk}{2|\lambda|} + O\left(\frac{R^2}{\lambda^2}\right) \tag{6.2.4}$$

where:

$$k = 1 + S(\alpha_y^2 - \alpha_x^2) \tag{6.2.4'}$$

The general solution of equation (6.2.2) is obtained by a continuous superposition of plane waves (the operator T being linear) and the addition of a harmonic function. This representation is not valid for the Alfvén flow ($S = 1$, $\alpha_x = 1$, $\alpha_y = 0$) since in this case operators T and Δ are no longer independent. A study of this case is presented in [18] (see also Section 7.3). Accordingly, we have:

$$\begin{aligned} v_y^\pm(x, y) &= \int_{-\infty}^{+\infty} A_\pm(\lambda) \exp(-\mathrm{i}\lambda x + s_\pm y)\,\mathrm{d}\lambda + \frac{\partial\varphi}{\partial y}, \quad \Delta\varphi = 0 \\ b_y^\pm(x, y) &= \int_{-\infty}^{+\infty} B_\pm(\lambda) \exp(-\mathrm{i}\lambda x + s_\pm y)\,\mathrm{d}\lambda + \frac{\partial\psi}{\partial y}, \quad \Delta\psi = 0 \end{aligned} \tag{6.2.5}$$

From equations (5.2.8) and (5.2.9) we get:

$$\begin{aligned} v_x^\pm(x, y) &= -\int_{-\infty}^{+\infty} r_\pm A_\pm \exp(-\mathrm{i}\lambda x + s_\pm y)\,\mathrm{d}\lambda + \frac{\partial\varphi}{\partial x} \\ b_x^\pm(x, y) &= -\int_{-\infty}^{+\infty} r_\pm B_\pm \exp(-\mathrm{i}\lambda x + s_\pm y)\,\mathrm{d}\lambda + \frac{\partial\psi}{\partial x} \end{aligned} \tag{6.2.6}$$

the upper sign again denoting the solution valid in the upper half-plane ($y > 0$) and the lower sign the solution valid in the lower half-plane ($y < 0$). According to the first remark we find that the integrals in (6.2.5) and (6.2.6) vanish for $y \to \pm\infty$. According to a theorem of Lebesgue on the behaviour at infinity of the Fourier-type integrals [G. 36], it results that the above integrals also vanish for $x \to \pm\infty$. Accordingly, in order to satisfy condition (5.1.3), it is necessary to have:

$$\lim_{x^2+y^2\to\infty}\left(\frac{\partial\varphi}{\partial x}, \frac{\partial\varphi}{\partial y}, \frac{\partial\psi}{\partial x}, \frac{\partial\psi}{\partial y}\right) = 0 \tag{6.2.7}$$

From equations (5.2.10) and (6.2.1) we have:

$$A_{\pm} = S(\alpha_x + \alpha_y r_{\pm})\, B_{\pm}$$

$$\frac{\partial\psi}{\partial y} = \alpha_x \frac{\partial\varphi}{\partial y} - \alpha_y \frac{\partial\varphi}{\partial x}, \quad \frac{\partial\psi}{\partial x} = \frac{\mathrm{d}\varphi}{\mathrm{d}\alpha} \tag{6.2.8}$$

This shows again that the general solution may be represented by means of three functions, namely $B_{\pm}(\lambda)$ and $\varphi\,(x, y)$.

Finally, from the equations of motion (5.2.5) and (5.2.6) we get:

$$p^{\pm}(x, y) = \int_{-\infty}^{+\infty} (\alpha_x r_{\pm} - \alpha_y)\, SB_{\pm} \exp(-\mathrm{i}\lambda x + s_{\pm} y)\, \mathrm{d}\lambda - \frac{\partial\varphi}{\partial x} \tag{6.2.9}$$

We notice that the harmonic part of the fields is of the same form as for the perfectly conducting and non-conducting fluids. What differs from one case to another is the rotational part of the fields, which for non-conducting fluids is zero, for perfectly conducting fluids is a hyperbolic wave and for finite electrical conductivity fluids has a rather complex structure. It is worthwhile observing that for aligned fields the harmonic part of the magnetic field coincides with that of the velocity field, while the rotational parts differ only by the factor S.

6.2.1 Boundary conditions

Using conditions (5.2.22), (5.2.24) and the general solution given above we get:

$$\int_{-\infty}^{+\infty} (\alpha_x + \alpha_y r_+)\, SB_+ \mathrm{e}^{-\mathrm{i}\lambda x} \mathrm{d}\lambda + \left.\frac{\partial\varphi}{\partial y}\right|_{y=+0} = h'_+(x), \quad |x| \leqslant 1. \tag{6.2.10}$$

$$\int_{-\infty}^{+\infty} S(\alpha_x[B] + \alpha_y[rB])\mathrm{e}^{-\mathrm{i}\lambda x}\mathrm{d}\lambda + \left[\frac{\partial\varphi}{\partial y}\right] = \eta[h'], \ \forall x \tag{6.2.11}$$

$$-\int_{-\infty}^{+\infty}[rB]\mathrm{e}^{-i\lambda x}\mathrm{d}\lambda+\alpha_x\left[\frac{\partial\varphi}{\partial x}\right]+\alpha_y\left[\frac{\partial\varphi}{\partial y}\right]=0,\ \forall x \tag{6.2.12}$$

$$\int_{-\infty}^{+\infty}[B]\mathrm{e}^{-i\lambda x}\mathrm{d}y+\alpha_x\left[\frac{\partial\varphi}{\partial y}\right]-\alpha_y\left[\frac{\partial\varphi}{\partial x}\right]=0,\ \forall x \tag{6.2.13}$$

Using the notations:

$$\left[\frac{\partial\varphi}{\partial x}\right]=f(x),\quad \frac{1}{2\pi}\int_{-\infty}^{+\infty}f(x)\,\mathrm{e}^{i\lambda x}\mathrm{d}x=F$$

$$\left[\frac{\partial\varphi}{\partial y}\right]=g(x),\quad \frac{1}{2\pi}\int_{-\infty}^{+\infty}g(x)\,\mathrm{e}^{i\lambda x}\mathrm{d}x=G \tag{6.2.14}$$

$$\frac{1}{2\pi}\int_{-1}^{+1}[h']\,\mathrm{e}^{i\lambda x}\mathrm{d}\lambda=I$$

and Fourier's inversion theorem, from (6.2.11) — (6.2.13) we get:

$$S(\alpha_x[B]+\alpha_y[rB])=I-G \tag{6.2.15}$$

$$[rB]=\alpha_xF+\alpha_yG \tag{6.2.16}$$

$$[B]=-\alpha_xG+\alpha_yF \tag{6.2.17}$$

This system yields:

$$kG=I-2\alpha_x\alpha_y\,SF \tag{6.2.18}$$

$$k[rB]=\alpha_x(1-S)\,F+\alpha_yI \tag{6.2.19}$$

$$k[B]=\alpha_y(1+S)\,F-\alpha_xI \tag{6.2.20}$$

Equation (6.2.18) is equivalent to:

$$k\left[\frac{\partial\varphi}{\partial y}\right]=\varepsilon\,[h']-2\alpha_x\alpha_y\,Sf(x) \tag{6.2.18'}$$

Finally, from (6.2.9) we get:

$$k[p]=-(S^2-1+2k)f+2\alpha_x\alpha_y\,\eta S\,[h'] \tag{6.2.21}$$

Obviously $f(x)\equiv 0$ and $g(x)\equiv 0$ for $|x|>1$.

These equations enable us to express the rotational part of the solution by means of the unknown function f. Obviously, this is possible (as in perfectly conducting fluids) only if $k \neq 0$. The singular case $k = 0$ requires a separate examination. It may be observed that $k = 0$ only if $\alpha_y < \alpha_x$.

6.2.2 The determination of function f

To determine the function f we proceed as in the previous chapter and introduce the holomorphic function $\Phi(z)$. Taking into account equations (6.2.14) and (6.2.18′) we have:

$$\Phi(z) = \frac{1}{2\pi i} \int_{-1}^{+1} \left\{ f(t) + \frac{2i\,\alpha_x \alpha_y}{k} Sf(t) - \frac{i}{k} [h']_t \right\} \frac{dt}{t - z} \tag{6.2.22}$$

From this equation we obtain the harmonic part of the general solution using function f.

The function f may be determined from (6.2.10) by using equations (6.2.19) and (6.2.20) and noticing that from (6.2.22) we get:

$$\left. \frac{\partial \varphi}{\partial y} \right|_{+0} = -\frac{\alpha_x \alpha_y}{k} Sf + \frac{1}{2k} [h'] + \frac{1}{2\pi} \int_{-1}^{\prime +1} \frac{f(t)}{t - x} dt \tag{6.2.22'}$$

Thus, we obtain the integral equation:

$$-2\alpha_x \alpha_y Sf + \frac{k}{\pi} \int_{-1}^{\prime +1} \frac{f(t)}{t - x} dt + S \int_{-1}^{+1} f(t) M(t - x)\, dt \tag{6.2.23}$$

$$= h'_+ + h'_- + 2S(\alpha_y^2 - \alpha_x^2) h'_+ - S \int_{-1}^{+1} [h']_t N(t - x)\, dt, \; |x| < 1$$

where:

$$\begin{Bmatrix} M \\ N \end{Bmatrix} (t - x) = \frac{1}{\pi} \int_{-\infty}^{+\infty} \begin{Bmatrix} m \\ n \end{Bmatrix} e^{i\lambda(t - x)}\, d\lambda \tag{6.2.24}$$

$$m = \frac{(\alpha_x + \alpha_y r_+)\{\alpha_x(1 - S) - \alpha_y r - (1 + S)\}}{r_+ - r_-} \tag{6.2.25}$$

$$n = \frac{(\alpha_x + \alpha_y r_+)\,(\alpha_y + \alpha_x r_-)}{r_+ - r_-}$$

Taking into account equations (6.2.4) it follows that for large λ we have:

$$m = \alpha_x\alpha_y + \frac{1}{2}(\alpha_x^2 - \alpha_y^2 - S)\,\mathrm{i}\,\mathrm{sign}\,\lambda + Rm^*, \quad m^* = O(1/|\lambda|)$$

$$n = -\frac{1}{2}(\alpha_x^2 - \alpha_y^2) + \mathrm{i}\,\alpha_x\alpha_y\,\mathrm{sign}\,\lambda + \alpha_x\alpha_y\,Rn^*, \quad n^* = O(1/|\lambda|) \tag{6.2.25'}$$

This proves that kernels M and N are generalized functions. It should be noticed that from (6.2.25) we get $n = -1/2$ for aligned fields and $n = 1/2$ for orthogonal fields. This observation proves that the last term in (6.2.25′) vanishes as the product $\alpha_x\alpha_y$.

Kernels M and N may be transformed by means of the following equations [G. 33]:

$$\int_{-\infty}^{+\infty} \mathrm{e}^{\mathrm{i}\lambda(t-x)}\,\mathrm{d}\lambda = 2\pi\,\delta(t-x) \tag{6.2.26}$$

$$\int_{-\infty}^{+\infty} \mathrm{sign}\,\lambda\,\mathrm{e}^{\mathrm{i}\lambda(t-x)}\,\mathrm{d}\lambda = \frac{2\mathrm{i}}{t-x} \tag{6.2.27}$$

δ being Dirac's distribution. Accordingly, we have:

$$M(t-x) = 2\alpha_x\alpha_y\,\delta(t-x) + \frac{\alpha_y^2 - \alpha_x^2 + S}{\pi(t-x)} + RM^*(t-x)$$

$$N(t-x) = (\alpha_y^2 - \alpha_x^2)\,\delta(t-x) - \frac{2\alpha_x\alpha_y}{\pi(t-x)} + \alpha_x\alpha_y\,RN^*(t-x)$$

such that the integral equation (6.2.23) becomes:

$$\frac{1}{\pi}\int_{-1}^{'+1} \frac{(S^2 - 1 + 2k)f(t) - 2\alpha_x\alpha_y\,S[h']_t}{t-x}\,\mathrm{d}t + RS\int_{-1}^{+1} f(t)\,M^*(t-x)\,\mathrm{d}t$$

$$= k(h'_+ + h'_-) - \alpha_x\alpha_y\,RS\int_{-1}^{+1} [h']_t\,N^*(t-x)\,\mathrm{d}t, \quad |x| < 1 \tag{6.2.28}$$

since [G. 33]:

$$\int_{-1}^{+1} f(t)\,\delta(t-x)\,\mathrm{d}t = f(x)$$

Kernels M^* and N^* have the same expression as M and N if m^* and n^* are substituted for m and n. According to Dirichlet's criterion integrals M and N are convergent.

The integral equation (6.2.28) can be finally reduced to a Fredholm type equation. In this sense we make use of the fact that the solution of an integral equation of the form:

$$\frac{1}{\pi}\int_{-1}^{\prime +1} \frac{v\ (t)}{t-x}\,\mathrm{d}t = P(x),\ |x| < 1 \tag{6.2.29}$$

is

$$v(x) = -\frac{1}{\pi}\left(\frac{1-x}{1+x}\right)^{1/2}\int_{-1}^{\prime +1}\left(\frac{1+t}{1-t}\right)^{1/2}\frac{P(t)\,\mathrm{d}t}{t-x} \tag{6.2.29'}$$

As noticed in the previous chapter this equation is characteristic for the thin airfoil theory of classical aerodynamics. Its solution may be found for instance in [G. 34].

With this result, equation (6.2.28) becomes:

$$(S^2 - 1 + 2k)f(x) = \tag{6.2.30}$$

$$RS\int_{-1}^{+1} f(\xi)\mathfrak{M}(x,\xi)\,\mathrm{d}\xi - \mathfrak{M}_0 + \alpha_x\alpha_y S\left\{2[h'] + R\int_{-1}^{+1}[h']\,\mathfrak{N}(x,\xi)\mathrm{d}\xi\right\}$$

where:

$$\begin{Bmatrix}\mathfrak{M}\\ \mathfrak{N}\end{Bmatrix} = \frac{1}{\pi}\left(\frac{1-x}{1+x}\right)^{1/2}\int_{-1}^{\prime +1}\left(\frac{1+t}{1-t}\right)^{1/2}\begin{Bmatrix}M^*(\xi - t)\\ N^*(\xi - t)\end{Bmatrix}\frac{\mathrm{d}t}{t-x}$$

$$\mathfrak{M}_0(x) = \frac{k}{\pi}\left(\frac{1-x}{1+x}\right)^{1/2}\int_{-1}^{\prime +1}\left(\frac{1+t}{1-t}\right)^{1/2}\frac{h'_+ + h'_-}{t-x}\,\mathrm{d}t$$

The solution of equation (6.2.30) may be found through successive approximations. As a first approximation, we may use the solution given in the classical aerodynamics ($RS = 0$).

Let us now consider some particular cases of the above solution.

6.2.3 Perfectly conducting fluid

By assuming $R = \infty$, from (6.2.3) (except for $\alpha_y = 0$) we have $r_{\pm} = -c_{\pm}$. Using the notation:

$$\int_{-\infty}^{+\infty} A_{\pm}\mathrm{e}^{-\mathrm{i}\lambda(x - c_{\pm}y)}\mathrm{d}\lambda = F_{\pm}(x - c_{\pm}y)$$

the general solution (6.2.5), (6.2.6) reduces to (5.2.14) (5.2.15). Then noticing that in this case:

$$2m = 2\alpha_x\alpha_y - A\left(1 + \frac{1}{A^2}\right)\alpha_y, \qquad M = 2m\,\delta(t - x)$$

$$2n = \alpha_y^2 - \alpha_x^2 + A\alpha_x, \qquad N = 2n\,\delta(t - x)$$

it follows that the integral equation (6.2.23) reduces to (5.2.37).

6.2.4 Aligned fields

By assuming $\alpha_x = 1$, $\alpha_y = 0$, we get:

$$r_\pm = \pm \mathrm{i}\sqrt{1 + \frac{R}{\mathrm{i}\lambda}(1 - S)} \tag{6.2.31}$$

and:

$$M(t - x) = \frac{S - 1}{2\pi \mathrm{i}}\int_{-\infty}^{+\infty}\sqrt{\frac{\lambda}{\lambda + \mathrm{i}R(S - 1)}}\,\mathrm{e}^{\mathrm{i}\lambda(t - x)}\mathrm{d}\lambda \tag{6.2.32}$$

Such integrals were explicitly calculated by N. Marcov [19].

For $S < 1$ we get:

$$M = -\frac{R(1 - S)^2}{2\pi}\left\{K_0\left(|t - x|\frac{R(1 - S)}{2}\right) + \operatorname{sign}(t - x)K_1\left(|t - x|\frac{R(1 - S)}{2}\right)\right\}$$
$$\times \exp\left\{-(t - x)\frac{R(1 - S)}{2}\right\} \tag{6.2.33}$$

where K_0 and K_1 are Bessel's functions.

In this case too, the solution of the integral equation (6.2.23) is laborious. However, the calculations simplify themselves if $0 < R(1 - S) \ll 1$. Indeed, under this assumption for M one can use the following asymptotic expression:

$$M(t - x) = \frac{R(1 - S)^2}{2\pi}(\Gamma + \ln|t - x|) - \frac{1 - S}{\pi}\frac{1}{t - x}$$
$$\Gamma = -2\ln 2 + C + 1 + \ln R(1 - S) \tag{6.2.34}$$

where C is Euler's constant ($= 0.577215$). Then equation (6.2.23) becomes ($S \neq 1$):

$$\frac{1}{\pi}\int_{-1}^{\prime +1} \frac{f(t)}{t - x}\,dt + \frac{\omega}{\pi}\int_{-1}^{+1} f(t)(\Gamma + \ln|t - x|)\,dt = 2h, \quad |x| < 1 \tag{6.2.35}$$

where: $2\omega = RS$, $2h = (h'_+ + h'_-)\,/\,(1 - S)$.

The solution of the integral equation (6.2.35) may be determined as follows [7]. We introduce the notation:

$$g(x) = -\frac{1}{\pi}\int_{-1}^{+1} f(t)(\Gamma + \ln|t - x|)\,dt \tag{6.2.36}$$

and obtain:

$$g'(x) = \frac{1}{\pi}\int_{-1}^{\prime +1} \frac{f(t)}{t - x}\,dt \tag{6.2.36'}$$

such that the integral equation (6.2.35) becomes:

$$g' - \omega g = 2h \tag{6.2.37}$$

whose general solution is:

$$g(x) = e^{\omega x}\left(\Gamma_0 + 2\int_0^x h(\xi)\,e^{-\omega\xi}\,d\xi\right) \tag{6.2.37'}$$

where Γ_0 is a constant.

By using (6.2.29) and (6.2.29'), $f(x)$ may be determined from (6.2.36') in the form:

$$f(x) = \Gamma_0 f_0(x) + f_1(x) \tag{6.2.38}$$

where:

$$f_0(x) = -\frac{\omega}{\pi}\left(\frac{1 - x}{1 + x}\right)^{1/2}\int_{-1}^{\prime +1}\left(\frac{1 + t}{1 - t}\right)^{1/2}\frac{e^{\omega t}}{t - x}\,dt$$

$$f_1(x) = -\frac{1}{\pi}\left(\frac{1 - x}{1 + x}\right)^{1/2}\int_{-1}^{\prime +1}\left(\frac{1 + t}{1 - t}\right)^{1/2}\frac{H(t)}{t - x}\,dt\,. \tag{6.2.38'}$$

$$H(t) = 2h(t) + \omega\int_0^t 2h(\xi)\,e^{\omega(t - \xi)}\,d\xi$$

The constant Γ_0 is determined by the relationship:

$$\pi\Gamma_1\Gamma_0 = (\ln 2 - \Gamma)\bar{f}_1 \tag{6.2.39}$$

$$+\int_{-1}^{+1} \sqrt{1 - t^2} H(t)\,\mathrm{d}t - \int_{-1}^{+1} \left(\frac{1+t}{1-t}\right)^{1/2} \left(\int_0^t 2h(\xi)\,\mathrm{e}^{\omega(t-\xi)}\,\mathrm{d}\xi\right)\mathrm{d}t$$

where:

$$\Gamma_1 = I_0 - I_1 + \omega(\ln 2 - \Gamma)(I_0 + I_1) \tag{6.2.39'}$$

$$\bar{f}_1 = -\int_{-1}^{+1} \left(\frac{1+t}{1-t}\right)^{1/2} H(t)\,\mathrm{d}t$$

$I_0(\omega)$ and $I_1(\omega)$ being the modified Bessel functions.

The lift is determined by the formula:

$$L = -\int_{-1}^{+1} [p]\,\mathrm{d}x = (1 - S)\int_{-1}^{+1} f(x)\,\mathrm{d}x = (1 - S)(\Gamma_0\bar{f}_0 + \bar{f}_1) \tag{6.2.40}$$

$$\bar{f}_0 = -\pi\omega(I_0 + I_1)$$

If the profile reduces itself to a flat plate at incidence ε, we get:

$$h'_+ + h'_- = -2\varepsilon,\ (1 - S)\bar{f}_1 = 2\pi\varepsilon(I_0 + I_1)$$

$$(1 - S)\Gamma_0 = 2\varepsilon\omega^{-1}(1 - \Gamma_1^{-1}),\ (1 - S)f(x) = -2\varepsilon\omega^{-1}\Gamma_1^{-1}f_0(x) \tag{6.2.41}$$

$$L = 2\pi\varepsilon(I_0 + I_1)\Gamma_1^{-1}$$

For $\omega = 0$ the lift reduces itself to $2\pi\varepsilon$, a well-known formula in classical aerodynamics. The table below shows some numerical values of the ratio $L/2\pi\varepsilon$.

$R =$	0	0.5	1	2	3	4	5	6
$S = 0.99$	1	0.45	0.32	0.21	0.16	0.13	0.11	0.100
$S = 0.96$	1	0.54	0.41	0.30	0.24			
$S = 0.92$	1	0.58	0.49					

It is obvious that the lift decreases as R increases.

6.2.5 Orthogonal fields

In this case ($\alpha_x = 0$, $\alpha_y = 1$) we obtain:

$$M(t-x) = \frac{1+S}{2\pi i}\int_{-\infty}^{+\infty}\sqrt{\frac{\lambda - iR}{\lambda + iRS}}\, e^{i\lambda(t-x)}\, d\lambda \tag{6.2.42}$$

$$= \frac{R(1+S)^2}{2}\left\{-K_0\left(|t-x|\frac{R(1+S)}{2}\right) + \text{sign}\,(t-x)K_1\left(|t-x,\frac{R(1+S)}{2}\right)\right\}$$

$$\times \exp\left\{-(t-x)\frac{R(S+1)}{2}\right\}$$

By using the asymptotic expressions of the functions K_0 and K_1, for $R \ll 1$ we get:

$$M(t-x) = \frac{R(1+S)^2}{2}(\Gamma^* + \ln|t-x|) + \frac{1+S}{\pi}\frac{1}{t-x}$$

$$\Gamma^* = -2\ln 2 + C - 1 + \ln R(1+S) \tag{6.2.43}$$

The integral equation (23) thus becomes:

$$\frac{1}{\pi}\int_{-1}^{+1}\frac{f(t)}{t-x}\,dt + \frac{\omega}{\pi}\int_{-1}^{+1} f(t)(\Gamma^* + \ln|t-x|)\,dt = 2h^*, \qquad |x| < 1 \tag{6.2.44}$$

where: $2h^* = (h'_+ + h'_-)/(1+S)$
Equation (6.2.44) is obtained from (6.2.35) by substituting h^* for h and Γ^* for Γ. Its solution is obtained from (6.2.38) with the same substitution.

Numerical computations show that for orthogonal fields the value of the lift is lower than for aligned fields. Thus for example if we consider the flat plate, for $R = 0.2$ and $S = 0.2$ we obtain $(L/2\pi\varepsilon) = 0.97$ in the case of aligned fields, while in the case of orthogonal fields, we have $(L/2\pi\varepsilon) = 0.94$.

6.3 Compressible fluids in aligned fields

6.3.1 General solution

In the case of compressible fluids in aligned fields the equations of motion are ($v_x = -p$):

$$\beta^2\frac{\partial v_x}{\partial x} + \frac{\partial v_y}{\partial y} = 0, \quad \beta^2 \overset{\text{def}}{=} 1 - M^2 \tag{6.3.1}$$

$$\frac{\partial v_y}{\partial x} - \frac{\partial v_x}{\partial y} = S\left(\frac{\partial b_y}{\partial x} - \frac{\partial b_x}{\partial y}\right) \tag{6.3.2}$$

$$\frac{\partial b_x}{\partial x} + \frac{\partial b_y}{\partial y} = 0 \tag{6.3.3}$$

$$\frac{\partial b_y}{\partial x} - \frac{\partial b_x}{\partial y} = R(b_y - v_y) \tag{6.3.4}$$

$$\lim_{x^2+y^2\to\infty} (v_x, v_y, b_x, b_y) = 0 \tag{6.3.5}$$

From (6.3.1), (6.3.2) and (6.3.3) we find

$$Hv_y = \beta^2 S\Delta b_y, \ H \overset{\text{def}}{=} \beta^2 \frac{\partial^2}{\partial x^2} + \frac{\partial^2}{\partial y^2} \tag{6.3.6}$$

and from (6.3.3) and (6.3.4)

$$\left(\Delta - R\frac{\partial}{\partial x}\right)b_y = -R\frac{\partial v_y}{\partial x} \tag{6.3.7}$$

Finally:

$$T(v_y, b_y) = 0 \tag{6.3.8}$$

where:

$$T = \left(\beta^2\frac{\partial^2}{\partial x^2} + \frac{\partial^2}{\partial y^2}\right)\left(\Delta - R\frac{\partial}{\partial x}\right) + \beta^2 RS\Delta\frac{\partial}{\partial x} \tag{6.3.8$'$}$$

For plane waves of the form exp $(-\mathrm{i}\lambda x + sy)$, $s = -\mathrm{i}\lambda r$, λ real, from (6.3.8) we get the following dispersion equation:

$$(\beta^2 + r^2)\left(1 + r^2 + \frac{R}{\mathrm{i}\lambda}\right) - \frac{R}{\mathrm{i}\lambda}\beta^2 S(1 + r^2) = 0 \tag{6.3.9}$$

The roots of equation (6.3.9) are distinct from one another and their imaginary parts different from zero. Indeed, if equation (6.3.9) had real roots then one would get:

$$(\beta^2 + r^2)(1 + r^2) = 0, \ \beta^2 + r^2 = \beta^2 S(1 + r^2)$$

But the real roots of the first equation ($r^2 = -\beta^2$) do not verify the second one.

Let us denote by r_j ($j = 1,2$) those roots of equation (6.3.9) whose real parts of the form $s_j = -\mathrm{i}\,\lambda r_j$ are negative for any λ (these parts depend on $|\lambda|$ only).

This is possible since (6.3.9) is a biquadratic equation; thus two roots will certainly have this property. The other two roots have opposite signs.

Now if a and b are real numbers we have:

$$\sqrt{a \pm \frac{b}{\mathrm{i}\lambda}} = \begin{cases} \sqrt{a} \pm \dfrac{b}{2\sqrt{a}} \dfrac{1}{\mathrm{i}\lambda} + O\left(\dfrac{b^2}{\lambda^2}\right), & \text{if } a > 0 \\[2ex] \mathrm{i}\sqrt{-a}\, \operatorname{sign} \lambda \mp \dfrac{b}{2\sqrt{-a}} \dfrac{1}{|\lambda|} + O\left(\dfrac{b^2}{\lambda^2}\right), & \text{if } a < 0 \end{cases}$$

Whence it follows that for large λ the roots r_j have the following behaviour:

$$r_1 = -\mathrm{i} \operatorname{sign} \lambda - \frac{R}{2|\lambda|} + O\left(\frac{R^2}{\lambda^2}\right)$$

$$r_2 = \begin{cases} -\mathrm{i}\sqrt{1 - M^2} \operatorname{sign} \lambda + \dfrac{RS\sqrt{1 - M^2}}{2|\lambda|} + O\left(\dfrac{R^2}{\lambda^2}\right), & \text{if } M < 1 \\[2ex] -\sqrt{M^2 - 1} + \dfrac{RS\sqrt{M^2 - 1}}{2\mathrm{i}\lambda} + O\left(\dfrac{R^2}{\lambda^2}\right), & \text{if } M > 1 \end{cases} \tag{6.3.10}$$

The general solution of equation (6.3.8) is:

$$\begin{Bmatrix} v_y^{\pm} \\ b_y^{\pm} \end{Bmatrix}(x, y) = \int_{-\infty}^{+\infty} \sum_j \begin{Bmatrix} A_j^{\pm} \\ B_j^{\pm} \end{Bmatrix} \exp(-\mathrm{i}\lambda x \pm s_j y)\, \mathrm{d}\lambda \tag{6.3.11}$$

From (6.3.1) and (6.3.3) we get ($M \neq 1$):

$$\begin{Bmatrix} \beta^2 v_x^{\pm} \\ b_x^{\pm} \end{Bmatrix}(x, y) = \mp \int_{-\infty}^{+\infty} \sum r_j \begin{Bmatrix} A_j^{\pm} \\ B_j^{\pm} \end{Bmatrix} \exp(-\mathrm{i}\lambda x \pm s_j y)\, \mathrm{d}\lambda \tag{6.3.12}$$

and from (6.3.4):

$$A_j^{\pm} = \left\{\frac{\mathrm{i}\lambda}{R}(1 + r_j^2) + 1\right\} B_j^{\pm}, \quad j = 1, 2. \tag{6.3.13}$$

The upper and lower signs indicate as usually the solutions valid in the upper ($y > 0$) and lower ($y < 0$) planes, respectively. In order to get (6.3.12) (12) and (6.3.13) (13) we noticed that a relationship of the form:

$$(*') \qquad \alpha_1 \mathrm{e}^{s_1 y} + \alpha_2 \mathrm{e}^{s_2 y} = 0, \quad \forall\, y$$

with distinct s_1 and s_2 entails $\alpha_1 = \alpha_2 = 0$. Indeed, from (*) it follows:

$$\alpha_1 e^{(s_1 - s_2)y} + \alpha_2 = 0$$

and by differentiating with respect to y we have $\alpha_1 = 0$.

6.3.2 Boundary conditions

Let the camber line of the profile be represented by the function $y = h(x)$, $|x| \leqslant 1$. Then, the linearized boundary conditions are:

$$v_y^+(x, 0) = h'(x), \text{ for } |x| \leqslant 1 \tag{6.3.14}$$

$$[b_x] = [b_y] = [v_y] = 0, \text{ for all } x \tag{6.3.15}$$

$$p^+(x, 0) = p^-(x, 0), \text{ for } |x| > 1 \tag{6.3.16}$$

Using the general solution and the inversion theorem of the Fourier integrals from (6.3.14) and (6.3.15) we have:

$$\int_{-\infty}^{+\infty} \Sigma \left\{\frac{i\lambda}{R}(1 + r_j^2) + 1\right\} B_j^+ e^{-i\lambda x} d\lambda = h'(x), \ |x| \leqslant 1 \tag{6.3.17}$$

$$\Sigma\, r_j(B_j^+ + B_j^-) = 0 \tag{6.3.18}$$

$$\Sigma\,(B_j^+ - B_j^-) = 0, \ \Sigma\,(1 + r_j^2)(B_j^+ - B_j^-) = 0 \tag{6.3.19}$$

Since the roots r_j are distinct from one another, from (6.3.19) it follows

$$B_j^+ = B_j^-, \ j = 1, 2 \tag{6.3.20}$$

and then from (6.3.18):

$$r_2 B_2^+ = - r_1 B_1^+ \tag{6.3.21}$$

so that (6.3.17) becomes:

$$\int_{-\infty}^{+\infty} k_1 B_1^+ e^{-i\lambda x} d\lambda = h'(x), \ |x| \leqslant 1$$

$$k_1 = \frac{r_2 - r_1}{r_2}\left\{\frac{i\lambda}{R}(1 - r_1 r_2) + 1\right\} \tag{6.3.17'}$$

Using the expression of pressure we find:

$$\beta^2[p] = \int_{-\infty}^{+\infty} k_2 B_1^+ e^{-i\lambda x} d\lambda \overset{\text{def}}{=} \beta^2 f(x) \tag{6.3.22}$$

where:

$$k_2 B_1^+ = \Sigma\, r_j(A_j^+ + A_j^-) = \frac{i\lambda}{R}\Sigma\, r_j(1 + r_j^2)(B_j^+ + B_j^-) = \frac{2i\lambda r_1}{R}(r_1^2 - r_2^2)$$

Condition (6.3.16) yields $f(x) \equiv 0$, for $|x| > 1$ and thus from (6.3.22) we have:

$$k_2 B_1^+ = \frac{\beta^2}{2\pi}\int_{-1}^{+1} f(t)\, e^{i\lambda t}\, dt \tag{6.3.23}$$

From conditions (6.3.17′) and (6.3.23) the following integral equation for f follows:

$$\int_{-1}^{+1} \beta^2 f(t) K(t - x)\, dt = h'(x), \; |x| \leqslant 1 \tag{6.3.24}$$

where:

$$K(t - x) = \frac{1}{2\pi}\int_{-\infty}^{+\infty} k(\lambda)\, e^{i\lambda(t-x)}\, d\lambda \tag{6.3.25}$$

$$k(\lambda) = \frac{k_1}{k_2} = \frac{-1 + r_1 r_2 - R/i\lambda}{2r_1 r_2(r_1 + r_2)} \tag{6.3.26}$$

This equation is to be used for the solution of the problem.

The kernel (6.3.25) represents a distribution. Using equations (6.3.10) we get:

$$k = \begin{cases} \dfrac{i\,\mathrm{sign}\,\lambda}{2\sqrt{1 - M^2}} + \dfrac{RS}{4\sqrt{1 - M^2}}\, k^*\left(\dfrac{1}{|\lambda|}\right), & M < 1 \\[2ex] \dfrac{1}{2\sqrt{M^2 - 1}} - \dfrac{RS}{\sqrt{4M^2 - 1}}\, k^{**}\left(\dfrac{1}{\lambda}\right), & M > 1 \end{cases} \tag{6.3.27}$$

Taking into account equations (6.2.26) and (6.2.27) we obtain:

$$K(t - x) = \begin{cases} \dfrac{1}{2\pi\sqrt{1 - M^2}}\dfrac{1}{(t - x)} + \dfrac{RS}{4\sqrt{1 - M^2}} K^*, & M < 1 \\[2ex] -\dfrac{1}{2\sqrt{M^2 - 1}}\delta(t - x) - \dfrac{RS}{4\sqrt{M^2 - 1}} K^{**}, & M > 1 \end{cases} \tag{6.3.28}$$

where K^* and K^{**} are convergent integrals.

Consequently, equation (6.3.24) reduces to the following ($|x| \leqslant 1$)

$$\frac{1}{\pi}\int_{-1}^{'+1}\frac{f(t)}{t-x}\,\mathrm{d}t = \omega\int_{-1}^{+1} f(t)K^{*}(t-x)\,\mathrm{d}t + H_1(x), \text{ if } M<1 \qquad (6.3.29)$$

$$f(x) + \omega\int_{-1}^{+1} f(t)K^{**}(t-x)\,\mathrm{d}t = H_2(x), \qquad \text{if } M>1 \qquad (6.3.30)$$

where:

$$\omega = \frac{RS}{2}, \; H_1 = \frac{-2h'(x)}{\sqrt{1-M^2}}, \quad H_2 = \frac{2h'(x)}{\sqrt{M^2-1}} \qquad (6.3.31)$$

Equation (6.3.29) may be reduced to an integral equation of the Fredholm type in a manner similar to that used in Section 6.2

$$f(x) = \omega\int_{-1}^{+1} f(\xi)\mathcal{K}(x,\xi)\,\mathrm{d}\xi + \mathcal{K}_1 \qquad (6.3.29')$$

$$\begin{Bmatrix}\mathcal{K}\\ \mathcal{K}_1\end{Bmatrix} = -\frac{1}{\pi}\left(\frac{1-x}{1+x}\right)^{1/2}\int_{-1}^{'+1}\left(\frac{1+t}{1-t}\right)^{1/2}\begin{Bmatrix}K^{*}\\ H_1\end{Bmatrix}\frac{\mathrm{d}t}{t-x}$$

Equations (6.3.29′) and (6.3.30) may be integrated using the method of succesive approximations, the first approximation being the classical aerodynamic solution ($\omega = 0$).

For $M = 0$ one obtains a solution valid for the case of incompressible fluids.

6.3.3 Weakly conducting fluids

Since air is a weakly conducting medium it is of interest in aerodynamics to obtain the solution under such a hypothesis. Neglecting the second and higher powers of R, the roots r_j have the following form:

$$r_1 = -\mathrm{i}\,\mathrm{sign}\,\lambda - \frac{R}{2|\lambda|} \qquad (6.3.32)$$

$$r_2 = \begin{cases}\left(-\mathrm{i}\,\mathrm{sign}\,\lambda + \dfrac{RS}{2|\lambda|}\right)\sqrt{1-M^2}, \text{ if } M<1\\[2ex] \left(-1 + \dfrac{RS}{2|\lambda|}\right)\sqrt{M^2-1}, \text{ if } M>1\end{cases}$$

so that:

$$k = \frac{\mathrm{i}\,\mathrm{sign}\,\lambda}{2\sqrt{1-M^2}} + \frac{RS}{4\sqrt{1-M^2}}\frac{1}{|\lambda|} + O\left(\frac{R^2}{\lambda^2}\right), \quad M < 1$$

$$k = -\frac{1}{2\sqrt{M^2-1}} - \frac{RS}{4\sqrt{M^2-1}}\frac{1}{\mathrm{i}\lambda} + O\left(\frac{R^2}{\lambda^2}\right), \quad M > 1 \tag{6.3.33}$$

Using (6.2.26) and (6.2.27) as well as the equations [G. 33]:

$$\int_{-\infty}^{+\infty} \frac{\mathrm{e}^{\mathrm{i}\lambda(t-x)}}{|\lambda|}\,\mathrm{d}\lambda = -2(C + \ln|t-x|) \tag{6.3.34}$$

$$\frac{1}{\pi}\int_{-\infty}^{+\infty} \frac{\mathrm{i}\lambda(t-x)}{\mathrm{i}\lambda}\,\mathrm{d}\lambda = \mathrm{sign}\,(t-x) \tag{6.3.35}$$

C being Euler's constant, we get:

$$K = -\frac{1}{2\pi\sqrt{1-M^2}}\frac{1}{t-x} - \frac{RS}{4\pi\sqrt{1-M^2}}(C + \ln|t-x|), \quad M < 1$$

$$K = -\frac{1}{2\sqrt{M^2-1}}\delta(t-x) - \frac{RS}{8\sqrt{M^2-1}}\,\mathrm{sign}\,(t-x), \quad M > 1 \tag{6.3.36}$$

so that the integral equation (6.3.24) becomes ($|x| \leqslant 1$)

$$\frac{1}{\pi}\int_{-1}^{\prime +1} \frac{f(t)}{t-x}\,\mathrm{d}t + \frac{\omega}{\pi}\int_{-1}^{+1} f(t)(C + \ln|t-x|)\,\mathrm{d}t = H_1(x), \quad M < 1 \tag{6.3.37}$$

$$f(x) + \frac{\omega}{2}\int_{-1}^{+1} f(t)\,\mathrm{sign}\,(t-x)\,\mathrm{d}t = H_2(x), \quad M > 1 \tag{6.3.38}$$

Equation (6.3.37) is solved in Section 6.2. Its solution is given in (6.2.38) where C is substituted for Γ and H_1 for H. For the flat plate of incidence ε we have:

$$\Gamma_0 = \frac{2\varepsilon}{\omega\sqrt{1-M^2}} \times \frac{1 - I_0 - \omega(\ln 2 - C)(I_0 + I_1)}{I_0 - I_1 + \omega(\ln 2 - C)(I_0 + I_1)}, \quad \lim_{\omega\to 0} \Gamma_0 = 0$$

$$L = \frac{2\varepsilon\pi(I_0 + I_1)(1 - I_1)}{\sqrt{1-M^2}\{I_0 - I_1 + \omega(\ln 2 - C)(I_0 + I_1)\}}, \quad \lim_{\omega\to 0} L = \frac{2\varepsilon\pi}{\sqrt{1-M^2}} \tag{6.3.39}$$

Equation (6.3.38) can be easily integrated if one notices that:

$$\int_{-1}^{+1} f(t)\,\mathrm{sign}\,(t-x)\,\mathrm{d}t = -\int_{-1}^{x} f(t)\,\mathrm{d}t + \int_{x}^{1} f(t)\,\mathrm{d}t$$

Indeed by differentiation this becomes:

$$f' - \omega f = H_2'$$

and has the solution:

$$f = \mathrm{e}^{\omega x}\left(\Gamma + \int_{-1}^{x} H_2' \mathrm{e}^{-\omega t}\,\mathrm{d}t\right) \tag{6.3.40}$$

The constant Γ is determined from the condition that f satisfies equation (6.3.38). After some simple calculations in which use is made of the following relationships:

$$\int_{-1}^{x} \mathrm{e}^{\omega t}\left(\int_{-1}^{t} H_2' \mathrm{e}^{-\omega\xi}\,\mathrm{d}\xi\right)\mathrm{d}t = \int_{-1}^{x} H_2' \mathrm{e}^{-\omega\xi}\left(\int_{\xi}^{x} \mathrm{e}^{\omega t}\,\mathrm{d}t\right)\mathrm{d}\xi$$

$$\int_{x}^{1} \mathrm{e}^{\omega t}\left(\int_{-1}^{t} H_2' \mathrm{e}^{-\omega\xi}\,\mathrm{d}\xi\right)\mathrm{d}t = \int_{-1}^{x} H_2' \mathrm{e}^{-\omega\xi}\left(\int_{x}^{1} \mathrm{e}^{\omega t}\,\mathrm{d}t\right)\mathrm{d}\xi + \int_{x}^{1} H_2' \mathrm{e}^{-\omega\xi}\left(\int_{\xi}^{1} \mathrm{e}^{\omega t}\,\mathrm{d}t\right)\mathrm{d}\xi \tag{6.3.41}$$

one obtains:

$$\Gamma \mathrm{ch}\omega = \frac{1}{2}\{H_2(1) + H_2(-1)\} - \frac{1}{2}\mathrm{e}^{\omega}\int_{-1}^{+1} H_2' \mathrm{e}^{-\omega t}\,\mathrm{d}t \tag{6.3.42}$$

The lift is determined from the following equation:

$$L = -2\Gamma\frac{\mathrm{sh}\omega}{\omega} - \int_{-1}^{+1} \mathrm{e}^{\omega x}\left(\int_{-1}^{x} H_2' \mathrm{e}^{-\omega t}\,\mathrm{d}t\right)\mathrm{d}x \tag{6.3.43}$$

For the flat plate of incidence ε one has:

$$f = -\frac{2\varepsilon}{\sqrt{M^2-1}}\frac{\mathrm{e}^{\omega x}}{\mathrm{ch}\,\omega} \tag{6.3.44}$$

$$L = \frac{4\varepsilon}{\sqrt{M^2-1}}\frac{\mathrm{th}\omega}{\omega}, \quad \lim_{\omega\to 0} L = \frac{4\varepsilon}{\sqrt{M^2-1}} \tag{6.3.45}$$

Equation (6.3.45) is of particular importance in aerodynamics since it demonstrates to a first approximation the influence on the lift of the magnetic field and resistivity of the medium. They both reduce the magnitude of the lift.

6.3.4 The symmetrical profile

If the airfoil is symmetrical about the Ox-axis the perturbed motion will also be symmetrical:

$$v_y^+(x, y) = -v_y^-(x, -y) \tag{6.3.46}$$

so that the Ox-axis will be a current line outside the airfoil. Consequently:

$$v_y^+(x, 0) = 0, \qquad \text{for } |x| > 1$$

$$v_y^+(x, 0) = h'(x), \text{ for } |x| \leqslant 1 \tag{6.3.47}$$

$y = h(x)$ being the equation of the upper surface of the body (the lower surface equation being $y = -h(x)$, $h(-1) = h(1) = 0$).

From (6.3.11) and (6.3.47) it follows that:

$$\sum A_j^+ = \frac{1}{2\pi}\int_{-1}^{+1} h'(x)\mathrm{e}^{\mathrm{i}\lambda x}\mathrm{d}x \overset{\text{def}}{=} I \tag{6.3.48}$$

$$\sum (A_j^+ + A_j^-) = 0 \tag{6.3.49}$$

also, from $[b_x] = [b_y] = 0$ one has:

$$\sum r_j(B_j^+ + B_j^-) = 0 \tag{6.3.50}$$

$$\sum (B_j^+ - B_j^-) = 0 \tag{6.3.51}$$

Equations (6.3.13) and (6.3.48) – (6.3.51) determine the solution.

From equations (6.3.13), (6.3.49) and (6.3.50) it follows:

$$B_j^+ = -B_j^- \, (A_j^+ = -A_j^-), \; j = 1, 2 \tag{6.3.52}$$

so that (6.3.51) and (6.3.48) become:

$$B_1^+ + B_2^+ = 0, \; (1 + r_1^2)\, B_1^+ + (1 + r_2^2)\, B_2^+ = RI/\mathrm{i}\lambda$$

To conclude:

$$B_1^+ = -\frac{RI}{i\lambda(r_2^2 - r_1^2)}, \quad B_2^+ = \frac{RI}{i\lambda(r_2^2 - r_1^2)} \tag{6.3.53}$$

For this case the lift vanishes. The symmetrical airfoil problem is of no importance in aerodynamics.

6.4 Compressible fluids in orthogonal fields

The following equations describe the motion in orthogonal fields:

$$M^2\frac{\partial p}{\partial x} + \frac{\partial v_x}{\partial x} + \frac{\partial v_y}{\partial y} = 0 \tag{6.4.1}$$

$$\frac{\partial v_x}{\partial x} + \frac{\partial p}{\partial x} = S\left(\frac{\partial b_x}{\partial y} - \frac{\partial b_y}{\partial x}\right), \quad \frac{\partial v_y}{\partial x} + \frac{\partial p}{\partial y} = 0, \tag{6.4.2}$$

$$\frac{\partial b_x}{\partial x} + \frac{\partial b_y}{\partial y} = 0 \tag{6.4.3}$$

$$\frac{\partial b_y}{\partial x} - \frac{\partial b_x}{\partial y} = R(b_y + v_x) \tag{6.4.4}$$

$$\lim_{x^2+y^2\to\infty} (v_x, v_y, b_x, b_y, p) = 0 \tag{6.4.5}$$

This system yields [13]:

$$T(v_y, b_x) = 0 \tag{6.4.6}$$

where:

$$T = H\frac{\partial}{\partial x}\left(\Delta - R\frac{\partial}{\partial x}\right) - RS\Delta\left(M^2\frac{\partial^2}{\partial x^2} - \frac{\partial^2}{\partial y^2}\right) \tag{6.4.6'}$$

H and Δ denoting the same operators as in Section 6.3.

For plane waves of the form $\exp(-i\lambda x + sy)$, $s = -i\lambda r$, from (6.4.6) we find the following dispersion equation:

$$-i\lambda(\beta^2 + r^2)(1 + r^2 + R/i\lambda) = RS(1 + r^2)(M^2 - r^2) \tag{6.4.7}$$

As usual we denote by r_j ($j = 1,2$) those roots of equation (6.4.7) for which the real part of expressions $s_j = -\mathrm{i}\lambda r_j$ are negative.

For large λ we have:

$$1 + r_1^2 = -\frac{R}{\mathrm{i}\lambda} + O\left(\frac{R^2 S}{\lambda^2}\right), \; 1 + r_2^2 = M^2 - \frac{RS}{\mathrm{i}\lambda} + O\left(\frac{R^2 S^2}{\lambda^2}\right)$$

and

$$r_1 = -\mathrm{i}\,\mathrm{sign}\,\lambda - \frac{R}{2|\lambda|} + O\left(\frac{R^2 S}{\lambda^2}\right) \tag{6.4.8}$$

$$r_2 = \begin{cases} -\mathrm{i}\sqrt{1 - M^2}\,\mathrm{sign}\,\lambda - \dfrac{RS}{2\sqrt{1 - M^2}}\dfrac{1}{|\lambda|} + O\left(\dfrac{R^2 S^2}{\lambda^2}\right), & M < 1 \\ -\sqrt{M^2 - 1} + \dfrac{RS}{2\sqrt{M^2 - 1}}\dfrac{1}{\mathrm{i}\lambda} + O\left(\dfrac{R^2 S^2}{\lambda^2}\right), & M > 1 \end{cases}$$

The general solution of system (6.4.1) — (6.4.5) may be written as:

$$\begin{Bmatrix} v_y^{\pm} \\ b_x^{\pm} \\ p^{\pm} \\ b_y^{\pm} \\ v_x^{\pm} \end{Bmatrix}(x, y) = \int_{-\infty}^{+\infty} \sum_j \begin{Bmatrix} \mp r_j A_j^{\pm} \\ \mp r_j B_j^{\pm} \\ A_j^{\pm} \\ B_j^{\pm} \\ (r_j^2 - M^2) A_j^{\pm} \end{Bmatrix} \exp(-\mathrm{i}\lambda x \pm s_j y)\,\mathrm{d}\lambda \tag{6.4.9}$$

with the relationship:

$$(\beta^2 + r_j^2) A_j^{\pm} + S(1 + r_j^2) B_j^{\pm} = 0, \; j = 1, 2 \tag{6.4.10}$$

obtained from (6.4.2).

From the boundary conditions (6.3.14)— (6.3.16) we have:

$$-\int_{-\infty}^{+\infty} \sum r_j A_j^{+} \mathrm{e}^{-\mathrm{i}\lambda x} \mathrm{d}\lambda = h'(x), \; |x| \leqslant 1 \tag{6.4.11}$$

$$\sum r_j (A_j^{+} + A_j^{-}) = 0 \tag{6.4.12}$$

$$\sum r_j (1 + r_j^2)^{-1} (A_j^{+} + A_j^{-}) = 0 \tag{6.4.13}$$

$$\sum (\beta^2 + r_j^2)(1 + r_j^2)^{-1} (A_j^{+} - A_j^{-}) = 0 \tag{6.4.14}$$

Using the fact that r_j are distinct roots from (6.4.12) and (6.4.13) we get:

$$A_j^+ + A_j^- = 0 \tag{6.4.15}$$

so that (6.4.14) yields:

$$A_2^+ = \alpha A_1^+, \ \alpha = -\frac{(\beta^2 + r_1^2)(1 + r_2^2)}{(\beta^2 + r_2^2)(1 + r_1^2)} \tag{6.4.16}$$

and (6.4.11) becomes:

$$-\int_{-\infty}^{+\infty} (r_1 + \alpha r_2) A_1^+ \mathrm{e}^{-\mathrm{i}\lambda x} \mathrm{d}\lambda = h'(x), \ |x| \leqslant 1 \tag{6.4.16'}$$

Now using the pressure expression we find:

$$[p] = 2\int_{-\infty}^{+\infty} (1 + \alpha) A_1^+ \mathrm{e}^{-\mathrm{i}\lambda x} \mathrm{d}\lambda \overset{\text{def}}{=} f(x) \tag{6.4.17}$$

From condition (6.3.16) it follows that $f \equiv 0$ for $|x| > 1$, so that from (6.4.17) we get:

$$2(1 + \alpha) A_1^+ = \frac{1}{2\pi}\int_{-1}^{+1} f(t)\, \mathrm{e}^{\mathrm{i}\lambda t}\, \mathrm{d}t \tag{6.4.17'}$$

Using (6.4.17′) the integral equation (6.4.11′) becomes:

$$\int_{-1}^{+1} f(t)\, K(t - x)\, \mathrm{d}t = h'(x), \ |x| \leqslant 1 \tag{6.4.18}$$

where:

$$K(t - x) = \frac{1}{2\pi}\int_{-\infty}^{+\infty} k(\lambda) \mathrm{e}^{\mathrm{i}\lambda(t-x)} \mathrm{d}\lambda$$

$$k = \frac{\beta^2(1 + r_1^2 + r_2^2) + r_1^2 r_2^2 - M^2 r_1 r_2}{2M^2(r_1 + r_2)} \tag{6.4.18'}$$

Equation (6.4.18) can also be reduced to an integral equation of the Fredholm type (as for aligned fields).

Let us consider a weakly conducting fluids. Neglecting the terms of the order of R^2 we obtain ($2\omega = RS$):

$$2k = \begin{cases} \mathrm{i}\,\mathrm{sign}\,\lambda \sqrt{1 - M^2} + \dfrac{\omega}{\sqrt{1 - M^2}} \dfrac{1}{|\lambda|}, & \text{if } M < 1 \\[2ex] \sqrt{M^2 - 1} - \dfrac{\omega}{\sqrt{M^2 - 1}} \dfrac{1}{\mathrm{i}\lambda}, & \text{if } M > 1 \end{cases} \tag{6.4.19}$$

Consequently:

$$2K = \begin{cases} -\dfrac{\sqrt{1-M^2}}{\pi}\dfrac{1}{t-x} - \dfrac{1}{\sqrt{1-M^2}}\dfrac{\omega}{\pi}(C + \ln|t-x|), & M < 1 \\ \sqrt{M^2-1}\,\delta(t-x) - \dfrac{1}{\sqrt{M^2-1}}\dfrac{\omega}{2}\,\mathrm{sign}\,(t-x), & M > 1 \end{cases} \qquad (6.4.19')$$

so that equation (6.4.18) becomes:

$$\frac{1}{\pi}\int_{-1}^{\prime +1} \frac{f(t)}{t-x}\,\mathrm{d}t + \frac{\omega'}{\pi}\int_{-1}^{+1} f(t)(C + \ln|t-x|)\,\mathrm{d}t = H_1,\ M < 1 \qquad (6.4.20)$$

$$f(x) + \frac{\omega'}{2}\int_{-1}^{+1} f(t)\,\mathrm{sign}\,(t-x)\,\mathrm{d}t = H_2,\ M > 1 \qquad (6.4.21)$$

where:

$$\omega' = \frac{\omega}{1-M^2} = \frac{RS}{2(1-M^2)} \qquad (6.4.22)$$

Therefore the solution will be the same as for aligned fields except for ω which will be replaced by ω'.

7

Theory of thin airfoils in ionized gases

7.1 Introduction

In problems of hypersonic aerodynamics and space flights the surrounding medium is an ionized gas. It is known that in such a case the Hall effect cannot be disregarded. The study of the problem of thin airfoils with consideration of this effect constitutes the subject of the present chapter.

The first considerations on this matter were made by Sonnerup [1] and W. R. Sears and E. L. Resler Jr. [6.9]. However, a quantitative discussion of the problem of thin airfoils was attempted independently by L. Dragoş [2] [G.24] and J. Tang and R. Seebass [3], in 1968. In [2] the incompressible fluid is studied the solution being obtained by means of generalized functions. In [3] the compressible fluid in aligned fields is considered but the solution is valid only for symmetrical airfoils.

A complete study of incompressible fluids with Hall effect was presented by L. Dragoş [4]. The problem was again discussed for aligned fields [5] and for orthogonal fields [6] in view of putting the results in a form from which the solutions for non-conducting fluids, perfectly conducting fluids and fluids with scalar finite conductivity could be obtained in a straightforward manner. The solutions given in [6] and [5] are included in Sections 2 and 3. The particular case of Alfvén flow in aligned fields [4], [5] is also included in Section 3.

The problem of compressible fluids is considered by L. Dragoş [7] for orthogonal fields and for aligned fields [8], [9]. These results are included in Sections 7.4 and 7.5. In all the considerations no restrictions concerning the airfoil shape are made.

7.1.1 Ohm's law

Ohm's law with the Hall effect included has the following form (chapter 14):

$$\boldsymbol{J} = \sigma(\boldsymbol{E} + \boldsymbol{V} \times \boldsymbol{B}) + \frac{\sigma}{nq} \boldsymbol{B} \times \boldsymbol{J} \tag{7.1.1}$$

$\boldsymbol{E}$ including the additional electric fields due to the electron pressure gradient. Using the notations (5.1.2) and linearizing equation (7.1.1) we obtain:

$$\boldsymbol{e}_0 + \boldsymbol{i}_1 \times \boldsymbol{\alpha} = 0$$

$$\boldsymbol{j} = R\boldsymbol{\mathscr{E}} + \nu(\boldsymbol{\alpha} \times \boldsymbol{j}) \tag{7.1.2}$$

$$\boldsymbol{\mathscr{E}} \overset{\text{def}}{=} \boldsymbol{e} + \boldsymbol{i}_1 \times \boldsymbol{b} + \boldsymbol{v} \times \boldsymbol{\alpha}, \quad \nu \overset{\text{def}}{=} \sigma B_0 (nq)^{-1}$$

where we used the fact that $\boldsymbol{j}_0 \equiv 0$ (5.2.1).

Taking into account the conditions under which the motion takes place it follows that for this case we have again:

$$\frac{\partial}{\partial t} = \frac{\partial}{\partial z} = 0 \tag{7.1.3}$$

But due to the Hall effect we may not conclude that $\boldsymbol{e}$ and b_z are vanishing quantities. From Maxwell's equation curl $\boldsymbol{e} = 0$ it follows that $e_z \equiv 0$ and

$$\frac{\partial e_x}{\partial y} - \frac{\partial e_y}{\partial x} = 0 \tag{7.1.4}$$

From (7.1.2) we have:

$$j_x = R\mathscr{E}_x + \nu\alpha_y j_z$$

$$j_y = R\mathscr{E}_y - \nu\alpha_x j_z$$

$$j_z = R\mathscr{E}_z + \nu(\alpha_x j_y - \alpha_y j_x)$$

so that

$$k j_z = \mathscr{E}_z + \nu(\alpha_x \mathscr{E}_y - \alpha_y \mathscr{E}_x)$$

$$k j_x = (1 + \nu^2\alpha_x^2)\mathscr{E}_x + \nu^2 \alpha_x\alpha_y\mathscr{E}_y + \nu\alpha_y\mathscr{E}_z$$

$$k j_y = \nu^2\alpha_x\alpha_y\mathscr{E}_x + (1 + \nu^2\alpha_y^2)\mathscr{E}_y - \nu\alpha_x\mathscr{E}_z \tag{7.1.5}$$

$$k \overset{\text{def}}{=} (1 + \nu^2)\, R^{-1}$$

Now using Maxwell's equation $\boldsymbol{j} =$ curl $\boldsymbol{b}$ and the expression of $\boldsymbol{\mathscr{E}}$ from (7.1.5) we obtain:

$$k\left(\frac{\partial b_y}{\partial x} - \frac{\partial b_x}{\partial y}\right) = \alpha_y v_x - \alpha_x v_y + b_y + \nu(v_z - \alpha_x b_z - \alpha_y e_z + \alpha_x e_y)$$

$$k\frac{\partial b_z}{\partial y} = \nu\alpha_y(\alpha_y v_x - \alpha_x v_y + b_y) - \alpha_y v_z - \nu^2\alpha_x\alpha_y(b_z - e_y) + (1 + \nu^2\alpha_x^2)e_x \tag{7.1.6}$$

$$k\frac{\partial b_z}{\partial x} = \nu\alpha_x(\alpha_y v_x - \alpha_x v_y + b_y) - \alpha_x v_z + (1 + \nu^2\alpha_y^2)(b_z - e_y) - \nu^2\alpha_x\alpha_y e_x$$

7.2 Incompressible fluids in orthogonal fields

7.2.1 Equations of motion

The system of equations of motion has in this case the following form:

$$\frac{\partial v_x}{\partial x} + \frac{\partial v_y}{\partial y} = 0 \tag{7.2.1}$$

$$\frac{\partial b_x}{\partial x} + \frac{\partial b_y}{\partial y} = 0 \tag{7.2.2}$$

$$\frac{\partial v_x}{\partial x} + \frac{\partial p}{\partial x} = S\left(\frac{\partial b_x}{\partial y} - \frac{\partial b_y}{\partial x}\right) \tag{7.2.3}$$

$$\frac{\partial v_y}{\partial x} + \frac{\partial p}{\partial y} = 0 \tag{7.2.4}$$

$$\frac{\partial v_z}{\partial x} = S\frac{\partial b_z}{\partial y} \tag{7.2.5}$$

$$k\left(\frac{\partial b_y}{\partial x} - \frac{\partial b_x}{\partial y}\right) = v_x + b_y + v(v_z - e_x) \tag{7.2.6}$$

$$e_x = -v(v_x + b_y) + v_z + k\,\partial b_z/\partial y \tag{7.2.7}$$

$$Re_y = (R - \partial/\partial x)\, b_z \tag{7.2.8}$$

to which equation (7.1.4) and the damping condition

$$\lim_{x^2+y^2\to\infty} (v_x, v_y, v_z, p, b_x, b_y, b_z, e_x, e_y) = 0 \tag{7.2.9}$$

should be added. One prefers to replace equations (7.2.6) and (7.2.7) by;

$$R(v_x + b_y) + \frac{\partial b_x}{\partial y} - \frac{\partial b_y}{\partial x} = v\frac{\partial b_z}{\partial y} \tag{7.2.6'}$$

$$Re_x = v\left(\frac{\partial b_x}{\partial y} - \frac{\partial b_y}{\partial x}\right) + Rv_z + \frac{\partial b_z}{\partial y} \tag{7.2.7'}$$

Again one can reduce the system of equations of motion to a single equation. Indeed from equations (7.2.1) —(7.2.4) we have as usually:

$$\Delta v_y + S\Delta b_x = 0 \tag{7.2.10}$$

From (7.2.7′), (7.2.8), (7.1.4) and (7.2.5) one obtains:

$$T_0 b_z + \nu \frac{\partial}{\partial x}\Delta b_x = 0 \tag{7.2.11}$$

$$T_0 \overset{\text{def}}{=} \Delta \frac{\partial}{\partial x} + RS\frac{\partial^2}{\partial y^2} - R\frac{\partial^2}{\partial x^2}$$

and finally from (7.2.6′), (7.2.1) and (7.2.2):

$$\left(\Delta - R\frac{\partial}{\partial x}\right)\frac{\partial}{\partial x} b_x = R\frac{\partial^2 v_y}{\partial y^2} + \nu\frac{\partial^3 b_z}{\partial x\,\partial y^2} \tag{7.2.12}$$

Equations (7.2.10)— (7.2.12) yield:

$$T\Delta(v_y, b_x) = 0,\ Tb_z = 0 \tag{7.2.13}$$

$$T \overset{\text{def}}{=} T_0^2 + \nu^2\frac{\partial^4}{\partial x^2 \partial y^2}\Delta$$

From (7.2.5), (7.2.8) and (7.1.4) it follows that $T(v_z, e_y, e_x) = 0$ and from the first equations $T\Delta(v_x, b_y, p) = 0$. It is of interest to point out this difference between the basic unknowns (v_x, v_y, p, b_x, b_y) and the unknowns (v_z, b_z, e_x, e_y) imposed by the Hall effect. For $\nu = 0$ operator T reduces itself to the operator T_0 which characterizes the motion of scalar conductivity fluids (Chapter 6).

7.2.2 General solution

A general solution of the form

$$v_y = \bar{v}_y + \frac{\partial\varphi}{\partial y},\ b_x = \bar{b}_x + \frac{\partial\psi}{\partial x}$$

$$T(\bar{v}_y, \bar{b}_x) = 0,\ \Delta(\varphi, \psi) = 0 \tag{7.2.14}$$

will be determined. With plane-wave solutions of the form exp $(-i\lambda x + sy)$, $s = -i\lambda r$ the operator T yields the following dispersion equation:

$$\{(i\lambda - RS)(1 + r^2) + R(1 + S)\}^2 = v^2\lambda^2 r^2(1 + r^2) \tag{7.2.15}$$

This equation has distinct and complex roots. In order to prove this let us write equation (7.2.16) in the form:

$$a(1 + r^2)^2 + 2b(1 + r^2) + c = 0$$

$$a = (i\lambda - RS)^2 - v^2\lambda^2 = a_1 + ia_2, \quad c = R^2(1 + S)^2 \tag{7.2.15'}$$

$$2b = 2R(1 + S)(i\lambda - RS) + v^2\lambda^2 = b_1 + ib_2$$

and notice that $b^2 \neq ac$. Also, if equation (7.2.15′) had real roots we should have had:

$$a_1(1 + r^2)^2 + 2b_1(1 + r^2) + c = 0, \; a_2(1 + r^2) + 2b_2 = 0$$

simultaneously. But the real roots of the latter equation ($r^2 = S^{-1}$) do not verify identically the former.

As usually we denote by r_j ($j = 1,2$) the roots of equation (7.2.15) for which $s_j = -i\lambda r_j$ have negative real parts. For large λ we have:

$$r_1 = -i \operatorname{sign} \lambda + O(R^2\lambda^{-2})$$

$$\sqrt{1 + v^2}\, r_2 = -i \operatorname{sign} \lambda - \left(1 + \frac{S}{1 + v^2}\right)\frac{R}{|\lambda|} + O\left(\frac{R^2}{\lambda^2}\right) \tag{7.2.16}$$

The general solution has the following form:

$$\begin{Bmatrix} v_y^{\pm} \\ b_x^{\pm} \end{Bmatrix}(x, y) = \mp \int_{-\infty}^{+\infty} \sum_{j=1,2} r_j \begin{Bmatrix} A_j^{\pm} \\ B_j^{\pm} \end{Bmatrix} \exp(-i\lambda x \pm s_j y)\, d\lambda + \begin{Bmatrix} \dfrac{\partial\varphi}{\partial y} \\ \dfrac{\partial\psi}{\partial x} \end{Bmatrix} \tag{7.2.17}$$

$$vb_z^{\pm}(x, y) = \int_{-\infty}^{+\infty} \sum_{j=1,2} C_j^{\pm} \exp(-i\lambda x \pm s_j y)\, d\lambda \tag{7.2.18}$$

From (7.2.1), (7.2.2), (7.2.4), (7.2.5) we find

$$\begin{Bmatrix} v_x^{\pm} \\ b_y^{\pm} \\ p^{\pm} \end{Bmatrix}(x, y) = \int_{-\infty}^{+\infty} \sum_{j=1,2} \begin{Bmatrix} r_j^2 A_j^{\pm} \\ B_j^{\pm} \\ A_j^{\pm} \end{Bmatrix} \exp(-i\lambda x \pm s_j y)\, d\lambda + \begin{Bmatrix} \dfrac{\partial \varphi}{\partial x} \\ \dfrac{\partial \psi}{\partial y} \\ -\dfrac{\partial \varphi}{\partial x} \end{Bmatrix} \quad (7.2.19)$$

$$v\, v_z^{\pm}(x, y) = \pm \int_{-\infty}^{+\infty} S \sum_{j=1,2} r_j C_j^{\pm} \exp(-i\lambda x \pm s_j y)\, d\lambda \quad (7.2.20)$$

$$\lim_{x^2 + y^2 \to \infty} \left(\frac{\partial \varphi}{\partial x}, \frac{\partial \varphi}{\partial y}, \frac{\partial \psi}{\partial x}, \frac{\partial \psi}{\partial y} \right) = 0 \quad (7.2.21)$$

Then taking into account equations (7.2.3) and (7.2.6′) we obtain:

$$A_j^{\pm} = -SB_j^{\pm},\ j = 1, 2 \quad (7.2.22)$$

$$s_j C_j^{\pm} = \pm P_j B_j^{\pm},\ P_j \overset{\text{def}}{=} (i\lambda - RS)(1 + r_j^2) + R(1 + S) \quad (7.2.23)$$

$$\frac{\partial \varphi}{\partial x} = -\frac{\partial \psi}{\partial y},\ \frac{\partial \varphi}{\partial y} = \frac{\partial \psi}{\partial x} \quad (7.2.24)$$

Finally, from (7.2.7′) we obtain:

$$v\, Re_x^{\pm}(x, y) = -\int_{-\infty}^{+\infty} \left(1 + \frac{R}{i\lambda}\right) \sum_{j=1,2} r_j^{-2} P_j B_j^{\pm} \exp(-i\lambda x \pm s_j y)\, d\lambda \quad (7.2.25)$$

The unknowns $B_j^{\pm}$ and φ are determined by the boundary conditions.

7.2.3 Boundary conditions

Let $y = h_{\pm}(x)$, $|x| \leqslant 1$ be the airfoil equation. Thus we have*:

$$v_y^{+}(x, 0) = h'_{+}(x),\ |x| \leqslant 1 \quad (7.2.26)$$

$$[v_y] = \eta(x)[h'],\ [b_x] = 0,\ \forall x \quad (7.2.27)$$

$$[b_z] = [b_y] = [e_x] = 0,\ \forall x \quad (7.2.28)$$

* If the airfoil is an insulator from the conditions:

$$j_y^{\pm}(x, 0) = -\left.\frac{\partial b_z}{\partial x}\right|_{y=\pm 0} = 0, \qquad \oint b_z dx = 0$$

the condition $b_z^{\pm}(x, 0) = 0$, $|x| \leqslant 1$ follows. This changes the boundary value problem.

so that

$$\int_{-\infty}^{+\infty} S \sum_j r_j B_j^+ e^{-i\lambda x} d\lambda + \left.\frac{\partial \varphi}{\partial y}\right|_{y=+0} = h'_+(x), \ |x| \leqslant 1 \qquad (7.2.26')$$

$$\int_{-\infty}^{+\infty} S \sum_j r_j (B_j^+ + B_j^-) e^{-i\lambda x} d\lambda + \left[\frac{\partial \varphi}{\partial y}\right] = \eta(x)\,[h'], \ \forall x$$

$$-\int_{-\infty}^{+\infty} \sum_j r_j (B_j^+ + B_j^-) e^{-i\lambda x} d\lambda + \left[\frac{\partial \varphi}{\partial y}\right] = 0, \ \forall x \qquad (7.2.27')$$

$$\sum P_j r_j^{-1} (B_j^+ + B_j^-) = 0$$

$$\sum (B_j^+ - B_j^-) = 2F, \ \sum P_j r_j^{-2} (B_j^+ - B_j^-) = 0 \qquad (7.2.28')$$

where

$$2F = \frac{1}{2\pi} \int_{-\infty}^{+\infty} f(t)\, e^{i\lambda t} dt, \ f = \left[\frac{\partial \varphi}{\partial x}\right] \qquad (7.2.29)$$

From equation (7.2.27′) we get:

$$\sum r_j (B_j^+ + B_j^-) = \frac{1}{2\pi(1+S)} \int_{-1}^{+1} [h']_t e^{i\lambda t} dt \stackrel{\text{def}}{=} 2I \qquad (7.2.30)$$

$$\left[\frac{\partial \varphi}{\partial y}\right] = \frac{\eta}{1+S}[h'], \qquad (7.2.31)$$

so that (7.2.28′) and (7.2.30) yield:

$$B_j^+ + B_j^- = 2\alpha_j I, \ B_j^+ - B_j^- = 2\beta_j F \qquad (7.2.32)$$

where:

$$\alpha_1 = r_1 P_2/\gamma, \ \alpha_2 = -r_2 P_1/\gamma$$

$$\beta_1 = r_1^2 P_2/\gamma, \ \beta_2 = -r_2^2 P_1/\gamma \quad (\beta_1 + \beta_2 = 1) \qquad (7.2.32')$$

$$\gamma = (i\lambda + R)(r_1^2 - r_2^2)$$

These equations determine the unknowns $B_j^{\pm}$.

We have also:

$$[p] = -(1+S) f(x) \qquad (7.2.33)$$

$$f(x) \equiv 0, \ |x| > 1 \qquad (7.2.34)$$

The last condition is required also by the continuity of the harmonic function φ.

For the holomorphic function:

$$\Phi(z) = \frac{\partial \varphi}{\partial x} - \mathrm{i}\frac{\partial \varphi}{\partial y} \tag{7.2.35}$$

we have:

$$[\Phi] = \eta\left(f - \frac{\mathrm{i}}{1+S}[h']\right), \ \forall x$$

such that:

$$\frac{\partial \varphi}{\partial x} - \mathrm{i}\frac{\partial \varphi}{\partial y} = \frac{1}{2\pi \mathrm{i}}\int_{-1}^{+1}\left\{f(t) - \frac{\mathrm{i}}{1+S}[h']_t\right\}\frac{\mathrm{d}t}{t-z} \tag{7.2.36}$$

and

$$\left.\frac{\partial \varphi}{\partial y}\right|_{y=+0} = \frac{1}{2(1+S)}[h'] + \frac{1}{2\pi}{\int_{-1}^{+1}}' \frac{f(t)}{t-x}\mathrm{d}t \tag{7.2.36'}$$

Consequently equation (7.2.26′) becomes:

$$\frac{1}{\pi}{\int_{-1}^{+1}}' \frac{f(t)}{t-x}\mathrm{d}t + S\int_{-1}^{+1} f(t)\,K(t-x)\,\mathrm{d}t = h'_+ + h'_-, \ |x| \leqslant 1 \tag{7.2.37}$$

where

$$K(t-x) = \frac{1}{2\pi}\int_{-1}^{+1} k(\lambda)\mathrm{e}^{\mathrm{i}\lambda(t-x)}\mathrm{d}\lambda$$

$$k(\lambda) = \frac{r_1^2 + r_1 r_2 + r_2^2 + r_1^2 r_2^2(\mathrm{i}\lambda - RS)(\mathrm{i}\lambda + R)}{r_1 + r_2} \tag{7.2.37'}$$

Equation (7.2.36) determines the harmonic parts of the fields and the integral equation (7.2.37) the function f.

Equation (7.2.37) may be also reduced to a Fredholm integral equation [6].

7.2.4 Weakly conducting fluids

For a weakly conducting fluid we have:

$$k = -\mathrm{i}\,\mathrm{sign}\,\lambda - \frac{R(1+S)}{\sqrt{1+v^2}\,(1+\sqrt{1+v^2})}\,\frac{1}{|\lambda|} \tag{7.2.38}$$

$$K = \frac{1}{\pi}\frac{1}{t-x} + \frac{R(1+S)}{\pi\sqrt{1+v^2}(1+\sqrt{1+v^2})}(C + \ln|t-x|)$$

such that equation (7.2.37) becomes:

$$\frac{1}{\pi}\int_{-1}^{'+1}\frac{f(t)}{t-x}\,dt+\frac{\omega}{\pi}\int_{-1}^{+1}f(t)(C+\ln|t-x|)\,dt=H(x) \qquad |x|\leqslant 1 \tag{7.2.39}$$

or

$$\frac{1}{\pi}\int_{-1}^{'+1}\frac{[p]}{t-x}\,dt+\frac{\omega}{\pi}\int_{-1}^{+1}[p](C+\ln|t-x|)\,dt=-(h'_{+}+h'_{-}) \tag{7.2.39'}$$

where

$$\omega=\frac{RS}{\sqrt{1+v^2(1+\sqrt{1+v^2})}},\quad H=\frac{h'_{+}+h'_{-}}{1+S} \tag{7.2.39''}$$

The solution of equation (7.2.39) is given in Section 6.2.

The lift has the following expression:

$$L=-\int_{-1}^{+1}[p]\,dx=(1+S)(\Gamma_0\bar{f}_0+\bar{f}_1) \tag{7.2.40}$$

For a flat plate we get:

$$L=\frac{2\varepsilon\pi(I_0+I_1)(1-I_1)}{I_0-I_1+\omega(\ln 2-C)(I_0+I_1)},\quad \lim_{\omega\to 0}L=2\varepsilon\pi \tag{7.2.41}$$

7.2.5 The camber problem

If $h'_{+}=h'_{-}=h'$ we get:

$$\left[\frac{\partial\varphi}{\partial y}\right]=0,\ \forall x \tag{7.2.42}$$

such that the harmonic function:

$$\frac{\partial\varphi}{\partial y}(x,y)-\frac{\partial\varphi}{\partial y}(x,-y)$$

is vanishing at infinity and on the $[-1, +1]$ segment. Consequently, it vanishes everywhere such that we have:

$$\frac{\partial\varphi}{\partial y}(x,y)=\frac{\partial\varphi}{\partial y}(x,-y),\quad \frac{\partial\varphi}{\partial x}(x,y)=-\frac{\partial\varphi}{\partial x}(x,-y) \tag{7.2.43}$$

From the last equation we have:

$$\left.\frac{\partial\varphi}{\partial x}\right|_{y=0} = 0, \ |x| > 1 \tag{7.2.44}$$

From (7.2.32) we also had $B_j^+ = -B_j^-$, $B_j^+ = \beta_j F$ such that condition (7.2.26′) becomes:

$$S\int_{-1}^{+1} \left.\frac{\partial\varphi}{\partial x}\right|_{\substack{x=t\\ y=+0}} K(t-x)\,\mathrm{d}t + \left.\frac{\partial\varphi}{\partial y}\right|_{y=+0} = h'(x), \ |x| \leqslant 1 \tag{7.2.45}$$

The boundary value problem (7.2.44) and (7.2.45) for the holomorphic function $\Phi(z)$ is solved by the Volterra-Signorini formula [6].

For a perfectly conducting fluid ($R \simeq \infty$) we get:

$$K(t-x) = -A\delta(t-x) \ \textit{where} \ A = \sqrt{S} \tag{7.2.46}$$

such that condition (7.2.45) becomes:

$$-\frac{1}{A}\left.\frac{\partial\varphi}{\partial x}\right|_{y=+0} + \left.\frac{\partial\varphi}{\partial y}\right|_{y=+0} = h'(x), \ |x| \leqslant 1 \tag{7.2.45`}$$

This condition is given by Sears and Resler [5.1]

7.2.6 Some particular cases

(1) *The scalar conductivity fluid.* If $\nu = 0$ the operator T reduces itself to T_0 such that the root of the dispersion equation is

$$r = -\sqrt{\frac{R + \mathrm{i}\lambda}{RS - \mathrm{i}\lambda}} = -\,\mathrm{i}\,\mathrm{sign}\,\lambda - \frac{R(1+S)}{2|\lambda|} + O\left(\frac{R^2}{\lambda^2}\right) \tag{7.2.47}$$

The general solution is given by (7.2.17), (7.2.19) without summation ($r_j = r$). Obviously $C_j^{\pm} = 0$. Equations (7.2.32) are replaced by:

$$r(B^+ + B^-) = 2I, \ B^+ - B^{-1} = 2F \tag{7.2.48}$$

and k by r.

(2) *Perfectly conducting fluids.* If $R \to \infty$ we have $r = -S^{-1/2} = -A$ such that:

$$k = -A, \ K = -A\delta(t-x) \tag{7.2.49}$$

Equation (7.2.37) becomes:

$$-\frac{1}{A}f(x)+\frac{1}{\pi}\int_{-1}^{\prime+1}\frac{f(t)}{t-x}\,\mathrm{d}t = h'_{+}+h'_{-} \tag{7.2.50}$$

With the notations:

$$\int_{-\infty}^{+\infty} B^{\pm}\,\mathrm{e}^{-\mathrm{i}\lambda(x\mp Ay)}\,\mathrm{d}\lambda = G_{\pm}(x\mp Ay) \tag{7.2.51}$$

the general solution (7.2.17), (7.2.19) reduces itself to that given in Section 5.2.

(3) *Non-conducting fluids*. If $S=0$ the integral equation (7.2.37) yields:

$$f(x)=-\frac{1}{\pi}\left(\frac{1-x}{1+x}\right)^{1/2}\int_{-1}^{\prime+1}\left(\frac{1+t}{1-t}\right)^{1/2}\frac{h'_{+}+h'_{-}}{t-x}\,\mathrm{d}t$$

such that (7.2.36) becomes:

$$\frac{\partial\varphi}{\partial x}-\mathrm{i}\frac{\partial\varphi}{\partial y}=-\frac{1}{2\pi}\int_{-1}^{+1}\frac{h'_{+}-h'_{-}}{t-z}\,\mathrm{d}t$$

$$-\frac{1}{2\pi\mathrm{i}}\left(\frac{z-1}{z+1}\right)^{1/2}\int_{-1}^{+1}\left(\frac{1+t}{1-t}\right)^{1/2}\frac{h'_{+}+h'_{-}}{t-z}\,\mathrm{d}t \tag{7.2.52}$$

which is a well known equation in classical aerodynamics (see for instance reference [G. 28], p. 664). From (7.2.22) it follows that $A_j^{\pm}=0$.

In order to obtain this equation we have used the fact that:

$$\frac{1}{\pi}\int_{-1}^{+1}\left(\frac{1-t}{1+t}\right)^{1/2}\frac{\mathrm{d}t}{(\xi-t)(t-z)}$$

$$=\frac{1}{(\xi-z)\pi}\int_{-1}^{+1}\left(\frac{1-t}{1+t}\right)^{1/2}\left(\frac{1}{\xi-t}+\frac{1}{t-z}\right)\mathrm{d}t=\frac{1}{\xi-z}\sqrt{\frac{z-1}{z+1}} \tag{7.2.53}$$

which is derived in the following way:

We consider the function

$$g(z)=\sqrt{\frac{z-1}{z+1}}-1 \tag{7.2.54}$$

where the square root is positive for $z = x > 1$ and notice that:

$$\mathcal{R}e(ig) = \begin{cases} 0 & , \text{ for } x > 1 \\ -\sqrt{\dfrac{1-t}{1+t}}, & \text{for } |x| \leqslant 1 \\ 0 & , \text{ for } x < -1 \end{cases} \tag{7.2.54'}$$

$$\lim_{z \to \infty} g(z) = 0$$

Thus according to a well known equation ([G. 28], p. 60) we obtain:

$$g(z) = \frac{1}{\pi} \int_{-1}^{+1} \sqrt{\frac{1-t}{1+t}} \frac{\mathrm{d}t}{t-z} \tag{7.2.55}$$

From (7.2.55) we have as well (using Plemelj's equation):

$$-1 = \frac{1}{\pi} \int_{-1}^{\prime +1} \sqrt{\frac{1-t}{1+t}} \frac{\mathrm{d}t}{t-x} \tag{7.2.55'}$$

Equations (7.2.55) and (7.2.55′) are then used in (7.2.53).

7.3 Incompressible fluids in aligned fields

In this case the equations of motion are:

$$\frac{\partial v_x}{\partial x} + \frac{\partial v_y}{\partial y} = 0 \quad (7.3.1), \qquad \frac{\partial b_x}{\partial x} + \frac{\partial b_y}{\partial y} = 0 \tag{7.3.2}$$

$$\frac{\partial v_y}{\partial x} - \frac{\partial v_x}{\partial y} = S\left(\frac{\partial b_y}{\partial x} - \frac{\partial b_x}{\partial y}\right) \tag{7.3.3}$$

$$v_x = -p \quad (7.3.4) \qquad v_z = Sb_z \tag{7.3.5}$$

$$R(v_y - b_y) + \frac{\partial b_y}{\partial x} - \frac{\partial b_x}{\partial y} + v\frac{\partial b_z}{\partial x} = 0 \tag{7.3.6}$$

$$Re_x = \frac{\partial b_z}{\partial y}, \quad Re_y = v\left(\frac{\partial b_y}{\partial x} - \frac{\partial b_x}{\partial y}\right) - \frac{\partial b_z}{\partial x} + R(1 - S)b_z \tag{7.3.7}$$

to which we add equation (7.1.4) and condition (7.2.9).

From (7.3.1), (7.3.2) and (7.3.3) we obtain:

$$\Delta v_y = S\Delta b_y \tag{7.3.8}$$

and from (7.3.6), (7.3.7), (7.1.4) and (7.3.2):

$$v\Delta b_y = T_0 b_z, \quad T_0 \overset{\text{def}}{=} \Delta - R(1 - S)\partial/\partial x \tag{7.3.9}$$

Finally, from (7.3.6), (7.2.3), (7.3.8) and (7.3.9) we obtain:

$$\begin{gathered} T\Delta(v_y, b_y) = 0, \quad Tb_z = 0 \\ T \overset{\text{def}}{=} T_0^2 + v^2\Delta\frac{\partial^2}{\partial x^2} \end{gathered} \tag{7.3.10}$$

T_0 being the scalar conductivity fluid operator.

For plane-wave solutions of the same type as those used in previous sections the operator T yields:

$$\left\{1 + r^2 + \frac{R}{\mathrm{i}\lambda}(1 - S)\right\}^2 + v^2(1 + r^2) = 0 \tag{7.3.11}$$

or

$$r_{1,2}^2 = -1 - \frac{v^2}{2} - \frac{R(1 - S)}{\mathrm{i}\lambda} \pm \frac{v^2}{2}\left\{1 + \frac{4R(1 - S)}{\mathrm{i}\lambda v^2}\right\}^{1/2}$$

For $v \neq 0$ r_1^2 and r_2^2 are obviously distinct roots. Now denoting as usually by r_j $(j = 1, 2)$ those roots for which $s_j = -\mathrm{i}\lambda r_j$ have negative real parts we have for large λ:

$$\begin{aligned} r_1 &= -\mathrm{i}\,\mathrm{sign}\,\lambda + R^2(1 - S)^2 O(\lambda^{-2}) \\ r_2 &= -\mathrm{i}\sqrt{1 + v^2}\,\mathrm{sign}\,\lambda - \frac{R(1 - S)}{\sqrt{1 + v^2}}\frac{1}{|\lambda|} + R^2(1 - S)^2 O(\lambda^{-2}) \end{aligned} \tag{7.3.12}$$

Taking into account the damping condition (7.2.9) and the equations (7.3.1), (7.3.2) and (7.3.7) we get the following general solution:

$$\begin{Bmatrix} v_y^{\pm} \\ b_y^{\pm} \\ v_x^{\pm} \\ b_x^{\pm} \end{Bmatrix}(x, y) = \int_{-\infty}^{+\infty} \sum_{j=1,2} \begin{Bmatrix} A_j^{\pm} \\ B_j^{\pm} \\ \mp r_j A_j^{\pm} \\ \mp r_j B_j^{\pm} \end{Bmatrix} \exp(-\mathrm{i}\lambda x \pm s_j y)\,\mathrm{d}\lambda + \begin{Bmatrix} \dfrac{\partial \varphi}{\partial y} \\ \dfrac{\partial \psi}{\partial y} \\ \dfrac{\partial \varphi}{\partial x} \\ \dfrac{\partial \psi}{\partial x} \end{Bmatrix} \tag{7.3.13}$$

$$v \begin{Bmatrix} b_z^{\pm} \\ Re_x^{\pm} \end{Bmatrix}(x, y) = \int_{-\infty}^{+\infty} \sum_j \begin{Bmatrix} C_j^{\pm} \\ \pm s_j C_j^{\pm} \end{Bmatrix} \exp(-\mathrm{i}\lambda x \pm s_j y)\,\mathrm{d}\lambda \tag{7.3.14}$$

$$\Delta\varphi = \Delta\psi = 0 \tag{7.3.15}$$

$$\lim_{x^2+y^2\to\infty} \left(\frac{\partial \varphi}{\partial x}, \frac{\partial \varphi}{\partial y}, \frac{\partial \psi}{\partial x}, \frac{\partial \psi}{\partial y} \right) = 0 \tag{7.3.16}$$

From (7.3.4) and (7.3.6) we obtain:

$$A_j^{\pm} = SB_j^{\pm} \ (S \neq 1), \qquad j = 1, 2 \tag{7.3.17}$$

$$\frac{\partial \varphi}{\partial y} = \frac{\partial \psi}{\partial y}, \ \frac{\partial \varphi}{\partial x} = \frac{\partial \psi}{\partial x} \tag{7.3.18}$$

$$C_j^{\pm} = -P_j B_j^{\pm}, \ P_j = 1 + r_j^2 + \mathrm{i}\lambda^{-1} R(S - 1) \tag{7.3.19}$$

The boundary conditions (7.2.26) – (7.2.28) yield:

$$\int_{-\infty}^{+\infty} S \Sigma B_j^{+} \mathrm{e}^{-\mathrm{i}\lambda x} \mathrm{d}\lambda + \left.\frac{\partial \varphi}{\partial y}\right|_{+0} = h'_{+}(x), \qquad |x| \leqslant 1 \tag{7.3.20}$$

$$\int_{-\infty}^{+\infty} S \Sigma (B_j^{+} - B_j^{-}) \mathrm{e}^{-\mathrm{i}\lambda x} \mathrm{d}\lambda + \left[\frac{\partial \varphi}{\partial y} \right] = \eta [h'], \qquad \forall x \tag{7.3.21}$$

$$\int_{-\infty}^{+\infty} \Sigma r_j (B_j^{+} + B_j^{-}) \mathrm{e}^{-\mathrm{i}\lambda x} \mathrm{d}\lambda - \left[\frac{\partial \varphi}{\partial x} \right] = 0, \qquad \forall x \tag{7.3.22}$$

$$\sum P_j(B_j^+ - B_j^-) = 0 \tag{7.3.23}$$

$$\int_{-\infty}^{+\infty} \sum (B_j^+ - B_j^-)\, e^{-i\lambda x}\, d\lambda + \left[\frac{\partial \varphi}{\partial y}\right] = 0, \ \forall x \tag{7.3.24}$$

$$\sum r_j P_j(B_j^+ + B_j^-) = 0 \tag{7.3.25}$$

In ref. [5] a surface current is also included.

With the notation

$$f = \left[\frac{\partial \varphi}{\partial x}\right], \quad (f \equiv 0 \text{ for } |x| > 1) \tag{7.3.26}$$

$$2F = \frac{1}{2\pi}\int_{-1}^{+1} f(x) e^{i\lambda x} dx, \ 2I = \frac{1}{2\pi(S-1)}\int_{-1}^{+1} [h'] e^{i\lambda x} dx \tag{7.3.27}$$

the conditions (7.3.21), (7.3.24) and (7.3.22) become:

$$\left[\frac{\partial \varphi}{\partial y}\right] = \frac{\eta}{1-S}[h'] \tag{7.3.28}$$

$$\sum (B_j^+ - B_j^-) = 2I \tag{7.3.29}$$

$$\sum r_j(B_j^+ + B_j^-) = 2F \tag{7.3.30}$$

From (7.3.29) and (7.3.23) we obtain:

$$B_j^+ - B_j^- = 2\alpha_j I$$

$$\alpha_1 = P_2/(r_2^2 - r_1^2), \ \alpha_1 + \alpha_2 = 1 \tag{7.3.31}$$

and from (7.3.30) and (7.3.25):

$$B_j^+ + B_j^- = 2\beta_j F$$

$$\beta_1 = P_2/r_1(r_2^2 - r_1^2), \ \beta_2 = -P_1/r_2(r_2^2 - r_1^2) \tag{7.3.32}$$

These equations determine $B_j^{\pm}$.

The discontinuity of pressure is given by the formula:

$$[p] = -(1 - S) f \tag{7.3.33}$$

The harmonic parts of the fields are determined by the equation:

$$\frac{\partial \varphi}{\partial x} - \mathrm{i}\frac{\partial \varphi}{\partial y} = \frac{1}{2\pi \mathrm{i}}\int_{-1}^{+1}\left\{f(t) - \frac{\mathrm{i}}{1-S}[h']_t\right\}\frac{\mathrm{d}t}{t-z} \tag{7.3.34}$$

and the function f by the integral equation:

$$\frac{1}{\pi}\int_{-1}^{\prime +1}\frac{f(t)}{t-x}\,\mathrm{d}t + S\int_{-1}^{+1} f(t)\,K(t-x)\,\mathrm{d}t = h'_+ + h'_- \qquad |x| \leqslant 1 \tag{7.3.35}$$

where:

$$\beta = \frac{1 - \mathrm{i}\lambda^{-1}R(1-S) + r_1^2 + r_2^2 + r_1 r_2}{r_1 r_2 (r_1 + r_2)}$$

$$= \mathrm{i}\,\mathrm{sign}\,\lambda - \frac{1-S}{S}\frac{\omega}{|\lambda|} - \frac{R^2(1-S)^2}{\lambda^2}(\gamma + \cdots), \quad \gamma = \mathrm{const}$$

$$K(t-x) = \frac{1}{2\pi}\int_{-\infty}^{+\infty}\beta\,\mathrm{e}^{\mathrm{i}\lambda(t-x)}\mathrm{d}\lambda$$

$$= \frac{1}{\pi}\frac{1}{t-x} + \frac{1-S}{S}\frac{\omega}{\pi}(C + \ln|t-x|) + \frac{R^2(1-S)^2|t-x|}{2}(\gamma + \cdots)$$

ω being given by (7.2.39″).

As usually, equation (7.3.35) may be reduced to an integral equation of the Fredholm type (ω parameter) [5].

For a weakly conducting fluid ($R(1-S) \ll 1$) equation (7.3.35) becomes:

$$\frac{1}{\pi}\int_{-1}^{\prime +1}\frac{[p]}{t-x}\,\mathrm{d}t + \frac{\omega}{\pi}\int_{-1}^{+1}[p](C + \ln|t-x|)\,\mathrm{d}t = -(h'_+ + h'_-) \qquad |x| \leqslant 1 \tag{7.3.36}$$

This equation coincides with (7.2.39′).
With this approximation the lift has therefore the same expression for both cases.

7.3.1 Alfvén flow

We have noticed above that the general solution (7.3.13) — (7.3.19) is not valid for $S = 1$. This particular case will be considered here. If in the system (7.3.1) — (7.3.7) we set $S = 1$ we have:

$$\Delta\left(\Delta + v^2\frac{\partial^2}{\partial x^2}\right)(v_y, b_y, b_z) = 0 \tag{7.3.37}$$

Using the notations:

$$r_1 = -\mathrm{i}\,\mathrm{sign}\,\lambda,\ r_2 = -\mathrm{i}\,\mu_2\,\mathrm{sign}\,\lambda,\ \mu_2 = \sqrt{1+v^2} \tag{7.3.38}$$

we obtain the following general solution ($v \neq 0$):

$$\begin{Bmatrix} v_y^{\pm} \\ b_y^{\pm} \\ v\,b_z^{\pm} \end{Bmatrix}(x, y) = \int_{-\infty}^{+\infty} \sum \begin{Bmatrix} A_j^{\pm} \\ B_j^{\pm} \\ C_j^{\pm} \end{Bmatrix} \exp(-\mathrm{i}\lambda\, x \pm s_j y)\,\mathrm{d}\lambda \tag{7.3.39}$$

$$\begin{Bmatrix} v_x^{\pm} \\ b_x^{\pm} \\ vRe_x^{\pm} \end{Bmatrix}(x, y) = \int_{-\infty}^{+\infty} \sum \begin{Bmatrix} \mp r_j A_j^{\pm} \\ \mp r_j B_j^{\pm} \\ \pm s_j C_j^{\pm} \end{Bmatrix} \exp(-\mathrm{i}\lambda\, x \pm s_j y)\,\mathrm{d}\lambda \tag{7.3.40}$$

$$A_2^{\pm} = B_2^{\pm},\ \mathrm{i}\lambda C_1^{\pm} = R(A_1^{\pm} - B_1^{\pm}),\ C_2^{\pm} = v^2 B_2^{\pm} \tag{7.3.41}$$

the unknowns being $A_1^{\pm}$, $B_1^{\pm}$, $B_2^{\pm}$.

The boundary conditions (7.2.26) — (7.2.28) provide the following relationships:

$$\begin{gathered} \int_{-\infty}^{+\infty} (A_1^{\pm} + B_2^{\pm})\,\mathrm{e}^{-\mathrm{i}\lambda x}\mathrm{d}\lambda = h'(x),\ |x| \leqslant 1 \\ A_1^{+} - A_1^{-} + B_2^{+} - B_2^{-} = \frac{1}{2\pi}\int_{-1}^{+1} [h']\,\mathrm{e}^{\mathrm{i}\lambda x}\mathrm{d}x \overset{\mathrm{def}}{=} 2I \end{gathered} \tag{7.3.42}$$

$$\begin{gathered} B_1^{+} + B_1^{-} + \mu_2(B_2^{+} - B_2^{-}) = 0 \\ \mu_1(A_1^{+} - A_1^{-}) - \mu_1(B_1^{+} - B_1^{-}) + B_2^{+} - B_2^{-} = 0,\ \mu_1 \overset{\mathrm{def}}{=} R/\mathrm{i}\lambda v^2 \\ B_1^{+} - B_1^{-} + B_2^{+} - B_2^{-} = 0 \\ \mu_1(A_1^{+} + A_1^{-}) - \mu_1(B_1^{+} + B_1^{-}) + \mu_2(B_2^{+} + B_2^{-}) = 0 \end{gathered} \tag{7.3.43}$$

From (7.3.43) we obtain:

$$\begin{gathered} A_1^{+} - A_1^{-} = 2(1+\mu_1)I,\ B_1^{+} - B_1^{-} = 2\mu_1 I,\ B_2^{+} - B_2^{-} = -2\mu_1 I \\ B_1^{+} = \frac{\mu_1}{1+\mu_1}A_1^{+},\ B_2^{+} = -\frac{\mu_1}{\mu_2(1+\mu_1)}A_1^{+} + \frac{\mu_1(1-\mu_2)}{\mu_2}I \end{gathered} \tag{7.3.44}$$

We also have:

$$[p] = 2\int_{-\infty}^{+\infty} r_1\{(1+\mu_1)^{-1}A_1^+ - I\}e^{-i\lambda x}d\lambda \overset{\text{def}}{=} f \tag{7.3.45}$$

$$f \equiv 0, \ |x| > 1$$

Equation (7.3.45) yields:

$$2A_1^+ = 2(1+\mu_1)I + \frac{1+\mu_1}{2\pi r_1}\int_{-1}^{+1} f(t)e^{i\lambda t}dt \tag{7.3.46}$$

such that (7.3.42) becomes:

$$\int_{-1}^{+1} f(t)\,K(t-x)\,dt = h'_+ + h'_-, \ |x| \leqslant 1 \tag{7.3.47}$$

$$K(t-x) = \frac{1}{2\pi}\int_{-\infty}^{+\infty}\left(i\,\mathrm{sign}\,\lambda + \frac{\omega}{|\lambda|}\right)e^{i\lambda(t-x)}d\lambda \tag{7.3.47$'$}$$

$$\omega = \frac{R}{v^2}\left(1 - \frac{1}{\sqrt{1+v^2}}\right) = \frac{R}{2} + O(v^2) \tag{7.3.47$''$}$$

Taking into account (6.2.27) and (6.3.34) we obtain:

$$K(t-x) = -\frac{1}{\pi(t-x)} - \frac{\omega}{\pi}(C + \ln|t-x|) \tag{7.3.48}$$

Accordingly, equation (7.3.47) becomes:

$$\frac{1}{\pi}\int_{-1}^{\prime +1}\frac{f(t)}{t-x}dt + \frac{\omega}{\pi}\int_{-1}^{+1} f(t)(C + \ln|t-x|)\,dt = -(h'_+ + h'_-)$$

$$|x| \leqslant 1 \tag{7.3.49}$$

In the general case this equation determines the solution for a weakly conducting fluid. In the Alfvén flow case it determines the solution for any arbitrary magnetic Reynolds number.

For a flat plate the lift is given by (7.2.41) with ω being given by (7.3.47$''$).

For $v = 0$ the unknowns v_x, v_y, b_x, b_y are bi-harmonic functions. The general solution is determined by means of Goursat's representation [6.18], [G. 24].

7.4 Compressible fluids in orthogonal fields

In this case the equations of motion are (7.2.2) — (7.2.5), (7.2.8), (7.2.6′), (7.2.7′) and (7.1.4), to which we add the continuity equation for compressible fluids:

$$M^2 \frac{\partial p}{\partial x} + \frac{\partial v_x}{\partial x} + \frac{\partial v_y}{\partial y} = 0 \tag{7.4.1}$$

The system of equations of motion reduces itself to [7]

$$T(v_y, b_x, b_z) = 0 \tag{7.4.2}$$

where

$$T = T_1 T_2 + v^2 \Delta H \frac{\partial^4}{\partial x^2 \partial y^2}, \quad T_1 = \Delta \frac{\partial}{\partial x} + RS \frac{\partial}{\partial y^2} - R \frac{\partial^2}{\partial x^2}$$

$$T_2 = H\left(\Delta - R\frac{\partial}{\partial x}\right)\frac{\partial}{\partial x} + RS\,\Delta\left(\frac{\partial^2}{\partial y^2} - M^2 \frac{\partial^2}{\partial x^2}\right),$$

$$H = \beta^2 \frac{\partial^2}{\partial x^2} + \frac{\partial^2}{\partial y^2}$$

the operator T_1 characterizing the incompressible fluids (with scalar conductivity) and the operator T_2 the compressible fluids.

The dispersion equation may be written as:

$$\left\{(1 + r^2)\left(1 - \frac{RS}{i\lambda}\right) + \frac{R(1 + S)}{i\lambda}\right\}\left\{(\beta^2 + r^2)\left(1 + r^2 + \frac{R}{i\lambda}\right)\right.$$

$$\left. - \frac{RS}{i\lambda}(1 + r^2)(r^2 - M^2)\right\} + v^2 r^2 (1 + r^2)(\beta^2 + r^2) = 0 \tag{7.4.3}$$

We denote by r_k ($k=0, 1, 2$) those roots r for which the expressions $s_k = -i\lambda r_k$ have negative real parts. It can be noticed that these roots depend only on $R/i\lambda$. Accordingly, their behaviour for large λ is the same as for small R. In order to point out this behaviour we write (7.4.3) as:

$$\left(1 + v^2 - \frac{2RS}{i\lambda}\right)(1 + r^2)^3 - \left\{v^2(1 + M^2) + M^2\right.$$

$$\left. - \frac{2R}{i\lambda}(1 + S + SM^2)\right\}(1 + r^2)^2$$

$$+ M^2\left\{v^2 - \frac{R}{i\lambda}(2 + S)\right\}(1 + r^2) + O\left(\frac{R^2}{\lambda^2}\right) = 0$$

Consequently:

$$1 + r_0^2 = O(R^2/\lambda^2)$$

$$1 + r_1^2 = M^2 + \frac{M^2RS}{i\lambda(\beta^2\nu^2 - M^2)} + O\left(\frac{R}{\lambda^2}\right)$$

$$(1+\nu^2)(1+r_2^2) = \nu^2 - \frac{2R\alpha}{i\lambda(\beta^2\nu^2 - M^2)} + O\left(\frac{R^2}{\lambda^2}\right)$$

$$\alpha \overset{\text{def}}{=} \nu^2\left(1 + \frac{S}{1+\nu^2}\right) - M^2\left\{1 + \nu^2 + \frac{S}{2}(1-\nu^2)\right\}$$

and

$$r_0 = -\mathrm{i}\,\mathrm{sign}\,\lambda + O(R^2/\lambda^2)$$

$$r_1 = \begin{cases} -\mathrm{i}\sqrt{1-M^2}\,\mathrm{sign}\,\lambda + \dfrac{M^2RS}{2\sqrt{1-M^2}(\beta^2\nu^2-M^2)\,|\lambda|} + O\left(\dfrac{R^2}{\lambda^2}\right), & M < 1 \\ -\sqrt{M^2-1} - \dfrac{M^2RS}{2\sqrt{M^2-1}(\beta^2\nu^2 - M^2)\,\mathrm{i}\lambda} + O\left(\dfrac{R^2}{\lambda^2}\right), & M > 1 \end{cases} \tag{7.4.4}$$

$$\sqrt{1+\nu^2}\,r_2 = -\,\mathrm{i}\,\mathrm{sign}\,\lambda - \frac{R\alpha}{\beta^2\nu^2 - M^2}\frac{1}{|\lambda|} + O\left(\frac{R^2}{\lambda^2}\right)$$

In the following neither r_1 nor r_2 will play a privileged role.

The general solution is given by:

$$\begin{Bmatrix} v_y^{\pm} \\ b_x^{\pm} \\ \nu b_z^{\pm} \end{Bmatrix}(x, y) = \int_{-\infty}^{+\infty} \sum_k \begin{Bmatrix} \mp r_k A_k^{\pm} \\ \mp r_k B_k^{\pm} \\ C_k^{\pm} \end{Bmatrix} \exp(-\mathrm{i}\lambda x \pm s_k y)\,\mathrm{d}\lambda \tag{7.4.5}$$

$$\begin{Bmatrix} b_y^{\pm} \\ p^{\pm} \\ v_x^{\pm} \\ \nu v_z^{\pm} \\ \nu Re_x^{\pm} \\ \nu Re_y^{\pm} \end{Bmatrix}(x, y) = \int_{-\infty}^{+\infty} \sum_k \begin{Bmatrix} B_k^{\pm} \\ A_k^{\pm} \\ (r_k^2 - M^2)A_k^{\pm} \\ \pm r_k S C_k^{\pm} \\ \pm(R + \mathrm{i}\lambda)r_k^{-1}C_k^{\pm} \\ (R + \mathrm{i}\lambda)C_k^{\pm} \end{Bmatrix} \exp(-\mathrm{i}\lambda x \pm s_k y)\,\mathrm{d}\lambda \tag{7.4.6}$$

From equations (7.2.4) and (7.2.6′) we obtain:

$$S(1 + r_k^2)B_k^{\pm} = -(\beta^2 + r_k^2)A_k^{\pm}, \qquad k = 0, 1, 2 \tag{7.4.7}$$

$$\pm r_k S(1 + r_k^2)C_k^{\pm} = Q_k A_k^{\pm}, \qquad k = 0, 1, 2 \tag{7.4.8}$$

where

$$Q_k = (\beta^2 + r_k^2)\left(1 + r_k^2 + \frac{R}{\mathrm{i}\lambda}\right) - \frac{RS}{\mathrm{i}\lambda}(1 + r_k^2)(r_k^2 - M^2)$$

If we also introduce the notation:

$$P_k = (1 + r_k^2)\left(1 - \frac{RS}{\mathrm{i}\lambda}\right) + \frac{R(1 + S)}{\mathrm{i}\lambda}$$

the dispersion equation (7.4.3) may be written as:

$$P_k Q_k + v^2 r_k^2(1 + r_k^2)(\beta^2 + r_k^2) = 0 \tag{7.4.3′}$$

such that (7.4.8) becomes:

$$SP_k C_k^{\pm} = \mp v^2 r_k(\beta^2 + r_k^2)A_k^{\pm}, \qquad k = 0, 1, 2 \tag{7.4.8′}$$

From the boundary conditions ($h'_+ = h'_- = h'$) we obtain:

$$-\int_{-\infty}^{+\infty} \Sigma\, r_k A_k^+ \mathrm{e}^{-\mathrm{i}\lambda x}\,\mathrm{d}\lambda = h'(x), \qquad |x| \leqslant 1 \tag{7.4.9}$$

$$\left.\begin{aligned} &\Sigma\, r_k(A_k^+ + A_k^-) = 0 \\ &\Sigma\, r_k(1 + r_k^2)^{-1}(A_k^+ + A_k^-) = 0 \\ &\Sigma\, r_k(\beta^2 + r_k^2)P_k^{-1}(A_k^+ + A_k^-) = 0 \end{aligned}\right\} \tag{7.4.10}$$

$$\left.\begin{aligned} &\Sigma\,(\beta^2 + r_k^2)(1 + r_k^2)^{-1}(A_k^+ - A_k^-) = 0 \\ &\Sigma\,(\beta^2 + r_k^2)P_k^{-1}(A_k^+ - A_k^-) = 0 \end{aligned}\right\} \tag{7.4.11}$$

Since the roots r_k are distinct from one another, from equations (7.4.10) we get:

$$A_k^+ + A_k^- = 0, \qquad k = 0, 1, 2 \tag{7.4.12}$$

such that (7.4.11) yields:

$$A_j^+ = \alpha_j A_0^+, \qquad j = 1, 2 \tag{7.4.13}$$

$$\alpha_1 = \frac{(\beta^2 + r_0^2)(1 + r_1^2)(r_0^2 - r_2^2)P_1}{(\beta^2 + r_1^2)(1 + r_0^2)(r_2^2 - r_1^2)P_0}, \quad \alpha_2 = \frac{(\beta^2 + r_0^2)(1 + r_2^2)(r_1^2 - r_0^2)P_2}{(\beta^2 + r_2^2)(1 + r_0^2)(r_2^2 - r_1^2)P_1}$$

Accordingly:

$$[p] = 2\int_{-\infty}^{+\infty} (1 + \alpha_1 + \alpha_2)A_0^+ \, e^{-i\lambda x}\, d\lambda \overset{\text{def}}{=} f \tag{7.4.14}$$

such that:

$$2(1 + \alpha_1 + \alpha_2)A_0^+ = \frac{1}{2\pi}\int_{-1}^{+1} f(t)\, e^{i\lambda t}\, dt \tag{7.4.15}$$

the condition of continuity of pressure ($f \equiv 0$, $|x| > 1$) being also imposed.

From (7.4.13), (7.4.15) and (7.4.9) we find the following integral equation:

$$\int_{-1}^{+1} f(t)K(t - x)\, dt = h'(x), \qquad |x| \leqslant 1 \tag{7.4.16}$$

where

$$K(t - x) = \frac{1}{2\pi}\int_{-\infty}^{+\infty} k(\lambda)\, e^{i\lambda(t-x)}\, d\lambda, \quad k = -\frac{r_0 + r_1\alpha_1 + r_2\alpha_2}{2(1 + \alpha_1 + \alpha_2)}$$

In [7] equation (7.4.16) is reduced to a Fredholm type equation.

7.4.1 Weakly conducting fluids

Taking into account equations (7.4.4) and (7.4.4′) we obtain:

$$k = \frac{i\sqrt{1 - M^2}\,\text{sign}\,\lambda}{2} + \frac{\omega\sqrt{1 - M^2}}{2|\lambda|} + O\left(\frac{R^2}{\lambda^2}\right), \quad \text{for } M < 1 \tag{7.4.17}$$

where:

$$\omega = -\frac{RS}{2} \times \frac{M^2(\beta^2 v^2 - M^2)\sqrt{1 + v^2} + 2v^2\sqrt{1 - M^2}(1 - \sqrt{1 - M^2}\sqrt{1 + v^2})}{(1 - M^2)\sqrt{1 + v^2}(\beta^2 v^2 - M^2)^2} \tag{7.4.18}$$

$$\lim_{v^2 \to 0} \omega = \frac{RS}{2(1 - M^2)} \tag{6.4.22}$$

$$\lim_{M^2 \to 0} \omega = \frac{RS}{\sqrt{1 + v^2}(1 + \sqrt{1 + v^2})} \tag{7.2.39''}$$

With the approximation $R^2 \simeq 0$ (7.4.16) is reduced to the following equation:

$$\frac{1}{\pi}\int_{-1}^{'+1}\frac{f(t)}{t-x}\,\mathrm{d}t+\frac{\omega}{\pi}\int_{-1}^{+1}f(t)(C+\ln|t-x|)\,\mathrm{d}t=-\frac{2h'}{\sqrt{1-M^2}} \quad (7.4.16')$$

This equation coincides with equation (6.4.20) for $v^2 = 0$ and equation (7.2.39′) for $M^2 = 0$.

Therefore for a flat plate the lift is given by equation (6.3.37) in which ω has the expression (7.4.18).

For $M > 1$ we get:

$$k=\frac{\sqrt{M^2-1}}{2}\left(1+\frac{\omega_1}{\mathrm{i}\lambda}-\frac{\omega_2}{|\lambda|}\right)+O\left(\frac{R^2}{\lambda^2}\right) \quad (7.4.19)$$

$$\omega_1=\frac{RS}{2}\times\frac{2\beta^2v^2-M^2(\beta^2v^2-M^2)}{\beta^2(\beta^2v^2-M^2)^2}, \quad \omega_2=\frac{RSv^2}{(\beta^2v^2-M^2)\sqrt{1+v^2}\sqrt{M^2-1}}$$

$$\lim_{v^2\to 0}\omega_1=\frac{RS}{\beta^2}, \quad \lim_{v^2\to 0}\omega_2=0$$

such that equation (7.4.16) becomes:

$$f(x)+\frac{\omega_1}{2}\int_{-1}^{+1}f(t)\,\mathrm{sign}\,(t-x)\,\mathrm{d}t$$

$$+\frac{\omega_2}{\pi}\int_{-1}^{+1}f(t)(C+\ln|t-x|)\,\mathrm{d}t=\frac{2h'}{\sqrt{M^2-1}}, \quad |x|\leqslant 1 \quad (7.4.20)$$

We have not succeeded in integrating this equation. Differentiating with respect to x we obtain:

$$f'(x)-\omega_1 f(x)-\frac{\omega_2}{\pi}\int_{-1}^{'+1}\frac{f(t)}{t-x}\,\mathrm{d}t=\frac{2h''}{\sqrt{M^2-1}}, \quad |x|\leqslant 1 \quad (7.4.20')$$

7.5 Compressible fluids in aligned fields

In this case the equations of motion are (6.3.1), (7.3.2) — (7.3.7), (7.1.4) and (7.2.9). This system yields [8]:

$$T(v_y, b_y, b_z)=0 \quad (7.5.1)$$

where

$$T = T_1 T_2 + v^2 H\Delta \frac{\partial^2}{\partial x^2}, \ T_1 = \Delta + R(S-1)\frac{\partial}{\partial x}$$

$$T_2 = \left(\Delta - R\frac{\partial}{\partial x}\right) H + \beta^2 RS\Delta \frac{\partial}{\partial x}$$

operators T_1 and T_2 corresponding to incompressible and compressible scalar conductivity fluids respectively. Equations (7.5.1) provide the following dispersion equation:

$$\left\{1 + r^2 - \frac{R(S-1)}{i\lambda}\right\}\left\{\left(1 + r^2 + \frac{R}{i\lambda}\right)(\beta^2 + r^2) - \frac{RS}{i\lambda}\beta^2(1 + r^2)\right\}$$
$$+ v^2(\beta^2 + r^2)(1 + r^2) = 0 \qquad (7.5.2)$$

or:

$$(1 + r^2)^3 + \left\{v^2 - M^2 + \frac{R}{i\lambda}(2 - S - S\beta^2)\right\}(1 + r^2)^2$$
$$-M^2\left\{v^2 + \frac{R}{i\lambda}(2 - S)\right\}(1 + r^2) + O\left(\frac{R^2}{\lambda^2}\right) = 0$$

Denoting by r_k ($k = 0, 1, 2$) those roots of equation (7.5.2) for which the expressions $s_k = -i\lambda r_k$ have negative real parts, we have:

$$1 + r_0^2 = O(R^2/\lambda^2)$$

$$1 + r_1^2 = M^2 + \frac{M^2 RS\beta^2}{i\lambda(v^2 + M^2)} + O\left(\frac{R^2}{\lambda^2}\right)$$

$$1 + r_2^2 = -v^2 - \frac{R}{i\lambda}\left(2 - S - \frac{S\beta^2 v^2}{v^2 + M^2}\right) + O\left(\frac{R^2}{\lambda^2}\right)$$

and:

$$r_0 = -i\,\mathrm{sign}\,\lambda + O(R^2/\lambda^2)$$

$$r_1 = \begin{cases} -i\sqrt{1-M^2}\,\mathrm{sign}\,\lambda + \dfrac{M^2 RS\beta^2}{2(v^2+M^2)\sqrt{1-M^2}}\dfrac{1}{|\lambda|} + O\left(\dfrac{R^2}{\lambda^2}\right), & M < 1 \\ -\sqrt{M^2-1} - \dfrac{M^2 RS\beta^2}{2(v^2+M^2)\sqrt{M^2-1}}\dfrac{1}{i\lambda} + O\left(\dfrac{R^2}{\lambda^2}\right), & M > 1 \end{cases} \qquad (7.5.3)$$

$$r_2 = -i\sqrt{1 + v^2}\,\mathrm{sign}\,\lambda - \left(2 - S - \frac{S\beta^2 v^2}{v^2 + M^2}\right)\frac{R}{2\sqrt{1+v^2}}\frac{1}{|\lambda|} + O\left(\frac{R^2}{\lambda^2}\right)$$

The general solution of the problem is:

$$\begin{Bmatrix} v_y^{\pm} \\ b_y^{\pm} \\ vb_z^{\pm} \end{Bmatrix}(x, y) = \int_{-\infty}^{+\infty} \sum_{k=0,1,2} \begin{Bmatrix} A_k^{\pm} \\ B_k^{\pm} \\ C_k^{\pm} \end{Bmatrix} \exp(-\mathrm{i}\lambda x \pm s_k y)\,\mathrm{d}\lambda \tag{7.5.4}$$

$$\begin{Bmatrix} \beta^2 v_x^{\pm} \\ b_x^{\pm} \\ vRe_x^{\pm} \end{Bmatrix}(x, y) = \int_{-\infty}^{+\infty} \sum \begin{Bmatrix} \mp r_k A_k^{\pm} \\ \mp r_k B_k^{\pm} \\ \pm s_k C_k^{\pm} \end{Bmatrix} \exp(-\mathrm{i}\lambda x \pm s_k y)\,\mathrm{d}\lambda \tag{7.5.5}$$

$$\left.\begin{aligned} &\beta^2 S(1 + r_k^2)B_k^{\pm} = (\beta^2 + r_k^2)A_k^{\pm} \\ &\beta^2 S P_k C_k^{\pm} = v^2(\beta^2 + r_k^2)A_k^{\pm} \end{aligned}\right. \tag{7.5.6}$$

$$P_k \overset{\text{def}}{=} 1 + r_k^2 + \mathrm{i}\lambda^{-1}R(S - 1) \tag{7.5.7}$$

The boundary conditions:

$$v_y^{+}(x, 0) = h'(x), \quad |x| \leqslant 1$$

$$[v_y] = [b_y] = [b_z] = [b_x] = [e_x] = 0, \quad \forall x$$

yield the following relationships:

$$\int_{-\infty}^{+\infty} \Sigma A_k^{+}\, \mathrm{e}^{-\mathrm{i}\lambda x}\,\mathrm{d}\lambda = h'(x), \qquad |x| \leqslant 1 \tag{7.5.8}$$

$$\left.\begin{aligned} \Sigma(A_k^{+} - A_k^{-}) = 0,\ \Sigma(1 + r_k^2)^{-1}(A_k^{+} - A_k^{-}) = 0 \\ \Sigma P_k^{-1}(\beta^2 + r_k^2)(A_k^{+} - A_k^{-}) = 0 \end{aligned}\right\} \tag{7.5.9}$$

$$\left.\begin{aligned} \Sigma r_k(\beta^2 + r_k^2)(1 + r_k^2)^{-1}(A_k^{+} + A_k^{-}) = 0 \\ \Sigma r_k P_k^{-1}(\beta^2 + r_k^2)(A_k^{+} + A_k^{-}) = 0 \end{aligned}\right\} \tag{7.5.10}$$

From (7.5.9) and (7.5.10) we get:

$$\begin{aligned} &A_k^{+} = A_k^{-}, \qquad k = 0, 1, 2 \\ &A_j^{+} = \alpha_j A_0^{+}, \qquad j = 1, 2 \end{aligned} \tag{7.5.11}$$

$$\alpha_1 = \frac{r_0 P_1(1 + r_1^2)(\beta^2 + r_0^2)(r_0^2 - r_2^2)}{r_1 P_0(1 + r_0^2)(\beta^2 + r_1^2)(r_2^2 - r_1^2)}, \quad \alpha_2 = \frac{r_0 P_2(1 + r_2^2)(\beta^2 + r_0^2)(r_1^2 - r_0^2)}{r_2 P_0(1 + r_0^2)(\beta^2 + r_2^2)(r_2^2 - r_1^2)}$$

From the expression of pressure we have:

$$\beta^2[p] = 2\int_{-\infty}^{+\infty} (r_0 + r_1\alpha_1 + r_2\alpha_2)A_0^+ \, e^{-i\lambda x} \, d\lambda \overset{\text{def}}{=} \beta^2 f(x)$$

$$f(x) \equiv 0, \qquad |x| > 1 \tag{7.5.12}$$

such that:

$$2(r_0 + r_1\alpha_1 + r_2\alpha_2)A_0^+ = \frac{\beta^2}{2\pi}\int_{-1}^{+1} f(t)\, e^{i\lambda t} \, dt \tag{7.5.12'}$$

Conditions (7.5.11), (7.5.12′) and equation (7.5.8) yield the following integral equation (for the determination of f):

$$\int_{-1}^{+1} \beta^2 f(t) K(t-x)\, dt = h'(x), \; |x| \leqslant 1$$

$$K(t-x) = \frac{1}{2\pi}\int_{-\infty}^{+\infty} k\, e^{i\lambda(t-x)} \, d\lambda, \; k = \frac{1+\alpha_1+\alpha_2}{2(r_0 + r_1\alpha_1 + r_2\alpha_2)} \tag{7.5.13}$$

This equation solves the problem.

For weakly conducting fluids we have:

$$k = \frac{i \operatorname{sign} \lambda}{2\sqrt{1-M^2}} + \frac{\omega}{2\sqrt{1-M^2}}\frac{1}{|\lambda|} + O\left(\frac{R^2}{\lambda^2}\right), \qquad M < 1 \tag{7.5.14}$$

where:

$$\omega = \frac{RS\beta^2\nu^2}{(\nu^2+M^2)^2}\left\{1 - \frac{\sqrt{1-M^2}}{\sqrt{1+\nu^2}} + \frac{M^2(\nu^2+M^2)}{2\nu^2\beta^2}\right\}$$

$$\lim_{\nu^2 \to 0} \omega = \frac{RS}{2}, \qquad \lim_{M^2 \to 0} \omega = \frac{RS}{\sqrt{1+\nu^2}(1+\sqrt{1+\nu^2})} \tag{7.5.15}$$

Neglecting R^2(7.9.13) reduces itself to (6.3.34) ω being given by (7.5.15). For $M = 0$ the resulting equation also coincides with (7.3.36). Consequently, the lift is given by equations (6.3.36) and (6.3.37) in which ω defined by (7.5.15) is substituted for ω defined by (6.3.31).

If $M > 1$ we get:

$$k = -\frac{1}{2\sqrt{M^2-1}} - \frac{\omega_1}{2\sqrt{M^2-1}}\frac{1}{\mathrm{i}\lambda} + \frac{\omega_2}{2\sqrt{M^2-1}}\frac{1}{|\lambda|} + O\left(\frac{R^2}{\lambda^2}\right)$$

$$\omega_1 = \frac{RS(M^4 + 2\nu^2 - \nu^2 M^2)}{2(\nu^2 + M^2)^2}, \quad \omega_2 = \frac{RS\nu^2\sqrt{M^2-1}}{(\nu^2 + M^2)\sqrt{1+\nu^2}}, \tag{7.5.16}$$

$$\lim_{\nu^2\to 0} k = -\frac{1}{2\sqrt{M^2-1}} - \frac{RS}{4\mathrm{i}\lambda\sqrt{M^2-1}}$$

such that (by neglecting R^2) (7.5.13) is reduced to the following equation:

$$f(x) + \frac{\omega_1}{2}\int_{-1}^{+1} f(t)\,\mathrm{sign}\,(t-x)\,\mathrm{d}t$$

$$+ \frac{\omega_2}{\pi}\int_{-1}^{+1} f(t)(C + \ln|t-x|)\,\mathrm{d}t = \frac{2h'}{\sqrt{M^2-1}}, \quad |x| \leqslant 1 \tag{7.5.17}$$

8

Viscous flow past thin bodies

8.1 Introduction

The motion of viscous fluids past a flat plate represents a classical problem in fluid mechanics. While studies on the flow past the aligned plate with unperturbed velocity have been performed in the thirties, the investigation of the flow past the incident plate was (due to the mathematical difficulties) carried out only in 1962. These investigations are mainly due to Ko Tamada and T. Miyagi [1] – [3].

The problem of the flow of conducting fluids past an aligned flat plate has been solved by H. Greenspan and G. Carrier [4] – [6]. Since one makes the hypothesis that the plate is oriented along the direction of the fluid velocity at infinity, this problem can be solved by means of the Fourier transform. However, this method cannot be used for the incident plate.

The incident plate problem has been solved by N. Marcov and the author in [7] and subsequently extended by N. Marcov to the case of any thin profile [8]. The solution is obtained through its potential representation and the solution of two integral equations. For the aligned plate one obtains the results of H. Greenspan and G. Carrier, and for the non-conducting fluid those of Ko Tamada and T. Miyagi. Due to its generality this chapter is based on this solution*. So in Section 8.2 the general problem for any thin body is presented, while in Section 8.3 and Section 8.4 the aligned plate and the incident plate problems, respectively, are presented. The perfectly conducting fluid case represents the subject of Section 8.5.

Different mathematical problems arise when the applied magnetic field is orthogonal to the plate. The case when the latter is aligned with the unperturbed velocity has been delt with by D. Hector [9] for the semi-infinite plate, by N. Marcov [10] for the case when the flow parameters obey the relationship $\varepsilon\mu\sigma = 1$ and by L. Dragoş [11] for the general case. Numerous investigations (Rossow [12], Hasimoto [13], Dix [14], etc.) use the boundary layer theory. The solution of [11] is presented in Section 8.6.

The results on the Hall effect are due to H. Hasimoto and G. Ianowitz [15] for an insulating plane, to L. Dragoş [16] for the conducting aligned plate and to S. Datta

* This chapter was written in cooperation with N. Marcov.

[17] for the orthogonal plate. Finally, the problem of the aligned plate in compressible fluids has been tackled in [18].

Some other papers are indicated in the bibliography but a proper systematization would be difficult at the present time.

8.2 Flow past thin bodies

8.2.1 The equations of motion

Let us consider a viscous incompressible fluid in uniform motion with velocity $\boldsymbol{V_0}$ past a thin cylindrical body whose cross-section contains the vector $\boldsymbol{V_0}$ and reduces itself to either a smooth arc Γ (or a straight line segment). Let us suppose that the motion takes place in the presence of a homogeneous magnetic field of induction $\boldsymbol{B_0}$ aligned with $\boldsymbol{V_0}$. One has to determine the perturbed motion of the fluid. It is obvious that this should be a steady plane motion. With the Ox-axis (of unit vector $\boldsymbol{i}_1$) along the direction of $\boldsymbol{V_0}$ the pressure, velocity and magnetic induction at some point will have the following forms:

$$\begin{aligned} & p_0 + \rho_0 V_0^2 p\,;\ p = p(x, y) \\ & V_0(\boldsymbol{i}_1 + \boldsymbol{v})\,;\ \boldsymbol{v} = \boldsymbol{v}(x, y),\ \boldsymbol{v} = (v_x, v_y) \\ & B_0(\boldsymbol{i}_1 + \boldsymbol{b})\,;\ \boldsymbol{b} = \boldsymbol{b}(x, y),\ \boldsymbol{b} = (b_x, b_y) \\ & \lim_{\infty}\,(p, \boldsymbol{v}, \boldsymbol{b}) = 0 \end{aligned} \tag{8.2.1}$$

Substituting these expressions in the equations of motion, linearizing these equations and assuming the perturbation to be small it follows that:

$$\frac{\partial v_x}{\partial x} = -\frac{\partial P}{\partial x} + \frac{1}{Re}\Delta v_x + Rh\frac{\partial b_x}{\partial x} \tag{8.2.2}$$

$$\frac{\partial v_y}{\partial x} = -\frac{\partial P}{\partial y} + \frac{1}{Re}\Delta v_y + Rh\frac{\partial b_y}{\partial x} \tag{8.2.3}$$

$$\frac{\partial v_x}{\partial x} + \frac{\partial v_y}{\partial y} = 0 \tag{8.2.4}$$

$$\frac{\partial b_x}{\partial x} + \frac{\partial b_y}{\partial y} = 0 \tag{8.2.5}$$

$$\frac{\partial b_y}{\partial x} - \frac{\partial b_x}{\partial y} = Rm(b_y - v_y) \tag{8.2.6}$$

$$P \overset{\text{def}}{=} p + Rhb_x \tag{8.2.7}$$

the electric field being constant and zero due to the boundary conditions.

From equations (8.2.2) and (8.2.3) one obtains:

$$\Delta P = 0, \ \Delta \overset{\text{def}}{=} \frac{\partial^2}{\partial x^2} + \frac{\partial^2}{\partial y^2} \tag{8.2.8}$$

so that we may introduce the function Q harmonically conjugated to P:

$$\frac{\partial P}{\partial x} = \frac{\partial Q}{\partial y}, \ \frac{\partial P}{\partial y} = -\frac{\partial Q}{\partial x} \tag{8.2.8'}$$

In the complex plane $z = x + iy$ the holomorphic function $P + iQ$ vanishes at infinity so that we have:

$$(P + iQ)_\infty = \frac{\alpha_1}{z} + \frac{\alpha_2}{z^2} + \ldots \tag{8.2.9}$$

If we now introduce the complex variable function

$$\Pi_1 + i\Pi_2 = Re \int (P + iQ)\, dz \tag{8.2.10}$$

we get:

$$P = \frac{1}{Re}\frac{\partial \Pi}{\partial y} \quad Q = \frac{1}{Re}\frac{\partial \Pi}{\partial x} \tag{8.2.11}$$

Equations (8.2.4) and (8.2.5) are satisfied by taking:

$$v_x = \frac{\partial \psi}{\partial y}, \ v_y = -\frac{\partial \psi}{\partial x}, \ \psi = \psi(x, y)$$

$$b_x = \frac{\partial \chi}{\partial y}, \ b_y = -\frac{\partial \chi}{\partial x}, \ \chi = \chi(x, y) \tag{8.2.12}$$

Taking into account (8.2.8') and (8.2.12) from (8.2.2) and (8.2.3) it follows that the first derivatives of the function:

$$\Delta\psi + Re\, Rh \frac{\partial \chi}{\partial x} - Re \frac{\partial \Pi}{\partial x} - ReQ$$

are vanishing, which proves that this function is a constant and since it is zero at infinity it will be so everywhere.

For the determination of functions Π, ψ and χ we have the following system of equations:

$$\Delta \Pi = 0$$

$$\Delta \psi = -Re\,Rh \frac{\partial \chi}{\partial x} + Re \frac{\partial \psi}{\partial x} + \frac{\partial \Pi}{\partial x}$$

$$\Delta \chi = Rm \left(\frac{\partial \chi}{\partial x} - \frac{\partial \psi}{\partial x} \right) \tag{8.2.13}$$

which by the transformation:

$$\Pi = 2\tilde{\chi}$$

$$\psi\delta = (a-b)\,\tilde{\chi} + (b-c)\,\tilde{\psi} + (c-a)\,\tilde{\Pi} \tag{8.2.14}$$

$$\chi\delta = (a-b)\,\tilde{\chi} + b\tilde{\psi} - a\tilde{\Pi}$$

is separated and the following equations are obtained:

$$\Delta\tilde{\chi} = 0, \; \Delta\tilde{\psi} - 2a \frac{\partial \tilde{\psi}}{\partial x} = 0, \; \Delta\tilde{\Pi} - 2b \frac{\partial \tilde{\Pi}}{\partial x} = 0 \tag{8.2.15}$$

where:

$$\begin{aligned} 4a &= Re + Rm - \sqrt{(Re - Rm)^2 + 4ReRmRh} \\ &= Re + Rm - \sqrt{(Re + Rm)^2 + 4Re\,Rm\,(Rh - 1)} \\ 4b &= Re + Rm + \sqrt{(Re + Rm)^2 + 4Re\,Rm(Rh - 1)} \\ 2c &= Re\,(1 - Rh), \; \delta = c(b - a) \end{aligned} \tag{8.2.16}$$

c and δ are assumed to be non-zero constants so that the transformation (8.2.14) does not apply to the Alfvén motions ($Rh = 1$). The constant b is positive in any case while a is positive and negative for the sub-alfvénian ($Rh < 1$) and super-alfvénian ($Rh > 1$) motions, respectively.

From the condition (8.2.1′) and from the representation (8.2.11) and (8.2.12) it follows that the first derivatives of the functions Π, ψ and χ are vanishing at infinity which property is also conveyed to the functions $\tilde{\Pi}$, $\tilde{\psi}$ and $\tilde{\chi}$. Consequently:

$$\lim_{\infty}\left(\frac{\partial}{\partial x}, \frac{\partial}{\partial y}\right)(\tilde{\Pi}, \tilde{\psi}, \tilde{\chi}) = 0 \tag{8.2.17}$$

The solution (8.2.15) has also been obtained by H. Yoshinobu [19] taking a particular case of the solution of the aligned field three-dimensional problem.

Let now:

$$x = x(s), \ \ y = y(s), \ \ s \in [0, l] \tag{8.2.18}$$

be the equations of the profile Γ. From the adherence condition of the viscous fluid it follows that:

$$v_x|_\Gamma = -1, \ v_y|_\Gamma = 0 \tag{8.2.19}$$

and from the condition of magnetic field continuity:

$$[b_x - ib_y]_\Gamma = 0 \tag{8.2.20}$$

the last relationship being imposed by the continuity of the tangential ($[\boldsymbol{b}] \cdot \boldsymbol{s} = 0$) and normal ($[\boldsymbol{b}] \cdot \boldsymbol{n} = 0$) components, respectively. In the Romanian edition of this work we considered the more general case when a surface current exists on the profile. In the following we arrange for the positive sense on Γ to correspond to increasing curvilinear coordinate s and will indicate by $(+)$ and $(-)$ the boundary values on the left and righthand side of Γ respectively.

8.2.2 Potential representation of the solution

Noticing that by a transformation of the form $\tilde{\psi} = F \exp(ax)$ the $\tilde{\psi}$ equation (8.2.15b) reduces itself to the Helmholtz equation $\Delta F - a^2 F = 0$ and knowing that the fundamental solution with respect to the origin of the latter is K_0 $(|a|\ r)$ we obtain that the fundamental solution of the $\tilde{\psi}$ equation is:

$$K_0(|a|r) \exp(ar \cos\theta)$$

where r and θ are the polar coordinates in the plane of motion and K_0 the imaginary argument Bessel function of the second kind and zero order. We shall also use in the following the (first order) Bessel function K_1 as well as the function Θ defined by the equation:

$$\Theta(ar, \theta) \exp(ar \cos\theta) = \theta + \sin\theta \int_0^{ar} K_0(|z|) \exp(z \cos\theta)\, dz$$

$$= \int_0^{\theta} r \{|a|\, K_1(|a|\, r) + a \cos\theta\, K_0(|a|\, r)\} \exp(ar \cos\theta)\, d\theta \tag{8.2.21}$$

With respect to K_0 the function Θ has the same role in the study of the Helmholtz equation as the function θ with respect to $-\ln r$ in the study of the Laplace equation. As a matter of fact in origin they have the following behaviour:

$$K_0 = \ln\frac{1}{r} + \ldots, \quad \Theta = \theta + \ldots \tag{8.2.22}$$

the terms left out being bounded. It is obvious that $\Theta(ar, \theta) \exp(ar \cos\theta)$ is a non-uniform function which obeys the Oseen equation for $\tilde{\psi}$. As a matter of fact we have the equations:

$$\frac{\partial \Theta}{\partial x} = \frac{\partial}{\partial y} K_0(|a|\, r|) - a\Theta \left(= -|a| \sin\theta\, K_1(|a|\, r) - a\Theta\right)$$

$$\frac{\partial \Theta}{\partial y} = -\frac{\partial}{\partial x} K_0(|a| r) + aK_0(|a| r) \left(= |a| \cos\theta\, K_1(|a| r) + aK_0(|a| r)\right) \tag{8.2.23}$$

Given two points s and t of the arc Γ we shall introduce the notation:

$$\xi + i\eta = x(s) - x(t) + i(y(s) - y(t)) = \rho e^{i\alpha} \operatorname{sign}(s - t) \tag{8.2.24}$$

Since Γ is a smooth arc it follows that the distance ρ between these points as well as the angle α between the chord which connects them and the Ox-axis are continuous functions. For any point on the place and for the point t of the body we have the notation:

$$\tilde{x} + i\tilde{y} = x - x(t) + i(y - y(t)) = \tilde{r} \exp(i\tilde{\theta}) \tag{8.2.25}$$

We have to assume in the following that $\tilde{\theta}$ is a continuous function of x, y and t. In this respect we shall consider a cross-cut $P_\infty P_0 P_t$ in the xOy-plane. Particularly, if $P(x, y)$ tends to a point s of the obstacle it follows that

$$\tilde{\theta} = \begin{cases} \alpha, & s > t \\ \alpha \pm \pi, & s < t \end{cases} \tag{8.2.26}$$

the (+) and (−) signs corresponding to the left- and right-hand side of the obstacle, respectively.

The equations (8.2.15) should be satisfied within the region occupied by the fluid, that is, in the plane xOy with the cross-cut Γ. The conditions (8.2.17) determine the following solution:

$$\tilde{\chi} = \int_0^l \left\{ A_1(t)\,\tilde{\theta} + B_1(t)\ln\frac{1}{\tilde{r}} \right\} \mathrm{d}t$$

$$\tilde{\psi} = \int_0^l \left\{ A_2(t)\,\Theta\,(a\tilde{r}, \tilde{\theta}) + B_2(t)\,K_0(|a|\tilde{r}) \right\} \exp\,(a\tilde{x})\,\mathrm{d}t \tag{8.2.27}$$

$$\tilde{\Pi} = \int_0^l \left\{ A_3(t)\,\Theta\,(b\tilde{r}, \tilde{\theta}) + B_3(t)\,K_0(b\tilde{r}) \right\} \exp\,(b\tilde{x})\,\mathrm{d}t$$

where $A_i(t)$ and $B_i(t)$ are functions to be determined from the boundary conditions. Taking into account equations (8.2.23), by differentiation one obtains:

$$\left(\frac{\partial}{\partial y} + \mathrm{i}\frac{\partial}{\partial x}\right)\tilde{\chi} = \int_0^l \overline{C_1(t)} \mathrm{e}^{-\mathrm{i}\tilde{\theta}}\frac{1}{\tilde{r}}\,\mathrm{d}t = \frac{1}{2\pi\mathrm{i}}\int_0^l \{-2\pi\mathrm{i}\overline{C_1(t)}\mathrm{e}^{-\mathrm{i}\theta(t)}\}\frac{\mathrm{d}z(t)}{z(t)-z}$$

$$\left(\frac{\partial}{\partial y} + \mathrm{i}\frac{\partial}{\partial x}\right)\tilde{\psi} = \int_0^l \{\overline{C_2(t)}\mathrm{e}^{-\mathrm{i}\tilde{\theta}}\,|a|\,K_1(|a|\tilde{r}) + aC_2(t)\,K_0(|a|\tilde{r})\}\mathrm{e}^{a\tilde{x}}\mathrm{d}t$$

$$\left(\frac{\partial}{\partial y} + \mathrm{i}\frac{\partial}{\partial x}\right)\tilde{\Pi} = \int_0^l \{\overline{C_3(t)}\mathrm{e}^{-\mathrm{i}\tilde{\theta}}\,bK_1(b\tilde{r}) + bC_3(t)\,K_0(b\tilde{r})\}\mathrm{e}^{b\tilde{x}}\mathrm{d}t \tag{8.2.28}$$

where:

$$C_j = A_j + \mathrm{i}B_j,\ j = 1, 2, 3 \tag{8.2.29}$$

Taking the limit in (8.2.28):

$$x + \mathrm{i}y \to x(s) + \mathrm{i}y(s),\ s \in (0, l)$$

one obtains the following values to the left/right-hand side:

$$\left(\frac{\partial\tilde{\chi}}{\partial y} + \mathrm{i}\frac{\partial\tilde{\chi}}{\partial x}\right)(s) = \mp\pi\mathrm{i}\,\overline{C_1(s)}\,\mathrm{e}^{-\mathrm{i}\theta(s)} + \int_0^{\prime l} \overline{C_1(t)}\frac{\mathrm{d}t}{\xi + \mathrm{i}\eta}$$

$$\left(\frac{\partial\tilde{\psi}}{\partial y} + \mathrm{i}\frac{\partial\tilde{\psi}}{\partial x}\right)(s) = \mp\pi\mathrm{i}\,\overline{C_2(s)}\mathrm{e}^{-\mathrm{i}\theta(s)}$$

$$+\int_0^{\prime l}\left\{\overline{C_2(t)}\frac{|a|\rho}{\xi+i\eta}K_1(|a|\rho)+aC_2(t)\,K_0(|a|\rho)\right\}e^{a\xi}\,dt \tag{8.2.30}$$

$$\left(\frac{\partial\tilde{\Pi}}{\partial y}+i\frac{\partial\tilde{\Pi}}{\partial x}\right)(s)=\mp\pi i\,\overline{C_3(s)}\,e^{-i\theta(s)}$$

$$+\int_0^{\prime l}\left\{\overline{C_3(t)}\frac{b\rho}{\xi+i\eta}K_1(b\rho)+bC_3(t)\,K_0(b\rho)\right\}e^{b\xi}\,dt$$

the integrals being considered in the Cauchy principal value. The first equation (8.2.30) is given by the Plemelj formulae. In order to get the other two equations one uses the fact that K_0 has a logarithmic singularity at the origin (8.2.22) and K_1 may be written as follows:

$$|a|\,K_1(|a|\tilde{r})=\frac{1}{\tilde{r}}+O\left(\tilde{r}\ln\frac{1}{\tilde{r}}\right)$$

In this way last two equations are derived from the first.

Using the equations (8.2.12), (8.2.14) and (8.2.28) for the velocity field and the magnetic induction one obtains the following representation:

$$w\delta=(a-b)\int_0^l\overline{C_1(t)}\,e^{-i\tilde{\theta}}\,\frac{1}{\tilde{r}}\,dt \tag{8.2.31}$$

$$+(b-c)\int_0^l\{\overline{C_2(t)}e^{-i\tilde{\theta}}|a|K_1(|a|\tilde{r})+C_2(t)\,aK_0(|a|\tilde{r})\}e^{a\tilde{x}}dt$$

$$+(c-a)\int_0^l\{\overline{C_3(t)}\,e^{-i\tilde{\theta}}bK_1(b\tilde{r})+C_3(t)\,bK_0(b\tilde{r})\}\,e^{b\tilde{x}}dt$$

$$B\delta=(a-b)\int_0^l\overline{C_1(t)}\,e^{-i\tilde{\theta}}\,\frac{1}{\tilde{r}}\,dt \tag{8.2.32}$$

$$+b\int_0^l\{\overline{C_2(t)}\,e^{-i\tilde{\theta}}\,|a|\,K_1(|a|\tilde{r})+C_2(t)\,aK_0(|a|\tilde{r})\}\;e^{a\tilde{x}}dt$$

$$-a\int_0^l\{\overline{C_3(t)}\,e^{-i\tilde{\theta}}\,bK_1(b\tilde{r})+C_3(t)\,bK_0(b\tilde{r})\}\;e^{b\tilde{x}}\,dt$$

with the notations:

$$w = v_x - iv_y \ (8.2.31'), \quad B = b_x - ib_y \ (8.2.32')$$

Now imposing the boundary conditions (8.2.19), (8.2.20) it follows that:

$$[w]\delta = -2\pi i\{(a-b)\overline{C_1(s)} + (b-c)\overline{C_2(s)} + (c-a)\overline{C_3(s)}\}e^{-i\theta(s)} = 0$$

$$[B]\delta = -2\pi i\{(a-b)\overline{C_1(s)} + b\overline{C_2(s)} - a\overline{C_3(s)}\}e^{-i\theta(s)} = 0$$

so that:

$$C_1(s) = C_2(s) = C_3(s) \overset{\text{def}}{=} C(s)$$

$$C(s) \overset{\text{def}}{=} A + iB \qquad (8.2.33)$$

With these expressions for C_1, C_2 and C_3, ψ and χ are uniform functions since the integrands in the representations of these functions (derived from (8.2.14), (8.2.27) and (8.2.23)) are periodic functions with respect to $\tilde{\theta}$.

8.2.3 Integral equations

Imposing the boundary condition (8.2.19) on the velocity (8.2.31) and taking into account (8.2.30) and (8.2.33) we have the following integral equation:

$$(a-b)\int_0^l \overline{C(t)}\frac{dt}{\xi + i\eta} + (b-c)\int_0^l \left\{\overline{C(t)}\frac{|a|\rho}{\xi + i\eta}K_1(|a|\rho) + C(t)\,aK_0(|a|\rho)\right\} e^{a\xi}\,dt$$

$$+ (c-a)\int_0^l \left\{\overline{C(t)}\frac{b\rho}{\xi + i\eta}K_1(b\rho) + C(t)\,bK_0(b\rho)\right\} e^{b\xi}\,dt = 0 \qquad (8.2.34)$$

the functions ξ, η and ρ being dependent on s and t through the formulae (8.2.24). If the arc equations are given with respect to some parameter t_1 and not with respect to the curvilinear coordinate t then the factor dt/dt_1 may be included in the unknown function C.

Since the kernel of the integral equation (8.2.34) has a logarithmic singularity this equation may be reduced to a second-kind Freedholm equation. Indeed, with the notation:

$$F(s) = -\frac{1}{\pi}\int_0^l C(t)\ln|s-t|\,dt \qquad (8.2.35)$$

and reversing [G.34] we obtain:

$$C(t) = \frac{1}{\pi\sqrt{t(l-t)}}\int_0^{\prime l} F'(\tau)\sqrt{\tau(l-\tau)}\,\frac{\mathrm{d}\tau}{t-\tau} + \frac{K}{\pi\sqrt{t(l-t)}} \tag{8.2.35$'$}$$

where:

$$K = \int_0^l C(t)\,\mathrm{d}t = -\frac{1}{\ln(l/4)}\int_0^l \frac{F(s)\,\mathrm{d}s}{\sqrt{s(l-s)}} \tag{8.2.35$''$}$$

Taking into account these results equation (8.2.34) becomes:

$$\pi F(s) - \int_0^l \{M(s,\tau)\,F'(\tau) + N(s,\tau)\,\overline{F'(\tau)}\}\,\mathrm{d}\tau$$

$$= -1 - KG(s) - \overline{K}H(s) \tag{8.2.36}$$

where:

$M(s,\tau) =$

$$\frac{1}{\pi\delta}\int_0^{\prime l}\{a(b-c)\,K_0(|a|\rho)\mathrm{e}^{a\xi} + b(c-a)K_0(b\rho)\mathrm{e}^{b\xi} + \delta\ln|s-t|\}\sqrt{\frac{\tau(l-\tau)}{t(l-t)}}\,\frac{\mathrm{d}t}{\tau-t}$$

$N(s,\tau) =$

$$\frac{1}{\pi\delta}\int_0^{\prime l}\left\{|a|(b-c)\,K_1(|a|\rho)\,\mathrm{e}^{a\xi} + b(c-a)\,K_1(b\rho)\,\mathrm{e}^{b\xi} + \frac{a-b}{\rho}\right\}\sqrt{\frac{\tau(l-\tau)}{t(l-t)}}\,\frac{\rho\,\mathrm{d}t}{(\xi+\mathrm{i}\eta)(\tau-t)}$$

$$G(s) = \frac{1}{\pi\delta}\int_0^l\{a(b-c)\,K_0(|a|\rho)\mathrm{e}^{a\xi} + b(c-a)\,K_0(b\rho)\mathrm{e}^{b\xi}$$

$$+\,\delta\ln|s-t|\}\frac{\mathrm{d}t}{\sqrt{t(l-t)}}$$

$$H(s) = \frac{1}{\pi\delta}\int_0^l\left\{|a|(b-c)\,K_1(|a|\rho)\,\mathrm{e}^{a\xi} + b(c-a)\,K_1(b\rho)\mathrm{e}^{b\xi}\right.$$

$$\left.+\,\frac{a-b}{\rho}\right\}\frac{\rho\,\mathrm{d}t}{(\xi+\mathrm{i}\eta)\sqrt{t(l-t)}} \tag{8.2.36$'$}$$

From (8.2.36) it follows that function F depends in a linear way on the constants K and $\bar{K}$. Once having determined the function F the constants K and $\bar{K}$ are determined from (8.2.35″). In this manner the proposed problem is reduced to finding the solution of the integral equation (8.2.34) or (8.2.36).

For $b \ll 1$ by expanding in series both the Bessel functions K_0 and K_1 as well as the exponential functions one obtains (up to an additive term of the order of $b \ln b$):

$$M(s,\tau) = \frac{1}{\pi}\int_0^{\prime l} \ln\frac{|s-t|}{\rho}\sqrt{\frac{\tau(l-\tau)}{t(l-t)}}\,\frac{dt}{\tau - t}$$

$$N(s,\tau) = \frac{1}{\pi}\int_0^{\prime l} \frac{\xi}{\xi + i\eta}\sqrt{\frac{\tau(l-\tau)}{t(l-t)}}\,\frac{dt}{\tau - t} \tag{8.2.36″}$$

$$G(s) = \frac{1}{\pi}\int_0^{l} \ln\frac{|s-t|}{\rho}\,\frac{dt}{\sqrt{t(l-t)}} - \delta_1, \quad H(s) = \frac{1}{\pi}\int_0^{l} \frac{\xi\, dt}{(\xi + i\eta)\sqrt{t(l-t)}}$$

$$\delta_1 = \gamma + \frac{1}{\delta}\left\{a(b-c)\ln\frac{|a|}{2} + b(c-a)\ln\frac{b}{2}\right\}, \quad \gamma = 0.577\ldots$$

8.2.4 Fluid action on the field

In order to calculate the magnetofluid dynamic (MFD) action we shall use the expression of the total tensor (2.4.17) in which the constants will be neglected (being of no importance in MFD action) and linearize the corresponding term according to the structure (8.2.1). In this manner we get:

$$\mathcal{T}_{ij} = \rho_0 V_0^2\, t_{ij} \tag{8.2.37}$$

where:

$$t_{11} = -P + \frac{2}{Re}\frac{\partial v_x}{\partial x} + 2Rhb_x = t_{11}^{(m)} + Rhb_x$$

$$t_{12} = \frac{1}{Re}\left(\frac{\partial v_x}{\partial y} + \frac{\partial v_y}{\partial x}\right) + Rhb_y = t_{12}^{(m)} + Rhb_y$$

$$t_{22} = -P + \frac{2}{Re}\frac{\partial v_y}{\partial y} = t_{22}^{(m)} - Rhb_x \tag{8.2.37′}$$

Let $\boldsymbol{n} = (n_1, n_2)$ be the normal to Γ oriented to the left when one goes along the arc in the positive sense. Then the projections on the coordinate axes of the MFD action are given by the formulae:

$$\mathscr{R}_x = \oint (\mathscr{T}_{11}n_1 + \mathscr{T}_{21}n_2)\,\mathrm{d}s = \rho_0 V_0^2 l \int_0^l \left(-[t_{11}]\frac{\mathrm{d}y}{\mathrm{d}s} + [t_{21}]\frac{\mathrm{d}x}{\mathrm{d}s}\right)\mathrm{d}s$$

$$\mathscr{R}_y = \oint (\mathscr{T}_{12}n_1 + \mathscr{T}_{22}n_2)\,\mathrm{d}s = \rho_0 V_0^2 l \int_0^l \left(-[t_{12}]\frac{\mathrm{d}y}{\mathrm{d}s} + [t_{22}]\frac{\mathrm{d}x}{\mathrm{d}s}\right)\mathrm{d}s$$

$$[t_{ij}] \stackrel{\text{def}}{=} t_{ij}^+ - t_{ij}^- \tag{8.2.38}$$

Due to condition (8.2.20) the magnetic field action is vanishing. This is no longer true for a body with surface currents.

For the calculation of the mechanical action one uses the continuity equation and the vortex expression:

$$\omega = \frac{\partial v_y}{\partial x} - \frac{\partial v_x}{\partial y} = -\Delta\psi = -Re\left(Rhb_y - v_y + \frac{1}{Re}\frac{\partial \Pi}{\partial y}\right) \tag{8.2.39}$$

We have:

$$t_x \stackrel{\text{def}}{=} -t_{11}^{(\mathrm{m})}\frac{\mathrm{d}y}{\mathrm{d}s} + t_{21}^{(\mathrm{m})}\frac{\mathrm{d}x}{\mathrm{d}s} = p\frac{\mathrm{d}y}{\mathrm{d}s} + \frac{2}{Re}\frac{\mathrm{d}v_y}{\mathrm{d}s} - \frac{\omega}{Re}\frac{\mathrm{d}x}{\mathrm{d}s}$$

$$t_y \stackrel{\text{def}}{=} -t_{12}^{(\mathrm{m})}\frac{\mathrm{d}y}{\mathrm{d}s} + t_{22}^{(\mathrm{m})}\frac{\mathrm{d}x}{\mathrm{d}s} = -p\frac{\mathrm{d}x}{\mathrm{d}s} - \frac{2}{Re}\frac{\mathrm{d}v_x}{\mathrm{d}s} - \frac{\omega}{Re}\frac{\mathrm{d}y}{\mathrm{d}s}$$

so that

$$[t_x + \mathrm{i}t_y] = -\left[\mathrm{i}p + \frac{\omega}{Re}\right]\left(\frac{\mathrm{d}x}{\mathrm{d}s} + \mathrm{i}\frac{\mathrm{d}y}{\mathrm{d}s}\right)$$

But from (8.2.7), (8.2.11) and (8.2.39) it follows:

$$[p] = \frac{1}{Re}\left[\frac{\partial \Pi}{\partial y}\right],\quad [\omega] = -\left[\frac{\partial \Pi}{\partial x}\right]$$

so that also using (8.2.30) we have:

$$[t_x + \mathrm{i}t_y] = -\frac{\mathrm{i}}{Re}\left[\frac{\partial \Pi}{\partial y} + \mathrm{i}\frac{\partial \Pi}{\partial x}\right]\mathrm{e}^{\mathrm{i}\vartheta(s)}$$

$$= -\frac{2\mathrm{i}}{Re}\left[\frac{\partial \tilde{\chi}}{\partial y} + \mathrm{i}\frac{\partial \tilde{\chi}}{\partial x}\right]\mathrm{e}^{\mathrm{i}\vartheta(s)} = -\frac{4\pi}{Re}\overline{C(s)}$$

Consequently:

$$\mathcal{R}_x - i\mathcal{R}_y = \rho_0 V_0^2 l(R_x - iR_y)$$

$$R_x - iR_y = -\frac{4\pi}{Re}\int_0^l C(s)\,ds = -\frac{4\pi K}{Re}. \tag{8.2.40}$$

8.3 The aligned flat plate

8.3.1 The general solution

In this section we shall assume that the profile Γ is a straight line segment on the Ox axis of equation:

$$x = t,\ y = 0,\ t \in [0, 1] \tag{8.3.1}$$

(the plate length is taken as the characteristic length L_0 in dimensionless variables). We obviously have the following boundary conditions:

$$v_x(t, 0) = -1,\ v_y(t, 0) = 0,\ t \in [0, 1] \tag{8.3.2}$$

$$[b_x]_\Gamma = 0,\ [b_y]_\Gamma = 0 \tag{8.3.3}$$

In this particular case the general solution of the previous section for the velocity and induction fields yields the following representations:

$$w\delta = (a - b)\int_0^1 f(\tau)\,e^{-i\tilde{\theta}}\frac{d\tau}{\tilde{r}} \tag{8.3.4}$$

$$+ (b - c)\int_0^1 f(\tau)\{|a|K_1(|a|\tilde{r})\,e^{-i\tilde{\theta}} + aK_0(|a|\tilde{r})\}\,e^{a(x-\tau)}\,d\tau$$

$$+ (c - a)\int_0^1 f(\tau)\{bK_1(b\tilde{r})\,e^{-i\tilde{\theta}} + bK_0(b\tilde{r})\}\,e^{b(x-\tau)}\,d\tau$$

$$B\delta = (a - b)\int_0^1 f(\tau)\,e^{-i\tilde{\theta}}\frac{d\tau}{\tilde{r}} \tag{8.3.5}$$

$$+ b\int_0^1 f(\tau)\{|a|K_1(|a|\tilde{r})\,e^{-i\tilde{\theta}} + aK_0(|a|\tilde{r})\}\,e^{a(x-\tau)}\,d\tau$$

$$- a\int_0^1 f(\tau)\{bK_1(b\tilde{r})\,e^{-i\tilde{\theta}} + bK_0(b\tilde{r})\}\,e^{b(x-\tau)}\,d\tau$$

where:

$$\tilde{r}\exp(i\tilde{\theta}) = x - \tau + iy$$

Also, the total pressure and the vortex have the following expressions:

$$P = \frac{2}{Re}\int_0^1 \frac{(x-\tau)f(\tau)\,d\tau}{(x-\tau)^2 + y^2} \tag{8.3.6}$$

$$\omega = -ReRhb_y + v_y + 2\int_0^1 \frac{yf(\tau)\,d\tau}{(x-\tau)^2 + y^2} \tag{8.3.7}$$

Either differentiating and taking the limit in (8.3.4) or directly from (8.3.7) we get:

$$\frac{\partial v_x}{\partial y}(x, \pm 0) = \begin{cases} \mp 2\pi f(x), & x \in (0, 1) \\ 0 & , x \in \complement[0, 1] \end{cases} \tag{8.3.8}$$

The drag and the lift are given by:

$$R_x = -\frac{4\pi}{Re}\int_0^1 f(\tau)\,d\tau, \; R_y = 0 \tag{8.3.9}$$

Finally, from the first condition (8.3.2) one obtains the following integral equation for the determination of function $f(t)$:

$$\int_0^1 f(\tau)\mathscr{K}(t-\tau)\,d\tau = -\delta \tag{8.3.10}$$

where:

$$\mathscr{K}(\xi) = (a-b)/\xi + a(b-c)\{K_1(|a\xi|)\operatorname{sign} a\xi + K_0(|a\xi|)\}\exp(a\xi)$$
$$+ b(c-a)\{K_1(b|\xi|)\operatorname{sign}\xi + K_0(b|\xi|)\}\exp(b\xi) \tag{8.3.10'}$$

Equation (8.3.10) cannot be solved in the general case. For a semi-infinite plate the solution could be obtained using a technique due to Wiener-Hopf.

8.3.2 Asymptotic solutions

For extreme characteristics of the fluid the kernel (8.3.10′) may be replaced by simpler kernels so that equation (8.3.10) can be integrated. In this respect we shall notice that Re, Rm and Rh are non-negative parameters and that we have:

$$\begin{aligned} &0 \leqslant a < c < b, && \text{if } 0 \leqslant Rh < 1 \\ &-b \leqslant c < a \leqslant 0, && \text{if } 1 < Rh < 2\,(1 + Rm/Re) \\ &c \leqslant -b < a \leqslant 0, && \text{if } 2\,(1 + Rm/Re) \leqslant Rh \end{aligned} \tag{8.3.11}$$

and $|a| < b$.

For the non-conducting fluid ($Rm = 0$) we get:

$$a = 0,\ 2b = Re,\ 2c = Re(1 - Rh),\ \delta = bc \tag{8.3.12}$$

and for the perfectly conducting fluid ($Rm = \infty$):

$$a = \delta/b = c = Re(1 - Rh)/2,\ b = \infty \tag{8.3.13}$$

It is thus sufficient to consider the following cases:
(1). $|a| < b \ll 1$ (weakly conducting fluid with small Re)
(2). $1 \ll |a| < b$ (good conductor fluid with large Re)
(3). $|a| \ll 1,\ 1 \ll b$ (weakly conducting fluid with large Re or well conducting fluid with small Re).
For the approximation of the kernel K the following expansions around the origin:

$$K_0(2z) = -\sum_0^\infty \frac{z^{2m}}{(m!)^2}\{\ln z - \psi(m+1)\}$$

$$K_1(2z) = \frac{1}{2z} + \sum_0^\infty \frac{z^{2m+1}}{m!(m+1)!}\left\{\ln z - \psi(m+1) - \frac{1}{2(m+1)}\right\} \tag{8.3.14}$$

$$\psi(1) = -\gamma,\ \psi(m+1) = -\gamma + 1 + \frac{1}{2} + \ldots + \frac{1}{m},\ \gamma = 0.577\,215\,7$$

(γ being Euler's constant) and for large z

$$K_\nu(z) = \sqrt{\frac{\pi}{2z}}\,e^{-z}\left\{1 + O\left(\frac{1}{z}\right)\right\},\ z > 0 \tag{8.3.15}$$

are used.

(1). For the weakly conducting fluid ($Rm \ll 1$) with $Re \ll 1$ ($b \ll 1$) the kernel $\mathscr{K}(\xi)$ may be approximated by:

$$\mathscr{K}_1(\xi) = \frac{a-b}{\xi} + a(b-c)\left(\frac{1}{a\xi} - \ln\frac{|a\xi|}{2} - \gamma + 1\right)$$

$$+ b(c-a)\left(\frac{1}{b\xi} - \ln\frac{b|\xi|}{2} - \gamma + 1\right)$$

so that the integral equation (8.3.10) becomes:

$$\int_0^1 f(\tau)(\ln 4|t - \tau| + c_0 - 1)\,d\tau = 1$$

where:

$$c_0 = \gamma + \frac{1}{\delta}\left\{a(b-c)\ln\frac{|a|}{8} + b(c-a)\ln\frac{b}{8}\right\}$$

and has the solution:

$$f(\tau) = \frac{1}{\pi(c_0 - 1)}\frac{1}{\sqrt{\tau(1-\tau)}} \tag{8.3.16}$$

For this case the drag is:

$$R_x = -\frac{4\pi}{(c_0 - 1)Re}, \qquad \lim_{Rm\to 0} c_0 = \gamma + \ln\frac{Re}{16} \tag{8.3.17}$$

(2). For the well conducting fluid ($Rm \gg 1$) with $Re \gg 1$ ($1 \ll |a|$) the kernel $\mathscr{K}(\xi)$ may be approximated by:

$$\mathscr{K}_2(t-\tau) = \frac{a-b}{t-\tau} + a(b-c)\{\text{sign}\, a(t-\tau) + 1\}\sqrt{\frac{\pi}{2|a(t-\tau)|}}\,\mathrm{e}^{-|a(t-\tau)|}\,\mathrm{e}^{a(t-\tau)}$$

$$+ b(c-a)\{\text{sign}\,(t-\tau) + 1\}\sqrt{\frac{\pi}{2b|t-\tau|}}\,\mathrm{e}^{-b|t-\tau|}\,\mathrm{e}^{b(t-\tau)} \tag{8.3.18}$$

For $a \gg 1$ the integral equation becomes:

$$c_2\sqrt{\pi}\int_0^t \frac{f(\tau)\,\mathrm{d}\tau}{\sqrt{t-\tau}} + c_3\sqrt{\pi}\int_0^t \frac{f(\tau)\,\mathrm{d}\tau}{\sqrt{t-\tau}} - \frac{1}{c}\int_0^{\prime 1} \frac{f(\tau)\,\mathrm{d}\tau}{t-\tau} = -1 \tag{8.3.19}$$

and for $a \ll -1$:

$$c_2\sqrt{\pi}\int_t^1 \frac{f(\tau)\,\mathrm{d}\tau}{\sqrt{\tau-t}} + c_3\sqrt{\pi}\int_0^t \frac{f(\tau)\,\mathrm{d}\tau}{\sqrt{t-\tau}} - \frac{1}{c}\int_0^{\prime 1} \frac{f(\tau)\,\mathrm{d}\tau}{t-\tau} = -1 \tag{8.3.20}$$

where:

$$c_2\delta = (b-c)\sqrt{2|a|}, \quad c_3\delta = (c-a)\sqrt{2b}$$

In order to solve these equations one writes:

$$f(\tau) = f_1(\tau) + f_2(\tau) + \ldots, \text{ if } a \gg 1 \tag{8.3.19$'$}$$

$$f(\tau) = f_{-1}(\tau) + f_{-2}(\tau) + \ldots, \text{ if } a \ll -1 \tag{8.3.20$'$}$$

the functions $f_{\pm 1}$ satisfying the equations:

$$(c_2 + c_3)\sqrt{\pi}\int_0^t \frac{f_1(\tau)}{\sqrt{t-\tau}}\,d\tau = -1 \tag{8.3.19''}$$

$$c_2\sqrt{\pi}\int_t^1 \frac{f_{-1}(\tau)}{\sqrt{\tau - t}}\,d\tau + c_3\sqrt{\pi}\int_0^t \frac{f_{-1}(\tau)}{\sqrt{t-\tau}}\,d\tau = -1 \tag{8.3.20''}$$

and the functions $f_{\pm 2}$ satisfying equations corresponding to (8.3.19″) and (8.3.20″) but with the right-hand side terms substituted by:

$$\frac{1}{c}\int_0^{\prime 1} \frac{f_{\pm}(\tau)}{t-\tau}\,d\tau$$

One obtains:

$$f_1(\tau) = -\frac{c_4}{\sqrt{\tau}}, \quad c_4 = \frac{1}{\pi^{3/2}(c_2 + c_3)} \tag{8.3.21}$$

The function $f_2(\tau)$ satisfies an Abel-type equation. Since on account of solution (8.3.21) we have:

$$\int_0^{\prime 1} \frac{f_1(\tau)}{t-\tau}\,d\tau = -\frac{c_4}{\sqrt{t}}\ln\frac{1+\sqrt{t}}{1-\sqrt{t}}$$

$$\int_0^t \left\{\int_0^{\prime 1} \frac{f_1(\tau)}{t'-\tau}\,d\tau\right\}\frac{dt'}{\sqrt{t-t'}} = -2c_4\int_0^{\pi/2} \ln\frac{1+\sqrt{\tau}\sin\sigma}{1-\sqrt{\tau}\sin\sigma}\,d\sigma$$

it follows that:

$$f_2(\tau) = \frac{1}{\pi}\frac{d}{d\tau}\int_0^\tau \left\{\frac{1}{c(c_2+c_3)\sqrt{\pi}}\int_0^{\prime 1} \frac{f_1(\tau)\,d\tau}{t'-\tau}\right\}\frac{dt'}{\sqrt{\tau - t'}}$$

$$= -\frac{2c_4^2}{c}\frac{d}{d\tau}\int_0^{\pi/2} \ln\frac{1+\sqrt{\tau}\sin\sigma}{1-\sqrt{\tau}\sin\sigma}\,d\sigma = -\frac{2c_4^2}{c}\frac{\arcsin\sqrt{\tau}}{\tau\sqrt{1-\tau}} \tag{8.3.22}$$

Now noticing that:

$$f(\tau) = f_1(\tau) + f_2(\tau) = -\frac{c_4}{\sqrt{\tau}} - \frac{2c_4^2}{c\sqrt{\tau(1-\tau)}}\frac{\arcsin\sqrt{\tau}}{\sqrt{\tau}}, \quad a \gg 1$$

the following formula for the drag is derived:

$$R_x = \frac{8\pi c_4}{Re}\left\{1 + \frac{2c_4}{c}\int_0^{\pi/2}\frac{\sigma\,\mathrm{d}\sigma}{\sin\sigma}\right\} = \frac{8\pi c_4}{Re}\left(1 + 3.663\,862\,\frac{c_4}{c}\right) \tag{8.3.23}$$

$$\lim_{Rm\to\infty} c_4 = \frac{1}{2\pi\sqrt{\pi}}\sqrt{Re\,(1 - Rh)},\quad \lim_{Rm\to\infty}\frac{c_4}{c} = \frac{1}{\pi\sqrt{\pi}\sqrt{Re(1 - Rh)}}$$

This is the solution for $Rh < 1$ (the sub-alfvénian motion).

For super-alfvénian motion $Rh > 1$ we have to determine $f_{-1}(\tau)$ from equation (8.3.20″). To this end we multiply (according to the Abel procedure) this equation by $\mathrm{d}t'/\sqrt{t - t'}$ and integrate over the interval $(0, t)$. By changing the order of integration [G.23] it follows (Figure 23):

$$\int_0^t\left\{\int_0^{t'}\frac{f(\tau)\,\mathrm{d}\tau}{\sqrt{t' - \tau}}\right\}\frac{\mathrm{d}t'}{\sqrt{t - t'}} = \int_0^t f(\tau)\left\{\int_\tau^t\frac{\mathrm{d}t'}{\sqrt{(t - t')(t' - \tau)}}\right\}\mathrm{d}\tau = \pi\int_0^t f(\tau)\,\mathrm{d}\tau$$

and (Figure 24)

$$\int_0^t\left\{\int_{t'}^1\frac{f(\tau)\,\mathrm{d}\tau}{\sqrt{\tau - t'}}\right\}\frac{\mathrm{d}t'}{\sqrt{t - t'}} = \int_0^t f(\tau)\left\{\int_0^\tau\frac{\mathrm{d}t'}{\sqrt{(t - t')\,(\tau - t')}}\right\}\mathrm{d}\tau$$

$$+\int_t^1 f(\tau)\left\{\int_0^t\frac{\mathrm{d}t'}{\sqrt{(t - t')\,(\tau - t')}}\right\}\mathrm{d}\tau = \int_0^t f(\tau)\ln\left|\frac{\sqrt{\tau} + \sqrt{t}}{\sqrt{\tau} - \sqrt{t}}\right|\mathrm{d}\tau$$

$$+\int_t^1 f(\tau)\ln\left|\frac{\sqrt{\tau} + \sqrt{t}}{\sqrt{\tau} - \sqrt{t}}\right|\mathrm{d}\tau = \int_0^1 f(\tau)\ln\left|\frac{\sqrt{\tau} + \sqrt{t}}{\sqrt{\tau} - \sqrt{t}}\right|\mathrm{d}t$$

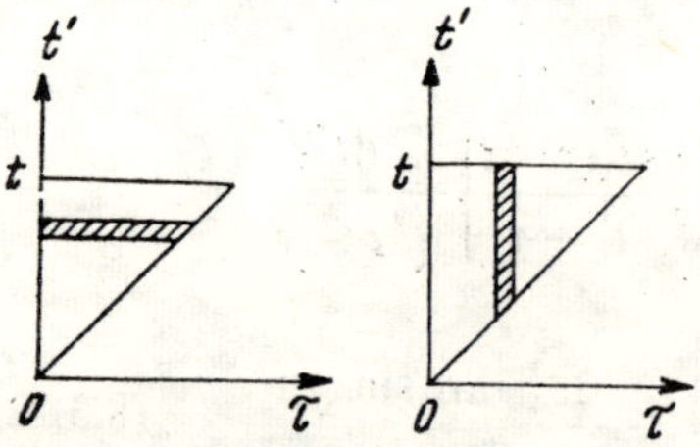

Fig. 23

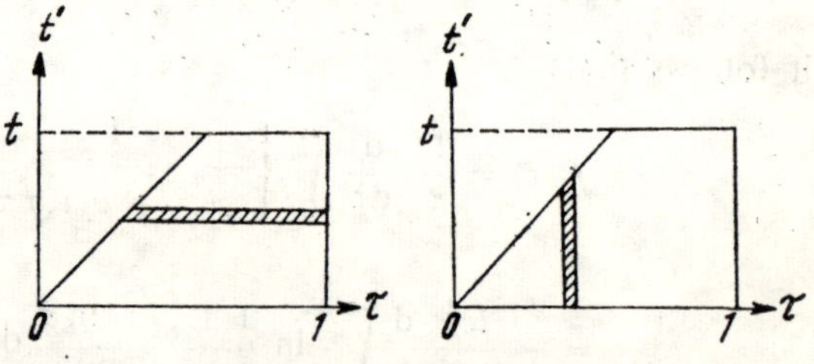

Fig. 24

In this manner equation (8.3.20″) becomes:

$$\pi\sqrt{\pi}c_3\int_0^t f_{-1}(\tau)\,\mathrm{d}\tau + c_2\sqrt{\pi}\int_0^1 f_{-1}(\tau)\ln\left|\frac{\sqrt{\tau} + \sqrt{t}}{\sqrt{\tau} - \sqrt{t}}\right|\mathrm{d}\tau = -2\sqrt{t}$$

and through differentiation:

$$\pi\sqrt{\pi}c_3 f_{-1}(t) + c_2\sqrt{\pi}\int_0^{\prime 1} f_{-1}(\tau)\sqrt{\frac{\tau}{t}}\frac{dt}{\tau - t} = -\frac{1}{\sqrt{t}}$$

If we now use the notation:

$$f_{-1}(\tau) = -\frac{c_{-4}}{\sqrt{\tau}}k(\tau),\ c_3 + ic_2 = \frac{\exp i\beta}{\pi\sqrt{\pi}\,c_{-4}},\ \beta \in \left[0, \frac{\pi}{2}\right] \tag{8.3.24}$$

it follows that function $k(\tau)$ verifies the equation:

$$\pi\cos\beta\, k(t) + \sin\beta\int_0^{\prime 1}\frac{k(\tau)}{\tau - t}d\tau = \pi,\ \beta \in \left[0, \frac{\pi}{2}\right], \tag{8.3.25}$$

This is an equation of the kind investigated by Carleman. According to Carleman's theory, equation (8.3.25) has an unique solution if the restriction of having only integrable singularities is imposed on the function $k(\tau)$. The solution of this equation may be found, e.g. in [G.34]. We used it in Chapter 5. This solution can be derived directly in the following way. Let us consider the function:

$$F(z) = \frac{1}{2\pi i}\int_0^1\frac{k(\tau)}{\tau - z}d\tau \tag{8.3.26}$$

holomorphic outside the segment [0, 1]. Using Plemelj's formulae and denoting by $F_+(t)$ the boundary values on the segment [0, 1] of the upper half-plane and by $F_-(t)$ the boundary values on the lower half-plane, we have:

$$k(t) = F_+(t) - F_-(t),\ \int_0^{\prime 1}\frac{k(\tau)}{\tau - t}d\tau = \pi i\{F_+(t) + F_-(t)\} \tag{8.3.26'}$$

so that (8.3.25) becomes:

$$e^{i\beta}F_+(t) - e^{-i\beta}F_-(t) = 1,\ \beta \in [0, \pi/2],\ t \in (0,1) \tag{8.3.25'}$$

The holomorphic function satisfying this equation and having singularities of the order < 1/2 has the following form:

$$F(z) = -\frac{1}{2\pi i}\left\{\left(1 - \frac{1}{z}\right)^{-\frac{\beta}{\pi}} - 1\right\}\frac{\pi}{\sin\beta} \tag{8.3.27}$$

For $|z| > 1$ we have:

$$F(z) = -\frac{1}{2\pi i}\frac{\beta}{\sin\beta}\frac{1}{z} + \ldots \tag{8.3.27'}$$

From (8.3.27) it follows that:

$$F_{\pm}(t) = -\frac{1}{2\pi i}\left\{\left(\frac{t}{1-t}\right)^{\beta/\pi} e^{\mp i\beta} - 1\right\}\frac{\pi}{\sin\beta}$$

so that (8.3.26′) yields:

$$k(t) = \left(\frac{t}{1-t}\right)^{\beta/\pi} \tag{8.3.28}$$

Comparing the expression of $F(z)$ given by (8.3.26) for $|z| > 1$ with equation (8.3.27′) it follows that:

$$\int_0^1 k(\tau)\,d\tau = \frac{\beta}{\sin\beta} \tag{8.3.29}$$

From (8.3.24) and (8.3.28) we have:

$$f(\tau) = f_-(\tau) + \ldots = -\frac{c_{-4}}{\sqrt{\tau}}\left(\frac{\tau}{1-\tau}\right)^{\beta/\pi} + O\left(\frac{c_{-4}^2}{c}\right), a \ll -1 \tag{8.3.30}$$

It is therefore obvious that the singularity on the leading edge of the plate is $\tau^{\frac{\beta}{\pi}-\frac{1}{2}}$ while that on the trailing edge is $(1-\tau)^{-\frac{\beta}{\pi}}$ and that the latter is preponderent if:

$$-\frac{\beta}{\pi} > \frac{\beta}{\pi} - \frac{1}{2} \Rightarrow \beta < \frac{\pi}{4}$$

From (8.3.25) it follows that:

$$\operatorname{tg}\beta = -\frac{c_3}{c_2} = \frac{(a-c)\sqrt{b}}{(b-c)\sqrt{|a|}} \overset{\text{def}}{=} \Lambda_0(\varepsilon, Rh),\ \varepsilon \overset{\text{def}}{=} Rh/Re$$

The equation $\Lambda_0 = 1$ (corresponding to $\beta = \pi/4$) divides the plane (ε, Rh) into two regions (Figure 25). In the region $\Lambda_0 < 1$ the downstream wake is more important (the trailing edge singularity is preponderent), while in the region $\Lambda_0 > 1$ the upstream wake predominates.

The following expression for the drag is obtained:

$$R_x \simeq \frac{4\pi}{Re}\int_0^1 \frac{c_{-4}}{\sqrt{\tau}}\left(\frac{\tau}{1-\tau}\right)^{\beta/\pi} d\tau = \frac{4\pi c_{-4}}{Re} B\left(\frac{1}{2}+\frac{\beta}{\pi}, 1-\frac{\beta}{\pi}\right)$$

$$\lim_{Rm\to\infty} c_{-4} = \frac{\sqrt{Re(Rh-1)}}{2\pi\sqrt{\pi}}, \quad \lim_{Rm\to\infty} \beta = \frac{\pi}{2} \qquad (8.3.31)$$

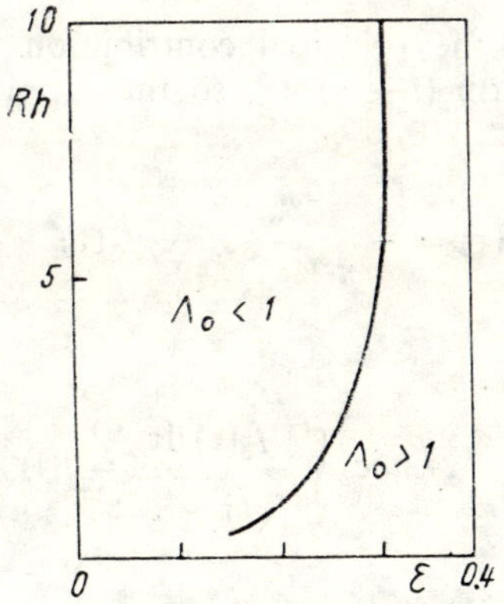

Fig. 25

(3) Finally, for weakly conducting fluids ($Rm \ll 1$) with $Re \gg 1$ or for the well conducting fluids ($Rm \gg 1$) with $Re \ll 1$ the following kernel:

$$\mathscr{K}_3(\xi) = \frac{a-b}{\xi} + a(b-c)\left\{\frac{1}{a\xi} - \gamma - \ln\frac{|a\xi|}{2} + 1\right\}$$

$$+ b(c-a)(\operatorname{sign}\xi + 1)\frac{\sqrt{\pi}}{\sqrt{2b|\xi|}} e^{-b|\xi|}e^{b\xi}$$

and integral equation

$$(a-c)\int_0^1 \frac{f(\tau)}{t-\tau} d\tau - a(b-c)\int_0^1 f(\tau)\left\{\gamma - 1 + \ln\frac{|a(t-\tau)|}{2}\right\} d\tau$$

$$+ (c-a)\sqrt{2\pi b}\int_0^t \frac{f(\tau)}{\sqrt{t-\tau}} d\tau = -\delta \qquad (8.3.32)$$

may be used. In this case the parameters a, b, c have the following approximate values

$$a \simeq \frac{Re\,Rm\,(1 - Rh)}{2(Re + Rm)}, \quad b = \frac{Re + Rm}{2}, \quad c = \frac{Re(1 - Rh)}{2}$$

so that:

$$\frac{a - c}{8} \simeq -\frac{2Re}{(Re + Rm)^2}, \quad \frac{a(b - c)}{\delta} \simeq \frac{Rm(Re+Rm)}{(Re + Rm)^2}, \quad \frac{c - a}{\delta}\sqrt{b} \simeq \frac{\sqrt{2}Re}{(Re + Rm)^{3/2}}$$

For $Re \gg 1$ and $Rm \ll 1$ the principal contribution to the kernel of equation (8.3.32) is given by the term $(t - \tau)^{-1/2}$, so that applying again the perturbation method we obtain:

$$f(\tau) = -\frac{1}{\pi^{3/2} c_3 \sqrt{\tau}} + f_3(\tau) + \dots \tag{8.3.33}$$

and

$$\sqrt{\pi}\delta c_3 \int_0^t \frac{f_3(\tau)\,\mathrm{d}\tau}{\sqrt{t - \tau}} = j(t) \tag{8.3.34}$$

where c_3 is defined by (8.3.21) and

$$j(t) = -\int_0^1 \left\{\frac{c - a}{t - \tau} + a(b - c)\left(\gamma - 1 + \ln\frac{|a(t - \tau)|}{2}\right)\right\}\frac{\mathrm{d}\tau}{\pi^{3/2} c_3 \sqrt{\tau}}$$

$$= -c_3'\left\{\gamma - 3 + \ln\frac{|a|}{2} + \ln(1 - t) + \left(\frac{c_3''}{\sqrt{t}} + \sqrt{t}\right)\ln\frac{1 + \sqrt{t}}{1 - \sqrt{t}}\right\} \tag{8.3.34'}$$

$$c_3' = \frac{2a(b - c)}{\pi^{3/2} c_3}, \quad c_3'' = \frac{c - a}{2a(b - c)}, \quad c_3' c_3'' = \frac{\delta}{\pi\sqrt{2\pi b}}$$

From (8.3.34) we get:

$$\pi^{3/2}\delta c_3 f_3(\tau) = \frac{\mathrm{d}}{\mathrm{d}\tau}\int_0^\tau \frac{j(t)\,\mathrm{d}t}{\sqrt{\tau - t}} = -c_3'\left\{\frac{\gamma - 5 + \ln(|a|/2)}{\sqrt{\tau}}\right.$$

$$\left. + 2\frac{1 + c_3'' - \tau}{\tau\sqrt{1 - \tau}} \arcsin\sqrt{\tau} + 2\int_0^{\pi/2} \sin^2\sigma \ln\frac{1 + \sqrt{\tau}\sin\sigma}{1 - \sqrt{\tau}\sin\sigma}\,\mathrm{d}\sigma\right\} \tag{8.3.35}$$

If we now take into account the fact that:

$$\int_0^1 \frac{\arcsin\sqrt{\tau}}{\tau\sqrt{1-\tau}}\,d\tau = 2\int_0^{\pi/2} \frac{\sigma\, d\sigma}{\sin\sigma} = 3.663\ 862$$

$$\int_0^1 \frac{\arcsin\sqrt{\tau}}{\sqrt{1-\tau}}\,d\tau = 2\int_0^{\pi/2} \sigma \sin\sigma\, d\sigma = 2$$

$$I(\tau) \overset{\text{def}}{=} \int_0^{\pi/2} \sin^2\sigma \ln\frac{1+\sqrt{\tau}\sin\sigma}{1-\sqrt{\tau}\sin\sigma}\,d\sigma = \sum_0^{\infty} \frac{2}{2n+1}\frac{(2n+2)\,!!}{(2n+3)\,!!}\tau^{n+1/2}$$

$$\int_0^1 I(\tau)\,d\tau = \sum_0^{\infty} \frac{2}{2n+1}\frac{(2n+2)!!}{(2n+3)!!}\frac{2}{2n+3} = 4\sum_0^{\infty}\frac{(2n)!!}{(2n+3)!!}\frac{(2n+2)}{(2n+1)(2n+3)}$$

$$= 1.137_{25}^{40}$$

$$\frac{d}{d\tau}\tau\int_0^{\pi/2} \sin^2\sigma \ln\frac{1+\sqrt{\tau}\sin\sigma}{1-\sqrt{\tau}\sin\sigma}\,d\sigma = -\frac{1}{\sqrt{\tau}} + \frac{\arcsin\sqrt{\tau}}{\tau\sqrt{1-\tau}}$$

$$+\int_0^{\pi/2} \sin 2\sigma \ln\frac{1+\sqrt{\tau}\sin\sigma}{1-\sqrt{\tau}\sin\sigma}\,d\sigma$$

we obtain the following formula for the drag:

$$R_x = -\frac{4\pi}{Re}\int_0^1 f(\tau)\,d\tau \simeq \frac{8}{\sqrt{\pi c_3 Re}}\left\{1+\frac{3.663\ 86}{\pi\sqrt{2\pi b}} + \frac{c_3'}{8}\left(\gamma + \ln\frac{|a|}{2} - 2.2088\right)\right\} \tag{8.3.36}$$

where:

$$\frac{c_3'}{8} \simeq \frac{1}{\pi\sqrt{\pi}}\frac{Rm(Rm + ReRh)}{Re\sqrt{Re+Rm}}$$

For $Rm = 0$ we have:

$$c_3\sqrt{Re} = 2,\ 2b = Re,\ c_3' = 0$$

and thus:

$$R_x = \frac{4}{\sqrt{\pi Re}} + \frac{14.6554}{\pi^2 Re} = \frac{4}{\sqrt{\pi Re}} + \frac{1.4863}{Re} \tag{8.3.37}$$

which result differs from that given by Carrier:

$$C_D = \frac{4}{\sqrt{\pi Re}} + \frac{1.484}{Re}$$

For $Re \ll 1$, $Rm \gg 1$ the principal contribution to the kernel of equation (8.3.32) is given by the logarithmic term. Applying again the perturbation method we obtain:

$$f(\tau) = \frac{1}{\pi\sqrt{\tau(1-\tau)}}\frac{\delta}{a(b-c)c_0'} + O\left(\frac{Re}{Rm^{3/2}}\right) \tag{8.3.38}$$

$$c_0' = \gamma - 1 + \ln\frac{|a|}{8}, \quad \frac{\delta}{a(b-c)} \simeq 1 + \frac{Re(Rh-1)}{Rm}, \quad a \simeq \frac{Re(1-Rh)}{2}$$

The drag having the expression:

$$R_x = -\frac{4\pi}{Re}\int_0^1 f(\tau)\,d\tau = -\frac{4\pi\delta}{a(b-c)c_0' Re} + O(Rm^{-3/2}) \tag{8.3.39}$$

For $Rm \to \infty$ we obtain:

$$R_x = -\frac{4\pi}{Re}\left\{\gamma - 1 + \ln\frac{Re(1-Rh)}{16}\right\}^{-1} \tag{8.3.40}$$

8.4 The incident flat plate

8.4.1 The general solution

Let Γ be a flat plate making an angle α with the Ox-axis (which is aligned with the unperturbed velocity and magnetic induction). For any point of the plate we can write:

$$x + iy = t\,e^{i\alpha}, \quad t \in [0, 1] \tag{8.4.1}$$

The general solution obtained from (8.2.31) and (8.2.32) has the form:

$$w\delta = (a-b)\int_0^1 \overline{C(\tau)}\,\mathrm{e}^{-\mathrm{i}\tilde{\theta}}\frac{\mathrm{d}\tau}{\tilde{r}}$$

$$+ (b-c)\int_0^1 \{\overline{C(\tau)}\,\mathrm{e}^{-\mathrm{i}\tilde{\theta}}|a|K_1(|a|\tilde{r}) + C(\tau)aK_0(|a|\tilde{r})\}\,\mathrm{e}^{a(x-\tau\cos\alpha)}\mathrm{d}\tau \qquad (8.4.2)$$

$$+ (c-a)\int_0^1 \{\overline{C(\tau)}\,\mathrm{e}^{-\mathrm{i}\tilde{\theta}}bK_1(b\tilde{r}) + C(\tau)bK_0(b\tilde{r})\}\,\mathrm{e}^{b(x-\tau\cos\alpha)}\,\mathrm{d}\tau$$

$$B\delta = (a-b)\int_0^1 \overline{C(\tau)}\,\mathrm{e}^{-\mathrm{i}\tilde{\theta}}\frac{\mathrm{d}\tau}{\tilde{r}}$$

$$+ b\int_0^1 \{C(\tau)\,\mathrm{e}^{-\mathrm{i}\tilde{\theta}}\,|a|K_1(|a|\tilde{r}) + C(\tau)aK_0(|a|\tilde{r})\}\,\mathrm{e}^{a(x-\tau\cos\alpha)}\,\mathrm{d}\tau \qquad (8.4.3)$$

$$- a\int_0^1 \{\overline{C(\tau)}\,\mathrm{e}^{-\mathrm{i}\tilde{\theta}}bK_1(b\tilde{r}) + C(\tau)bK_0(b\tilde{r})\}\,\mathrm{e}^{b(x-\tau\cos\alpha)}\mathrm{d}\tau$$

where:

$$\tilde{r}\exp(\mathrm{i}\tilde{\theta}) = x - \tau\cos\alpha + \mathrm{i}(y - \tau\sin\alpha)$$

$$C = A + \mathrm{i}B \qquad (8.4.4)$$

the constants a, b, c having the same expressions as in the previous sections. In the Romanian edition of this work [G.24] a more general solution for the plate having a surface current is given.

With the boundary condition (8.2.19) and since the integrand has a logarithmic singularity one obtains the following integral equation:

$$-\delta = (a-b)\int_0^{\prime 1} \overline{C(\tau)}\frac{\mathrm{e}^{-\mathrm{i}\alpha}}{t-\tau}\,\mathrm{d}\tau \qquad (8.4.5)$$

$$+ (b-c)\int_0^{\prime 1} \{\overline{C(\tau)}\mathrm{e}^{-\mathrm{i}\alpha}\,\mathrm{sign}\,(t-\tau)|a|K_1(|a(t-\tau)|) + C(\tau)aK_0(|a(t-\tau)|)\}\,\mathrm{e}^{a(t-\tau)\cos\alpha}\,d\tau$$

$$+ (c-a)\int_0^{\prime 1} \{\overline{C(\tau)}\,\mathrm{e}^{-\mathrm{i}\alpha}\,\mathrm{sign}\,(t-\tau)bK_1(b|t-\tau|) + C(\tau)bK_0(b|t-\tau|)\}\,\mathrm{e}^{b(t-\tau)\cos\alpha}\,\mathrm{d}\tau$$

Now, introducing the notation:

$$C(\tau)\exp\frac{i\alpha}{2}=f(\tau)\cos\frac{\alpha}{2}+i\,g(\tau)\sin\frac{\alpha}{2} \tag{8.4.6}$$

multiplying the integral equation (8.4.5) by exp (iα/2) and separating the real and imaginary parts, the following two integral equations for the determination of the real functions $f(\tau)$ and $g(\tau)$ follow:

$$\int_0^1 f(\tau)\mathscr{K}_+(t-\tau)\,d\tau=-\,\delta \tag{8.4.7}$$

$$\int_0^1 g(\tau)\mathscr{K}_-(t-\tau)\,d\tau=\delta \tag{8.4.8}$$

where:

$$\mathscr{K}_\pm(\zeta)=\frac{a-b}{\zeta}+a(b-c)\{K_1(|a\zeta|)\,\text{sign}\,(a\zeta)\pm K_0(|a\zeta|)\}\,e^{a\zeta\cos\alpha}$$

$$+b(c-a)\{K_1(b|\zeta|)\,\text{sign}\,\zeta\pm K_0(b|\zeta|)\}\,e^{b\zeta\cos\alpha} \tag{8.4.7'}$$

It is sufficient to consider for α the interval $(0, \pi)$ only.

If in (8.4.7) we make the substitution:

$$\tau, t, \alpha\rightarrow 1-\tau, 1-t, \pi-\alpha \tag{8.4.9}$$

it follows that:

$$\int_0^1 f(1-\tau)\mathscr{K}_-(t-\tau)\,d\tau=\delta$$

Comparing this equation with (8.4.8) we find:

$$g(\tau;\alpha)=f(1-\tau;\pi-\alpha),\ \alpha\in[0,\pi) \tag{8.4.10}$$

Consequently for the solution of the problem it is sufficient to solve equation (8.4.7).

Taking into account (8.4.6) and (8.2.40) it follows:

$$R_x-iR_y=-\frac{2\pi}{Re}\int_0^1\{f(\tau)+g(\tau)+(f(\tau)-g(\tau))\,e^{-i\alpha}\}\,d\tau \tag{8.4.11}$$

8.4.2 Approximate solution for small values of the parameter $b (Re \ll 1, Rm \ll 1)$

If $b \ll 1$ one may use the expansion of the kernel $\mathscr{K}_+$ about the origin and obtain the following approximation kernel:

$$\mathscr{K}_+(\zeta) = \frac{a-b}{\zeta} + a(b-c)\left\{\frac{1}{a\zeta} + \cos\alpha - \gamma - \ln\frac{|a\zeta|}{2}\right\}$$

$$+ b(c-a)\left\{\frac{1}{b\zeta} + \cos\alpha - \gamma - \ln\frac{|b\zeta|}{2}\right\}$$

In this manner equation (8.4.7) becomes:

$$\int_0^1 f(\tau)\{\ln 4|t-\tau| + c_0 - \cos\alpha\}\,d\tau = 1$$

and has the solution:

$$f(\tau) = \frac{K'}{\pi\sqrt{\tau(1-\tau)}}, \quad K' \stackrel{\text{def}}{=} \frac{1}{c_0 - \cos\alpha} \tag{8.4.12}$$

Using the same approximation it follows that:

$$g(\tau) = \frac{K''}{\pi\sqrt{\tau(1-\tau)}}, \quad K'' \stackrel{\text{def}}{=} \frac{1}{c_0 + \cos\alpha} \tag{8.4.13}$$

and then:

$$R_x - iR_y = -\frac{4\pi}{Re}\,\frac{c_0 + \bar{e}^{i\alpha}\cos\alpha}{c_0^2 - \cos^2\alpha} \tag{8.4.14}$$

For $\alpha = 0$ one obtains the results of the previous section.

In order to compare these results with the classical ones we shall introduce again the notations:

$$\varepsilon = Rm/Re, \ \lambda = \sqrt{(1+\varepsilon)^2 + 4\varepsilon(Rh - 1)} \tag{8.4.15}$$

In this way one obtains:

$$4a = Re(1 + \varepsilon - \lambda), \ 4b = Re(1 + \varepsilon + \lambda), \ c_0 = \gamma + \ln\frac{Re}{16} - \Lambda(\varepsilon, Rh) \tag{8.4.16}$$

where:

$$\Lambda(\varepsilon, Rh) = -\frac{1}{2\lambda}\left\{(\lambda+\varepsilon-1)\ln\frac{|1+\varepsilon-\lambda|}{2} + (\lambda-\varepsilon+1)\ln\frac{1+\varepsilon+\lambda}{2}\right\}$$

$$\lim_{Rm\to 0} \Lambda = \lim_{\varepsilon\to 0} \Lambda = 0 \qquad (8.4.16')$$

With the classical notations for the drag and lift equation (8.4.14) becomes:

$$C_D + iC_L = -\frac{8\pi}{Re}\,\frac{c_0 + e^{i\alpha}\cos\alpha}{c_0^2 - \cos^2\alpha} \qquad (8.4.14')$$

The influence of the external magnetic field upon the drag and lift coefficients manifests itself by the presence of the coefficient Rh and the influence of the fluid conductivity by the presence of the coefficient Rm (that is, through ε). These coefficients appear only in the expression of Λ. In Figure 26 the values of $C_D = C_D(Re)$ for various values of the parameter Λ are indicated both for zero and $\pi/2$ incidence (the latter corresponding to the plate being normal to the unperturbed current). For fixed Λ the curves corresponding to some given incidence lie between the calculated curves for $\alpha = 0$ and $\alpha = \pi/2$. It follows that with increasing angle of incidence the drag coefficient increases up to some maximum value corresponding

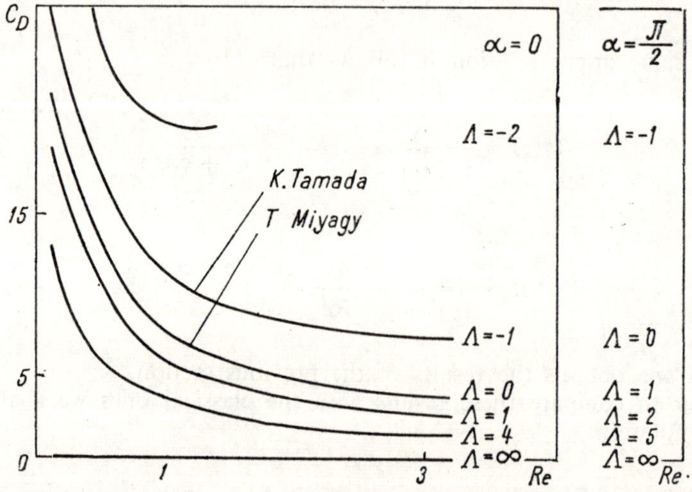

Fig. 26

to the normal plate and then decreases down to the aligned plate value. For a non-conducting fluid ($\Lambda = 0$) the curves superpose themselves on those given by Ko Tamada and T. Miyagi.

The variations of the lift coefficient C_L for $\alpha = -\pi/4$ and $\alpha = -\pi/8$ are given in Figure 27. For zero incidence, the lift coefficient vanishes, then increases with increasing angle of incidence up to a maximum value, then decreases to zero, this value corresponding to the plate being normal to the unperturbed current.

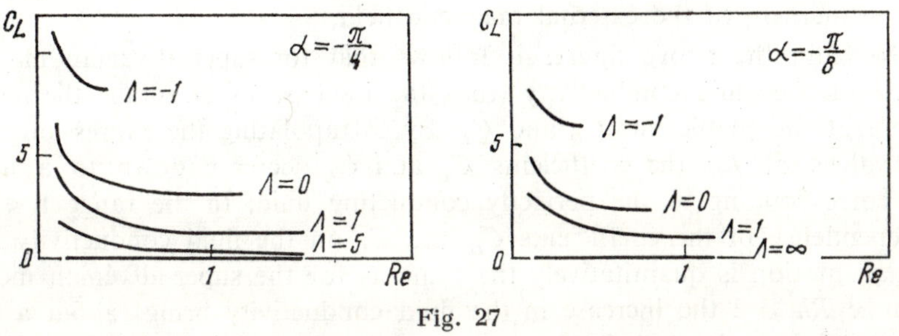

Fig. 27

Finally, the dependence of the parameter Λ on ε and Rh is presented in Figure 28. For the non-conducting fluid ($\varepsilon = 0$) or in the absence of an external magnetic field ($Rh = 0$) one obtains $\Lambda = 0$ which relationship corresponds to the $C_D = C_D^0(Re)$ dependence given by Ko Tamada and T. Miyagi. The parameter Λ also vanishes on a curve dividing the (ε, Rh) plane into two regions: $\Lambda > 0$ and $\Lambda < 0$ in the lower and upper regions, respectively. This curve is indicated in Figure 28. Figures 26 and 27 show that for $\Lambda > 0$ the values of the coefficients C_D and C_L are lower than for the classical fluid while for $\Lambda < 0$ they are higher than the classical values.

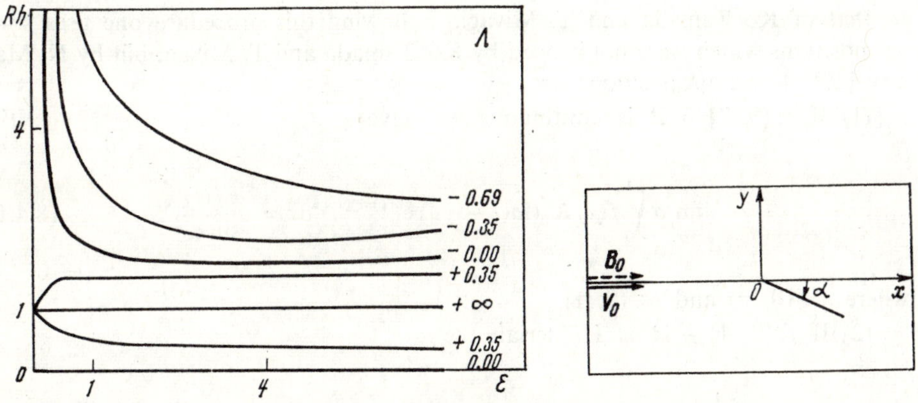

Fig. 28

Some more conclusions can be drawn from the above figures. So, for super-alfvénian motions ($Rh < 1$) the increase of Rh entails the monotonic decrease of the coefficients C_D and C_L tending to zero (the zero value cannot be reached since we supposed $Rh \neq 1$). For sub-alfvénian motion ($Rh > 1$) on the contrary the increase of Rh entails the increment of the coefficients C_D and C_L from zero up to the classical values (considered for Rh situated on the $\Lambda = 0$ curve) and so on. This conclusion results from the expression of Λ and is true for finite values of Rh. For large Rh it follows that for incidences different from zero and for strictly conducting fluids ($Rm \neq 0$) the coefficients C_D and C_L increase indefinitely with the increase of the external magnetic field.

Also from the above figures it follows that for super-alfvénian motion the increase of the fluid conductivity (thus the increase of ε) entails the monotonic decrease of the coefficients C_D and C_L. By extrapolating the expression of c_0 for large values of Rm the coefficients C_D and C_L decrease down to a minimum value corresponding to the perfectly conducting fluid. In the range $1 < Rh \lessapprox 2$ the dependence of the coefficients C_D and C_L on the fluid conductivity for sub-alfvénian motion is quantitatively the same as for the super-alfvénian motion. In the range $Rh \gtrapprox 2$ the increase in the fluid conductivity brings about a decrease of the coefficients C_D and C_L down to minimum values close to the classical ones. After these thresholds the coefficients C_D and C_L increase surpassing the classical values reach a maximum and by extrapolation tend (for $\varepsilon \to \infty$) to the perfectly-conducting fluid values.

8.4.3 Asymptotic solutions for large α

For $|a| \gg 1$ one can obtain an approximate integral equation in a manner similar to that of Ko Tamada and T. Miyagi. Following this procedure one uses some propositions which were not proved by Ko Tamada and T. Miyagi but by N. Marcov [20]. These propositions are:

(1) If $f\colon [0, 1] \to \mathbb{R}$ is continuous we have:

$$\lim_{a\to\infty} a \int_0^1 f(\tau)\, K_0(|a(t-\tau)|)\, e^{a(t-\tau)\cos\alpha} d\tau = \frac{\pi f(t)}{\sin\alpha} \tag{8.4.17}$$

where $\alpha \in (0, \pi)$ and $t \in (0, 1)$.

(2) If $f\colon [0, 1] \to \mathbb{R}$ is Hölderian then:

$$\lim_{a\to\infty} a \int_0^{\prime 1} f(\tau)\, K_1(|a(t-\tau)|) e^{a(t-\tau)\cos\alpha} \operatorname{sign} a(t-\tau)\, d\tau = \pi f(t) \operatorname{ctg}\alpha \tag{8.4.18}$$

the integral being taken in the Cauchy principal value.

Equations (8.4.17) and (8.4.18) are also true when $f(\tau)$ has integrable singularities at the ends of the interval.

(3) If $f(\tau)$ also depends on a, if this dependence is uniformly Hölderian and if the limits $\lim_{a\to\infty} f_a(\tau) = f_\infty(\tau)$ Hölderian on (0, 1) exist and if $f(\tau)$ has integrable singularities at the ends of the interval then:

$$\lim_{a\to\infty} a \int_0^1 f_a(\tau) e^{a(t-\tau)\cos\alpha} K_1(|a(t-\tau)|) \operatorname{sign} a(t-\tau)\, d\tau - \pi f_\infty(t) \operatorname{cotg}\alpha$$

$$= \lim_{a\to\infty} a \int_0^1 \{f_a(\tau) - f_\infty(\tau)\} e^{a(t-\tau)\cos\alpha} K_1(|a(t-\tau)|) \operatorname{sign} a(t-\tau)\, d\tau = 0 \quad (8.4.19)$$

Using these results one may approximately write (for $a \gg 1$):

$$a \int_0^1 f_a(\tau) K_1(|a(t-\tau)|) e^{a(t-\tau)\cos\alpha} \operatorname{sign} a(t-\tau)\, d\tau$$

$$\simeq \pi f_\infty(t) \operatorname{ctg}\alpha \simeq \pi f_a(t) \operatorname{ctg}\alpha$$

$$a \int_0^1 f_a(\tau) K_0(|a(t-\tau)|) e^{a(t-\tau)\cos\alpha}\, d\tau \simeq \pi f_a(t)/\sin\alpha$$

In this manner the integral equation (8.4.7) is approximated by the equation:

$$(a-b)\int_0^1 \frac{f(\tau)\, d\tau}{t-\tau} + (b-a) f(t) \left(\pi \operatorname{ctg}\alpha + \frac{\pi}{\sin\alpha}\right) = -\delta$$

and then by:

$$\pi \cos\frac{\alpha}{2} f(t) + \sin\frac{\alpha}{2} \int_0^1 \frac{f(\tau)\, d\tau}{\tau - t} = -c \sin\frac{\alpha}{2}, \quad \alpha \in (0, \pi) \quad (8.4.20)$$

The solution of this equation was found in the previous section. We thus have:

$$f(\tau) = -\frac{c}{\pi}\left(\frac{\tau}{1-\tau}\right)^{\frac{\alpha}{2\pi}} \sin\frac{\alpha}{2}, \quad \int_0^1 f(\tau)\, d\tau = -\frac{c\alpha}{2\pi} \quad (8.4.21)$$

From (8.4.10) we obtain to the same degree of approximation:

$$g(\tau) = -\frac{c}{\pi}\left(\frac{\tau}{1-\tau}\right)^{\frac{\alpha}{2\pi}-\frac{1}{2}} \cos\frac{\alpha}{2}, \quad \int_0^1 g(\tau)\, d\tau = -\frac{c(\pi-\alpha)}{2\pi} \quad (8.4.22)$$

Consequently:

$$R_x - \mathrm{i}R_y = \frac{1}{2}(1 - Rh)\{\pi - (\pi - 2\alpha)\mathrm{e}^{-\mathrm{i}\alpha}\}$$

$$C_D + \mathrm{i}C_L = (1 - Rh)\{\pi - (\pi - 2\alpha)\mathrm{e}^{\mathrm{i}\alpha}\} \tag{8.4.23}$$

where $\alpha \in (0, \pi)$, $Rh \in [0, 1)$.

For $a \ll -1$ the integral equation (8.4.7) is approximated by:

$$(a - b)\int_0^1 \frac{f(\tau)\,\mathrm{d}\tau}{t - x} + (2c - a - b)f(t)\left(\pi \operatorname{ctg} \alpha + \frac{\pi}{\sin \alpha}\right) = -\delta$$

whence:

$$\pi(a + b - 2c)\cos\frac{\alpha}{2}\, f(t) + (b - a)\sin\frac{\alpha}{2}\int_0^1 \frac{f(t)\,\mathrm{d}\tau}{t - \tau} = \delta \sin\frac{\alpha}{2}$$

$$\alpha \in (0, \pi) \tag{8.4.24}$$

Since in this case we have $c < 0$, $a + b - 2c > 0$ and $b - a > 0$ we introduce the notation:

$$c_1(a + b - 2c)\cos\frac{\alpha}{2} = -\delta \cos\frac{\alpha_1}{2}$$

$$c_1(b - a)\sin\frac{\alpha}{2} = -\delta \sin\frac{\alpha_1}{2} \tag{8.4.25}$$

$$\alpha_1 \in (0, \pi),\ c_1 > 0$$

so that equation (8.4.24) becomes:

$$\pi \cos\frac{\alpha_1}{2}\, f(t) - \sin\frac{\alpha_1}{2}\int_0^1 \frac{f(\tau)\,\mathrm{d}\tau}{\tau - t} = c \sin\frac{\alpha_1}{2} \tag{8.4.26}$$

This equation can be obtained from equation (8.4.20) with the substitution τ, t, $c \to 1 - \tau$, $1 - t$, $-c$. To conclude, the solution is:

$$f(\tau) = \frac{c}{\pi}\left(\frac{1 - \tau}{\tau}\right)^{\frac{\alpha_1}{2\pi}} \sin\frac{\alpha_1}{2}, \quad \int_0^1 f(\tau)\,\mathrm{d}\tau = \frac{c\alpha_1}{2\pi} \tag{8.4.27}$$

From (8.4.10) we obtain:

$$g(\tau) = \frac{c}{\pi}\left(\frac{\tau}{1-\tau}\right)^{\frac{\alpha_1}{2\pi}} \sin\frac{\alpha_2}{2}, \quad \int_0^1 g(\tau)\,d\tau = \frac{c\alpha_2}{2\pi} \tag{8.4.28}$$

where:

$$\alpha_2 = \alpha_1(\pi - \alpha) = 2\,\mathrm{arc\,tg}\left(\frac{b-a}{a+b-2c}\,\mathrm{ctg}\,\frac{\alpha}{2}\right), \; \alpha_2 \in (0, \pi)$$

To conclude:

$$C_D + iC_L = (Rh - 1)\{\alpha_1 + \alpha_2 + (\alpha_1 - \alpha_2)\,e^{i\alpha}\}$$
$$Rh > 1, \; \alpha, \alpha_1, \alpha_2 \in (0, \pi) \tag{8.4.29}$$

For a very strong applied magnetic field the solution of the problem is again (8.4.27) where:

$$\alpha_1 \simeq 2\sqrt{\frac{\varepsilon}{Rh}}\,\mathrm{tg}\,\frac{\alpha}{2}, \; \alpha_2 \simeq 2\sqrt{\frac{\varepsilon}{Rh}}\,\mathrm{ctg}\,\frac{\alpha}{2}$$

Consequently in this case:

$$C_D + iC_L \simeq 4\sqrt{\varepsilon Rh}\,(\sin\alpha - i\cos\alpha), \; \alpha \in (0, \pi) \tag{8.4.30}$$

In the Romanian edition of the present work [G.24] as well as in [7] one can find solutions of the problem valid for quite large (though finite) values of the parameter *a*. In this case the kernel is approximated by:

$$\mathcal{K}_+(\zeta) = \frac{a-b}{\zeta} + a(b-c)(\mathrm{sign}\,a\zeta + 1)\frac{\sqrt{\pi}}{\sqrt{2|a\zeta|}}\,e^{-|a\zeta| + a\zeta\cos\alpha}$$
$$+ b(c-a)(\mathrm{sign}\,\zeta + 1)\frac{\sqrt{\pi}}{\sqrt{2|a\zeta|}}\,e^{-b|\zeta| + b\zeta\cos\alpha}$$

and the integral equation (for $a \gg 1$) by:

$$(a-b)\int_0^{1}\frac{f(\tau)\,d\tau}{t-\tau} + (b-c)\sqrt{2\pi a}\int_0^t \frac{f(\tau)}{\sqrt{t-\tau}}\,e^{-a(1-\cos\alpha)(t-\tau)}d\tau$$
$$+ (c-a)\sqrt{2\pi b}\int_0^t \frac{f(\tau)}{\sqrt{t-\tau}}\,e^{-b(1-\cos\alpha)(t-\tau)}d\tau = -\delta = -c(b-a)$$

The integration of this equation as well as its various solutions are presented in the above mentioned works. For a small incidence C_D and C_L are plotted in Figure 29.

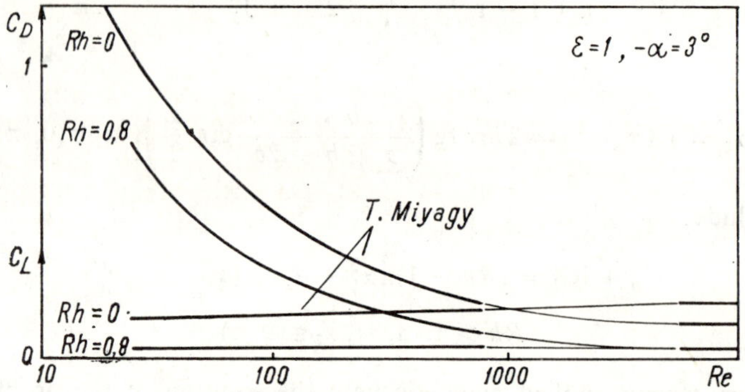

Fig. 29

8.5 Perfectly conducting fluids

For perfectly conducting fluids equations (8.2.2)–(8.2.5) are still valid. But equation (8.4.6) is replaced by the equation:

$$b_y - v_y \equiv 0 \tag{8.5.1}$$

derived from Ohm's law for $Rm \simeq \infty$. From equations (8.2.4) (8.2.5) and (8.5.1) as well as from the condition for the perturbed motion to vanish at infinity it also follows that:

$$b_x - v_x \equiv 0 \tag{8.5.2}$$

Consequently, in this case the motion will be described by the system:

$$(1 - Rh)\frac{\partial v_x}{\partial x} = -\frac{\partial P}{\partial x} + \frac{1}{Re}\Delta v_x$$

$$(1 - Rh)\frac{\partial v_y}{\partial x} = -\frac{\partial P}{\partial y} + \frac{1}{Re}\Delta v_y \tag{8.5.3}$$

$$\frac{\partial v_x}{\partial x} + \frac{\partial v_y}{\partial y} = 0$$

with the boundary condition (8.2.19). The solution $\boldsymbol{b} \equiv \boldsymbol{v}$ which follows from (8.5.1) and (8.5.2) satisfies the boundary conditions (8.2.20), the conditions (8.2.19) being taken into account. It follows that the perfectly-conducting fluid theory reduces to the classical fluid theory with a modified Reynolds number. Indeed the coefficient $(1 - Rh)$ may be included in the expressions of P and Re. For sub-alfvénian motions ($Rh > 1$) a perturbation of opposite direction to the classical fluid perturbation follows (for the same obtacle and the same velocity at infinity — but with reverse direction).

The motion of classical viscous fluids past a thin body has been investigated by N. Marcov [21].

8.6 The flat plate in an orthogonal magnetic field

.6.1 The general solution

Let us now consider a flat plate aligned with the unperturbed velocity $\boldsymbol{V}_0$ (and with the Ox-axis) in the presence of an orthogonal magnetic field $\boldsymbol{H}_0$. Using the same definitions of the dimensionless variables as in Section 8.2 the velocity and induction at some point within the fluid have the expressions:

$$\boldsymbol{V} = (1 + v_x)\,\boldsymbol{i}_1 + v_y\boldsymbol{i}_2,\ \boldsymbol{B} = b_x\boldsymbol{i}_1 + (1 + b_y)\,\boldsymbol{i}_2 \tag{8.6.1}$$

$\boldsymbol{v}$ and $\boldsymbol{b}$ being functions of (x, y) only. In this case the linearized system of equations reduces itself to:

$$P_1 v_x - \frac{\partial p}{\partial x} + Rh\left(\frac{\partial b_x}{\partial y} - \frac{\partial b_y}{\partial x}\right) = 0 \tag{8.6.2}$$

$$P_1 v_y - \frac{\partial p}{\partial y} = 0,\ P_1 \overset{\text{def}}{=} Re^{-1}\Delta - \frac{\partial}{\partial x},\ \Delta \overset{\text{def}}{=} \frac{\partial^2}{\partial x^2} + \frac{\partial^2}{\partial y^2} \tag{8.6.3}$$

$$\frac{\partial v_x}{\partial x} + \frac{\partial v_y}{\partial y} = 0 \tag{8.6.4}$$

$$\frac{\partial b_x}{\partial x} + \frac{\partial b_y}{\partial y} = 0 \tag{8.6.5}$$

$$Rm(v_x + b_y) + \frac{\partial b_x}{\partial y} - \frac{\partial b_y}{\partial x} = 0 \tag{8.6.6}$$

$$\lim_{x^2 + y^2 \to \infty} (v_x, v_y, p, b_x, b_y) = 0 \tag{8.6.7}$$

This linear system with constant coefficients can be reduced to a higher order equation by the following procedure. We eliminate the pressure from equations (8.6.2) and (8.6.3) and then from the new equation we eliminate v_x and b_x using equations (8.6.4) and (8.6.5). We obtain:

$$\Delta\left(P_1 v_y + Rh\frac{\partial}{\partial y} b_y\right) = 0 \tag{8.6.8}$$

From (8.6.6), (8.6.5) and (8.6.4) we obtain:

$$P_2 b_y + \frac{\partial}{\partial y} v_y = 0,\ P_2 \overset{\text{def}}{=} Rm^{-1}\Delta - \frac{\partial}{\partial x} \tag{8.6.9}$$

so that eventually:

$$\Delta L(v_y, b_y) = 0,\ L \overset{\text{def}}{=} P_1 P_2 - Rh\frac{\partial^2}{\partial y^2} \tag{8.6.10}$$

The unknowns v_x, b_x and p obey the same equation (8.6.10).

The linear differential operator L with constant coefficients cannot unfortunately be factorized as for the aligned magnetic field. Due to this fact the general solution cannot be represented by potentials as in Section 8.2. We can however reduce the boundary value problem to an integral equation if we use the same method as in Chapter 6. In this respect the solution is represented in the following form:

$$v_y = \bar{v}_y + \frac{\partial \varphi}{\partial y},\ b_y = \bar{b}_y + \frac{\partial \psi}{\partial y} \tag{8.6.11}$$

where:

$$L(\bar{v}_y, \bar{b}_y) = 0,\ \Delta(\varphi, \psi) = 0 \tag{8.6.11'}$$

For plan-wave solutions of the form $\exp(-\mathrm{i}\lambda x + sy)$, $s = -\mathrm{i}\lambda r$ the first equation (8.6.11') yields the following dispersion equation:

$$\{\mathrm{i}\lambda Re^{-1}(1 + r^2) + 1\}\{\mathrm{i}\lambda Rm^{-1}(1 + r^2) + 1\} - Rh\, r^2 = 0 \tag{8.6.12}$$

The roots r of this equation are distinct from one another and their imaginary parts are different from zero. We denote by r_j ($j = 1, 2$) those roots for which the expressions $s_j = -\mathrm{i}\lambda r_j$ have negative imaginary parts. Since (8.6.12) is a biquadratic equation two of the roots will certainly have this property no matter the positive or negative real number λ (the other two roots will be equal to each other and of opposite sign). For large λ we have:

$$r_1 = -\mathrm{i}\,\mathrm{sign}\,\lambda + O(\lambda^{-2})$$

$$r_2 = -\mathrm{i}\,\mathrm{sign}\,\lambda - \frac{Re + Rm}{2|\lambda|} + O\left(\frac{1}{\lambda^2}\right) \tag{8.6.13}$$

Taking into account the damping condition (8.6.7) it follows that in the upper ($y > 0$) and lower ($y < 0$) half-planes perturbation waves of the form $\exp(-\mathrm{i}\lambda x + s_j y)$ and $\exp(-\mathrm{i}\lambda x - s_j y)$, respectively, will propagate. Making use of the fact that L is a linear operator and of the equations (8.6.3), (8.6.4) and (8.6.5) the following general solution is derived:

$$\begin{Bmatrix} v_y^{\pm} \\ v_x^{\pm} \\ b_y^{\pm} \\ b_x^{\pm} \\ p^{\pm} \end{Bmatrix}(x, y) = \int_{-\infty}^{+\infty} \sum_j \begin{Bmatrix} A_j^{\pm} \\ \mp r_j A_j^{\pm} \\ B_j^{\pm} \\ \mp r_j B_j^{\pm} \\ \mp r_j^{-1} P_{1j} A_j^{\pm} \end{Bmatrix}(\lambda)\exp(-\mathrm{i}\lambda x \pm s_j y)\mathrm{d}\lambda + \begin{Bmatrix} \dfrac{\partial\varphi}{\partial y} \\ \dfrac{\partial\varphi}{\partial x} \\ \dfrac{\partial\psi}{\partial y} \\ \dfrac{\partial\psi}{\partial x} \\ -\dfrac{\partial\varphi}{\partial x} \end{Bmatrix} \tag{8.6.14}$$

$$\lim_{\infty}\left(\frac{\partial\varphi}{\partial x}, \frac{\partial\varphi}{\partial y}, \frac{\partial\psi}{\partial x}, \frac{\partial\psi}{\partial y}\right) = 0 \tag{8.6.14'}$$

Since this solution also satisfies equations (8.6.2) and (8.6.6) and taking into account the dispersion equation as well as the fact that φ and ψ are harmonic functions which obey the conditions (8.6.14') it follows that:

$$Rh r_j B_j^{\pm} = \pm P_{1j} A_j^{\pm}, \quad P_{1j} \overset{\mathrm{def}}{=} \mathrm{i}\lambda Re^{-1}(1 + r_j^2) + 1 \tag{8.6.15}$$

$$\frac{\partial\psi}{\partial x} = \frac{\partial\varphi}{\partial y}, \quad \frac{\partial\psi}{\partial y} = -\frac{\partial\varphi}{\partial x} \tag{8.6.16}$$

Like in Chapter 6 the upper and lower signs indicate the solutions in the upper and lower half-planes, respectively. Therefore the general solution is represented by means of five functions $A_j^{\pm}(\lambda)$ and $\varphi(x, y)$, the latter being a harmonic function within the fluid. In order to find these functions one uses the boundary conditions in the form:

$$v_x^+(x,0) = -1,\ v_y^+(x,0) = 0,\ x \in [0,1] \tag{8.6.17}$$

$$[v_x] = [b_y] = [v_y] = [b_x] = 0,\ \forall x \tag{8.6.18}$$

with the usual notations.

8.6.2 The boundary value problem and the integral equation

Using the general solution and the inversion theorem for Fourier-type integrals from (8.6.17) and (8.6.18) we obtain:

$$\int_{-\infty}^{+\infty} \Sigma r_j A_j^+ \mathrm{e}^{-\mathrm{i}\lambda x} \mathrm{d}\lambda - \left.\frac{\partial \varphi}{\partial x}\right|_{y=+0} = 1,\ x \in [0,1] \tag{8.6.19}$$

$$\int_{-\infty}^{+\infty} \Sigma A_j^+ \mathrm{e}^{-\mathrm{i}\lambda x} \mathrm{d}\lambda + \left.\frac{\partial \varphi}{\partial y}\right|_{y=+0} = 0,\ x \in [0,1] \tag{8.6.20}$$

$$\Sigma r_j(A_j^+ + A_j^-) = 2F,\ \Sigma r_j^{-1} P_{1j}(A_j^+ + A_j^-) = 2RhF \tag{8.6.21}$$

$$\Sigma(A_j^+ - A_j^-) = -2G,\ \Sigma P_{1j}(A_j^+ - A_j^-) = 2RhG \tag{8.6.22}$$

with the notations:

$$2F = \frac{1}{2\pi}\int_{-\infty}^{+\infty} f(x)\mathrm{e}^{\mathrm{i}\lambda x}\mathrm{d}x,\ f = \left[\frac{\partial \varphi}{\partial x}\right] \tag{8.6.23}$$

$$2G = \frac{1}{2\pi}\int_{-\infty}^{+\infty} g(x)\mathrm{e}^{\mathrm{i}\lambda x}\mathrm{d}x,\ g = \left[\frac{\partial \varphi}{\partial y}\right] \tag{8.6.24}$$

From (8.6.21) and (8.6.22) it follows that:

$$A_j^+ + A_j^- = 2\alpha_j F,\ A_j^+ - A_j^- = 2\beta_j G \tag{8.6.25}$$

where:

$$\alpha_1 = \frac{r_1(P_{12} - Rhr_2^2)}{(\mathrm{i}\lambda Re^{-1} + 1)(r_1^2 - r_2^2)},\ \alpha_2 = \frac{r_2(Rhr_1^2 - P_{11})}{(\mathrm{i}\lambda Re^{-1} + 1)(r_1^2 - r_2^2)}$$

$$\beta_1 = \frac{P_{12} + Rh}{\mathrm{i}\lambda Re^{-1}(r_1^2 - r_2^2)},\ \beta_2 = -\frac{P_{11} + Rh}{\mathrm{i}\lambda Re^{-1}(r_1^2 - r_2^2)} \tag{8.6.25'}$$

Using the general solution it follows also that:

$$[p] = -(1 + Rh)f(x) \tag{8.6.26}$$

$$\left[\frac{\partial v_x}{\partial y}\right] = \int_{-\infty}^{+\infty} \Sigma i\lambda r_j^2(A_j^+ - A_j^-)\,e^{-i\lambda x}\,d\lambda + \frac{\partial}{\partial x}\left[\frac{\partial \varphi}{\partial y}\right] = (1 + Rh)Re\,g(x) \tag{8.6.27}$$

which are useful formulas for finding the lift and drag. The continuity of pressure and vortex outside the plate entails:

$$f(x) \equiv 0,\ g(x) \equiv 0,\ x \in \complement\,[0, 1] \tag{8.6.28}$$

In order to express the harmonic part of the solution as a function of f and g we shall consider the complex plane $z = x + iy$ and

$$\Phi(z) = \frac{\partial \varphi}{\partial x} - i\frac{\partial \varphi}{\partial y} \tag{8.6.29}$$

a holomorphic function outside the plate. Since:

$$[\Phi] = f(x) - i\,g(x),\ \forall x \tag{8.6.30}$$

$$\lim_{z\to\infty} \Phi(z) = 0$$

it follows that:

$$\Phi(z) = \frac{1}{2\pi i}\int_0^1 \frac{f(t) - ig(t)}{t - z}\,dt \tag{8.6.31}$$

This solution determines the derivatives of function φ by means of f and g.

In order to derive the integral equation of the problem we shall use the conditions (8.6.19) and (8.6.20). In this respect we shall take the limit in (8.6.31) with z tending from the upper half-plane to a point $x + i0$ on the plate. Using Plemelj's formulae one obtains:

$$\begin{aligned} \left.\frac{\partial \varphi}{\partial x}\right|_{y=+0} &= \frac{1}{2}f(x) - \frac{1}{2\pi}\int_0^{\prime 1} \frac{g(t)}{t - x}\,dt,\ x \in (0, 1) \\ \left.\frac{\partial \varphi}{\partial y}\right|_{y=+0} &= \frac{1}{2}\,g(x) + \frac{1}{2\pi}\int_0^{\prime 1} \frac{f(t)}{t - x}\,dt,\ x \in (0, 1) \end{aligned} \tag{8.6.32}$$

the integrals being taken in the Cauchy principal value. Using all these results from (8.6.19) and (8.6.20) it follows that:

$$-f(x)+\frac{1}{\pi}\int_0^{\prime 1}\frac{g(t)}{t-x}\,dt+\int_0^1\{f(t)M^*(t-x)+g(t)M(t-x)\}\,dt=2$$

$$g(x)+\frac{1}{\pi}\int_0^{\prime 1}\frac{f(t)}{t-x}\,dt+\int_0^1\{f(t)N(t-x)+g(t)N^*(t-x)\}\,dt=0 \qquad (8.6.33)$$

$$x\in(0,1)$$

where:

$$M^*(t-x)=-N^*(t-x)=\frac{1}{2\pi}\int_{-\infty}^{+\infty}e^{i\lambda(t-x)}\,d\lambda=\delta(t-x)$$

$$\begin{Bmatrix}M\\N\end{Bmatrix}(t-x)=\frac{1}{2\pi}\int_{-\infty}^{+\infty}\begin{Bmatrix}m\\n\end{Bmatrix}e^{i\lambda(t-x)}\,d\lambda$$

$$m=\Sigma r_j\beta_j=\frac{1-r_1r_2-Re(1+Rh)i\lambda^{-1}}{r_1+r_2}=i\,\mathrm{sign}\,\lambda+m_0$$

$$n=\Sigma\alpha_j=\frac{1-r_1r_2(i\lambda-ReRh)(i\lambda+Re)^{-1}}{r_1+r_2}=i\,\mathrm{sign}\,\lambda+n_0$$

where δ is the Dirac distribution. Using this definition of δ the system (8.6.33) is reduced to the following two equations:

$$\frac{1}{\pi}\int_0^{\prime 1}\frac{g(t)}{t-x}\,dt+\int_0^1 g(t)M(t-x)\,dt=2,\ x\in(0,1) \qquad (8.6.34)$$

$$\frac{1}{\pi}\int_0^{\prime 1}\frac{f(t)}{t-x}\,dt+\int_0^1 f(t)N(t-x)\,dt=0,\ x\in(0,1) \qquad (8.6.35)$$

These equations solve the problem.

From the behaviour of m and n for large λ it follows that M and N are distributions. Using the relationship (6.2.27) equations (8.6.34) and (8.6.35) become:

$$\int_0^1 g(t)M_0(t-x)\,dt=2,\ x\in(0,1) \qquad (8.6.34')$$

$$\int_0^1 f(t)N_0(t-x)\,dt=0,\ x\in(0,1) \qquad (8.6.35')$$

M_0 and N_0 being obtained from M and N by replacing m and n by m_0 and n_0. According to the Dirichlet criterion M_0 and N_0 are convergent kernels.

Obviously if the functions g and f are found from (8.6.31) and (8.6.35) then all the other functions appearing in the general solution will be determined. From equation (8.2.38), from the continuity conditions (8.6.18) and from equation (8.6.26) and (8.6.27) it follows that:

$$R_x = \int_0^1 [t_{21}]\,\mathrm{d}x = \frac{1}{Re}\int_0^1 \left[\frac{\partial v_x}{\partial y}\right]\mathrm{d}x = (1 + Rh)\int_0^1 g(x)\,\mathrm{d}x$$

$$R_y = \int_0^1 [t_{22}]\,\mathrm{d}x = -\int_0^1 [p]\,\mathrm{d}x = (1 + Rh)\int_0^1 f(x)\,\mathrm{d}x \qquad (8.6.36)$$

8.6.3 The approximate solution for the weakly conducting fluid, ($Rm \ll 1$) with low Reynolds number ($Re \ll 1$)

For this case neglecting the products $ReRm$, Re^2 and Rm^2 we have:

$$r_1 = -\,\mathrm{i}\,\mathrm{sign}\,\lambda, \quad r_2 = -\left(-1 + \frac{2\varepsilon \mathrm{i}}{\lambda}\right)^{1/2}$$

$$2\varepsilon \overset{\mathrm{def}}{=} Re + Rm \qquad (8.6.37)$$

So that:

$$m_0 = \frac{\omega}{2\varepsilon}\left\{-\mathrm{i}\,\mathrm{sign}\,\lambda + \left(-1 + \frac{2\varepsilon \mathrm{i}}{\lambda}\right)^{1/2}\right\}, \quad \omega \overset{\mathrm{def}}{=} Re(1 + Rh) \qquad (8.6.38)$$

Using the results of M. Marcov [22] we obtain:

$$\int_{-\infty}^{+\infty}\left(-1 + \frac{2\varepsilon \mathrm{i}}{\lambda}\right)^{1/2} \mathrm{e}^{-\mathrm{i}\lambda(x-t)}\,\mathrm{d}\lambda$$

$$= 2\varepsilon\{K_0(\varepsilon|x - t|) + \mathrm{sign}\,(x - t)K_1(\varepsilon|x - t|)\}\exp\{\varepsilon(x - t)\}$$

K_0 and K_1 ($= -K_0'$) being Bessel functions of imaginary argument. But for small values of the argument we have:

$$K_0(z) = -\ln z + \ln z - \gamma + O(z^2)$$

γ being Euler's constant. Using then this result equation (8.6.34′) reduces itself to:

$$-\frac{\omega}{2\pi}\int_0^1 g(t)(\ln|x - t| + \Gamma_1)\,\mathrm{d}t = 2, \quad x \in (0, 1)$$

$$\Gamma_1 \overset{\mathrm{def}}{=} \ln\frac{\varepsilon}{2} + \gamma - 1 \qquad (8.6.39)$$

Equations of this type were solved in Chapters 6 and 7. Theorems of existence and uniqueness have been proved by Carleman. The solution of (8.6.39) has the form:

$$g(t) = \frac{\Gamma}{\sqrt{t(1-t)}} \tag{8.6.40}$$

and it is found by differentiating (8.6.39) with respect to x. The constant Γ is determined from (8.6.39) and has the value:

$$\Gamma = \frac{4}{\omega(\ln 8/\varepsilon + 1 - \gamma)} \tag{8.6.41}$$

For the drag one has the expression:

$$R_x = (1 + Rh)\pi\, \Gamma$$

which reduces itself to the classical one for $Rm = 0$. In this case $f = 0$.

The above method can be used both for the case of Hall effect presence [16] and for compressible fluids [18]. Recently, it has been extended by J. Vacca [23] to the case of non-stationary motion.

9

The theory of small perturbations. The Cauchy problem

9.1 Introduction

9.1.1 Incompressible fluids

As mentioned already in the introduction to this book the problem of the production and propagation of perturbation waves in a conducting fluid has occurred for the first time in astrophysics. As early as 1942 Alfvén * [G.1] proved that a perturbation in a incompressible non-dissipating fluid at rest gives rise to a wave propagating with velocity $B_0/\sqrt{\mu\rho}$ in the direct and reverse sense along the direction of the external magnetic field. Such a wave was not known at that time either in hydrodynamics or magnetodynamics. Alfvén's argument was based on an analogy with the transverse vibrations of a string but Wallen gave in 1944 a rigorous proof. We present below Wallen's derivation [G.2].

Let us consider an incompressible unbounded fluid at rest in a homogeneous magnetic field $\boldsymbol{H}_0$. The state of the perturbed fluid at time t is characterized by the functions $\boldsymbol{v}(t, \boldsymbol{x})$ and $\boldsymbol{H}_0 + \mathbf{h}(t, \boldsymbol{x})$. Supposing that $\boldsymbol{v}$ and $\mathbf{h}$ are small perturbations the equations of motion can be linearized so that we have:

$$\operatorname{div} \boldsymbol{v} = 0, \quad \operatorname{div} \boldsymbol{h} = 0 \tag{9.1.1}$$

$$\rho_0 \partial_t \boldsymbol{v} = \rho_0 \boldsymbol{F} - \operatorname{grad} p + \operatorname{curl} \boldsymbol{h} \times \mu \boldsymbol{H}_0 \tag{9.1.2}$$

$$\partial_t \boldsymbol{h} = \operatorname{curl} (\boldsymbol{v} \times \boldsymbol{H}_0) \tag{9.1.3}$$

Assuming that the mechanical force is a conservative one ($\boldsymbol{F} = -\operatorname{grad} \Pi$) and using the identity (A.26′) equations (9.1.2) and (9.1.3) become:

$$\rho_0 \partial_t \boldsymbol{v} = -\operatorname{grad} P + (\mu \boldsymbol{H}_0 \cdot \nabla) \boldsymbol{h} \tag{9.1.2′}$$

$$\partial_t \boldsymbol{h} = (\boldsymbol{H}_0 \cdot \nabla) \boldsymbol{v} \tag{9.1.3′}$$

$$P \overset{\text{def}}{=} \rho_0 \Pi + p + \boldsymbol{H}_0 \cdot \boldsymbol{h}$$

---

* For these results the Swedish scientist received the Nobel prize in 1970.

Applying the *div* operator in (9.1.2′) and using (9.1.1) it follows that $\Delta P = 0$. But the harmonic function P reduces itself to a constant (ρ_0, Π_∞) at infinity and thus we have all over the space:

$$\rho_0 \Pi + p + \boldsymbol{H}_0 \cdot \boldsymbol{h} = \rho_0 \Pi_\infty \tag{9.1.4}$$

In this manner equations (9.1.2′) and (9.1.3′) yield:

$$\left(\frac{\partial^2}{\partial t^2} - b^2 \frac{\mathrm{d}^2}{\mathrm{d}\alpha^2}\right)(\boldsymbol{v}, \boldsymbol{h}) = 0 \tag{9.1.5}$$

where:

$$b = H_0 \sqrt{\frac{\mu}{\rho_0}} = V_A, \quad \frac{\mathrm{d}}{\mathrm{d}\alpha} = (\boldsymbol{\alpha} \cdot \nabla)$$

$\boldsymbol{\alpha}$ being the unit vector along the external magnetic field direction ($\boldsymbol{H}_0 = H_0 \boldsymbol{\alpha}$).

Equation (9.1.5) indicates that the perturbation propagates along the magnetic field direction in the direct and reverse sense with velocity V_A (the Alfvén velocity). From (9.1.1) it follows that the fluid motion is performed without volume changes and that the vibrations are perpendicular to the direction of propagation (as for the vibrating strings).

9.1.2 Compressible fluids

The propagation of small perturbations in a compressible fluid was studied for the first time by Herlofson [2] in 1950. He showed that three types of waves (Alfvén, fast and slow waves) can propagate in the plane motion of a non-dissipating fluid. A large number of works have been devoted to this subject since 1950 so that we cannot mention here all of them for lack of space. Two main directions of investigation occur: the studies of the propagation of small perturbations based on the linearized magnetofluid dynamic (MFD) equations on the one hand and the studies of non-linear phenomena on the other. In this chapter the linearized theory will be presented while the non-linear problems will be the subject of the following chapter.

Even in the linearized problem one has more than one point of view. We have those studies concerned with the perturbation wave structure only without taking into account the initial conditions (van de Hulst [3], Anderson (4], Baños [5], Ludford [6], etc.) on the one hand and there are studies on the more complex problem of the initial conditions and the fluid motions consistent with them (the Cauchy problem) on the other. In this chapter we shall deal with the latter problems as this is the most important since the perturbation motion of a fluid is produced by a continuous superposition of MFD waves and it depends in a fundamental way on the initial conditions. The study of wave structure is only one aspect of the problem. The first paper adopting this causal point of view is due to Friedlander [7].

Using John's method of spherical media Friedlander solved the initial pressure problem. A fundamental asymptotic study of the initial condition problem was given by Lighthill [8], this study being afterwards used by Broadbent [9] for the investigation of the Hall effect and by Agostineli [G.2] and M. Ignat [10] for some other problems. Using the plane wave (Fourier) method we have taken up again in two notes [11] the general Cauchy problem but here again the solution does not look quite natural. That is why we have resumed its study for the plane [12] and for the three-dimensional [13] case. In these later papers we succeeded in constructing a Cartesian basis (a complete system of eigenvectors) and to express the solution of the general Cauchy problem in terms of it. These results will be presented in Sections 9.3 and 9.4. In Section 9.2 we shall prove that the non-dissipating MFD system is of hyperbolic type (which is a fundamental result for the solution of the Cauchy problem). Finally, in section 9.5 some quantitative results [13] [14] regarding the general case of the dissipating fluid with Hall effect will be presented.

9.2 Hyperbolic characteristics of the (non-dissipating) MFD equations

It is important in the following to establish that the MFD system of equations is of the hyperbolic type. Indeed, according to a well-known theorem [G.30, vol. 2], if this is the case then the Cauchy problem is well posed, i.e. the solution of the problem is represented by continuous functions of the initial conditions. To conclude, if the initial (perturbations) conditions are small enough the solution (the perturbation at some instant t) will also be small enough so that the linearized equations of motion may be used.

Let us therefore consider a compressible fluid obeying the perfect gas law ($p = \rho RT$) and whose dissipating effects (that is, its viscosity and electrical resistivity) may be neglected. If moreover the motion is adiabatic from the energy equation it follows that this is isentropic so that:

$$\frac{\mathrm{d}}{\mathrm{d}t}\left(\ln\frac{p}{\rho^{\gamma}}\right) = 0 \tag{9.2.1}$$

and then

$$\frac{\mathrm{d}p}{\mathrm{d}t} = \gamma\frac{p}{\rho}\frac{\mathrm{d}\rho}{\mathrm{d}t} = a^2\frac{\mathrm{d}\rho}{\mathrm{d}t} \tag{9.2.2}$$

Using:

$$\frac{\partial p}{\partial x_i} = \frac{\mathrm{d}p}{\mathrm{d}\rho}\frac{\partial \rho}{\partial x_i} = a^2\frac{\partial \rho}{\partial x_i} \tag{9.2.3}$$

it follows that the MFD equations can be written in the following form ($\boldsymbol{B} = \mu \boldsymbol{H}$):

$$\frac{\mathrm{d}\rho}{\mathrm{d}t} + \rho \frac{\partial V_j}{\partial x_j} = 0 \tag{9.2.4}$$

$$\rho \frac{\mathrm{d}V_j}{\mathrm{d}t} + a^2 \frac{\partial \rho}{\partial x_i} + B_j \frac{\partial H_j}{\partial x_i} - B_j \frac{\partial H_i}{\partial x_j} = 0, \; i = 1, 2, 3 \tag{9.2.5}$$

$$\frac{\mathrm{d}H_i}{\mathrm{d}t} + H_i \frac{\partial V_j}{\partial x_j} - H_j \frac{\partial V_i}{\partial x_j} = \;, \; i = 1, 2, 3 \tag{9.2.6}$$

j being a summation index with values 1, 2, 3. As usual we have the notation:

$$\frac{\mathrm{d}}{\mathrm{d}t} = \frac{\partial}{\partial t} + V_j \frac{\partial}{\partial x_j}$$

If we pose the Cauchy problem as in Section 10.1 it follows that the characteristic surface $G(t, x_1, x_2, x_3) = 0$ is determined by equation (10.1.6′) which for the system (9.2.4)—(9.2.6) becomes:

$$\begin{vmatrix}
\rho\dot{G} & 0 & 0 & -B_2\alpha_2 - B_3\alpha_3 & B_2\alpha_1 & B_3\alpha_1 & a^2\alpha_1 \\
0 & \rho\dot{G} & 0 & B_1\alpha_2 & -B_1\alpha_1 - B_3\alpha_3 & B_3\alpha_2 & a^2\alpha_2 \\
0 & 0 & \rho\dot{G} & B_1\alpha_3 & B_2\alpha_3 & -B_1\alpha_1 - B_2\alpha_2 & a^2\alpha_3 \\
-H_2\alpha_2 - H_3\alpha_3 & H_1\alpha_2 & H_1\alpha_3 & \dot{G} & 0 & 0 & 0 \\
H_2\alpha_1 & -H_1\alpha_1 - H_3\alpha_3 & H_2\alpha_3 & 0 & \dot{G} & 0 & 0 \\
H_3\alpha_1 & H_3\alpha_2 & -H_1\alpha_1 - H_2\alpha_2 & 0 & 0 & \dot{G} & 0 \\
\rho\alpha_1 & \rho\alpha_2 & \rho\alpha_3 & 0 & 0 & 0 & \dot{G}
\end{vmatrix} = 0$$

The above determinant takes a symmetrical form if the factors μ, ρ and $a^2\mu^{-1}$ are taken out of the first three lines, the last line and the last column, respectively. One further introduces the unit vector $\boldsymbol{n}$ normal to the characteristic surface:

$$\boldsymbol{n} = \frac{\alpha}{|\alpha|} \tag{9.2.7}$$

by dividing by $|\alpha|$ each line of the determinant. Also noticing that we have:

$$|\alpha|^{-1}\dot{G} = |\alpha|^{-1}\{\partial_t G + (\boldsymbol{V}.\nabla)G\} = -d + \boldsymbol{V}.\boldsymbol{n} = -P \tag{9.2.8}$$

d being the displacement (or drift) velocity and P the propagation velocity (Section 1.5) we have:

$$\begin{vmatrix} -P/v & 0 & 0 & H_1n_1-H_n & H_2n_1 & H_3n_1 & n_1 \\ 0 & -P/v & 0 & H_1n_2 & H_2n_2-H_n & H_3n_2 & n_2 \\ 0 & 0 & -P/v & H_1n_3 & H_2n_3 & H_3n_3-H_n & n_3 \\ H_1n_1-H_n & H_1n_2 & H_1n_3 & -P & 0 & 0 & 0 \\ H_2n_1 & H_2n_2-H_n & H_2n_3 & 0 & -P & 0 & 0 \\ H_3n_1 & H_3n_2 & H_3n_3-H_n & 0 & 0 & -P & 0 \\ n_1 & n_2 & n_3 & 0 & 0 & 0 & -vP/a^2 \end{vmatrix} = 0$$

where v is defined by the relationship $\rho v = \mu$, μ being the magnetic permeability and ρ the variable fluid density.

In this latter form the determinant is symmetrical so that all its roots P are real the system of equations being of hyperbolic type. Expanding this determinant one obtains:

$$P(P^2 - b_n^2)\{P^4 - (a^2 + b^2)P^2 + a^2b_n^2\} = 0 \tag{9.2.9}$$

where:

$$a^2 = \gamma\frac{p}{\rho}, \quad b^2 = \frac{\boldsymbol{B}.\boldsymbol{H}}{\rho}, \quad b_n = \frac{B_nH_n}{\rho} \tag{9.2.9'}$$

The hyperbolic-type symmetrical systems have been studied by Friedrichs [10.1].

9.3 The plane problem

9.3.1 The equations of motion

The solution of the Cauchy problem in this case can be derived from the general three-dimensional solution given in the next section. However since some aspects which are not so apparent in the general case can be outlined in the plane problem we prefer to present here its elementary solution.

The problem can be stated in the following manner: we consider an unbounded compressible fluid at rest in the presence of a homogeneous magnetic field $\boldsymbol{H}_0$; at some given instant — the initial time — a perturbation characterized by the functions $p^0(\boldsymbol{x})$, $\boldsymbol{v}^0(\boldsymbol{x})$ and $\boldsymbol{h}^0(\boldsymbol{x})$ (div $\boldsymbol{h}^0 = 0$) alters the state of some region D of the fluid; one is required to determine the fluid motion caused by this perturbation. We shall consider the fluid to be a non-dissipating simple medium which obeys the perfect gas law.

Taking into account both the unperturbed and the perturbed states of the fluid it follows that at some time t the fluid motion will be characterized by the following variables:

$$p_0 + p(t, \boldsymbol{x}),\ \boldsymbol{v}(t, \boldsymbol{x}),\ \boldsymbol{H}_0 + \boldsymbol{h}(t, \boldsymbol{x}),\ \rho_0 + \rho(t, \boldsymbol{x}) \tag{9.3.1}$$

so that in D we have:

$$p(0, \boldsymbol{x}) = p^0(\boldsymbol{x}),\ \boldsymbol{v}(0, \boldsymbol{x}) = \boldsymbol{v}^0(\boldsymbol{x}),\ \boldsymbol{h}(0, \boldsymbol{x}) = \boldsymbol{h}^0(\boldsymbol{x}) \tag{9.3.2}$$

By linearization of the equations of motion we have:

$$\partial_t \rho + \rho_0 \operatorname{div} \boldsymbol{v} = 0 \tag{9.3.3}$$

$$\rho_0 \partial_t \boldsymbol{v} = -\operatorname{grad} p + \operatorname{curl} \boldsymbol{h} \times \mu \boldsymbol{H}_0 \tag{9.3.4}$$

$$\partial_t \boldsymbol{h} = (\boldsymbol{H}_0 \,.\, \nabla)\, \boldsymbol{v} - \boldsymbol{H}_0 \operatorname{div} \boldsymbol{v} \tag{9.3.5}$$

$$a^2 \partial_t \rho = \partial_t p,\ a^2 = \gamma p_0/\rho_0,\ \gamma = c_p/c_v \tag{9.3.6}$$

the equation div $\boldsymbol{h} = 0$ being a consequence of equation (9.3.5).

A first observation is that from the initial condition (9.3.2) and system (9.3.3) — (9.3.6) it follows that the first-order derivatives:

$$\partial_t \rho|_{t=0},\ \partial_t \boldsymbol{v}|_{t=0},\ \partial_t \boldsymbol{h}|_{t=0},\ \partial_t p|_{t=0}$$

are known in D. Now differentiating with respect to t the equations (9.3.3) — (9.3.6) and using the initial conditions it follows that the second order derivatives are known functions as well. In this manner it follows that one can determine the derivatives with respect to t of any order at the initial time and therefore express the analytical solution of the Cauchy problem.

Let us now consider the case when the initial perturbations depend on two variables only $\boldsymbol{x} = (x, y)$ and assume that this applies to the resulting motion too. Choosing the reference system such that the Ox-axis is along the direction of $\boldsymbol{H}$

from the system (9.3.3) — (9.3.6) we have:

$$\rho_0 \frac{\partial v_z}{\partial t} = \mu H_0 \frac{\partial h_z}{\partial x}, \quad \frac{\partial h_z}{\partial t} = H_0 \frac{\partial v_z}{\partial x} \tag{9.3.7}$$

$$\left.\begin{aligned} &\frac{\partial p}{\partial t} + \gamma p_0 \left(\frac{\partial v_x}{\partial x} + \frac{\partial v_y}{\partial y} \right) = 0, \ \rho_0 \frac{\partial v_x}{\partial t} = - \frac{\partial p}{\partial x} \\ &\rho_0 \frac{\partial v_y}{\partial t} = - \frac{\partial p}{\partial y} + \mu H_0 \left(\frac{\partial h_y}{\partial x} - \frac{\partial h_x}{\partial y} \right) \\ &\frac{\partial h_x}{\partial t} = -H_0 \frac{\partial v_y}{\partial y}, \ \frac{\partial h_y}{\partial t} = H_0 \frac{\partial v_y}{\partial x} \end{aligned}\right\} \tag{9.3.8}$$

From (9.3.7) it follows that:

$$\left(\frac{\partial^2}{\partial t^2} - b^2 \frac{\partial^2}{\partial x^2} \right) (v_z, h_z) = 0, \ b \overset{\text{def}}{=} H_0 \sqrt{\frac{\mu}{\rho_0}} \tag{9.3.7'}$$

which indicates that the Alfvén waves affect the components v_z and b_z. The solution of equation (9.3.7′) is determined by the classical (Fourier or d'Alembert) methods in terms of the initial conditions v_z^0 and b_z^0. Particularly if $v_z^0 \equiv b_z^0 \equiv 0$ no transverse waves will appear ($v_z \equiv b_z \equiv 0$).

The system (9.3.8) can also be reduced to a high-order equation.

In this respect in the last three equations one eliminates h_x and h_y and obtains:

$$\left(\frac{\partial^2}{\partial t^2} - b^2 \Delta \right) v_y = - \frac{1}{\rho_0} \frac{\partial^2 p}{\partial t \, \partial y}, \ \Delta = \frac{\partial^2}{\partial x^2} + \frac{\partial^2}{\partial y^2} \tag{9.3.9}$$

Combining this equation with the first two we have:

$$T(p, v_x, v_y) = 0$$

$$T \overset{\text{def}}{=} \frac{\partial^4}{\partial t^4} - (a^2 + b^2) \Delta \frac{\partial^2}{\partial t^2} + a^2 b^2 \Delta \frac{\partial^2}{\partial x^2} \tag{9.3.10}$$

Equation (9.3.10) is also verified by h_x and h_y.

T is a hyperbolic-type operator. Indeed the equation of the characteristic lines $G(t, x, y) = 0$ reduces itself to:

$$E(s) = s^4 - \Omega_1 s^2 + \Omega_2 \tag{9.3.11}$$

where:

$$s = \partial G/\partial t, \ \alpha = \partial G/\partial x, \ \beta = \partial G/\partial y$$

$$\Omega_1 = (a^2 + b^2)(\alpha^2 + \beta^2), \ \Omega_2 = a^2 b^2 \alpha^2 (\alpha^2 + \beta^2)$$

which has real s roots for any real parameters α and β. This statement is a consequence of the following two inequalities:

$$\Omega_1^2 - 4\Omega_2 = \alpha^4(a^2 - b^2)^2 + \beta^4(a^2 + b^2)^2 + 2\alpha^2\beta^2(a^4 + b^4) > 0$$

$$s_1^2 + s_2^2 = \Omega_1 > 0, \ s_1^2 s_2^2 = \Omega_2 > 0$$

As a matter of fact we have:

$$2s_1 = \sqrt{2(\Omega_1 + \Omega)} = \sqrt{\alpha^2 + \beta^2}(\sqrt{a^2 + b^2 + 2ab_1} + \sqrt{a^2 + b^2 - 2ab_1})$$

$$2s_2 = \sqrt{2(\Omega_1 - \Omega)} = \sqrt{\alpha^2 + \beta^2}(\sqrt{a^2 + b^2 + 2ab_1} - \sqrt{a^2 + b^2 - 2ab_1}) \quad (9.3.12)$$

$$s_3 = -s_1, \ s_4 = -s_2$$

with the notations:

$$\Omega = \sqrt{\Omega_1^2 - 4\Omega_2}, \ b_1 = b\alpha/\sqrt{\alpha^2 + \beta^2} \quad (9.3.12')$$

Expressions (9.3.12) will be met in the following chapter also. They provide the displacement (drift) velocities of the fast ($\pm s_1$) and slow ($\pm s_2$) waves. For $H_0 = 0$ the Alfvén and slow waves disappear while the fast one is transformed into an acoustic wave of velocity $\pm a$. As a matter of fact it will be latter shown that an entropy wave also develops out of a perturbation. It does not appear at the present time since it does not affect the pressure, velocity and magnetic field.

9.3.2 The Cauchy problem

Equation (9.3.10) is satisfied by plane-wave solutions of the form exp i ($\alpha x + \beta y - st$), if s is determined from (9.3.12). Since T is a linear operator it follows that the general solution is obtained by a continuous superposition of waves:

$$\begin{Bmatrix} p \\ v_x \\ v_y \\ h_x \\ h_y \end{Bmatrix} (t, x, y) = \iint_{-\infty}^{+\infty} \sum_{j=1}^{4} \begin{Bmatrix} A_j \\ B_j \\ C_j \\ D_j \\ E_j \end{Bmatrix} (\alpha, \beta) e^{-is_j t} e^{i(\alpha x + \beta y)} d\alpha d\beta \quad (9.3.13)$$

Let us now notice that from a relation of the form:

$$\sum_j a_j \exp(s_j t) = 0$$

satisfied for any t (the roots s_j being distinct from one another) it follows that $a_j = 0$ (Section 6.3) [12]. Hence (9.3.8) and (9.3.13) yield ($j = 1, 2, 3, 4$):

$$\begin{aligned} \alpha A_j &= \rho_0 s_j B_j \\ \alpha\beta a^2 C_j &= (s_j^2 - a^2\alpha^2)B_j \\ \alpha a^2 s_j D_j &= H_0(s_j^2 - a^2\alpha^2)B_j \\ \beta a^2 s_j E_j &= -H_0(s_j^2 - a^2\alpha^2)B_j \end{aligned} \tag{9.3.14}$$

By imposing the boundary conditions (9.3.2) on (9.3.13) and using the inversion theorem of Fourier-type integrals we have:

$$\sum_j \begin{Bmatrix} A_j \\ B_j \\ C_j \\ D_j \\ E_j \end{Bmatrix} = \left(\frac{1}{2\pi}\right)^2 \iint_D \begin{Bmatrix} p^0 \\ v_x^0 \\ v_y^0 \\ h_x^0 \\ h_y^0 \end{Bmatrix} \mathrm{e}^{-\mathrm{i}(\alpha x + \beta y)} \mathrm{d}x\, \mathrm{d}y \overset{\text{def}}{=} \begin{Bmatrix} A \\ B \\ C \\ D \\ E \end{Bmatrix} \tag{9.3.15}$$

Using equations (9.3.14) the system (9.3.15) determines the functions $B_j(\alpha, \beta)$. Two observations regarding this determination follow immediately. The first is that the last relation of (9.3.15) is contained in the last but one equation since one has the condition div $\boldsymbol{h}^0 = 0$. Indeed we have:

$$\operatorname{div} \boldsymbol{h}|_{t=0} = \iint \mathrm{i} \sum_j (\alpha D_j + \beta E_j) \mathrm{e}^{\mathrm{i}(\alpha x + \beta y)} \mathrm{d}\alpha\, \mathrm{d}\beta$$

so that $\alpha D_j + \beta E_j = 0$. Consequently we shall use the first four equations of (9.3.15) for the determination of the four functions B_j. The second observation refers to the fact that the determination of the unknowns B_j is possible. Indeed with the notation $B_j = s_j B_j^*$ system (9.3.15) reduces itself to

$$\Sigma\, s_j^2 B_j^* = \alpha A/\rho_\infty, \quad \Sigma\, s_j B_j^* = B, \tag{9.3.16}$$

$$\sum s_j^3 B_j^* = \alpha\beta a^2 C + a^2\alpha^2 B, \qquad \sum B_j^* = A/\alpha\gamma p_\infty - D/\alpha H_\infty$$

It is now obvious that the determinant of the unknowns B_j^* is not vanishing (it is a Vandermonde determinant).

9.3.3 The point perturbation of pressure

Let us consider as an application a perturbation P_0 of the pressure at some point taken as the origin of the reference system. Denoting by δ the Dirac distribution we have in this case:

$$P^0 = P_0\delta(\boldsymbol{x}),\ \boldsymbol{v}^0 \equiv 0,\ \boldsymbol{h}^0 \equiv 0 \tag{9.3.17}$$
$$A = \left(\frac{1}{2\pi}\right)^2 P_0,\ B = C = D = 0$$

Consequently:

$$B_1^* = B_3^* = -\frac{A}{2\gamma p_0}\frac{s_2^2 - a^2\alpha^2}{\alpha\Omega},\ B_2^* = B_4^* = \frac{A}{2\gamma p_0}\frac{s_1^2 - c^2\alpha^2}{\alpha\Omega}$$

and

$$A_1 = A_3 = \frac{A}{4}\left\{1 - \frac{(b^2 - a^2)(\alpha^2 + \beta^2)}{\Omega}\right\},\ A_2 = A_4 = \frac{A}{4}\left\{1 + \frac{(b^2 - a^2)(\alpha^2 + \beta^2)}{\Omega}\right\}$$

$$B_1 = -B_3 = -\frac{A}{2\gamma p_\infty}\frac{s_1(s_2^2 - a^2\alpha^2)}{\alpha\Omega},\ B_2 = -B_4 = \frac{A}{2\gamma p_\infty}\frac{s_2(s_1^2 - a^2\alpha^2)}{\alpha\Omega} \tag{9.3.18}$$

$$C_1 = -C_3 = \frac{A\beta s_1}{2\rho_0\Omega},\ C_2 = -C_4 = -\frac{A\beta s_2}{2\rho_0\Omega}$$

$$D_1 = D_3 = -D_2 = -D_4 = \frac{AH_0\beta^2}{2\rho_0\Omega},\ E_1 = E_3 = -E_2 = -E_4 = -\frac{AH_0\alpha\beta}{2\rho_0\Omega}$$

The general form of the solution is:

$$p(t, x, y) = 2\iint_{-\infty}^{+\infty}(A_1 \cos s_1 t + A_2 \cos s_2 t)e^{i(\alpha x + \beta y)}d\alpha\, d\beta$$

$$v_x(t, x, y) = -2i\iint_{-\infty}^{+\infty}(B_1 \sin s_1 t + B_2 \sin s_2 t)e^{i(\alpha x + \beta y)}d\alpha d\beta$$

. . . .

$$h_y(t, x, y) = 2\iint_{-\infty}^{+\infty}E_1 (\cos s_1 t - \cos s_2 t)e^{i(\alpha x + \beta y)}d\alpha\, d\beta \tag{9.3.19}$$

where one has still to perform the study of the integrals. For the asymptotic behaviour one can use the stationary phase principle [15] — [17].

9.3.4 The one-dimensional case

If the initial perturbation does not depend on the variable y we may assume this to be also true for the perturbation at time t so that the system (9.3.8) is separated into the following two systems:

$$\frac{\partial p}{\partial t} + \gamma p_0 \frac{\partial v_x}{\partial x} = 0, \; \rho_0 \frac{\partial v_x}{\partial t} + \frac{\partial p}{\partial x} = 0 \tag{9.3.20}$$

$$\rho_0 \frac{\partial v_y}{\partial t} - \mu H_0 \frac{\partial h_y}{\partial x} = 0, \; \frac{\partial h_y}{\partial t} - H_\infty \frac{\partial v_y}{\partial x} = 0 \tag{9.3.21}$$

$h_x = h_x^0 = \text{const.}$

Looking for a solution of system (9.3.20) of the form:

$$\begin{Bmatrix} p \\ v_x \end{Bmatrix} = \int_{-\infty}^{+\infty} \begin{Bmatrix} A \\ B \end{Bmatrix} e^{i(\alpha x - st)} \, d\alpha \tag{9.3.22}$$

we obtain:

$$-sA + \gamma p_0 \alpha B = 0, \; \alpha A - s\rho_0 B = 0 \tag{9.3.23}$$

In order that system (9.3.23) has a non-vanishing solution one should have $s = \pm a|\alpha|$ so that:

$$p(t, x) = a\rho_0 \int_{-\infty}^{+\infty} |\alpha| \, (A_1 e^{-ia|\alpha|t} - A_2 e^{ia|\alpha|t}) e^{i\alpha x} d\alpha$$

$$v_x(t, x) = \int_{-\infty}^{+\infty} \alpha \, (A_1 e^{-ia|\alpha|t} + A_2 e^{ia|\alpha|t}) e^{i\alpha x} d\alpha$$

By imposing the initial conditions we get:

$$a\rho_0 |\alpha| \, (A_1 - A_2) = \frac{1}{2\pi} \int_{-\infty}^{+\infty} p^0(x) e^{-i\alpha x} dx$$

$$\alpha(A_1 + A_2) = \frac{1}{2\pi} \int_{-\infty}^{+\infty} v_x^0(x) e^{-i\alpha x} dx$$

so that:

$$p(t, x) = \int_{-\infty}^{+\infty} p^0(\xi) \, K_1(t, x - \xi) \, d\xi - ia\rho_0 \int_{-\infty}^{+\infty} v_x^0(\xi) \, K_2(t, x - \xi) \, d\xi$$

$$v_x(t, x) = -\frac{i}{a\rho_0} \int_{-\infty}^{+\infty} p^0(\xi) \, K_2 d\xi + \int_{-\infty}^{+\infty} v_x^0(\xi) \, K_1 d\xi$$

But [G. 31]:

$$2K_1 \stackrel{\text{def}}{=} \frac{1}{\pi}\int_{-\infty}^{+\infty} \cos \alpha a t\, e^{i\alpha(x-\xi)} d\alpha = \delta(x-\xi-at) + \delta(x-\xi+at)$$

$$2K_2 \stackrel{\text{def}}{=} \frac{1}{\pi}\int_{-\infty}^{+\infty} \sin \alpha a t\, e^{i\alpha(x-\xi)} d\alpha = i\{\delta(x-\xi-at) - \delta(x-\xi+at)\}$$

such that:

$$2p(t, x) = p^0(x-at) + p^0(x+at) + a\rho_0\{v_x^0(x-at) - v_x^0(x+at)\}$$

$$2v_x(t, x) = v_x^0(x-at) + v_x^0(x+at) - \{p^0(x-at) - p^0(x+at)\}/a\rho_\infty$$

In a similar way we have:

$$2v_y(t, x) = v_y^0(x-bt) + v_y^0(x+bt) - \{h_y^0(x-bt) - h_y^0(x+bt)\}b/H_\infty$$

$$2h_y(t, x) = h_y^0(x-bt) + h_y^0(x+bt) - \{v_y^0(x-bt) - v_y^0(x+bt)\}H_\infty/b$$

This represents the final form of the solution. Its interpretation as a superposition of waves determined by the initial perturbation is obvious.

9.4 The three-dimensional problem

9.4.1 The construction of a base

When the perturbation depends on the z-variable as well the system (9.3.3) — (9.3.6) reduces itself to:

$$DU = 0 \tag{9.4.1}$$

where we used the notation:

$$D = \begin{pmatrix} \partial_t & a\partial_x & a\partial_y & a\partial_z & 0 & 0 & 0 \\ a\partial_x & \partial_t & 0 & 0 & 0 & 0 & 0 \\ a\partial_y & 0 & \partial_t & 0 & b\partial_y & -b\partial_x & 0 \\ a\partial_z & 0 & 0 & \partial_t & b\partial_z & 0 & -b\partial_x \\ 0 & 0 & b\partial_y & b\partial_z & \partial_t & 0 & 0 \\ 0 & 0 & -b\partial_x & 0 & 0 & \partial_t & 0 \\ 0 & 0 & 0 & -b\partial_x & 0 & 0 & \partial_t \end{pmatrix}, U = \begin{pmatrix} p^* \\ v_x \\ v_y \\ v_z \\ h_x^* \\ h_y^* \\ h_z^* \end{pmatrix}$$

$$\partial_x = \partial/\partial x, \ldots ; \; a\rho_0 p^* = p; \; H_0 h^* = bh$$

For plane-wave solutions of the form:

$$U = \bar{U} \exp i(\boldsymbol{\alpha} \cdot \boldsymbol{x} - st)$$
$$\boldsymbol{\alpha} = (\alpha, \beta, \gamma), \ \boldsymbol{x} = (x, y, z) \tag{9.4.2}$$

system (9.4.1) yields:

$$(A_0 - sE)\bar{U} = 0 \tag{9.4.3}$$

where:

$$A_0 - sE = \begin{pmatrix} -s & a\alpha & a\beta & a\gamma & 0 & 0 & 0 \\ a\alpha & -s & 0 & 0 & 0 & 0 & 0 \\ a\beta & 0 & -s & 0 & b\beta & -b\alpha & 0 \\ a\gamma & 0 & 0 & -s & b\gamma & 0 & -b\alpha \\ 0 & 0 & b\beta & b\gamma & -s & 0 & 0 \\ 0 & 0 & -b\alpha & 0 & 0 & -s & 0 \\ 0 & 0 & 0 & -b\alpha & 0 & 0 & -s \end{pmatrix}$$

E being the unit matrix.

Since A_0 is a symmetrical matrix it follows that its eigenvalues s_j are real (the system is of hyperbolic type) and the associated eigenvectors $\bar{U}_j$ are real and orthogonal.

An elementary calculation which uses the fact that the algebraic complements of the first line are:

$$\begin{pmatrix} s(s^2 - b^2\omega^2) \\ a\alpha(s^2 - b^2\omega^2) \\ a\beta s^2 \\ a\gamma s^2 \\ ab(\beta^2 + \gamma^2)s \\ -ab\,\alpha\beta\, s \\ -ab\,\alpha\gamma\, s \end{pmatrix} \times s(s^2 - b^2\alpha^2)$$

yields:

$$\det(A_0 - sE) = -s(s^2 - b^2\alpha^2)\{s^4 - (a^2 + b^2)\omega^2 s^2 + a^2 b^2 \alpha^2 \omega^2\} \tag{9.4.4}$$

where:

$$\omega = \sqrt{\alpha^2 + \beta^2 + \gamma^2}$$

Consequently one obtains the following eigenvalues:

$$s_1 = 0, \quad s_2 = b\alpha, \quad s_3 = -b\alpha \tag{9.4.5}$$

$$s_4 = \left(\sqrt{a^2 + b^2 + 2ab\alpha/\omega} + \sqrt{a^2 + b^2 - 2ab\alpha/\omega}\right)\omega/2, \quad s_6 = -s_4$$

$$s_5 = \left(\sqrt{a^2 + b^2 + 2ab\alpha/\omega} - \sqrt{a^2 + b^2 - 2ab\alpha/\omega}\right)\omega/2, \quad s_7 = -s_5$$

and the following eigenvectors:

$$\bar{U}_1 = \frac{1}{\omega}\begin{pmatrix} 0 \\ 0 \\ 0 \\ 0 \\ \alpha \\ \beta \\ \gamma \end{pmatrix}, \quad \bar{U}_2 = \frac{1}{\sqrt{2(\beta^2 + \gamma^2)}}\begin{pmatrix} 0 \\ 0 \\ -\gamma \\ \beta \\ 0 \\ \gamma \\ -\beta \end{pmatrix}, \quad \bar{U}_3 = \frac{1}{\sqrt{2(\beta^2 + \gamma^2)}}\begin{pmatrix} 0 \\ 0 \\ -\gamma \\ \beta \\ 0 \\ -\gamma \\ \beta \end{pmatrix}$$

$$\bar{U}_j = \frac{1}{N_j}\begin{pmatrix} s_j(s_j^2 - b^2\omega^2) \\ a\alpha(s_j^2 - b^2\omega^2) \\ a\beta s_j^2 \\ a\gamma s_j^2 \\ ab(\beta^2 + \gamma^2)s_j \\ ab\alpha\beta s_j \\ -ab\alpha\gamma s_j \end{pmatrix} \qquad j = 4, 5, 6, 7 \tag{9.4.6}$$

$$N_j^2 = (s_j^2 - b^2\omega^2)^2(s_j^2 + a^2\alpha^2) + a^2(\beta^2 + \gamma^2)s_j^2(s_j^2 + b^2\omega^2)$$

The norm N_j of the vector $\bar{U}_j$ can be simplified by using the relation:

$$(s_j^2 - a^2\alpha^2)(s_j^2 - b^2\omega^2) = a^2(\beta^2 + \gamma^2)s_j^2 \tag{9.4.7}$$

derived from (9.4.4). One has:

$$N_j^2 = 2\omega^2(s_j^2 - b^2\omega^2)\{(a^2 + b^2)s_j^2 - 2a^2b^2\alpha^2\} \tag{9.4.6'}$$

One can easily check the orthogonality of the eigenvectors ($\bar{U}_k \cdot \bar{U}_l = \delta_{kl}$). Therefore they form a base.

9.4.2 The Cauchy problem

Since (9.4.1) represents a linear system it follows that the general solution is obtained as a continuous superposition of plane waves of the form (9.4.2). Consequently we obtain:

$$U(t, \boldsymbol{x}) = \iiint_{-\infty}^{+\infty} \Sigma\, c_k(\alpha, \beta, \gamma)\overline{U}_k\, \mathrm{e}^{-\mathrm{i}s_k t}\, \mathrm{e}^{\mathrm{i}\boldsymbol{\alpha}\cdot\boldsymbol{x}}\, \mathrm{d}\boldsymbol{\alpha} \tag{9.4.8}$$

$$\mathrm{d}\boldsymbol{\alpha} = \mathrm{d}\alpha\, \mathrm{d}\beta \mathrm{d}\,\gamma$$

the functions c_k being determined by the initial conditions. To this end one calculates $U(O, \boldsymbol{x})$ in D from (9.3.2) (in $\complement D$ $U(0, \boldsymbol{x})$ is vanishing). Substituting these values in (9.4.8) and using the Fourier inversion theorem we get:

$$\Sigma\, c_k \overline{U}_k = (2\pi)^{-3} \iiint_D U(0, \boldsymbol{\xi})\, \mathrm{e}^{-\mathrm{i}\boldsymbol{\alpha}\cdot\boldsymbol{\xi}}\, \mathrm{d}\boldsymbol{\xi}$$

$$\boldsymbol{\xi} = (\xi_1, \xi_2, \xi_3), \;\; \mathrm{d}\boldsymbol{\xi} = \mathrm{d}\xi_1\, \mathrm{d}\xi_2\, \mathrm{d}\xi_3$$

and then (using the orthonormality conditions) it follows that

$$c_k = (2\pi)^{-3} \iiint_D \overline{U}_k \cdot U(0, \boldsymbol{\xi})\, \mathrm{e}^{-\mathrm{i}\boldsymbol{\alpha}\cdot\boldsymbol{\xi}} \mathrm{d}\boldsymbol{\xi} \tag{9.4.9}$$

In this way the general solution takes the following form:

$$U(t, \boldsymbol{x}) = \iiint_D M(t, \boldsymbol{x} - \boldsymbol{\xi}) U(0, \boldsymbol{\xi})\, \mathrm{d}\boldsymbol{\xi} \tag{9.4.10}$$

where $M = \Sigma M_k$,

$$M_k(t, \boldsymbol{x}) = (2\pi)^{-3} \iiint_{-\infty}^{+\infty} \overline{U}_k \overline{U}_k\, \mathrm{e}^{\mathrm{i}(\boldsymbol{\alpha}\cdot\boldsymbol{x} - s_k t)}\, \mathrm{d}\boldsymbol{\alpha} \tag{9.4.10$'$}$$

$\overline{U}_k \overline{U}_k$ being the dyadic product.

A potential representation of the kernels M_1, M_2, M_3 is given in [13]. One obtains:

$$\iiint_D M_1(t, \boldsymbol{x} - \boldsymbol{\xi})\, U(0, \boldsymbol{\xi})\, \mathrm{d}\boldsymbol{\xi} = \left(\begin{array}{c|ccc} 0 & & 0 & \\ \hline & \partial^2_{xx} & \partial^2_{xy} & \partial^2_{xz} \\ 0 & \partial^2_{yx} & \partial^2_{yy} & \partial^2_{yz} \\ & \partial^2_{zx} & \partial^2_{zy} & \partial^2_{zz} \end{array}\right) \frac{1}{4\pi} \iiint_D \frac{U(0, \boldsymbol{\xi})}{|\boldsymbol{x} - \boldsymbol{\xi}|}\, \mathrm{d}\boldsymbol{\xi}$$

$$\iiint_D M_2(t, \boldsymbol{x} - \boldsymbol{\xi}) U(0, \boldsymbol{\xi})\, \mathrm{d}\boldsymbol{\xi} = \left(\begin{array}{c|ccccc} 0 & & & 0 & & \\ \hline & \partial^2_{zz} & -\partial^2_{yz} & 0 & -\partial^2_{zz} & \partial^2_{yz} \\ 0 & -\partial^2_{yz} & \partial^2_{yy} & 0 & \partial^2_{yz} & -\partial^2_{yy} \\ & 0 & 0 & 0 & 0 & 0 \\ & -\partial^2_{zz} & \partial^2_{yz} & 0 & \partial^2_{zz} & -\partial^2_{yz} \\ & \partial^2_{yz} & -\partial^2_{yy} & 0 & -\partial^2_{yz} & \partial^2_{yy} \end{array}\right)$$

$$\times \frac{1}{4\pi} \iint U(0;\ x - bt, \eta, \zeta) \lg \frac{1}{\sqrt{(y-\eta)^2 + (z-\zeta)^2}}\, \mathrm{d}\eta\, \mathrm{d}\zeta$$

$$\iiint_D M_3(t, \boldsymbol{x} - \boldsymbol{\xi}) U(0, \boldsymbol{\xi})\, \mathrm{d}\boldsymbol{\xi} = \left(\begin{array}{c|ccccc} 0 & & & 0 & & \\ \hline & \partial^2_{zz} & -\partial^2_{yz} & 0 & \partial^2_{zz} & -\partial^2_{yz} \\ 0 & -\partial^2_{yz} & \partial^2_{yy} & 0 & -\partial^2_{yz} & \partial^2_{yy} \\ & 0 & 0 & 0 & 0 & 0 \\ & \partial^2_{zz} & -\partial^2_{yz} & 0 & \partial^2_{zz} & -\partial^2_{yz} \\ & -\partial^2_{yz} & \partial^2_{yy} & 0 & -\partial^2_{yz} & \partial^2_{yy} \end{array}\right)$$

$$\times \frac{1}{4\pi} \iint U(0;\ x + bt, \eta, \zeta) \lg \frac{1}{\sqrt{(y-\eta)^2 + (z-\zeta)^2}}\, \mathrm{d}\eta\, \mathrm{d}\zeta$$

9.4.3 The point perturbation of pressure

Let us consider here again the pressure perturbation at the origin of the system of coordinates:

$$p^0(\boldsymbol{x}) = a\rho_0 P_0 \delta(\boldsymbol{x}) \tag{9.4.11}$$

Supposing $\boldsymbol{v}^0 \equiv 0$, $\boldsymbol{h}^0 \equiv 0$ it follows that:

$$U(0, \boldsymbol{\xi}) = \begin{pmatrix} P_0 \delta(\boldsymbol{\xi}) \\ 0 \\ 0 \\ 0 \\ 0 \\ 0 \\ 0 \end{pmatrix}$$

so that (9.4.10) becomes:

$$U(t, \boldsymbol{x}) = P_0 M_1(\xi = 0)$$

$$= \frac{P_0}{(2\pi)^3} \iiint_{-\infty}^{+\infty} \sum_j \begin{pmatrix} s_j(s_j^2 - b^2\omega^2) \\ a\alpha(s_j^2 - b^2\omega^2) \\ a\beta s_j^2 \\ a\gamma s_j^2 \\ ab(\beta^2 + \gamma^2)s_j \\ -ab\alpha\beta s_j \\ -ab\alpha\gamma s_j \end{pmatrix} \times \frac{s_j(s_j^2 - b^2\omega^2)}{N_j^2} \, \mathrm{e}^{\mathrm{i}(\boldsymbol{\alpha}\cdot\boldsymbol{x} - s_j t)} \, \mathrm{d}\boldsymbol{\alpha} \qquad (9.4.12)$$

Taking into account (9.4.7) we have:

$$p(t, \boldsymbol{x}) = \frac{a^2 P_0}{(2\pi)^3} \iiint_{-\infty}^{+\infty} \sum_{l=4,5} \frac{s_l^2 - b^2\alpha^2}{(a^2 + b^2)s_l^2 - 2a^2b^2\alpha^2} \cos(s_l t) \, \mathrm{e}^{\mathrm{i}\boldsymbol{\alpha}\cdot\boldsymbol{x}} \, \mathrm{d}\boldsymbol{\alpha}$$

$$v_x(t, \boldsymbol{x}) = \frac{a^3 \mathrm{i} P_0}{(2\pi)^3} \iiint_{-\infty}^{+\infty} \sum_{l=4,5} \frac{\alpha(s_l^2 - b^2\alpha^2)}{(a^2 + b^2)s_l^3 - 2a^2b^2\alpha^2 s_l} \sin(s_l t) \, \mathrm{e}^{\mathrm{i}\boldsymbol{\alpha}\cdot\boldsymbol{x}} \, \mathrm{d}\boldsymbol{\alpha} \qquad (9.4.13)$$

One still has to perform the study of these integrals.

9.4.4 Non-conducting fluids

In order to outline the general aspect of the solution let us consider the particular case of the non-conducting fluid. In this case equations (9.3.3) and (9.3.4) become the acoustics equations:

$$\partial_t p + \gamma p_0 \operatorname{div} \boldsymbol{v} = 0, \quad \rho_0 \partial_t \boldsymbol{v} = -\operatorname{grad} p \qquad (9.4.14)$$

and are easily integrated since they imply:

$$\frac{\partial^2 p}{\partial t^2} - a^2 \Delta p = 0, \quad \Delta = \frac{\partial^2}{\partial x^2} + \frac{\partial^2}{\partial y^2} + \frac{\partial^2}{\partial z^2} \qquad (9.4.15)$$

Given p and $\boldsymbol{v}$ at the initial time from (9.4.14) we get $\partial_t p$ at the initial time so that equation (9.4.15) determines the pressure.

Afterwards the velocity is derived from equation (9.4.14). However the solution does not have a general form.

Applying the above method we shall look for a solution of the system (9.4.14) of the form (9.4.2). The following eigenvalues and eigenvectors are correspondingly obtained:

$$s_1 = 0, \quad \bar{U}_1 = \frac{1}{\sqrt{\alpha^2+\beta^2}}\begin{pmatrix} 0 \\ -\beta \\ \alpha \\ 0 \end{pmatrix}; \quad s_2 = 0, \quad \bar{U}_2 = \frac{1}{\omega\sqrt{\alpha^2+\beta^2}}\begin{pmatrix} 0 \\ -\alpha\gamma \\ -\beta\gamma \\ \alpha^2+\beta^2 \end{pmatrix};$$

$$s_3 = a\omega, \quad \bar{U}_3 = \frac{1}{\omega\sqrt{2}}\begin{pmatrix} \omega \\ \alpha \\ \beta \\ \gamma \end{pmatrix}; \quad s_4 = -a\omega, \quad \bar{U}_4 = \frac{1}{\omega\sqrt{2}}\begin{pmatrix} -\omega \\ \alpha \\ \beta \\ \gamma \end{pmatrix} \tag{9.4.16}$$

$\bar{U}_{1,2,3,4}$ being an orthogonal base. Thus the general solution has the form:

$$U(t, \boldsymbol{x}) = \iiint_{-\infty}^{+\infty} \Sigma\, c_j \bar{U}_j\, e^{i(\boldsymbol{\alpha}\cdot\boldsymbol{x} - s_j t)}\, d\boldsymbol{\alpha} \tag{9.4.17}$$

where:

$$c_j = (2\pi)^{-3} \iiint_D \bar{U}_j \cdot U(0, \xi)\, e^{-i\boldsymbol{\alpha}\cdot\xi}\, d\xi \tag{9.4.17'}$$

For an initial pressure perturbation of the form (9.4.11) we have:

$$c_1 = c_2 = 0, \quad c_3 = -c_4 = -P_0/\sqrt{2}(2\pi)^3$$

such that:

$$p = \frac{a\rho_0 P_0}{(2\pi)^3} \iiint_{-\infty}^{+\infty} \cos a\omega t\, e^{i\boldsymbol{\alpha}\cdot\boldsymbol{x}}\, d\boldsymbol{\alpha}$$

$$\boldsymbol{v} = -\frac{iP_0}{(2\pi)^3} \iiint_{-\infty}^{+\infty} \frac{\boldsymbol{\alpha}}{\omega} \sin a\omega t\, e^{i\boldsymbol{\alpha}\cdot\boldsymbol{x}}\, d\boldsymbol{\alpha} \tag{9.4.18}$$

The integrals (9.4.18) are distributions (generalized functions) obtained from the fundamental solution of the wave equation. Indeed we have [G. 33, Ch. 2, Section 3.4]:

$$\varphi \overset{\text{def}}{=} \frac{P_0}{(2\pi)^3} \iiint_{-\infty}^{+\infty} \frac{\sin a\omega t}{\omega} e^{i\boldsymbol{\alpha}\cdot\boldsymbol{x}} d\boldsymbol{\alpha} = \frac{P_0}{2\sqrt{2\pi}} \frac{\delta(r-at)}{at} \tag{9.4.19}$$

$$r = \sqrt{x^2 + y^2 + z^2} \quad \left(\frac{\partial^2 \varphi}{\partial t^2} - a^2 \Delta\varphi = 0 \right)$$

such that:

$$p = \rho_0 \frac{\partial \varphi}{\partial t}, \qquad \boldsymbol{v} = -\operatorname{grad} \varphi \tag{9.4.20}$$

The solution (9.4.19), (9.4.20) verifies the system (9.4.14). Indeed, the last equation of (9.4.14) is satisfied by a representation of the form (9.4.20) while the first equation (9.4.14) leads to a wave equation for φ.

9.5 The general problem of the non-zero resistivity fluid with Hall effect

9.5.1 The eigenvalue equation

If we adopt as usual Ohm's law in its general form (3.3.1) we obtain by linearization:

$$\operatorname{curl} \boldsymbol{h} = \sigma(\boldsymbol{e} + \boldsymbol{e}_e + \boldsymbol{v} \times \boldsymbol{B}_0) + \frac{\sigma}{nq} \boldsymbol{B}_0 \times \operatorname{curl} \boldsymbol{h} \tag{9.5.1}$$

$\boldsymbol{e}_e$ being the electric field due to the electron pressure gradient. Applying the *curl* operator to equation (9.5.1) and using the equations:

$$\operatorname{div} \boldsymbol{h} = 0, \ \operatorname{curl} \boldsymbol{e} = -\mu\partial_t \boldsymbol{h}, \ \operatorname{curl} \boldsymbol{e}_e = 0$$

we obtain:

$$P_0 \boldsymbol{h} = H_0 \partial_x \boldsymbol{v} - \boldsymbol{H}_0 \operatorname{div} \boldsymbol{v} - \nu \operatorname{curl} \partial_x \boldsymbol{h} \tag{9.5.2}$$

where:

$$P_0 = \frac{\partial}{\partial t} - \eta\left(\frac{\partial^2}{\partial x^2} + \frac{\partial^2}{\partial y^2} + \frac{\partial^2}{\partial z^2}\right), \ \eta = \frac{1}{\sigma\mu}, \ \nu = \frac{H_0}{nq} \tag{9.5.2'}$$

the Ox-axis being as in previous cases oriented along the direction of $\boldsymbol{H}_0$.

For the determination of the fluid motion equations (9.3.3), (9.3.4) and (9.3.6) are still valid, equation (9.3.5) being replaced by the equation (9.5.2) obtained above. Consequently system (9.4.1) becomes:

$$\mathscr{D}U = 0 \tag{9.5.3}$$

where:

$$\mathscr{D} = \begin{pmatrix} \partial_t & a\partial_x & a\partial_y & a\partial_z & 0 & 0 & 0 \\ a\partial_x & \partial_t & 0 & 0 & 0 & 0 & 0 \\ a\partial_y & 0 & \partial_t & 0 & b\partial_y & -b\partial_x & 0 \\ a\partial_z & 0 & 0 & \partial_t & b\partial_z & 0 & -b\partial_x \\ 0 & 0 & b\partial_y & b\partial_z & P_0 & -\nu\partial^2_{xz} & \nu\partial^2_{xy} \\ 0 & 0 & -b\partial_x & 0 & \nu\partial^2_{xz} & P_0 & -\nu\partial^2_{xx} \\ 0 & 0 & 0 & -b\partial_x & -\nu\partial^2_{xy} & \nu\partial^2_{xx} & P_0 \end{pmatrix}$$

In this case equation (9.4.3) is replaced by the equation:

$$(A - sE)\bar{U} = 0 \tag{9.5.4}$$

where:

$$A - sE = \begin{pmatrix} -s & a\alpha & a\beta & a\gamma & 0 & 0 & 0 \\ a\alpha & -s & 0 & 0 & 0 & 0 & 0 \\ a\beta & 0 & -s & 0 & b\beta & -b\alpha & 0 \\ a\gamma & 0 & 0 & -s & b\gamma & 0 & -b\alpha \\ 0 & 0 & b\beta & b\gamma & -s-i\eta\omega^2 & -i\nu\alpha\gamma & i\nu\alpha\beta \\ 0 & 0 & -b\alpha & 0 & i\nu\alpha\gamma & -s-i\eta\omega^2 & -i\nu\alpha^2 \\ 0 & 0 & 0 & -b\alpha & -i\nu\alpha\beta & i\nu\alpha^2 & -s-i\eta\omega^2 \end{pmatrix}$$

In order to calculate the determinant det $(A - sE)$ we use the notation

$$\sigma = s + i\eta\omega^2 \tag{9.5.5}$$

and denote by $C_1, \ldots, C_7$ the algebraic complements of the first line elements. Using also the notation:

$$C = \begin{pmatrix} C_1 \\ C_2 \\ C_3 \\ C_4 \\ C_5 \\ C_6 \\ C_7 \end{pmatrix}, \quad V_0 = \begin{pmatrix} s(s\sigma - b^2\omega^2) \\ a\alpha(s\sigma - b^2\omega^2) \\ a\beta\, s\sigma \\ a\gamma\, s\sigma \\ ab(\beta^2 + \gamma^2)s \\ -ab\,\alpha\beta\, s \\ -ab\,\alpha\gamma\, s \end{pmatrix}, \quad V_1 = \begin{pmatrix} 0 \\ 0 \\ -b\alpha\gamma \\ b\alpha\beta \\ 0 \\ \gamma s \\ -\beta s \end{pmatrix}, \quad V_2 = \begin{pmatrix} s \\ a\alpha \\ a\beta \\ a\gamma \\ 0 \\ 0 \\ 0 \end{pmatrix}$$

one obtains:

$$C = \sigma(s\sigma - b^2\alpha^2)V_0 + i\nu\, ab\,\alpha\omega^2 s\sigma V_1 - \nu^2\alpha^2\omega^2 s^2\sigma V_2$$

and then:

$$\det(A - sE)$$

$$= \sigma(s\sigma - b^2\alpha^2)\{-s^3\sigma + (b^2 s + a^2\sigma)s\omega^2 - a^2b^2\alpha^2\omega^2\} + \nu^2\omega^2 s^2\sigma(s^2 - a^2\omega^2) \qquad (9.5.6)$$

This equation determines the displacement velocities of the perturbation waves (the eigenvalues of the matrix A). It can be noticed that $\sigma = 0$ ($s = -i\eta\omega^2$) is an eigenvalue whose corresponding eigenvector is $(0, 0, 0, 0, \alpha, \beta, \gamma)$. For $\nu = 0$, V_1 is the eigenvector attached to the eigenvalue $s\sigma = b^2\alpha^2$ and V_0 the eigenvector attached to the roots of the equation:

$$-s^3\sigma + (a^2\sigma + b^2 s)s\omega^2 - a^2b^2\alpha^2\omega^2 = 0$$

In this case the explicit determination of the eigenvectors is very difficult.

9.5.2 A qualitative study

We include here some qualitative results concerning the eigenvalue (and eigenvector) structure [14]. In this respect let us search for solutions of the system (9.6.3) of the form:

$$U = u \exp ik(\boldsymbol{\omega} \cdot \boldsymbol{x} - st) \qquad (9.5.7)$$

and take $k = |\boldsymbol{\omega}|^{-1}$. Now using the notations:

$$\boldsymbol{\alpha} = \frac{\boldsymbol{\omega}}{|\boldsymbol{\omega}|}, \quad \lambda = \frac{s}{|\boldsymbol{\omega}|}, \quad \boldsymbol{\alpha} = (\alpha_1, \alpha_2, \alpha_3) \qquad (9.5.8)$$

we shall have $|\boldsymbol{\alpha}| = 1$. The system (9.5.3) can be written in the following way:

$$Du = \lambda u \tag{9.5.9}$$

where:

$$D = \begin{pmatrix} A & B \\ B' & -i\eta E + i\nu C \end{pmatrix}, \quad C = \alpha_1 \begin{pmatrix} 0 & -\alpha_3 & \alpha_2 \\ \alpha_3 & 0 & -\alpha_1 \\ -\alpha_2 & \alpha_1 & 0 \end{pmatrix}$$

$$A = a \begin{pmatrix} 0 & \alpha_1 & \alpha_2 & \alpha_3 \\ \alpha_1 & 0 & 0 & 0 \\ \alpha_2 & 0 & 0 & 0 \\ \alpha_3 & 0 & 0 & 0 \end{pmatrix}, \quad B = b \begin{pmatrix} 0 & 0 & 0 \\ 0 & 0 & 0 \\ \alpha_2 & -\alpha_1 & 0 \end{pmatrix}$$

E being the unit matrix of three lines by three columns and B' the transposed matrix of B.

We now breakdown the column vector $\boldsymbol{u}$ in the following way:

$$u = \begin{pmatrix} u_1 \\ u_2 \end{pmatrix}, \quad u_1 = \begin{pmatrix} p^* \\ v_x \\ v_y \\ v_z \end{pmatrix}, \quad u_2 = \begin{pmatrix} h^* \\ h_y^* \\ h_z^* \end{pmatrix} \tag{9.5.10}$$

With these notations equation (9.5.9) becomes:

$$Au_1 + Bu_2 = \lambda u_1 \tag{9.5.11}$$

$$B'u_1 + i(\nu C - \eta E)u_2 = \lambda u_2 \tag{9.5.12}$$

Generally λ is a complex quantity $\lambda' + i\lambda''$ and u a complex vector. Indicating by an asterisk the adjoint (transposed and conjugated) matrix from (9.5.11) and (9.5.12) we obtain:

$$u_1^* A u_1 + u_1^* B u_2 = \lambda(u_1, u_1)$$

$$u_1^* B + u_2^* i(\nu C + \eta E) = \bar{\lambda} u_2^* \qquad (C' = -C) \tag{9.5.12'}$$

$$u_1^* B u_2 + i\nu u_2^* C u_2 + i\eta(u_2, u_1) = \bar{\lambda}(u_2, u_2)$$

where:

$$(u_1, u_1) = u_1^* \cdot u_1, \quad (u_2, u_2) = u_2^* \cdot u_2$$

From (9.5.11′) and (9.5.12′) $u_1^* B u_2$ is eliminated. In the derived relation one separates the imaginary and real parts and obtains:

$$u_1^* A u_1 - i\nu u_2^* C u_2 = \lambda' \{(u_1, u_1) - (u_2, u_2)\} \tag{9.5.13}$$

$$-\eta (u_2, u_2) = \lambda'' \{(u_1, u_1) + (u_2, u_2)\} \tag{9.5.14}$$

A first conclusion can be drawn from (9.5.14): *the imaginary parts of the complex eigenvalues are negative and their moduli are bounded by* η (if $\eta = 0$ the eigenvalues are real). From the mechanical point of view this means that the waves propagating through a non-zero resistivity medium will be damped in time while in a perfectly conducting fluid (even with Hall effect) this property will not be observed.

Considering now the general case ($\eta \neq 0$) let us see if the existence of real eigenvalues is any longer possible. If $\lambda'' = 0$ from (9.5.14) it follows that $u_2 = 0$ so that equations (9.5.11) and (9.5.12) become:

$$A u_1 = \lambda u_2, \quad B' u_1 = 0$$

From the first equation we have:

$$\det(A - \lambda E) = \lambda^4 - a^2 \lambda^2 = 0$$

To the root $\lambda = 0$ the vector $u_{10} = (u_{10}^{(i)})$ corresponds and the latter should according to (9.5.15) satisfy the following relations:

$$\begin{aligned} &\alpha_1 u_{10}^{(2)} + \alpha_2 u_{10}^{(3)} + \alpha_3 u_{10}^{(4)} = 0 \qquad u_{10}^{(1)} = 0 \\ &\alpha_2 u_{10}^{(3)} + \alpha_3 u_{10}^{(4)} = 0, \ \alpha_1 u_{10}^{(3)} = 0, \ \alpha_1 u_{10}^{(4)} = 0 \end{aligned} \tag{9.5.15}$$

If $\alpha_1 \neq 0$ it follows that $u_{10} = 0$ so that $\lambda = 0$ is not an eigenvalue. If $\alpha_1 = 0$ then the eigenvectors:

$$\begin{aligned} u_{10} &= (0, 1, 0, 0, 0, 0, 0) \\ u_{10} &= (0, 0, -\alpha_3, \alpha_2, 0, 0, 0) \end{aligned} \tag{9.5.16}$$

corresponding to the double value $\lambda = 0$ exist. From the mechanical point of view the condition $\alpha_1 = 0$ means that the perturbation U does not depend on the x-variable (such waves are possible only if the initial perturbation has this property).

Using equations (9.5.15) one can show that all the components of the eigenvectors u_{1a} which should correspond to the eigenvalues $\lambda = \pm a$ are necessarily zero. Consequently, for *the general case* $\eta \neq 0$ *there can be no real eigenvalues unless* $\alpha_1 = 0$; *if this is so* $\lambda = 0$ *is a double value having the corresponding vectors* (9.5.16).

Let us now see whether it is possible for only a magnetic field perturbation to exist. If $u_1 = 0$ from (9.5.14) it follows that $\lambda'' = -\eta$ and from (9.5.12) that:

$$Cu_2 = \mu u_2, \quad \nu\mu \stackrel{\text{def}}{=} \eta - i\lambda = -i\lambda' \tag{9.5.17}$$

which shows that μ is an eigenvalue of C. But the eigenvalues of C can be easily calculated. Indeed we have:

$$\det(C - \mu E) = -\mu(\mu^2 + \alpha_1^2) = 0 \Rightarrow \mu = 0, \mu = \pm i\alpha_1$$

However equation (9.5.11) ($Bu_2 = 0$) implies:

$$u_2 = (\alpha_1, \alpha_2, \alpha_3)$$

and this vector satisfies equation (9.5.17) only if $\mu = 0$ ($\lambda = -i\eta$). The values $\mu = \pm i\alpha_1$ are not possible. Therefore, *there exists only one eigenvalue* ($\lambda = -i\eta$) *consistent with* $u_1 = 0$.

These considerations can be further developed (in the above mentioned work some other results are also derived) but the important problem remains that of actually determining the eigenvalues and eigenvectors of equations (9.5.9). While for $\eta = 0$, $\nu = 0$ the problem was solved in the previous chapter it still remains an unsolved problem even for the particular cases $\eta = 0$, $\nu \neq 0$ (perfectly conducting fluid with Hall effect) and $\eta \neq 0$, $\nu = 0$ (electrically resistive fluid without Hall effect).

10

The non-linear theory of waves

10.1 General considerations on quasi-linear systems

10.1.1 System classification

It is well known that the discontinuities that may appear in a fluid affect either only the derivatives of the variables of motion in which case they are termed weak discontinuities, or the variables themselves in which case they are called strong discontinuities or shock waves. The theory of weak discontinuities can be developed from the differential form of the equations of motion but shock wave theory should start from the integral form of these equations. We have established in Chapter 2 that the differential equations of motion are only valid for continuous motions.

The MFD differential equations are quasi-linear equations, that is, the higher-order derivatives occur in a linear way. This is why the presentation of the weak discontinuity theory which represents the subject of this chapter should be preceded by some considerations on the quasi-linear systems. This theory is largely due to Friedrichs who in a series of papers [1], [2] derived its fundamental results.

A first order quasi-linear system has the following general form:

$$\sum_{k=0}^{m} \sum_{j=1}^{n} a_{ij}^{k} \frac{\partial u_j}{\partial x_k} + b_i = 0, \quad i = 1, 2, \ldots, n \tag{10.1.1}$$

n being the number of unknowns and $m + 1$ the number of independent variables the time being denoted by x_0. The coefficients a_{ij}^k and b_i may be functions of $x_0 \ldots\ldots$ $\ldots x_m, u_1 \ldots\ldots\ldots u_n$. Using the notation:

$$A_k = \begin{pmatrix} a_{11}^k & a_{12}^k \ldots a_{1n}^k \\ a_{21}^k & \ldots & \\ & \ldots & \\ a_{n1}^k \ldots & & a_{nn}^k \end{pmatrix}, \quad U = \begin{pmatrix} u_1 \\ \cdot \\ \cdot \\ \cdot \\ u_n \end{pmatrix}, \quad B = \begin{pmatrix} b_1 \\ \cdot \\ \cdot \\ \cdot \\ b_n \end{pmatrix}$$

system (10.1.1) can be written as follows:

$$\sum A_k \frac{\partial U}{\partial x_k} + B = 0 \tag{10.1.1'}$$

Now indicating by A the vector of components $A_1 \ldots . A_m$ and by ∇ the operator $\left(\frac{\partial}{\partial x_1}, \ldots, \frac{\partial}{\partial x_m}\right)$ system (10.1.1') becomes:

$$A_0 \frac{\partial U}{\partial x_0} + (A \cdot \nabla)U + B = 0 \tag{10.1.1''}$$

The system in question being of the first order, one needs, in order to solve the Cauchy problem, to know the matrix U on a manifold Σ of the $(m+1)$-dimensional space. Let us write:

$$U|_{\Sigma} = F \tag{10.1.2}$$

F being a column matrix. If from conditions (10.1.2) together with the system of equations (10.1.1) one can determine all the derivatives of U then one can solve the Cauchy problem (the solution of the system is determined as a series). Otherwise Σ will be called a characteristic manifold.

Let us now determine the condition for Σ to be a characteristic manifold. It is obvious that taking into account the conditions (10.1.2) one can calculate the derivatives of U along any tangential direction the derivative normal to Σ remaining to be determined from the system (10.1.1). Let:

$$G(x_0, x_1, \ldots, x_m) = 0 \tag{10.1.3}$$

be the equation of the manifold Σ and let T be the following change of variables:

$$\begin{aligned} \xi_0 &= G(x_0, x_1, \ldots, x_m) \\ \xi_l &= G_l(x_0, x_1, \ldots, x_m) \end{aligned} \tag{10.1.4}$$

the only restriction imposed on the function G_l being that T be a reversible transformation. In terms of the new variables system (10.1.1') becomes:

$$\sum_{k=0}^{m} \sum_{j=0}^{m} A_k \frac{\partial \xi_j}{\partial x_k} \frac{\partial U}{\partial \xi_j} + B = \sum_{k=0}^{m} A_k \frac{\partial G}{\partial x_k} \frac{\partial U}{\partial \xi_0} + \ldots = 0 \tag{10.1.5}$$

where we pointed out the normal derivative U_k. This derivative can be determined only if:

$$\sum_{k=0}^{m} A_k \frac{\partial G}{\partial x_k}$$

is a non-singular matrix. Consequently, the condition that Σ be a characteristic manifold reduces itself to the partial derivative equation:

$$\det\left(\sum_{k=0}^{m} A_k \frac{\partial G}{\partial x_k}\right) = 0 \tag{10.1.6}$$

which should be satisfied on $G = 0$.

If we introduce the notations:

$$\alpha_k = \partial G/\partial x_k, \ k = 0, 1, \ldots, m \tag{10.1.7}$$

equation (10.1.6) is reduced to the equation $Q = 0$ where:

$$Q = \det\left(\sum_k a_{ij}^k \alpha_k\right)$$

Q being a n-degree homogeneous polynomial expressed in terms of the variables $\alpha_0, \ldots, \alpha_m$. If in some domain D of the $(x_0, \ldots, x_m)$ – space the equation $Q = 0$ is verified only by the solution $\alpha_0 = \alpha_1 \ldots = \alpha_m = 0$, then the system will have no real characteristic manifolds (one says that the system is of elliptic type). If all the roots of the algebraic equation for α_0 are real for any real values given to the parameters $\alpha_1, \ldots, \alpha_m$, then the system will have real characteristic manifolds (it is a *completely hyperbolic* system) [G. 30].

10.1.2 Discontinuity waves

A fluid perturbation propagates itself as a discontinuity wave (surface). Let Σ be such a surface. It is obvious that a discontinuity surface represents at the same time a characteristic surface (the values of the unknowns on such a surface do not uniquely determine the solution). It follows that the theory of discontinuity waves is closely related to the characteristic-surface theory.

Returning to the Cauchy problem (10.1.2), we shall notice that from the Cauchy conditions the continuity of the tangential derivatives of the unknown U is derived, the discontinuities being possible only for the normal derivative. Then writing system (10.1.5) on each side of the surface Σ and subtracting one has:

$$\left(\sum A_k \frac{\partial G}{\partial x_k}\right) [U_{,\xi_0}] = 0 \tag{10.1.8}$$

This equation yields the so-called *dynamic conditions of consistency*. Writing in the form:

$$\left\{A_0 \frac{\partial G}{\partial x_0} + (A \cdot \nabla) G\right\} [U_{,G}] = 0 \tag{10.1.8'}$$

dividing this equation by $|\nabla G|$ and using the expression of the drift velocity (Section 1.5) it follows :

$$(-A_0 d + \boldsymbol{A} \cdot \boldsymbol{n})[U_{,G}] = 0 \tag{10.1.8''}$$

$\boldsymbol{n}$ being the unit vector normal to the surface $G = 0$ ($\boldsymbol{n} = \nabla G/|\nabla G|$). In MFD $A_0 = E$, E being the unit matrix. For the system (10.1.8'') to have a non-zero solution (Σ to be a discontinuity surface) it is necessary and sufficient to have:

$$\det(-Ed + \boldsymbol{A} \cdot \boldsymbol{n}) = 0 \tag{10.1.9}$$

which equation is equivalent to (10.1.6).

From (10.1.9) and (10.1.8'') it follows that the diift velocities of the discontinuity surface are the eigenvalues of the matrix $\boldsymbol{A} \cdot \boldsymbol{n}$ while the discontinuities $[U_{,G}]$ are the associated eigenvectors. Whence the following classification is obtained: if, for any $\boldsymbol{n}$, the matrix $\boldsymbol{A} \cdot \boldsymbol{n}$ has n real and distinct eigenvalues then the system will be of (strictly) *hyperbolic type*; if, for any $\boldsymbol{n}$, the matrix $\boldsymbol{A} \cdot \boldsymbol{n}$, does not have real eigenvalues the system will be of *elliptic type*; finally if for some direction $\boldsymbol{n}$ part of eigenvalues of $\boldsymbol{A} \cdot \boldsymbol{n}$ are imaginary the system will be *ultra-hyperbolic.*

For the construction of the *wave front* one can use either the ray velocity or the normal velocity method [G. 30], [3], [4]. The wave-front construction by means of the normal velocity surface is performed by starting from the definition of the discontinuity-surface drift velocity. Indeed, the following construction (for the wave front determined by a point perturbation) follows from the observation that the discontinuity surface is at any of its points, normal to the drift velocity vector: one first determines the drift velocities from equation (10.1.9); for the hyperbolic system they are all real velocities (a wave front corresponds to each velocity); a certain velocity d depends on $\boldsymbol{n}$; when $\boldsymbol{n}$ changes the points of polar coordinates d, $\boldsymbol{n}$ (with the origin at the perturbation source) determine a surface called the *normal velocity surface*; the wave front (after one unit time) is obtained by taking the envelope of the planes normal to the position vector at each point of the normal velocity surface.

10.1.3 Constant state propagation

A difficulty appears if one applies the above theory to quasi-linear systems: the functions A_k also depend on the (a priori unknown) solution of the system. However, if the discontinuity propagation is produced in a *constant state* of the fluid (the fluid is at rest or in uniform motion in a homogeneous magnetic field) the problem of the determination of propagation velocities and discontinuities can be solved substituting U in A_k by its constant state value*. In this case the discontinuity surfaces will be spatially-variable wave-fronts.

---

* As a matter of fact this substitution implies the linearization of the equations of motion about the constant state.

In this case we can give a geometrical meaning to the normal discontinuities. We shall now consider the distances ε on the normal to the discontinuity surface taken both in the direct (towards the unperturbed fluid) and reverse sense. Noticing that:

$$U(G+\varepsilon) = U(G)$$

$$U(G-\varepsilon) = U(G) - \varepsilon U_{,G}$$

it follows that:

$$\delta U \overset{\text{def}}{=} U(G+\varepsilon) - U(G-\varepsilon) = \varepsilon[U_{,G}] \tag{10.1.10}$$

Consequently system (10.1.8″) becomes:

$$(-dE + \boldsymbol{A}\cdot\boldsymbol{n})\delta U = 0 \tag{10.1.11}$$

This equation will determine up to a constant factor the discontinuities δU.

It is useful in what follows to notice [3] that equation (10.1.11) can be formally obtained in a direct manner from the system of equations of motion (10.1.1″) by the following substitution:

$$\partial_t \to -d\partial, \quad \nabla \to \boldsymbol{n}\delta \tag{10.1.12}$$

10.2 Incompressible fluids

Let us now consider a discontinuity surface propagating through a non-dissipating fluid. The system of equations of motion for this case is of the first order and thus the above theory may be applied. As usual we assume that ε and μ are constant so that we have $\varepsilon\mu \simeq \varepsilon_0\mu_0$ ($\boldsymbol{B} = \mu\boldsymbol{H}$). For an incompressible fluid the equations of motion can be written as follows:

$$\nabla\cdot\boldsymbol{V} = 0, \quad \nabla\cdot\boldsymbol{H} = 0 \tag{10.2.1}$$

$$\rho(\partial_t + \boldsymbol{V}\cdot\nabla)\boldsymbol{V} = -\nabla p + (\boldsymbol{B}\cdot\nabla)\boldsymbol{H} - \nabla\left(\frac{1}{2}\boldsymbol{B}\cdot\boldsymbol{H}\right) \tag{10.2.2}$$

$$(\partial_t + \boldsymbol{V}\cdot\nabla)\boldsymbol{H} = (\boldsymbol{H}\cdot\nabla)\boldsymbol{V} \tag{10.2.3}$$

the last being obviously the induction equation. Using the formalism of (10.3.12) we derive the following equations for the discontinuities:

$$\delta V_n = 0, \ \delta H_n = 0 \tag{10.2.1'}$$

$$-\rho P \, \delta \boldsymbol{V} = -\boldsymbol{n} \, \delta p + B_n \delta \boldsymbol{H} - (\boldsymbol{B} \cdot \delta \boldsymbol{H})\boldsymbol{n} \tag{10.2.2'}$$

$$-P \delta \boldsymbol{H} = H_n \delta \boldsymbol{V} \tag{10.2.3'}$$

$\boldsymbol{n}$ being the unit vector normal to the wave front ($V_n = \boldsymbol{V} \cdot \boldsymbol{n}$, $H_n = \boldsymbol{H} \cdot \boldsymbol{n}$) and P the propagation velocity (1.5.19).

Equations (10.2.1') indicate that the normal components of the velocity and magnetic field are continuous across the wave front. Only the tangential components and pressure present discontinuities. Denoting by $\boldsymbol{v}$ and $\boldsymbol{h}$ the velocity and magnetic field components tangential to the wave front from (10.2.2') and (10.2.3') we obtain:

$$\delta p + \mu \boldsymbol{h} \cdot \delta \boldsymbol{h} = 0 \tag{10.2.4}$$

$$aP \, \delta \boldsymbol{v} + \mu H_n \, \delta \boldsymbol{h} = 0, \ H_n \, \delta \boldsymbol{v} + P \, \delta \boldsymbol{h} = 0 \tag{10.2.5}$$

If system (10.2.5) has a non-zero solution then we necessarily have:

$$\rho P^2 = \mu H_n^2 \Rightarrow P = \pm P_t, \quad P_t \overset{\text{def}}{=} H_n \sqrt{\mu/\rho} \tag{10.2.6}$$

which fact indicates that the discontinuity surface propagates with the Alfvén velocity. With this value of P system (10.2.4) determines the following discontinuities:

$$\delta \boldsymbol{H} = \delta \boldsymbol{h} = \varepsilon H_n \boldsymbol{\tau}, \ \delta \boldsymbol{V} = \delta \boldsymbol{v} = -\varepsilon P \boldsymbol{\tau} \tag{10.2.7}$$

ε being an infinitesimal parameter characterizing the magnitude of the jump. Taking into account (10.1.10) we can write:

$$\left[\frac{\partial \boldsymbol{H}}{\partial n}\right] = \left[\frac{\partial \boldsymbol{h}}{\partial n}\right] = \varepsilon H_n \boldsymbol{\tau}, \ \left[\frac{\partial \boldsymbol{V}}{\partial n}\right] = \left[\frac{\partial \boldsymbol{v}}{\partial n}\right] = -\varepsilon P \boldsymbol{\tau} \tag{10.2.7'}$$

The pressure jump is determined by (10.2.4) while the total pressure jump is zero since we have:

$$\delta\left(p + \frac{1}{2} \boldsymbol{B} \cdot \boldsymbol{H}\right) = \delta p + \boldsymbol{B} \cdot \delta \boldsymbol{H} = 0 \tag{10.2.8}$$

Denoting by θ the angle between the applied magnetic field and the normal to the wave front we have $H_n = H \cos\theta$ and

$$P_t = b\,|\cos\theta| \text{ where } b = H\sqrt{\nu},\ \nu \overset{\text{def}}{=} \mu/\rho \tag{10.2.9}$$

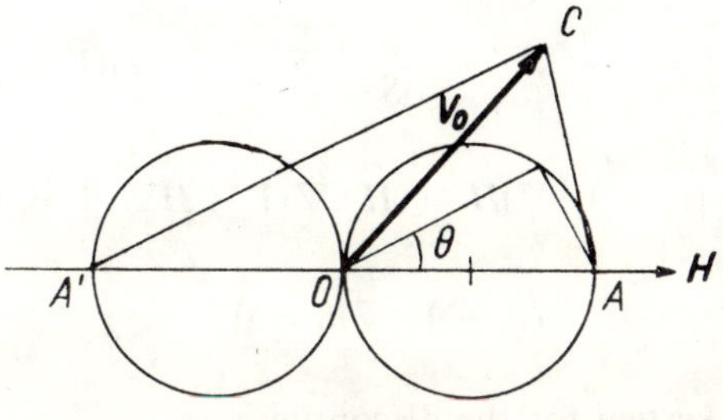

Fig. 30

For varying θ, P_t will generate the normal velocity surface. Let us now consider the source 0 and the applied magnetic field $\boldsymbol{H}$ (Figure 30). It is obvious that when θ changes in the intervals $[0, \pi/2, \pi)$ and $[-0, -\pi/2, -\pi]$ the end of P_t generates two circles having their centers along the direction of $\boldsymbol{H}$ at a distance $b/2$. Constructing the wave front according to the above-mentioned method one finds that for $t = 1$ this reduces to the point A and A'.

If the constant-state fluid has velocity V_0 then one will take into account the fact that on its way from 0 to A and A' the perturbation continuously emits waves so that the perturbed fluid region after one unit time will be $AA'C$.

10.3 Compressible fluids

10.3.1 Determination of propagation velocities

Let us now search for the discontinuity surfaces which may appear in a non-dissipating, compressible fluid obeying the perfect gas law ($p = \rho RT$). Neglecting any heat exchange it follows that the motion is isentropic. The calculations can be simplified if we take as independent variables the density and entropy instead of density and pressure as usual. Then the pressure becomes a function of density and entropy (2.3.18) such that we have:

$$\nabla p = \frac{\partial p}{\partial \rho}\nabla\rho + \frac{\partial p}{\partial S}\nabla S = a^2\nabla\rho + \frac{\partial p}{\partial S}\nabla S \tag{10.3.1}$$

The MFD equations in this form have for the first time been used by Lundquist [2.3] (being thus called the Lundquist equations). Let us write these equations in the following way:

$$(\partial_t + \boldsymbol{V}\cdot\nabla)\,\rho + \rho\nabla\cdot\boldsymbol{V} = 0, \ \nabla\cdot\boldsymbol{H} = 0 \tag{10.3.2}$$

$$\rho(\partial_t + \boldsymbol{V}.\nabla)\boldsymbol{V} = -a^2\nabla\rho - \frac{\partial p}{\partial S}\nabla S + (\boldsymbol{B}\cdot\nabla)\boldsymbol{H} - \nabla\left(\frac{1}{2}\boldsymbol{B}\cdot\boldsymbol{H}\right) \tag{10.3.3}$$

$$(\partial_t + \boldsymbol{V}\cdot\nabla)\boldsymbol{H} = (\boldsymbol{H}\cdot\nabla)\boldsymbol{V} - \boldsymbol{H}\nabla\cdot\boldsymbol{V} = 0 \tag{10.3.4}$$

$$(\partial_t + \boldsymbol{V}\cdot\nabla)S = 0 \tag{10.3.5}$$

We get the following system for the discontinuities:

$$-P\delta\rho + \rho\delta V_n = 0, \ \delta H_n = 0 \tag{10.3.2'}$$

$$-P\rho\delta\boldsymbol{V} + \left(a^2\delta\rho + \frac{\partial p}{\partial S}\delta S\right)\boldsymbol{n} - B_n\delta\boldsymbol{H} + (\boldsymbol{B}\cdot\delta\boldsymbol{H})\boldsymbol{n} = 0 \tag{10.3.3'}$$

$$-P\,\delta\boldsymbol{H} - H_n\delta\boldsymbol{V} + \boldsymbol{H}\delta V_n = 0 \tag{10.3.4'}$$

$$P\,\delta S = 0 \tag{10.3.5'}$$

the notations being those of the previous section.

Let us choose the reference system ($\boldsymbol{\tau}_1$, $\boldsymbol{\tau}_2$, $\boldsymbol{n}$) at some point on the wave front, $\boldsymbol{n}$ being as usual the normal to the wave front and $\boldsymbol{\tau}_1$ and $\boldsymbol{\tau}_2$ being unit vectors in the tangential plane ($\boldsymbol{\tau}_1$ lying in the plane determined by the magnetic field direction of the constant state and by the normal $\boldsymbol{n}$). In this reference system equations (10.3.2') — (10.3.5') become:

$$\left.\begin{aligned} &-P\,\delta\rho + \rho\delta V_n = 0 \\ &a^2\delta\rho - \rho P\,\delta V_n + \mu H_1\,\delta H_1 + \frac{\partial p}{\partial S}\delta S = 0 \\ &\rho P\,\delta V_1 + B_n\delta H_1 = 0 \\ &H_1\delta V_n - H_n\delta V_1 - P\delta H_1 = 0 \end{aligned}\right\} \tag{10.3.6}$$

$$\left.\begin{aligned} &\rho P\delta V_2 + B_n\delta H_2 = 0 \\ &H_n\delta V_2 + P\delta H_2 = 0 \end{aligned}\right\} \tag{10.3.7}$$

$$P\,\delta S = 0 \tag{10.3.8}$$

As already indicated above the coefficients ρ, a^2, μH_1, B_n, $\partial p/\partial S$ appearing in the system (10.3.6) and (10.3.7) have the values corresponding to the constant state. The system (10.3.6) — (10.3.8) (from which the unknowns $\delta\rho$, δV_n, δV_1, δH_1, δV_2, δH_2 and δS are to be determined) will have a non-trivial solution only if the determinant of its coefficients is zero. Writing down this condition and expanding the determinant the following equation is obtained:

$$P(P^2 - b_n^2)\{P^4 - (a^2 + b^2)P^2 + a^2 b_n^2\} = 0 \tag{10.3.9}$$

with the notations:

$$b_n = H_n\sqrt{v}, \quad b = H\sqrt{v} \ (b_n = b|\cos\theta|) \tag{10.3.10}$$

H and H_n being the absolute value of the unperturbed magnetic field and of its projection on the normal, respectively, and θ the angle between $\boldsymbol{H}$ and the normal direction.

Equation (10.3.9) has the following solution:

$$P = P_0, \pm P_t, \pm P_f, \pm P_s \tag{10.3.11}$$

where:

$$P_0 = 0, P_t = b_n$$

$$2P_f = \sqrt{a^2 + b^2 + 2ab_n} + \sqrt{a^2 + b^2 - 2ab_n} \tag{10.3.12}$$

$$2P_s = \sqrt{a^2 + b^2 + 2ab_n} - \sqrt{a^2 + b^2 - 2ab_n}$$

to whom the following drift velocities correspond:

$$d_0 = V_n, \ d_t = V_n \pm P_t$$
$$d_f = V_n \pm P_f, \ d_s = V_n \pm P_s \tag{10.3.13}$$

Therefore four types of waves are possible in a compressible fluid: one wave drifting with velocity d_0 (thus moving with the fluid) which will be called an *entropy wave*, one wave of velocity d_t called a *transverse* or *Alfvén wave*, one wave of velocity d_f called a *fast wave* and one wave drifting with velocity d_s which will be called a *slow wave*, the last two being also called *magneto-acoustic* waves. The entropy wave—the only one appearing in conventional gasdynamics—is also called an *acoustic* wave since as will be later shown it represents a pressure discontinuity.

Taking into account equation:

$$P^4 - (a^2 + b^2)P^2 + a^2 b_n = 0 \tag{10.3.14}$$

which determines the propagation velocities of the magneto-acoustic waves we obtain:

$$(P^2 - a^2)(P^2 - b_n^2) = P^2(b^2 - b_n^2) \geqslant 0 \tag{10.3.15}$$

such that:

$$P_s^2 \leqslant a^2 \leqslant P_f^2,\ P_s^2 \leqslant b_n^2 \leqslant P_f^2 \tag{10.3.16}$$

From these inequalities it follows that the fast-wave velocity is higher than the sound and Alfvén-wave velocities and the slow-wave propagation velocity is smaller than these two velocities. The relations become equalities for $b = b_n$, i.e. when the normal direction is along the unperturbed magnetic field direction. In this case if $a > b$ it follows that the fast wave is superposed on the sound wave and the slow wave on the Alfvén wave ($P_f = a$, $P_s = b$). On the contrary for an intense magnetic field ($b > a$) the fast wave degenerates into the Alfvén wave while the slow one coincides with the sound wave ($P_f = b$, $P_s = a$). These considerations justify the name of magneto-acoustic waves.

Another particular case is that of the propagation perpendicular to the magnetic field ($\theta = \pi/2$). In this case we have:

$$P_t = P_s = 0,\ P_f = \sqrt{a^2 + b^2} \tag{10.3.17}$$

Finally, for a vanishing magnetic field we have $P_t = P_s = 0$, $P_f = a$.

We should mention again that for the wave propagation through a constant-state fluid the propagation velocities can be calculated in terms of the constant-state parameters assumed to be known quantities. As a matter of fact one can also present graphical methods [5] for the determination of these velocities. So, if we construct the vector $\boldsymbol{OB} = \sqrt{\nu}\boldsymbol{H}$ ($\boldsymbol{H}$ being the unperturbed field) at the point of interest 0 on the wave front (Figure 31) then the orthogonal projection of $\boldsymbol{OB}$ on $\boldsymbol{n}$ determines the magnitude of b_n.

If now one takes the distances $OB = \nu H^2/2$, $BA = a^2/2$ along the normal $\boldsymbol{n}$, pointing outwards from the point 0 and if one denotes by D the intersection of the circle of centre B and radius BO with the magnetic field direction (Figure 32) and by C_1 and C_2 the points of intersection of the circle of centre A and radius AD with the normal then one has $P_s^2 = OC_1$ and $P_f^2 = OC_2$.

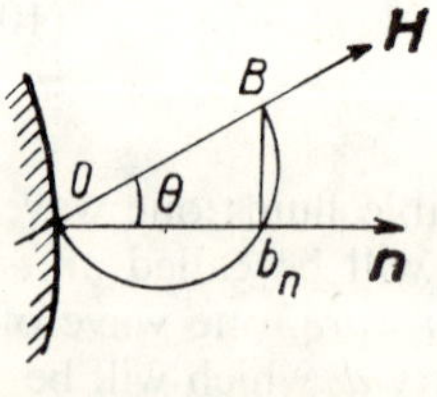

Fig. 31

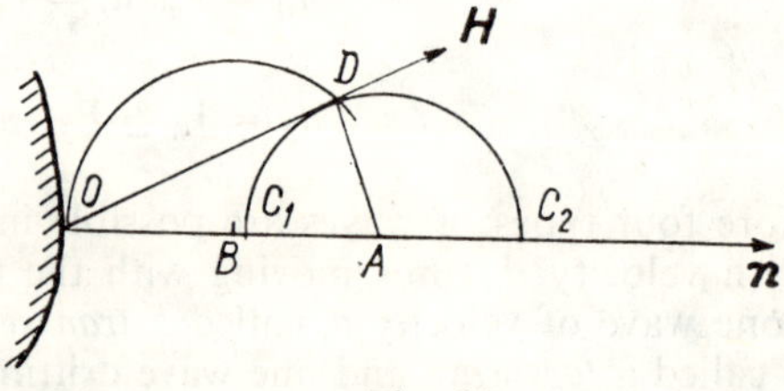

Fig. 32

10.3.2 Determination of discontinuities

Let us firstly consider the magneto-acoustic waves. The discontinuities determined by them are obtained from the systems (10.3.6) — (10.3.8) with the substitutions $P = \pm P_f$ and $P = \pm P_s$. Assuming that P_f and P_s are non-vanishing and different from b_n we can notice that the systems (10.3.7) and (10.3.8) have only the zero solution:

$$\delta H_2 = \delta V_2 = \delta S = 0 \tag{10.3.18}$$

On the contrary the system (10.3.6) having a vanishing determinant has a non-zero solution which can be determined up to a multiplying factor ε. So taking $\delta\rho = \varepsilon\rho$ we obtain:

$$\delta V_n = \varepsilon P,\ B_1\delta H_1 = \varepsilon\rho(P^2 - a^2),\ B_1 P\delta V_1 = -\varepsilon B_n(P^2 - a^2) \tag{10.3.19}$$

The discontinuities lying in the $(\boldsymbol{\tau}_1, \boldsymbol{n})$-plane only can be represented by means of the vectors $\boldsymbol{n}$ and $\boldsymbol{H}$. Using equation (10.3.14) we have:

$$\delta \boldsymbol{V} = \frac{\varepsilon P(\rho P^2\boldsymbol{n} - B_n\boldsymbol{H})}{\rho(P^2 - b_n^2)},\ \delta\boldsymbol{H} = \frac{\varepsilon P^2(\boldsymbol{H} - H_n\boldsymbol{n})}{P^2 - b_n^2} \tag{10.3.20}$$

Since the pressure depends on ρ and S we obtain $\delta p = a^2\varepsilon\rho$. These waves are therefore made up of velocity, magnetic field and pressure discontinuities which again justify their name.

For *transverse* waves we substitute $P = \pm P_t$ in the systems (10.3.6) — (10.3.8). Since system (10.3.6) has in this case a non-zero determinant it will only have a vanishing solution such that:

$$\delta\boldsymbol{V} = \varepsilon P\,\boldsymbol{n}\times\boldsymbol{H},\ \delta\boldsymbol{H} = -\varepsilon H_n\,\boldsymbol{n}\times\boldsymbol{H} \tag{10.3.21}$$

$$\delta S = \delta\rho = \delta p = 0$$

ε being as usual an infinitesimal parameter characterizing the jump. Since the density, pressure and normal component of the velocity do not change across the wave front it follows that we have a non-compressive wave. This wave is also consistent with the incompressible fluid.

Since $\delta H^2 = 2\boldsymbol{H}\,.\,\delta\boldsymbol{H} = 0$ it follows that the magnitude of the magnetic field does not change during the wave propagation. The wave is called a transverse one since the normal components of the velocity and magnetic field do not change. These waves are not consistent with the plane motion model since the normal components to the plane $(\boldsymbol{n}, \boldsymbol{H})$ are the ones affected by the motion.

Finally for the *entropy* waves $(P = 0,\ H_n \neq 0)$ we obtain:

$$\delta\boldsymbol{V} = \delta\boldsymbol{H} = 0,\ \delta S = \varepsilon,\ \delta\rho = -\frac{\varepsilon}{a^2}\frac{\partial p}{\partial S} = -\frac{\varepsilon}{a^2}\frac{p}{c_v} = -\frac{\varepsilon\rho}{\gamma c_v} \tag{10.3.22}$$

Their name is justified by the fact that they are the only waves for which the entropy is discontinuous. The entropy waves also appear in non-conducting aerodynamics but while in that case the tangential component of the velocity may be discontinuous in this case the total velocity is continuous across the wave front. The pressure is continuous as well $(\delta p = 0)$.

Let us now consider the two particular cases $b = b_n$ ($H_1 = 0$) and $b = b_1$ ($H_n = 0$). If $b = b_n$ one of the two velocities P_f and P_s becomes equal to the Alfvén velocity so that in this case we can no longer conclude that the solution of system (10.3.7) is vanishing. Substituting $H_1 = 0$ into (10.3.6) this separates itself into two systems:

$$-P\delta\rho + \rho\delta V_n = 0 \qquad \rho P\delta V_1 + B_n\delta H_1 = 0$$
$$a^2\delta\rho - \rho P\delta V_n = 0 \quad , \quad H_n\delta V_1 + P\delta H_1 = 0 \tag{10.3.23}$$

The first system has a non-zero solution for the magneto-acoustic wave degenerating into an acoustic wave such that this wave affects the normal component of the velocity and the density and pressure.

If $b = b_1$ ($H_n = 0$) $P_s = 0$ such that an entropy discontinuity ($\delta S = \varepsilon$) becomes possible. Substituting $H_n = 0$ into the systems (10.3.6) and (10.3.7) we notice that for the slow wave one has $\delta V_n = 0$ while $\delta\rho$, δV_1, δV_2 and δH_2 have arbitrary values. We also have the relation:

$$a^2\delta\rho + B\delta H_1 + (\delta p/\partial S)\varepsilon = 0 \tag{10.3.24}$$

which determines δH_1 for a given $\delta\rho$ and *vice versa*. Taking into account equation (10.3.24) it follows that the total pressure is continuous across the wave front:

$$\delta(p + \mu H^2/2) = 0 \tag{10.3.25}$$

The solution can be easily obtained for the fast wave $P = \pm\sqrt{a^2 + b^2}$.

10.3.3 Friedrichs wave-front diagram

Let us consider the perturbation produced by a source 0 in a fluid at rest. In such a case the propagation velocities determined by the formulae (10.3.12) are just the displacement velocities of the wave fronts. It also follows that only three waves emerge from the source: the transverse wave and the two magneto-acoustic waves. The normal velocity curves for the three waves are determined by the polar coordinates $(d_t/b, \theta)$, $(d_f/b, \theta)$ and $(d_s/b, \theta)$ connected together by the following equations:

$$d_t/b = |\cos\theta|$$
$$2d_f/b = \sqrt{s^2 + 1 + 2s|\cos\theta|} + \sqrt{s^2 + 1 - 2s|\cos\theta|}$$
$$2d_s/b = \sqrt{s^2 + 1 + 2s|\cos\theta|} - \sqrt{s^2 + 1 - 2s|\cos\theta|} \tag{10.3.26}$$

s being a dimensionless parameter ($= a/b$) determined by the constant state values of a and b and θ varying between $(0, \pi)$ and $(0, -\pi)$. These equations for three particular values of the parameter s ($= 1.2, 1.2$) are plotted in Figure 33 [3].

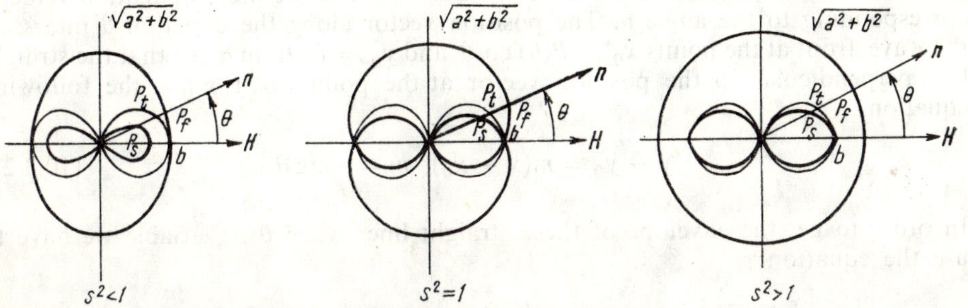

Fig. 33

The curves of the first diagram are characteristic for $s < 1$ and those of the last diagram for $s > 1$. The normal-velocity surfaces are obtained by rotating these curves about the direction of $\boldsymbol{H}$.

Having the normal-velocity curves the curves generating the wave-fronts can be constructed by taking the envelope of the straight lines perpendicular to the position vectors at the intersection of the latter with the normal velocity curves. The diagrams of Figure 34 called the Friedrichs diagrams are obtained. *The wave fronts after one time-unit are obtained by rotating these diagrams about the direction of* $\boldsymbol{H}$. One notices that the Alfvén wave reduces itself to the points A and A' which drift in the direct and reverse sense with velocity b along the

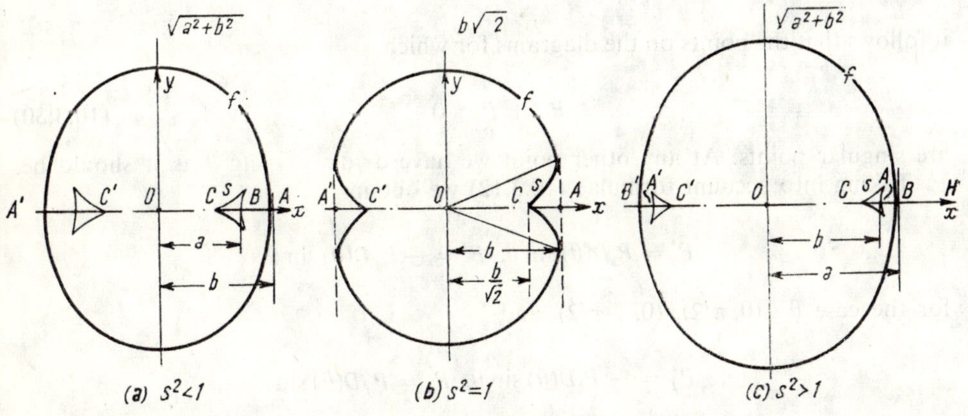

Fig. 34

direction of the unperturbed magnetic field (indeed, in such a case, the normal velocity curves reduce themselves to two circles tangent to one another at 0 so that the straight lines perpendicular to the position vectors will pass through A and A').

We can also determine analytically the Friedrichs diagram equations. In this respect we shall define a reference system with the origin at 0, having the Ox-axis along the direction of the unperturbed magnetic field and the Oy-axis perpendicular to Ox and lying in the plane $(\boldsymbol{H}, \boldsymbol{n})$. Let $P = P(\theta)$ be the propagation velocity corresponding to the angle θ. The position vector along the direction θ intersects the wave front at the points $x_0 = P(\theta)\cos\theta$ and $y_0 = P(\theta)\sin\theta$, so that the straight line perpendicular to the position vector at the point (x_0, y_0) has the following equation:

$$y - y_0 = m(x - x_0), \; m = -\operatorname{ctg}\theta \tag{10.3.27}$$

In order to get the envelope of these straight lines when θ is variable we have to use the equation:

$$-y_0' = m'(x - x_0) - mx_0' \tag{10.3.28}$$

where the prime sign indicates the derivative with respect to the parameter θ. From (10.3.27) and (10.3.28) we obtain the parametric equations of the Friedrichs diagrams:

$$\begin{aligned} x &= P(\theta)\cos\theta - P'(\theta)\sin\theta \\ y &= P(\theta)\sin\theta + P'(\theta)\cos\theta \end{aligned} \tag{10.3.29}$$

Since

$$\frac{dx}{d\theta} = -(P'' + P)\sin\theta, \; \frac{dy}{d\theta} = (P'' + P)\cos\theta$$

it follows that the points on the diagrams for which

$$P'' + P = 0 \tag{10.3.30}$$

are singular points. At any other point we have $dy/dx = -\operatorname{ctg}\theta$ as it should be.

Taking into account formulae (10.3.12) we obtain

$$P_f' = P_s D(\theta)\sin\theta, \; P_s' = -P_f D(\theta)\sin\theta$$

for the case $\theta \in (0, \pi/2)$, $(0, -\pi/2)$ and

$$P_f' = -P_s D(\theta)\sin\theta, \; P_s' = P_f D(\theta)\sin\theta$$

for the case $\theta \in (\pi/2, \pi)$, $(-\pi/2, -\pi)$, where we used the notations:

$$D(\theta) = \frac{ab}{\sqrt{(a^2 + b^2)^2 - 4a^2b^2\cos^2\theta}} \tag{10.3.31}$$

Consequently the parametric equations of the fast waves have the form:

$$x = P_f \cos\theta \mp P_s D \sin^2\theta$$
$$y = P_f \sin\theta \pm P_s D \sin\theta\cos\theta \tag{10.3.32}$$

while for the slow waves have the form:

$$x = P_s \cos\theta \pm P_f D \sin^2\theta$$
$$y = P_s \sin\theta \mp P_f D \sin\theta\cos\theta \tag{10.3.33}$$

the (+) and (−) signs being valid for the first and second case, respectively.

For transverse waves we get $x = \pm b$, $y = 0$.

Let us now consider the slow wave in the first case. A simple calculation yields:

$$P_s'' + P_s = P_s F(\theta)\{M(\theta) - G(\theta)\} \tag{10.3.34}$$

where:

$$F(\theta) = 1 - 4D^2\sin^2\theta = \frac{(s^2-1)^2}{(s^2+1)^2 - 4s^2\cos^2\theta} \geqslant 0$$

$$M(\theta) = \frac{1 - D^2\sin^2\theta}{1 - 4D^2\sin^2\theta} = 1 + \frac{3s^2}{(s^2-1)}\sin\theta$$

$$G(\theta) = \frac{P_f D}{P_s}\cos\theta = \frac{1}{2}\left\{1 + \frac{1+s^2}{\sqrt{(1+s^2)^2 - 4s^2\cos^2\theta}}\right\}$$

Since for $s \neq 1$ $F(\theta) > 0$ it follows that the singular points will correspond to both $\theta = \pm\pi/2$ (where $P_s = 0$) and to the points at which $M(\theta) = G(\theta)$. Plotting the functions $M(\theta)$ and $G(\theta)$ it follows that two roots $\theta = \pm\theta_1$ with corresponding singular points exist on the first case intervals. Therefore the slow wave diagram has in this case three singular points. For $\theta = \pm\pi/2$ one obtains:

$$C: x = ab/\sqrt{a^2 + b^2}, \quad y = 0 \tag{10.3.35}$$

From equation (10.3.33) it follows that this diagram intersects the Ox-axis either at point a if $a < b$ $(s < 1)$ or b if $a > b$ $(s > 1)$. These points correspond to the angle $\theta = 0$. Expanding equations (10.3.33) in series for small θ we obtain:

$$x = b\left(1 + \frac{b^2\theta^2}{2(a^2 - b^2)} + \ldots\right), \; y = \frac{b^3\theta}{a^2 - b^2}$$

which shows that at these points the convexity of the diagram is oriented towards the small values of x. Taking into account these results the slow wave diagram for $s \neq 1$ is obtained. This is symmetrical about the Oy-axis.

For the fast wave diagram equation (10.3.30) has no root (except for $s = 1$) so that one has a closed curve with no singular points. This result indicates that the fast waves propagate like the acoustic ones (except for the normal variation of the velocity) while the slow and transverse waves propagate in a different manner. This diagram intersects the Ox-axis either at point b if $a < b$ or at point a if $a > b$ and the convexity of the curve at these points is oriented towards the large values of x. The diagram is again symmetrical about the Oy-axis.

For $s = 1$, $F(\theta) = 0$. The fast wave diagram has minima at the points $\theta = \pm 0$ and $\theta = \pm \pi$. From (10.3.32) it follows that:

$$\lim_{\theta \to 0, \pi} y = b, \quad \lim_{\theta \to -\theta, -\pi} y = -b$$

At these points the fast wave diagram matches the slow wave one (the common tangent is vertically oriented) at those former singular points of the slow wave diagram which were not lying on the Ox-axis. For $a \to b$ the points C and C' still remain singular points ($\pm b/\sqrt{2}$).

One can notice from the Friedrichs diagram that the slow wave propagates only within a cone of apex 0 (the perturbation source) and axis $\boldsymbol{H}$ and with the opening determined by the distance between the two singular points $\theta = \pm \theta_1$.

10.3.4 Wave front diagrams for an arbitrary perturbation

Let us now assume that the initial perturbation is a plane of equation:

$$x \cos \theta + y \sin \theta = R_0 \tag{10.3.36}$$

in a reference system whose xOy-plane is coincident with the $(\boldsymbol{H}, \boldsymbol{n})$-plane and whose Ox-axis has the direction of $\boldsymbol{H}$. Since the wave-front drift velocity depends on θ (Figure 35) it follows that all the points of the plane (10.3.36) will have the

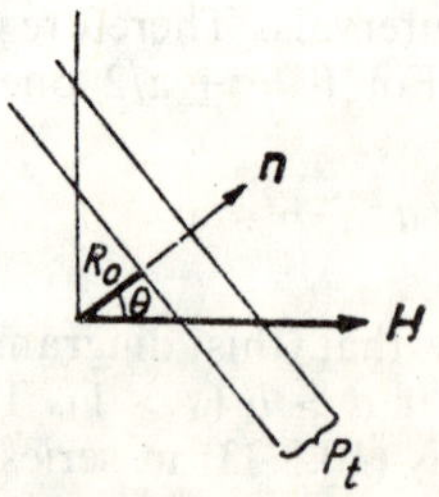

Fig. 35

same velocity $P(\theta)$ so that at time t the initial perturbation will lie in the plane of equation

$$x \cos \theta + y \sin \theta = R_0 + P(\theta)t \tag{10.3.37}$$

where $P(\theta)$ has the value of $\pm P_t$, $\pm P_f$ and $\pm P_s$ for the transverse fast and slow waves, respectively. The perturbation is propagated as three plane waves.

Let us now suppose that the initial perturbation represents a cylinder of radius R_0 obtained as the envelope about θ of the planes (10.3.36). It follows that the wave front at time t will again be a cylinder, namely a cylinder obtained as the envelope of the planes (10.3.37). The equations of this envelope are obtained from equation (10.3.37) and its derivative

$$-x\sin\theta + y\cos\theta = P'(\theta)t$$

We finally get:

$$\begin{aligned} x &= R_0\cos\theta + (P\cos\theta - P'\sin\theta)t \\ y &= R_0\sin\theta + (P\sin\theta + P'\cos\theta)t \end{aligned} \tag{10.3.38}$$

Equations (10.3.29) represent a particular case ($R_0 = 0$) of the equations (10.3.38). For transverse waves equations (10.3.38) become:

$$x = R_0\cos\theta \pm bt, \; y = R_0\sin\theta \tag{10.3.39}$$

which represent two circles of radius R_0 in the xOy plane moving in the positive and negative sense of the Ox-axis. For the limiting case of the source ($R_0 = 0$) these circles reduce themselves to the points A and A' already mentioned above. For slow waves equations (10.3.38) become:

$$\begin{aligned} x &= R_0\cos\theta \pm (P_s\cos\theta + P_f D\sin^2\theta)t \\ y &= R_0\sin\theta \pm (P_s\sin\theta - P_f D\cos\theta\sin\theta)t \end{aligned}$$

The slow wave drifting in the direct sense and determined by these equations is plotted in Figure 36 [3], [6]. One can see that its shape changes with time. Equations (10.3.40) can also be written in a form independent of θ[7].

If the initial perturbation is a sphere of centre 0 and radius R_0 then the wave front can be obtained by rotating the curve defined by equations (10.3.38) around

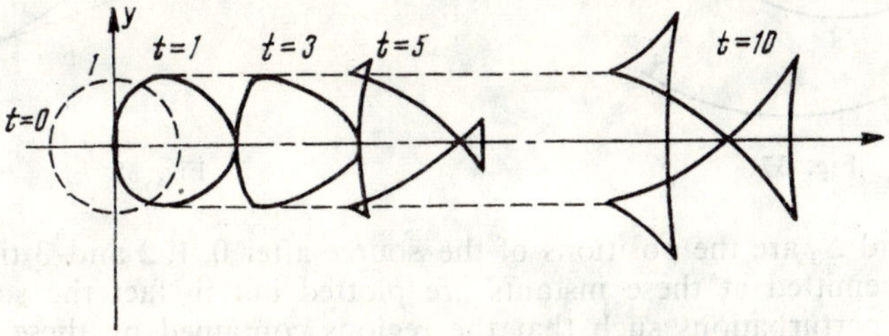

Fig. 36

the Ox-axis. Denoting by X, Y, Z the coordinates of an arbitrary point on the wave front we have:

$$X = x, \; Y = y\cos\varphi, \; Z = y\sin\varphi$$

x and y being defined by (10.3.38). For instance, for the transverse wave the wave front will be formed by two spheres whose centres move with velocity b in the direct and reverse sense along the direction of the magnetic field.

For the transverse and fast waves the wave front can be constructed starting from an arbitrary initial perturbation while the wave-front construction for the slow waves is more difficult.

10.4 Spatial discontinuities in steady motion

Besides the initial perturbations, discontinuity waves can also be produced by certain time-invariant conditions. Let us consider for instance a point source (e.g. the board of an aircraft) moving at constant velocity V_0. For a non-conducting fluid the waves propagate isotropically. We present in Figure 37 these waves for subsonic ($V_0 < a$) motion and in Figure 38 for supersonic motion ($V_0 > a$).

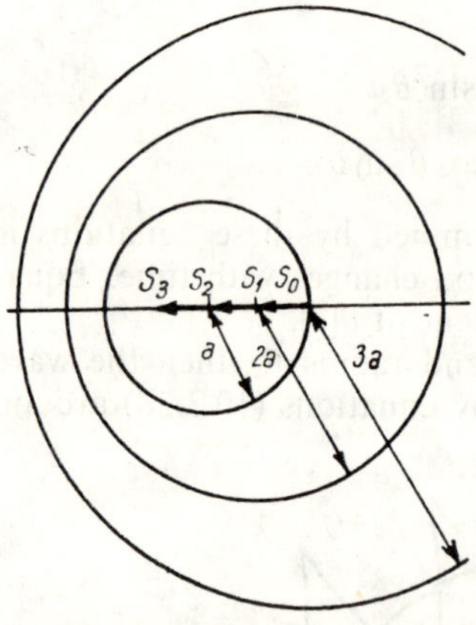

Fig. 37

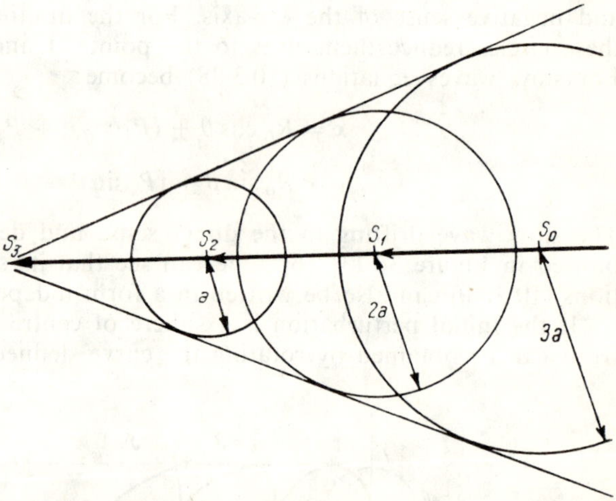

Fig. 38

S_0, S_1, S_2 and S_3 are the positions of the source after 0, 1, 2 and 3 time units. The wave-fronts emitted at these instants are plotted but in fact the source continuously emits perturbations such that the regions contained by these fronts receive perturbations in a continuous manner. For subsonic motion the perturbation leads in time the source motion while in the supersonic motion the source leads the perturbation. For supersonic motion the perturbed region is a cone (the Mach cone) having the apex at the source which is obtained by taking the envelope of the spherical wave fronts emitted at various successive times.

In MFD the problem is more complicated due to the anisotropy introduced by the magnetic field and by the large number of wave fronts which are produced. However, the principle of the wave-front construction remains the same. The perturbation region (at time $t = 1$) is obtained by drawing straight lines from the source tangent to the wave surface (and the Friedrichs diagram, respectively) corresponding to the same time. Their envelope will determine the front of the perturbation waves. If one can draw real tangent straightlines from the source S to the wave fronts then the characteristics (along which the perturbations propagate) are real.

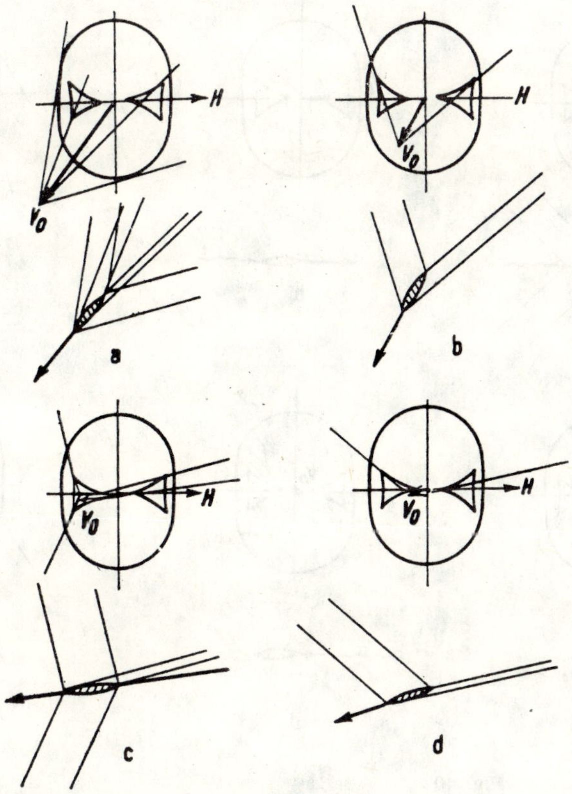

Fig. 39

For the two-dimensional case these constructions have been pointed out for the first time by Sears [8] and J. E. Mc Cune and E. L. Resler, Jr. [5.5].

These constructions are presented in Figure 39, 40 and 41 for arbitrary, aligned and orthogonal (V_0 and H) fields (H being the unperturbed magnetic field). One notices that the situation for fast waves is similar to that of classical aerodynamics (there exist real characteristics if and only if the point S is external to the wave front) while for the slow waves real characteristics exist even if the point S is inside the wave front.

For arbitrary fields (Figure 39 a) two fast and two slow characteristics emerge from any perturbation source; for the situation of Figure 39 b only two slow characteristics are possible while for that of Figure 39 c three slow characteristics are possible.

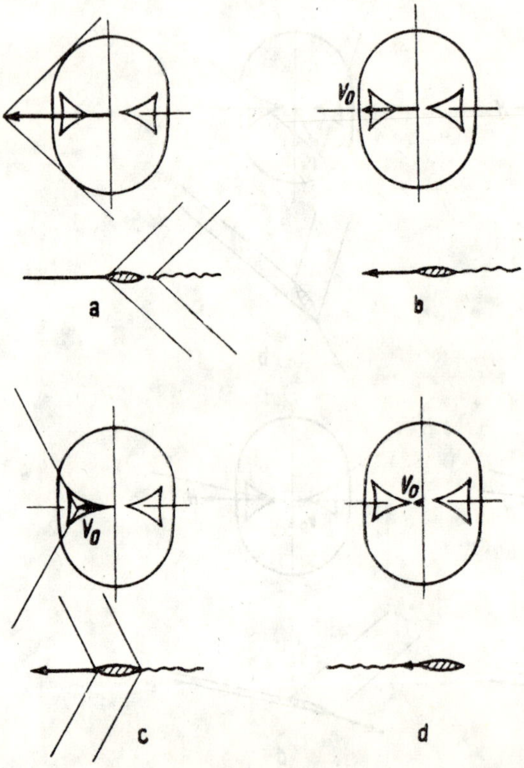

Fig. 40

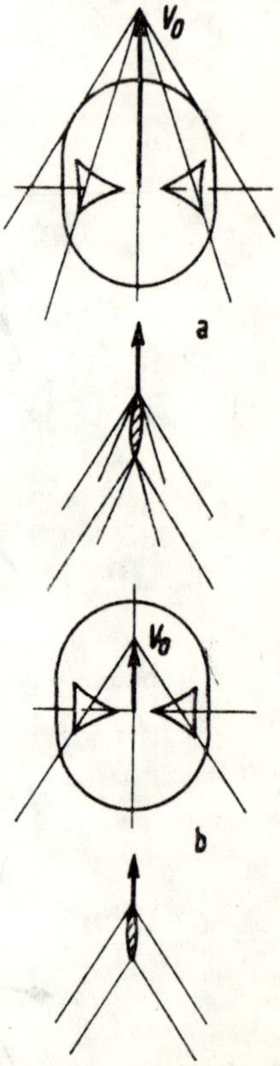

Fig. 41

Particularly, for orthogonal fields four (two fast and two slow) real characteristics exist if $V_0 > \sqrt{a^2 + b^2}$ and two (slow) real characteristics exist if $V_0 < \sqrt{a^2 + b^2}$.

For aligned fields and $V_0 > a$ (one assumes $a > b$) we have two real characteristics associated with the fast waves and oriented downstream and a downstream non-propagation layer corresponding to the slow waves (Figure 40 a). The non-propagation layer has the following explanation: the tangent drawn from S to the slow wave diagram passes through the singular point A corresponding to the values $\theta = \pi/2$ and $\theta = -\pi/2$; but at this point $P_s = 0$ such that one does not have a characteristic but a non-propagating layer (e.g. a tangential discontinuity

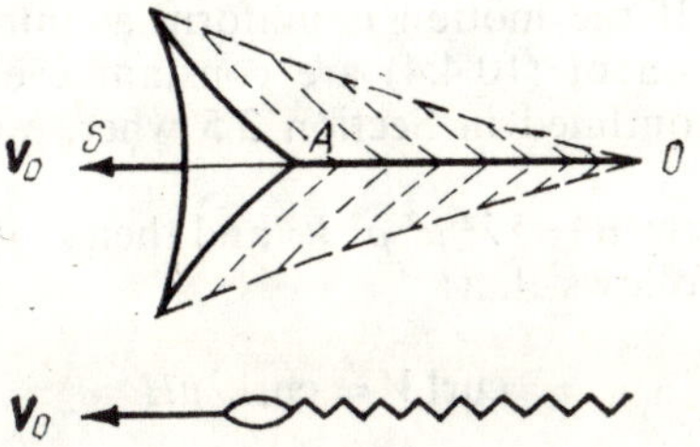

Fig. 42

or a vortex layer). Indeed starting at the moment at which the source is at 0 as t increases the slow wave diagram increases to reach the shape it should have after the lapsing of one unit time. During all this time interval point A remains behind S such that SA will be the envelope of certain cross-sectional fronts which does not propagate (non-propagating layer) and which remains behind (Figure 42). If $a > V_0 > b$ the characteristics reduce themselves to the non-propagating layer (Figure 40 b). If $b < V_0 < ab/\sqrt{a^2 + b^2}$ two real characteristics (envelopes of the slow fronts in cross-section) and the downstream layer appear (Figure 40 c). Finally, if $ab/\sqrt{a^2 + b^2} > V_0$ one has only the upstream layer (indeed, for this case, starting from 0 A is always in front of S). These results explain the theory of Chapter 5.

10.4.1 Analytical study of the aligned field problem

Let us now assume the existence of certain conditions such that at any point we have:

$$\boldsymbol{B} = \lambda\mu\rho\boldsymbol{V} \quad (\boldsymbol{V} \times \boldsymbol{B} = 0) \tag{10.4.1}$$

From equations

$$\operatorname{div} \rho\boldsymbol{V} = 0, \ \operatorname{div} \boldsymbol{B} = 0 \tag{10.4.2}$$

it follows that λ is a constant along the current lines ($\boldsymbol{V} . \nabla\lambda = 0$). Scalarly multiplying the equation of motion by $\boldsymbol{V}$ we obtain:

$$\rho\left(\operatorname{curl} \boldsymbol{V} \times \boldsymbol{V} + \frac{1}{2} \operatorname{grad} V^2\right) = -\operatorname{grad} p + \operatorname{curl} \boldsymbol{H} \times \boldsymbol{B} \tag{10.4.3}$$

and taking into account (10.4.1) it follows that:

$$\frac{1}{2}V^2+\int\frac{dp}{\rho} = \text{const} \tag{10.4.4}$$

along the current lines. If the motion is uniform at infinity one may assume that both λ and the expression of (10.4.4) are constant everywhere within the fluid. These results were also outlined in Section 2.5 where we dealt with the Bernoulli integral.

From (10.4.1) we obtain $B^2 = \lambda^2 \mu^2 \rho^2 V^2$ and then $\lambda^2 \mu\rho = b^2/V^2$. From (10.4.1), (10.4.3) and (10.4.4) it follows that:

$$\text{curl}\, \boldsymbol{V} = \text{curl}\, \lambda\mu\boldsymbol{H}$$

and then

$$\text{curl}\, \boldsymbol{V}^* = 0, \quad \boldsymbol{V}^* \overset{\text{def}}{=} \boldsymbol{V}(1 - b^2/V^2) \tag{10.4.5}$$

Noticing that:

$$(\boldsymbol{B}\cdot\nabla)\boldsymbol{H} = (\rho\boldsymbol{V}\cdot\nabla)\frac{b^2}{V^2}\,\boldsymbol{V}$$

from the equation of motion in the form:

$$(\rho\boldsymbol{V}\cdot\nabla)\boldsymbol{V} = -\text{grad}\, p^* + (\boldsymbol{B}\cdot\nabla)\boldsymbol{H}$$

$$p^* \overset{\text{def}}{=} p + \frac{1}{2}\boldsymbol{B}\cdot\boldsymbol{H} = p + \frac{1}{2}b^2\rho$$

we obtain

$$(\rho^*\boldsymbol{V}^*\cdot\nabla)\boldsymbol{V}^* = -\text{grad}\, p^*, \rho^* \overset{\text{def}}{=} \rho\Big/\left(1 - \frac{b^2}{V^2}\right) \tag{10.4.6}$$

Noticing that we have

$$\text{div}\, \rho^*\boldsymbol{V}^* = 0 \tag{10.4.7}$$

we obtained in (10.4.6), (10.4.7) and (10.4.5) the equations of potential motion of classical aerodynamics.

With the notations:

$$\boldsymbol{V}^* = \text{grad}\Phi, \quad a^* = dp^*/d\rho^* \tag{10.4.8}$$

we obtain for the plane motion that:

$$(V_x^{*2} - a^*)\frac{\partial^2\Phi}{\partial x^2} + 2V_x^*V_y^*\frac{\partial^2\Phi}{\partial x\,\partial y} + (V_y^{*2} - a^*)\frac{\partial^2\Phi}{\partial y^2} = 0 \tag{10.4.9}$$

The characteristic lines of equation (10.4.9) are determined by:

$$(V_x^{*2} - a^*)\,dy^2 - 2V_x^*V_y^*\,dx\,dy + (V_y^{*2} - a^*)\,dx^2 = 0 \qquad (10.4.10)$$

and are either real or imaginary according to whether the discriminant

$$\Delta = V_x^{*2}V_y^{*2} - (V_x^{*2} - a^*)(V_y^{*2} - a^*) = a^*(V^{*2} - a^*) \qquad (10.4.11)$$

is positive or negative. Since a^* can also be negative it follows that equation (10.4.9) is of hyperbolic type if $V^{*2} > a^* > 0$ and of elliptic type if $a^* < 0$ or $V^{*2} < a^*$ and $a^* > 0$.

Let us now calculate the expression of a^* as defined in (10.4.8). Taking into account the definitions of p^* and ρ^* it follows that:

$$a^* = \left(1 - \frac{b^2}{V^2}\right)^2 \frac{dp + \frac{1}{2}\rho\, db^2 + \frac{1}{2}b^2\, d\rho}{d\rho\left(1 - \frac{b^2}{V^2}\right) + \rho\, d\left(\frac{b^2}{V^2}\right)} \qquad (10.4.12)$$

From (10.4.4) we get:

$$\frac{dV^2}{dp} = -\frac{2}{\rho} \Rightarrow \frac{dV^2}{d\rho} = -\frac{2}{\rho}a^2\left(a^2 = \frac{dp}{d\rho}\right)$$

Since we also have:

$$b^2 = \frac{B^2}{\mu\rho} = \lambda^2\mu\rho V^2 \Rightarrow \frac{db^2}{d\rho} = \frac{b^2}{\rho} + \frac{b^2}{V^2}\frac{dV^2}{d\rho} = \frac{b^2}{\rho}\left(1 - \frac{2a^2}{V^2}\right)$$

and dividing the numerator and denominator of (10.4.12) by $d\rho$ we obtain:

$$a^* = \left(1 - \frac{b^2}{V^2}\right)^2\left(a^2 + b^2 - \frac{a^2b^2}{V^2}\right) \qquad (10.4.12')$$

The conclusions follow immediately from the Friedrichs diagram if we notice that:

$$\frac{a^*}{V^{*2}} = \frac{a^2 + b^2}{V^4}\left(V^2 - \frac{a^2b^2}{a^2 + b^2}\right) \qquad (10.4.12'')$$

Indeed for both cases ($s^2 < 1$, $s^2 > 1$) we have $a^* < 0$ on the segments $C'C$ ($V^2 < a^2b^2/(a^2 + b^2)$) so that the first ellipticity condition is satisfied (equation (10.4.9) is of elliptic type). Outside $C'C$ we have $a^* > 0$ so that one has to compare the ratio a^*/V^{*2} with unity. But:

$$1 - \frac{a^*}{V^{*2}} = \frac{1}{V^4}\{V^4 - (a^2 + b^2)V^2 + a^2b^2\}$$

and the equation:

$$V^4 - (a^2 + b^2)V^2 + a^2b^2 = 0 \tag{10.4.13}$$

has the roots $V_1^2 = a^2$, $V_2^2 = b^2$ which correspond to the points B', B and A', A on the diagrams. Studying the sign of the trinom (10.4.13) it follows that equation (10.4.9) is of elliptic type both on the segments $A'B'$ and BA for $s^2 < 1$ and on the segments $B'A'$ and AB for $s^2 > 1$. Outside the points A', A ($s^2 < 1$) and B', $B(s^2 > 1)$ as well as on the segments $B'C'$, CB ($s^2 < 1$) and $A'C'$, CA ($s^2 > 1$) the equation is of hyperbolic type. This result due to L. Dinu [9] is coincident with the one found graphically in Figure 40.

10.5 Simple waves

10.5.1 General considerations

The MFD equations admit some solutions for which all the unknown functions ρ, V, H, S depend on the problem variables through a single function $\varphi(t, \mathbf{x})$. The corresponding motions are called *simple-wave motions* and the surfaces of equation $\varphi(t, \mathbf{x}) =$ const. are called *simple waves*. Their particular importance follows from the following statement due to Friedrichs: "*the solution adjacent to a constant state is either a simple-wave solution or a constant*". Within the class of continuous functions only the simplewave motions may be connected to the constant states.

The investigations of simple waves in MFD are due to Friedrichs and Kranzer [2], Polovin [7] and Bohachevsky [10]. Some systematic reviews can be found in the works of Liubimov and Kulikowski [G.10], Jeffrey and Taniutti [3], Cabannes [G.23] and Jeffrey [G.21]. The problem is solved completely for the one-dimensional case. This case will also be considered here.

Let us consider the system of equations of motion written in the form:

$$U_{,t} + AU_{,x} = 0, \quad A = A(U) \tag{10.5.1}$$

U being the unknowns column matrix and A the square matrix known as a function of U. We consider the general case of n unknowns for the time being.

We assume that the system (10.5.1) is of *hyperbolic type* and therefore that the matrix A has n real eigenvalues $\lambda_i(U)$. They are solutions of the equation:

$$\det(A - \lambda E) = 0 \tag{10.5.2}$$

E being the unit matrix. We denote by $r^{(i)}$ the right-hand eigenvector associated to the eigenvalue λ_i:

$$Ar^{(i)} = \lambda_i r^{(i)} \tag{10.5.3}$$

$r^{(i)} = r^{(i)}(U)$ being a column matrix of components $r_k^{(i)}$.

The simple-wave solutions of the equation (10.5.1) are of the form $U(t, x) = U(\varphi)$, $\varphi = \varphi(t, x)$. From (10.5.1) it follows that:

$$\frac{\mathrm{d}U}{\mathrm{d}\varphi}\left(\frac{\partial\varphi}{\partial t} + A\frac{\partial\varphi}{\partial x}\right) = 0 \Rightarrow AU_{,\varphi} = -\frac{\varphi_{,t}}{\varphi_{,x}}U_{,\varphi} \tag{10.5.4}$$

which indicates that $U_{,\varphi}$ is the eigenvector associated to the eigenvalue

$$\lambda = -\frac{\varphi_{,t}}{\varphi_{,x}} \tag{10.5.5}$$

It follows that $U(\varphi)$ is determined by the equations:

$$\frac{\mathrm{d}U}{\mathrm{d}\varphi} = r^{(i)}(U) \Rightarrow \frac{\mathrm{d}u_1}{r_1^{(i)}} = \ldots = \frac{\mathrm{d}u_n}{r_n^{(i)}} = \mathrm{d}\varphi \tag{10.5.6}$$

which indicates that to each eigenvector $r^{(i)}$ (thus to each eigenvalue λ_i) one simple wave corresponds. On the whole n simple waves result. The function φ is determined by the equation (10.5.3):

$$\frac{\partial\varphi}{\partial t} + \lambda_i(U)\frac{\partial\varphi}{\partial x} = 0 \tag{10.5.7}$$

We shall call a *Riemann invariant* any scalar quantity J which remains constant during a simple-wave motion. We thus have:

$$\mathrm{d}J = \sum_k \frac{\partial J}{\partial u_k}\mathrm{d}u_k = 0 \tag{10.5.8}$$

and taking into account the proportionality of (10.5.6) we obtain the following equation of definition of the Riemann invariants:

$$\sum_k \frac{\partial J}{\partial u_k} r_k^{(i)} = 0, \; i = 1, \ldots, n \tag{10.5.9}$$

For fixed i, that is, for a given simple wave, equation (10.9.9) has $(n - 1)$ (and only $(n - 1)$) linearly independent solutions (determined by (10.9.6). Therefore $(n - 1)$ Riemann invariants can be associated to each simple wave; they are the only non-trivial constants associated to the motion.

10.5.2 Application to MFD

If we assume that the motion parameters depend on the variables t and x only from the equation div $\boldsymbol{H} = 0$ it follows that $(\partial H_x/\partial x) = 0$ and from the projection on the Ox-axis of the induction equation (2.2.10) it follows that $(\partial H_x/\partial t) = 0$ as well. Consequently the H_x component is a constant equal to its constant state value (H_x^0).

The other unknowns ρ, V_x, V_y, V_z, H_y, H_z, S satisfy the system (10.3.2)–(10.3.5) which can be written in the form (10.5.1) if we introduce the notation:

$$A = \begin{pmatrix} V_x & \rho & 0 & 0 & 0 & 0 & 0 \\ \dfrac{a^2}{\rho} & V_x & 0 & 0 & \dfrac{B_y}{\rho} & \dfrac{B_z}{\rho} & \dfrac{1}{\rho}\dfrac{\partial p}{\partial S} \\ 0 & 0 & V_x & 0 & -B_x^0/\rho & 0 & 0 \\ 0 & 0 & 0 & V_x & 0 & -B_x^0/\rho & 0 \\ 0 & H_y & -H_x^0 & 0 & V_x & 0 & 0 \\ 0 & H_z & 0 & -H_x^0 & 0 & V_x & 0 \\ 0 & 0 & 0 & 0 & 0 & 0 & V_x \end{pmatrix}, \quad U = \begin{pmatrix} \rho \\ V_x \\ V_y \\ V_z \\ H_y \\ H_z \\ S \end{pmatrix}$$

The eigenvalues λ_i are determined by using equation (10.5.2) and the expression of A. One has:

$$V_x - \lambda_1 = 0$$

$$V_x - \lambda_{2,3} = \pm P_t, \; P_t \overset{\text{def}}{=} b_x$$

$$V_x - \lambda_{4,5} = \pm P_f, \; 2P_f \overset{\text{def}}{=} \sqrt{a^2 + b^2 + 2ab_x} + \sqrt{a^2 + b^2 - 2ab_x} \tag{10.5.10}$$

$$V_x - \lambda_{6,7} = \pm P_s, \; 2P_s \overset{\text{def}}{=} \sqrt{a^2 + b^2 + 2ab_x} - \sqrt{a^2 + b^2 - 2ab_x}$$

where:

$$b_x = \sqrt{\frac{B_x^0 H_x^0}{\rho}}, \; b^2 = \frac{\boldsymbol{B}.\boldsymbol{H}}{\rho} = \frac{B_x^0 H_x^0 + B_y H_y + B_z H_z}{\rho} \tag{10.5.10'}$$

The eigenvectors are determined by the system:

$$
\begin{aligned}
(V_x - \lambda)r_1 + \rho r_2 &= 0 \\
a^2 r_1 + \rho(V_x - \lambda)r_2 + B_y r_5 + B_z r_6 + (\partial p/\partial S)r_7 &= 0 \\
\rho(V_x - \lambda)r_3 - B_x^0 r_5 &= 0 \\
\rho(V_x - \lambda)r_4 - B_x^0 r_6 &= 0 \\
H_y r_2 - H_x^0 r_3 + (V_x - \lambda)r_5 &= 0 \\
H_z r_2 - H_x^0 r_4 + (V_x - \lambda)r_6 &= 0 \\
(V_x - \lambda)r_+ &= 0
\end{aligned}
\tag{10.5.11}
$$

Substituting successively $(V_x - \lambda)$ by the values $0, \pm P_t, \pm P_f, \pm P_s$ we obtain the eigenvectors $r^{(1)}, r^{(2,3)}, r^{(4,5)}, r^{(6,7)}$:

$$
r^{(1)} = \begin{pmatrix} -\dfrac{1}{a^2}\dfrac{\partial p}{\partial S} \\ 0 \\ 0 \\ 0 \\ 0 \\ 0 \\ 1 \end{pmatrix}, \quad r^{(2,3)} = \begin{pmatrix} 0 \\ 0 \\ B_x^0 H_z \\ -B_x^0 H_y \\ \pm\rho H_z P_t \\ \mp\rho H_y P_t \\ 0 \end{pmatrix}, \quad r^{(i)}(P) = \begin{pmatrix} \rho^2(P^2 - b_x^2) \\ -\rho(P^2 - b_x^2)P \\ B_x^0 H_y P \\ B_x^0 H_z P \\ \rho H_y P^2 \\ \rho H_z P^2 \end{pmatrix}
$$

$$
r^{(i)}(\pm P_f) = r^{(4,5)}, \quad r^{(i)}(\pm P_s) = r^{(6,7)}
$$

10.5.3 Magnetoacoustic waves

Taking into account the eigenvectors $r^{(i)}$ we obtain that equation (10.5.9) which defines the Riemann invariants associated to the simple waves has the form:

$$
\rho^2(P^2 - b_x^2)\frac{\partial J}{\partial \rho} - \rho(P^2 - b_x^2)P\frac{\partial J}{\partial V_x} + B_x^0 H_y P\frac{\partial J}{\partial V_y} + B_x^0 H_z P\frac{\partial J}{\partial V_z}
$$

$$
+ \rho H_y P^2\frac{\partial J}{\partial H_y} + \rho H_z P^2\frac{\partial J}{\partial H_z} = 0 \tag{10.5.12}
$$

and the following characteristic system:

$$\frac{d\rho}{\rho^2(P^2 - b_x^2)} = \frac{dV_x}{-\rho(P^2 - b_x^2)P} = \frac{dV_y}{B_x^0 H_y P} = \frac{dV_z}{B_x^0 H_z P}$$

$$= \frac{dH_y}{\rho H_y P^2} = \frac{dH_z}{\rho H_z P^2} = \frac{dS}{0} \tag{10.5.13}$$

The following first integrals are obtained immediately:

$$\begin{aligned} &J_1(H_y, H_z) = H_z/H_y = K_1 \\ &J_2(V_y, V_z, H_y, H_z) = V_y - (H_y/H_z)V_z = K_2 \\ &J_3(p, \rho) = p/\rho^\gamma = K_3 \end{aligned} \tag{10.5.14}$$

K_1, K_2 and K_3 being constants everywhere within the fluid (and are determined by the adjacent constant motion). J_1, J_2, J_3 are the Riemann invariants.

In order to derive the fourth first integral (the fourth Riemann invariant) we consider the first, fifth and sixth ratios of (10.5.13). Using the properties of equal ratios we obtain:

$$\frac{d\rho}{\rho(P^2 - b_x^2)} = \frac{d(H_y^2 + H_z^2)}{2P^2(H_y^2 + H_z^2)} \tag{10.5.15}$$

In order to integrate this equation the following dimensionless parameters are introduced:

$$\alpha = \frac{P^2}{a^2},\ \beta = \frac{a^2}{b_x^2} = \frac{a^2\rho}{B_x^0 H_x^0} = \frac{\gamma p}{B_x^0 H_x^0} \tag{10.5.16}$$

β representing (up to the constant factor $\gamma/2$) the ratio of the fluid pressure and magnetic pressure along the Ox-axis direction. β is obviously the same both for the slow and the fast waves. Using the third first integral as well we get:

$$p = \bar{p}\beta,\ \rho = \bar{\rho}\beta^{1/\gamma},\ a^2 = \bar{a}^2\beta^{1-1/\gamma} \tag{10.5.17}$$

where we used the notations:

$$\bar{p} = \frac{B_x^0 H_x^0}{\gamma},\ \bar{\rho} = \rho_0\left(\frac{\bar{p}}{p_0}\right)^{\frac{1}{\gamma}},\ \bar{a}^2 = \gamma\frac{\bar{p}}{\bar{\rho}} \tag{10.5.17'}$$

Then from (10.5.16) we have:

$$P^2 = \bar{a}^2 \alpha \beta^{1-1/\gamma} \tag{10.5.18}$$

Now if we use the transformation (10.5.16) ($P^2 = b_x^2 \alpha\beta$, $a^2 = b_x^2 \beta$) in equation:

$$P^4 - P^2(a^2 + b^2) + a^2 b_x^2 = 0 \tag{10.5.19}$$

which determines the velocity P we have:

$$H_y^2 + H_z^2 = (\alpha - 1)\left(\beta - \frac{1}{\alpha}\right) H_x^{02} \tag{10.5.20}$$

With these results equation (10.5.15) becomes:

$$\alpha^2(\alpha - 1)\,\mathrm{d}\beta = \bar{\gamma}(\alpha^2\beta - 1)\,\mathrm{d}\alpha, \quad \bar{\gamma}(2 - \gamma) \overset{\text{def}}{=} \gamma \tag{10.5.21}$$

which is known as the Friedrichs equation.

Noticing that equation (10.5.21) has an integrating factor $(\alpha - 1)^{-\bar{\gamma}}$ it follows that:

$$J_4(H_y, H_z, \rho) = \frac{\beta}{(\alpha - 1)^{\bar{\gamma}}} + \bar{\gamma} \int \frac{\mathrm{d}\alpha}{\alpha^2(\alpha - 1)^{1+\bar{\gamma}}} = K_4 \tag{10.5.22}$$

J_4 being the fourth Riemann invariant (the magnetic invariant).

For certain values of γ, the adiabatic constant of the fluid, the invariant J_4 can be expressed by means of elementary functions:

$$\gamma = 1 \Rightarrow J_4 = \frac{\beta - 1}{\alpha - 1} - 2\log\left(1 - \frac{1}{\alpha}\right)$$

$$\gamma = \frac{5}{3} \Rightarrow J_4 = \frac{\beta - 1}{(\alpha - 1)^5} + \frac{5}{2}\frac{1}{(\alpha - 1)^4} - \frac{5}{(\alpha - 1)^3} + \frac{10}{(\alpha - 1)^2} - \frac{25}{\alpha - 1} - \frac{5}{\alpha} + 30\ln\left(1 - \frac{1}{\alpha}\right)$$

$$\gamma = 2 \Rightarrow J_4 = \alpha$$

Let us mention here some qualitative results regarding equation (10.5.21) due to Polovin. Using the notations $\alpha_+ = P_f^2/a^2$ and $\alpha_- = P_s^2/a^2$ and noticing that like in Section 10.3 from equation (10.9.19) we have:

$$P_s^2 \leqslant a^2 \leqslant P_f^2, \; P_s^2 \leqslant b_x^2 \leqslant P_f^2$$

it follows that:

$$\beta\alpha_+ \geqslant 1,\ \alpha_+ \geqslant 1,\ (\beta\alpha_+^2 > 1)$$
$$\beta\alpha_- \leqslant 1,\ \alpha_- \leqslant 1,\ (\beta\alpha_-^2 < 1) \tag{10.5.23}$$

with $\beta > 0$ and $\alpha_- \geqslant 0$. Using these inequalities from equation (10.5.21) we obtain:

$$\mathrm{d}\beta/\mathrm{d}\alpha_\pm > 0 \tag{10.5.24}$$

which indicates that the parameter $\alpha_\pm$ decreases (for both waves) with decreasing pressure. From the first inequalities, it follows that a maximum amplitude wave represented by the curve $\alpha\beta = 1$ appears. This curve together with $\alpha = 1$ separates the (α, β)-plane into two regions: they are denoted by the (+) and (−) signs corresponding to the fast and slow waves respectively. The relation $\alpha\beta = 1$ can be realized both by the fast (for $\alpha > 1$) and slow (for $\alpha < 1$) waves. For $\alpha\beta = 1$ and $\alpha = 1$ the transverse magnetic field is vanishing.

The point $\alpha = 1$, $\beta = 1$ where the two singular curves intersect each other represents a singular node point for the equation (10.5.21). Using the substitution $\tilde{\alpha} = \alpha - 1$, $\tilde{\beta} = \beta - 1$ in the neighbourhood of the singular point (for small α and $\tilde{\beta}$) equation (10.5.21) becomes:

$$\tilde{\alpha}\frac{\mathrm{d}\tilde{\beta}}{\mathrm{d}\tilde{\alpha}} = \bar{\gamma}(2\tilde{\alpha} + \tilde{\beta})$$

and has the solution:

$$\tilde{\beta} = -\frac{\gamma}{\gamma - 1}\tilde{\alpha} + C\tilde{\alpha}^{\bar{\gamma}}$$

C being a constant of integration. It follows that all the integral curves of equation (10.5.21) have the same tangent

$$\frac{\mathrm{d}\beta}{\mathrm{d}\alpha} = -\frac{\gamma}{\gamma - 1} < 0$$

at the singular point. At the points of intersection of the integral curves with the $(\alpha^2\beta = 1)$ curve the tangents are horizontal since $(\mathrm{d}\beta/\mathrm{d}\alpha = 0)$.

The general shape of the integral curves of the equation (10.5.21) is presented in Figure 43. The dotted parts of the integral curves are physically impossible since, as it follows from (10.5.20) in these regions the magnetic field would be imaginary. The pressure vanishes on the $\beta = 0$ curve and thus the cavitation phenomenon appears. Obviously the cavitation can appear only for a slow wave $(\alpha < 1)$.

From the above discussion it follows that the integral (10.5.22) may be written in the following form:

$$J_{4\pm} = \frac{\beta}{(\alpha_\pm - 1)^{\bar{\gamma}}} \pm \bar{\gamma} \int \frac{d\alpha_\pm}{\alpha_\pm^2 (\alpha_\pm - 1)^{1+\bar{\gamma}}} = K_{4\pm} \qquad (10.5.22')$$

the (+) and (−) signs corresponding to the fast and slow waves respectively.

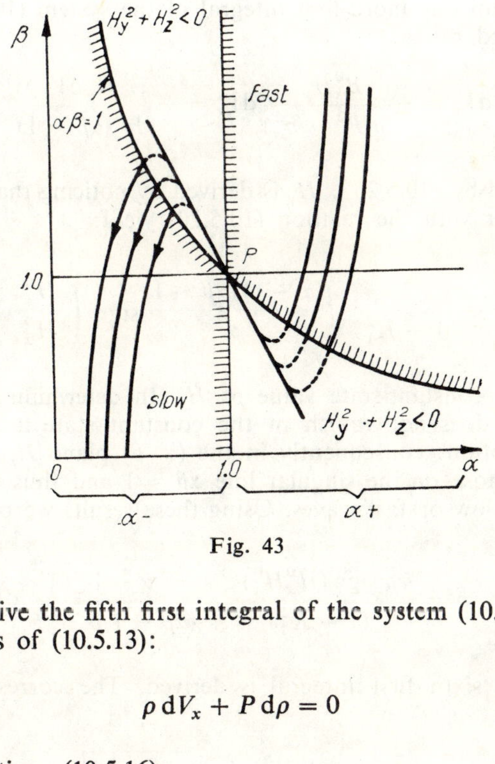

Fig. 43

In order to derive the fifth first integral of the system (10.5.13) one considers the first two ratios of (10.5.13):

$$\rho\, dV_x + P\, d\rho = 0 \qquad (10.5.25)$$

or, using the notations (10.5.16),

$$dV_x = -\frac{\varepsilon \bar{a}}{\gamma} \alpha^{\frac{1}{2}} \beta^{-\frac{1}{2}\left(1+\frac{1}{\gamma}\right)} d\beta \qquad (10.5.22'')$$

ε being equal to ± 1 for the direct waves ($P = +P_f$, $+P_s$) and to − 1 for the reverse waves ($P = -P_f$, $-P_s$). It is obvious that for the direct waves V_x increases or decreases as β (and thus p and ρ) decreases or increases while for the reverse waves V_x increases or decreases together with β. The velocity along the Ox-axis has therefore the same tendency to increase or decrease with respect to the compression or rarefaction of the wave as in classical aerodynamics. To conclude the expression:

$$J_5 = V_x + \frac{\varepsilon\bar{a}}{\gamma}\int \alpha^{\frac{1}{2}}\beta^{-\frac{1}{2}\left(1+\frac{1}{\gamma}\right)}\mathrm{d}\beta \tag{10.5.26}$$

together with the relationship between α and β determined by $J_4 = K_4$ yields the fifth Riemann invariant. In the magnetoacoustic simple-wave motion $J_5 = K_5$, K_5 being a constant determined by the constant state of the fluid.

In order to obtain one more first integral of the system (10.5.13) we consider the second and third ratios:

$$\mathrm{d}V_y = -\frac{B_x^0 H_y}{\rho(P^2 - b_x^2)}\mathrm{d}V_x = -\frac{H_y\,\mathrm{d}V_x}{H_x^0(\alpha\beta - 1)}$$

where $\mathrm{d}V_x$ is defined by (10.5.25′). H_y is derived by noticing that the first integral of (10.5.14) together with the relation (10.5.20) yield:

$$\frac{H_y}{H_x^0} = \frac{1}{1 + K_1^2}\sqrt{\frac{(\alpha - 1)(\alpha\beta - 1)}{\alpha}}\,\mathrm{sign}\left(\frac{H_y^0}{H_x^0}\right) \tag{10.5.27}$$

H_y^0 representing the constant-state value of H_y. In determining the sign we used the fact that if H_y does not vanish in the constant state it will not do so for the simple wave motion; consequently in the (α, β) plane H_y has no other zero values apart from those on the singular line $\alpha\beta = 1$ and thus it does not change the sign across the slow or fast waves. Using these results we obtain

$$\mathrm{d}V_y = \pm\frac{\varepsilon\bar{a}\,\mathrm{sign}\,(H_x^0 H_y^0)}{\gamma(1 + K_1^2)}\sqrt{\frac{\alpha - 1}{(\alpha\beta - 1)\beta^{1+1/\gamma}}}\,\mathrm{d}\beta$$

and integrating the sixth first integral is derived. The corresponding Riemann invariant will be

$$J_6 = V_y \mp \frac{\varepsilon\bar{a}\,\mathrm{sign}\,(H_x^0 H_y^0)}{\gamma(1 + K_1^2)}\int\sqrt{\frac{\alpha - 1}{(\alpha\beta - 1)\beta^{1+1/\gamma}}}\,\mathrm{d}\beta \tag{10.5.28}$$

the upper sign corresponding to the fast waves (in which case α_+ will be substituted for α) and the lower sign to the slow waves (substitution $\alpha = \alpha_-$).

Having determined the six first integrals we can express all the unknowns in terms of the constant values of the adjacent constant state $(K_1, \ldots K_6)$. From the first two invariants we obtain $H_z = K_1 H_y$, $V_z = K_1(V_y - K_2)$. If for instance one chooses the reference system such that the component H_z is vanishing in the constant state ($H_z^0 = 0$) then $K_1 = 0$ and $H_z = V_z = 0$. This choice simplifies the solution.

Let us now derive another important relation for the total pressure

$$p^* = p + \mu H^2/2$$

Taking into account equations (10.5.15) and (10.5.19) we have:

$$\frac{dp^*}{d\rho} = a^2 + \frac{P^2(b^2 - b_x^2)}{P^2 - b_x^2} = P^2 \tag{10.5.29}$$

which indicates that the simple wave is compressive.

10.5.4 Transverse simple waves and entropy waves

In this case the system (10.5.6) (the characteristic system associated with the equations (10.5.9)) reduces itself to:

$$\frac{d\rho}{0} = \frac{dV_x}{0} = \frac{dV_y}{B_x^0 H_z} = \frac{dV_z}{-B_x^0 H_y} = \frac{dH_y}{\rho H_z P} = \frac{dH_z}{\rho H_y P} = \frac{dS}{0}$$

whence the following first integrals are obtained straight forwardly:

$$\rho = \rho^0, \; V_x = V_x^0, \; S = S^0 \tag{10.5.30}$$

$$H_y^2 + H_z^2 = K_4^2, \; V_y \pm \sqrt{\nu} H_y \operatorname{sign} H_x^0 = K_5, \; V_z + \sqrt{\nu} H_z \operatorname{sign} H_x^0 = K_6$$

A first observation that can be made is that in this motion the magnitude of the magnetic field is conserved ($H^2 = H_0^2$). Since ρ is constant it follows that $P_t = P_t^0$. Since V_x is also constant it follows that $\lambda_{2,3}(= V_x \pm P_t)$ are constant which indicates that all the phases of this wave drift with the same velocity (the wave profile does not change).

Finally, for the entropy waves we have:

$$V = V^0 \; H = H^0, \; p = p^0 \tag{10.5.31}$$

ρ being an arbitrary function. Since V_x is constant all the phases drift again with the same velocity (the velocity of the unperturbed fluid).

11

The theory of shock waves

1.1 The equations of shock phenomena

1.1.1 A material derivative

So far we have investigated only continuous motions determined by the differential form of the MFD equations (the fields have been expressed in terms of some variables continuous throughout the fluid). However if the fluid is crossed by a surface S which changes the motion and the fields (a shock wave) then the formulae of Section 2.1 according to which the equations of motion in differential form were obtained are no longer valid. Let us then consider these formulae. The most important is the one providing the domain integral.

Let D be a material domain, $\mathscr{S}$ a surface discontinuity and D_1 and D_2 the two sub-domains determined by $\mathscr{S}$ in D at some arbitrary time $t(D_1 \cup D_2 = D)$. For each of the two domains D_1 and D_2 the formula (2.1.15) is valid such that for an arbitrary field F we have:

$$\frac{\mathrm{d}}{\mathrm{d}t}\int_{D_1} F\,\mathrm{d}v = \int_{D_1} \partial_t F\,\mathrm{d}v + \int_{\Sigma_1} F\boldsymbol{V}\cdot\boldsymbol{n}_1\,\mathrm{d}a - \int_S F_1\boldsymbol{d}\cdot\boldsymbol{n}\,\mathrm{d}a$$

$$\frac{\mathrm{d}}{\mathrm{d}t}\int_{D_2} F\,\mathrm{d}v = \int_{D_2} \partial_t F\,\mathrm{d}v + \int_{\Sigma_2} F\boldsymbol{V}\cdot\boldsymbol{n}_2\,\mathrm{d}a + \int_S F_2\boldsymbol{d}\cdot\boldsymbol{n}\,\mathrm{d}a$$

$$\Sigma_1 \bigcup \Sigma_2 = \partial D, \quad S = \mathscr{S} \bigcap D$$

where by F_1 and F_2 we denoted the limit values of F on S from D_1 and D_2 respectively. We also denoted by $\boldsymbol{n}$ the normal to S (oriented from D_2 to D_1), by $\boldsymbol{n}_1$ the external normal to Σ_1 and by $\boldsymbol{n}_2$ the external normal to Σ_2 (Figure 44). We used the fact that the particles situated on Σ_1 and Σ_2 had the velocity $\boldsymbol{V}$ (the fluid velocity) while the points of S had the velocity $\boldsymbol{d}$ (the discontinuity surface velocity). We now add up these formulae and take the limit as the volume of D tends to zero, the surface S remaining constant ($\Sigma_1 \to S$, $\Sigma_2 \to S$). Then the right-hand side volume integrals tend to zero (if $\partial_t F$ is bounded within D_1 and D_2) and the surface integrals tend to the integral on S:

$$\lim_{\Sigma_1 \to S} \int_{\Sigma_1} F \boldsymbol{V} \cdot \boldsymbol{n}_1 \, da = \int_S F_1 \boldsymbol{V}_1 \cdot \boldsymbol{n} \, da, \quad \lim_{\Sigma_2 \to S} \int_{\Sigma_2} F \boldsymbol{V} \cdot \boldsymbol{n}_2 \, da = -\int_S F_2 \boldsymbol{V}_2 \cdot \boldsymbol{n} \, da$$

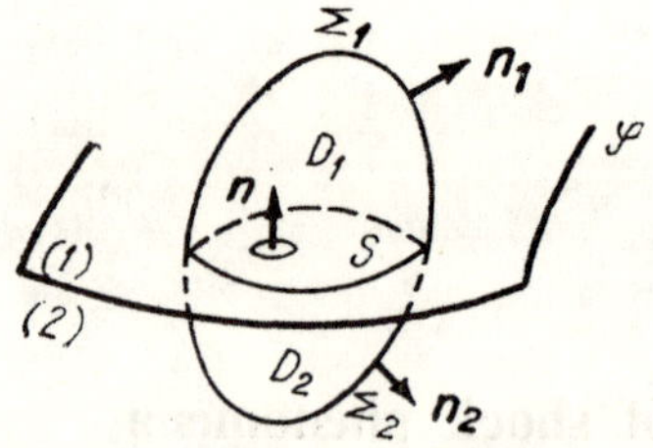

Fig. 44

Consequently we have:

$$\lim \frac{d}{dt} \int_D F \, dv = \int_S [F(V_n - d)] \, da \tag{11.1.1}$$

With the usual notation

$$[F(V_n - d)] = F_1(\boldsymbol{V}_1 - \boldsymbol{d}) \cdot \boldsymbol{n} - F_2(\boldsymbol{V}_2 - \boldsymbol{d}) \cdot \boldsymbol{n}$$

The formula (11.1) is also valid when F is a vector field.

11.1.2 The shock equations

We assume that the fluid is a simple, non-dissipating medium ($\tilde{\mu} = \varkappa = \sigma^{-1} = 0$) in the presence of a field of bounded mechanical forces. If we neglect the quantities of the order of the relativistic correction (the displacement current in the Maxwell equations and the electric field in the Maxwell tensor) then the equations of mass and momentum conservation reduce themselves to:

$$\frac{d}{dt} \int_D \rho \, dv = 0 \tag{11.1.2}$$

$$\frac{d}{dt} \int_D \rho \boldsymbol{V} \, dv + \int_{\partial D} \boldsymbol{\Pi} \, da = 0 \tag{11.1.3}$$

where

$$\boldsymbol{\Pi} = -\boldsymbol{n} \cdot (\underline{\boldsymbol{T}} + \underline{\boldsymbol{T}}^{(em)}) = \left(p + \frac{1}{2} \boldsymbol{B} \cdot \boldsymbol{H}\right) \boldsymbol{n} - (\boldsymbol{n} \cdot \boldsymbol{B}) \boldsymbol{H} \tag{11.1.3'}$$

the body mechanical forces being neglected since they do not interfere with the shock phenomena (they appear under the volume integrals which vanish as the volumes tend to zero as indicated above). Equation (11.3) is derived from (2.2.3) in which

$$F_i^{(\mathrm{em})} = T_{ji,j}^{(\mathrm{em})}, \quad T_{ji}^{(\mathrm{em})} = B_j H_i - \frac{1}{2}\boldsymbol{B}\cdot\boldsymbol{H}\delta_{ji}$$

For a perfectly conducting fluid the term $(\boldsymbol{E}' \times \boldsymbol{H}')$ disappears from the energy equation (2.3.35) since $\boldsymbol{E}' = 0$. Consequently this reduces to:

$$\frac{\mathrm{d}}{\mathrm{d}t}\int_D \left[\rho\left(\frac{1}{2}V^2 + U\right) + \frac{1}{2}\boldsymbol{B}\cdot\boldsymbol{H}\right]\mathrm{d}v + \int_{\partial D} \boldsymbol{\Pi}\cdot\boldsymbol{V}\mathrm{d}a = 0 \qquad (11.1.4)$$

The equations (11.2)—(11.4) are also valid if D is in the presence of a discontinuity surface. Using equation (11.1) as well as the fact that S is an arbitrary surface we obtain:

$$[\rho(V_n - d)] = 0 \qquad (11.1.5)$$

$$[\rho\boldsymbol{V}(V_n - d) + \boldsymbol{\Pi}] = 0 \qquad (11.1.6)$$

$$\left[\left(\frac{1}{2}\rho V^2 + \rho U + \frac{1}{2}\boldsymbol{B}\cdot\boldsymbol{H}\right)(V_n - d) + \boldsymbol{\Pi}\cdot\boldsymbol{V}\right] = 0 \qquad (11.1.7)$$

We used the fact that the surface integrals of (11.1.3) and (11.1.4) tend (as $\Sigma_1 \to S$ and $\Sigma_2 \to S$) towards the integrals on S of the integrand jumps.

The following equation

$$[\boldsymbol{H}]\cdot\boldsymbol{n} = 0 \qquad (11.1.8)$$

derived from (1.5.17) in the hypothesis that the fluid is characterized by the same constant μ in D_1 and D_2 can be used for the magnetic field. In order to obtain a complete system of shock equations we have to use the magnetic induction equation to derive another jump equation. To this end we shall integrate the induction equation over a material domain D. Using the Stokes formula we get:

$$\int_D \partial_t\boldsymbol{H}\,\mathrm{d}v = \int_{\partial D} \boldsymbol{n}\times(\boldsymbol{V}\times\boldsymbol{H})\,\mathrm{d}a = \int_{\partial D}\{\boldsymbol{V}(\boldsymbol{n}\cdot\boldsymbol{H}) - \boldsymbol{H}(\boldsymbol{n}\cdot\boldsymbol{V})\}\,\mathrm{d}a$$

Derived in the hypothesis of field continuity this equations may be used both in D_1 and D_2. Adding up the two equations and using (11.1.8) one has:

$$\int_D \partial_t\boldsymbol{H}\,\mathrm{d}v = \sum_{i=1,2}\int_{\Sigma_i}\{\boldsymbol{V}(\boldsymbol{n}_i\cdot\boldsymbol{H}) - \boldsymbol{H}(\boldsymbol{n}_i\cdot\boldsymbol{V})\}\,\mathrm{d}a + \int_S [\boldsymbol{H}]d\,\mathrm{d}a$$

(3) The stability conditions to be established in Section 11.4 for one-dimensional motion will show that only the fast and slow shocks are stable to small perturbations.

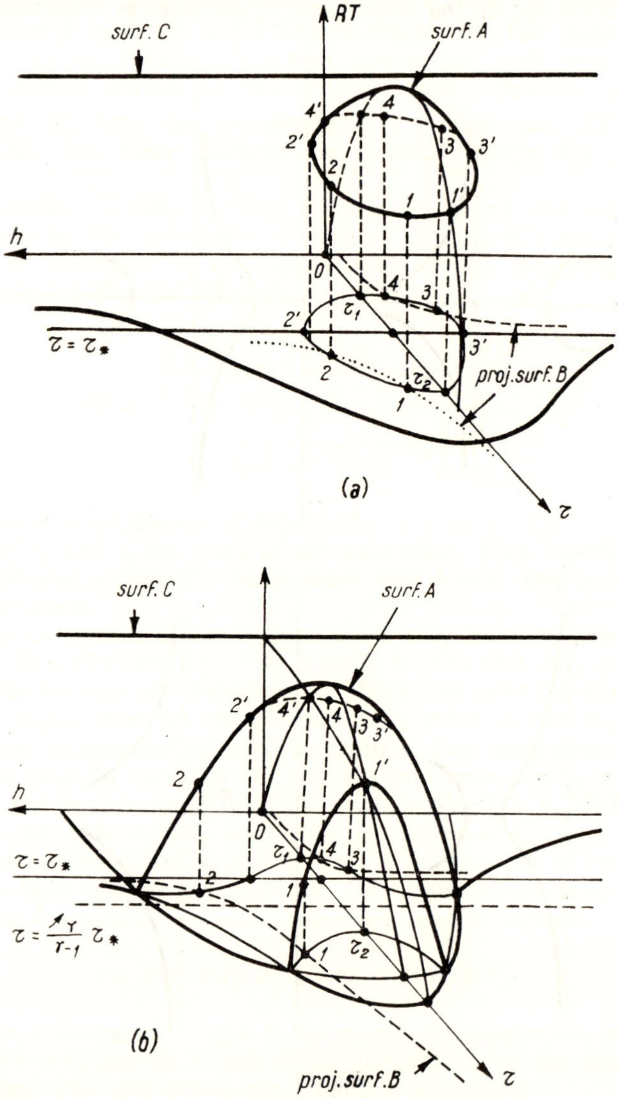

Fig. 48 a, b

(4) If $F_t \to 0$ the hyperbolic cylinder degenerates into the planes $\tau = \tau_*$ and $h = 0$ so that $1 \to 1'$, $2 \to 2'$, $3 \to 3'$, $4 \to 4'$. $1' \to 2'$ represents a switch-on shock while $3' \to 4'$ represents a switch-off shock. The shock $1' \to 2'$ is a common gas

dynamics shock in the presence of a normal magnetic field (if $B_n \neq 0$). As all the intermediate shocks this is a non-evolutionary shock.

In what follows we shall prove the inequalities (11.3.28). This has been done by Germain [5] and Shercliff [6] using different methods. Before proving (28) we have to define the Rayleigh line.

11.3.3 The Rayleigh line

If we eliminate h from equations (11.3.21), (11.3.22) and (11.3.23) we get the following two curves in the (τ, p)-plane:

$$p + m^2\tau + \frac{1}{2}\frac{\mu F_t^2}{(\tau - \tau_*)^2} = F_n \tag{11.3.29}$$

$$I + \frac{1}{2}m^2\tau^2 + \frac{\mu}{2\tau_*}\frac{\tau^2 F_t^2}{(\tau - \tau_*)^2} = K \tag{11.3.30}$$

The curve (11.3.29) is called the Rayleigh line [5] and (30) the K-line. The Rayleigh line is plotted in Figure 49 and the K-line in Figure 50. The intersection of the curves (11.3.29) and (11.3.30) provides the points that may be possibly connected by the shock.

The Rayleigh line has two asymptotes (one oblique of equation $p = -m^2\tau + F_n$ and another vertical of equation $\tau = \tau_*$) and two branches (a fast and a slow one). Given the values of m, F_n and F_t the fast branch has a maximum at the point (p', τ') determined by the equations:

$$p' = F_n + \frac{1}{2}m^2\tau_* - \frac{3}{2}m^2\tau' \tag{11.3.31}$$

$$\mu F_t = m^2(\tau' - \tau_*)^3 \tag{11.3.32}$$

For the fast branch to exist one should have $p' > 0$, that is,

$$0 < \frac{3}{2}m^2(\tau' - \tau_*) < F_n - m^2\tau_*$$

Using (11.3.32) this condition becomes:

$$0 < F_t < \frac{8(F_n - H_n B_n)^3}{27\,\mu m^4} \tag{11.3.33}$$

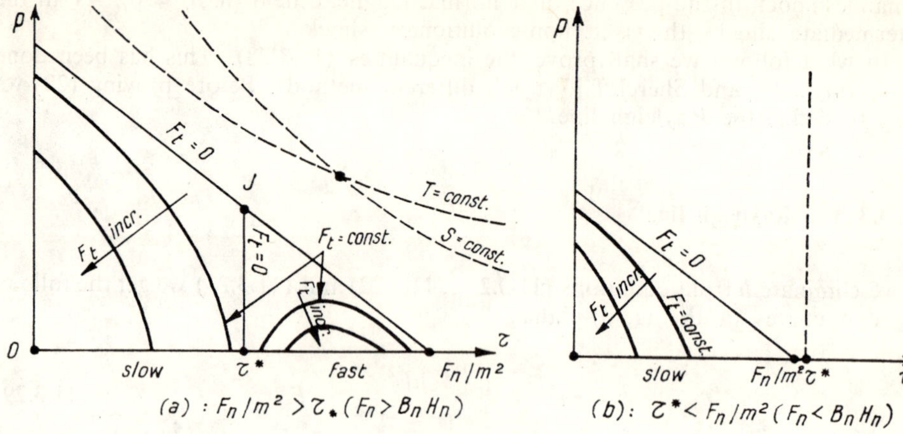

Fig. 49 a, b

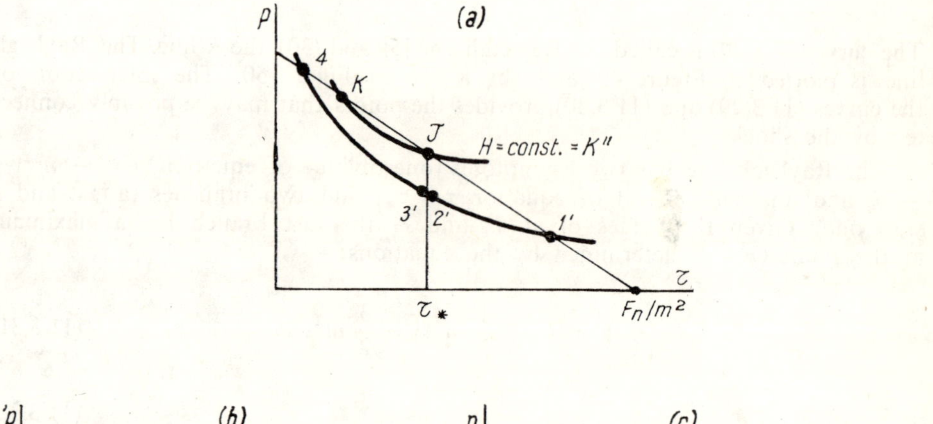

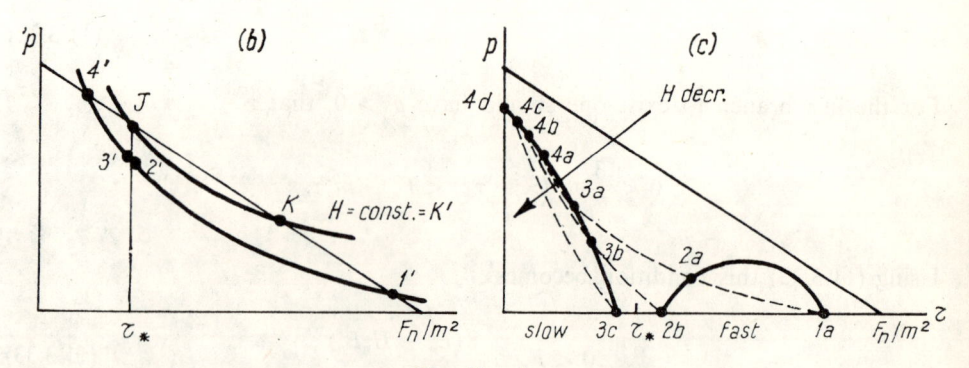

Fig. 50 a, b, c

Finally the existence condition for the slow branch is that $p_* = p|_{\tau=0} > 0$. Taking into account equation (11.3.29) this condition becomes:

$$0 < F_t^2 < 2F_n B_n H_n \qquad (11.3.34)$$

The K-line has a vertical asymptote for $\tau = \tau_*$. Let K'' and K' be the values for which the K-line passes through I the point of intersection of the Rayleigh lines $F_t = 0$. For K smaller than K'' or K' we shall have switch-on or switch-off shocks respectively. The shock $2' \to 3'$ is Alfvénian while $1' \to 4'$ is a common gasdynamic shock. From Figure 50 c one notices that some parts of the Rayleigh line are inaccessible to the shocks: no "post-shock" states can exist on the intervals (2*a*, 2*b*), (3*b*, 3*c*), (4*c*, 4*d*) since no corresponding "pre-shock"-states exist.

Differentiating (11.3.29) and (11.3.30) it follows that:

$$\mathrm{d}p + m^2\,\mathrm{d}\tau = \frac{\mu F_t^2}{(\tau - \tau_*)^3}\mathrm{d}\tau,\ \mathrm{d}I + m^2\tau\,\mathrm{d}\tau = \frac{\mu F_t^2}{(\tau - \tau_*)^3}\tau\,\mathrm{d}\tau \qquad (11.3.35)$$

So that at a point of tangency of the two lines we have:

$$\mathrm{d}I = \tau\,\mathrm{d}p \qquad (11.3.36)$$

I being the specific enthalpy. But in a non-dissipating process:

$$T\mathrm{d}S = \mathrm{d}I - \tau\,\mathrm{d}p$$

Consequently, at a point of tangency of the curves (11.3.29) and (11.3.30) both lines are tangent to an isentropic curve ($S = \text{const.}$) and at this point the entropy has a maximum.

Taking now into account the fact that an isentropic curve has at any of its points a positive curvature (Weyl's hypothesis) and negative slope it follows that for each branch of the Rayleigh line there exists only one entropy maximum and therefore only one point of tangency between the curves (11.3.29) and (11.3.30). It follows that the K-line can intersect each branch of the Rayleigh line at two points at most.

Let us now determine the Rayleigh line in the (T, S)-plane. Since for non-dissipating processes:

$$\mathrm{d}U = T\mathrm{d}S - p\,\mathrm{d}\tau$$

it follows that:

$$\frac{\partial}{\partial p}\left(T\frac{\partial S}{\partial \tau} - p\right) = \frac{\partial}{\partial \tau}\left(T\frac{\partial S}{\partial p}\right) \Rightarrow \frac{\partial(T, S)}{\partial(p, \tau)} = 1$$

The last relation shows that a transformation from the (τ, p)- to the (T, S)-plane is possible. In the (T, S)-plane the Rayleigh line becomes $T = T(S)$ which is represented [6] in Figure 51 M_1 and M_2 corresponding to an entropy maximum (they separate the points 1 and 2, and 3 and 4 respectively).

One notices that at the point of tangency of the curves (11.3.29) and (11.3.30) one has:

$$\left(\frac{dp}{d\tau}\right)_S = -m^2 + \frac{\mu F_t^2}{(\tau - \tau_*)^3} \tag{11.3.37}$$

Using now the definition of the parameters m, F_t and τ_* as well as the sound velocity definition $a^2 = -\tau^2(dp/d\tau)_S$ equation (11.3.37) is put into the form:

$$V_n^2 - (a^2 + b^2) V_n^2 + a^2 b_n^2 = 0 \tag{11.3.38}$$

This relation indicates that at the tangency points V_n becomes equal to the propagation velocity of the magnetoacoustic waves: the slow ones at the point of intersection with the slow branch and the fast ones at the point of intersection with the fast branch (Figure 52).

We are now able to present the proof of the inequalities (11.3.28) [5]. To this end we shall fix the constants m, τ_*, F_n and F_t and let K change its value. In such a case equation (11.3.30) yields $I = I(\tau, K)$ so that by differentiation we obtain:

$$dI + m^2\tau\, d\tau = \frac{\mu F_t^2}{(\tau - \tau_*)^3}\tau\, d\tau + dK \tag{11.3.39}$$

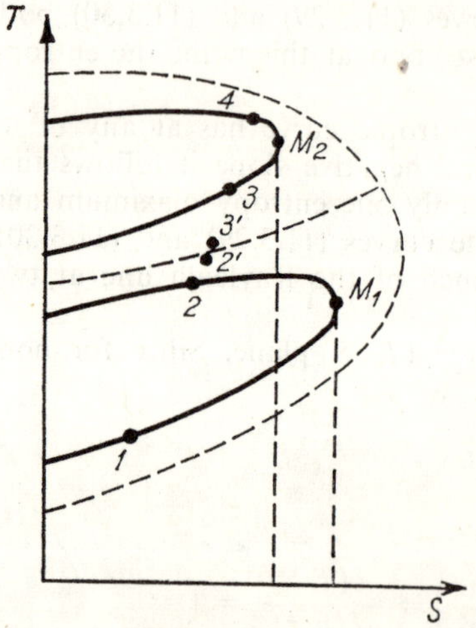

Fig. 51

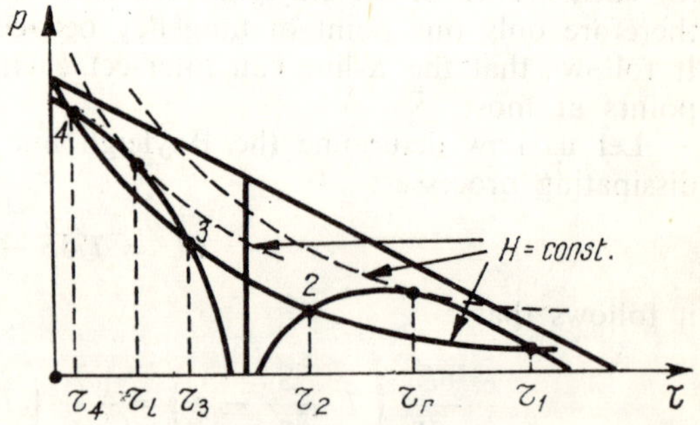

Fig. 52

Using also the first equation of (11.3.35) we get:

$$dK = T\, dS \tag{11.3.40}$$

Since two points (Q_1 and Q_2, say) connected by a shock lie on a K-line (thus $K =$ const.) it follows that:

$$0 = \int_{Q_1}^{Q_2} \mathrm{d}K = \int_{Q_1}^{Q_2} T(S)\,\mathrm{d}S = \int_{Q_1}^{M} T(S)\,\mathrm{d}S + \int_{M}^{Q_2} T(S)\,\mathrm{d}S$$

M being the point of entropy maximum ($M = M_1$ if Q_1 and Q_2 coincide with the points 1 and 2 respectively; $M = M_2$ if Q_1 and Q_2 coincide with 3 and 4 respectively). But $\int_{Q_1}^{M} T(S)\mathrm{d}S$ represents the area under the curve Q_1M. It follows that $S_1 \leqslant S_2$, $S_3 \leqslant S_4$. The proof of the inequality $S_2 \leqslant S_3$ is more complicated [3], [5].

As a matter of fact the inequalities (11.3.28) are required by the second law of thermodynamics according to which the entropy can only increase behind the shock.

11.4 The stability of shock waves

11.4.1 A stability criterion

One can show (and this was done by Lax in 1963) that both in classical gas dynamics and in magneto-gas dynamics an initial discontinuity may be consistent with more than one solution of the shock equations. If there could be found a criterion for choosing one physically acceptable solution from the one formally available then the *uniqueness* of the solution would be ensured. In classical gas dynamics the condition of entropy increase through a shock discontinuity ensures the uniqueness of the solution. Due to the large number of characteristics in magneto-gas dynamics (here one has the slow waves with no counterpart in classical gas dynamics) this condition is no longer sufficient. The uniqueness problem in this case is solved through the *evolutionary condition* for a wave or a sequence of waves. Therefore the class of solutions satisfying the condition of entropy increase by the shock is larger than that of the solutions satisfying the evolutionary conditions. Some authors (among them Friedrichs [10. 1, 2], Lax [10. 12, 17], Ghelfand [10. 15], Akhiezer, Liubarski and Polovin [7], Sirovatski [8], Polovin [9]) dealt with the definition of the evolutionary condition. The proof of the fact that the evolutionary condition is more restrictive than the entropy increase condition can be done on the shock polar curve.

We present below the solution of Akhiezer and Liubarski and Polovin for a steady plane shock wave separating two constant states of a conducting gas. The reference system is chosen such that the yOz-plane is coincident with the wave plane and the Ox-plane is normal to the shock wave and oriented so that the projection of the fluid velocity on this axis is positive. The shock wave separates the fluid into two regions D_1 and D_2 in which $x < 0$ and $x > 0$ respectively. The parameters of the motion are constant in both regions but their values in D_1 differ from those in D_2. We denote by U the column matrix of these parameters and by U_1 and U_2 its value in D_1 and D_2 respectively.

We now assume that at a given instant of time taken as the initial time the fluid state is slightly perturbed, the perturbation being a function of x only. After this time the fluid motion will be described by the matrix $U(t, x)$, the solution of the MFD equations:

$$U_{,t} + AU_{,x} = 0, \; A = A(U) \tag{11.4.1}$$

The solution in the two regions is of the form:

$$U_1(t, x) = U_1 + U_1'(t, x)$$
$$U_2(t, x) = U_2 + U_2'(t, x) \tag{11.4.2}$$

where U_1' and U_2' are the perturbations satisfying the following equations (derived from (11.4.1) by linearization):

$$U'_{1,t} + A(U_1)U'_{1,x} = 0$$
$$U'_{2,t} + A(U_2)U'_{2,x} = 0 \tag{11.4.3}$$

In order to determine the fluid motion we shall have to solve system (11.4.3) knowing that U_1' and U_2' are known at the initial time and that they vanish at infinity both upstream and downstream. Moreover the functions $U_1(t, x)$ and $U_2(t, x)$ should obey the shock equations on the shock wave ($x = 0$). We thus have a Cauchy problem with supplementary conditions. We shall call it the $\mathscr{P}$-problem.

For hyperbolic systems the uniqueness of the solution of the Cauchy problem (without supplementary conditions) also ensures its continuous dependence on the initial conditions. Taking into account this theorem we adopt the following criterion: "*the shock wave motion is stable (evolutionary) if the* $\mathscr{P}$*-problem has a unique solution*". Otherwise one has a so-called unstable motion.

Since (11.4.3) is a linear system it can be integrated using the Laplace transform:

$$\bar{U}'(s, x) = \int_0^\infty U'(t, x)\, e^{-st} dt \qquad s > 0 \tag{11.4.4}$$

From (11.4.3) it follows that:

$$A \frac{d\bar{U}'}{dx} + s\bar{U}' = U'(0, x) \tag{11.4.5}$$

in the D_k region A being substituted by $A(U_k)$, $\bar{U}'$ by $\bar{U}_k'$ and U' by U_k' ($k = 1, 2$). The homogeneous equation of (11.4.5) has a general solution of the form:

$$\bar{U}_0' = C_j e^{\lambda_j x} \tag{11.4.6}$$

C_j being a column matrix and λ_j a solution of the equation:

$$\det(\lambda_j A + sE) = 0 \tag{14.4.7}$$

The general solution of equation is of the form:

$$\bar{U}'(x, s) = \sum_j C_j e^{\lambda_j x} g_j(x) \tag{11.4.8}$$

where:

$$g_j(x) = g_j(0) + C_j^{-1} A^{-1} \int_0^x e^{-\lambda_j \xi} U'(0, \xi)\, d\xi \tag{11.4.8'}$$

If n is the number of unknowns, that is, if A is a square matrix of order n, then n roots λ_j will exist (and they are real since the system is hyperbolic). The general solution therefore depends on $2n$ arbitrary constants $g_j^{(k)}(0) = 0$. The solution depends on n constants in each region. In order that the Cauchy problem solution be unique the conditions at infinity together with the shock equations should make possible the complete determination of these constants. The initial conditions were posed in (11.4.5).

Let us now investigate the conditions for vanishing perturbations at infinity. In D_1 at infinity $x \to -\infty$ while in D_2 $x \to +\infty$. Therefore in (11.4.8) the waves for which we have

$$\lambda_j^{(1)} > 0, \ \lambda_j^{(2)} < 0 \tag{11.4.9}$$

are damping waves. The waves for which

$$\lambda_l^{(1)} < 0, \ \lambda_l^{(2)} > 0 \tag{11.4.10}$$

remain of finite amplitude only if

$$g_l^{(1)}(-\infty) = 0, \ g_l^{(2)}(+\infty) = 0 \tag{11.4.10'}$$

The conditions (11.4.10') will determine the constants $g_l^{(k)}(0)$. One is therefore left with the determination of constants $g_j^{(k)}(0)$ corresponding to the roots (11.4.9). Let us denote by m their number.

If the shock equations make possible the determination of these m constants then the solution is unique. But we have n shock equations which also contain the drift velocity. Eliminating this we get $(n - 1)$ equations. Consequently if $m = n - 1$ the solution of the $\mathscr{P}$-problem is unique, if $m < n - 1$ the solution of the $\mathscr{P}$-problem is impossible and if $m > n - 1$ the solution is undetermined. Therefore for the solution to be stable it is necessary to have $m = n - 1$, n being the number of roots of the characteristic equations (11.4.7) which satisfy the inequalities (11.4.9).

Let us now look for a physical interpretation of the number m. If the general solution of equations (11.4.3) is written as a continuous superposition of plane waves of the form $\exp i(kx - \omega t)$ one obtains:

$$\det(kA - \omega E) = 0$$

From a comparison of this equation with (11.4.7) it follows that between the drift velocities $V_j(= \omega/k_j)$ of the plane waves and the roots a relationship $V_j = -s/\lambda_j$ exists. Since s is positive it follows: in D_1 the inequality (11.4.9) is valid for V_j negative velocities, that is, for waves starting from the shock and propagating towards $-\infty$; in D_2 the inequality (4.4.9) is valid for positive velocities V_j, that is, also for waves starting on the shock and propagating towards infinity. To conclude, m is equal to the number of plane waves which emerge from the shock and propagate into the two regions.

11.4.2 Applications to MFD

In a MFD problem the number of unknowns is eight: density, pressure and the components of velocity and magnetic field. But using the fact that the variables depend on t and x only it follows from the equation

$$\partial_t \boldsymbol{H} = \operatorname{curl}(\boldsymbol{V} \times \boldsymbol{H})$$

that in the two domains we have:

$$\partial_t H_x = 0 \Rightarrow H_x(t, x) = H_x(0, x) \tag{11.4.11}$$

so that this quantity is no longer unknown. Therefore $n = 7$. In order to have an unique solution it is necessary to have $m = 6$.

As we already know the MFD waves may propagate with one of the following four velocities 0, P_t, P_f and P_s. Seven drift velocities correspond to these velocities in each domain: V, $V \mp P_t$, $V \mp P_f$, $V \mp P_s$, V being the fluid velocity in the direction of the Ox-axis.

The waves drifting with one of the velocities V, $V + P_t$, $V + P_f$, $V + P_s$ are always convergent in D_1 and divergent in D_2. On the other hand those drifting with one of the velocities $V - P_t$, $V - P_f$, $V - P_s$ may be either divergent or convergent. Their number is six: three in each region. Since we already have four divergent waves in the first category it follows that it is necessary to have only two such waves in the second one.

In order to determine the drifting direction of these waves (that is the sign of the velocities) a table having the velocities in D_1 indicated on the horizontal lines and those in D_2 on the vertical ones is constructed,

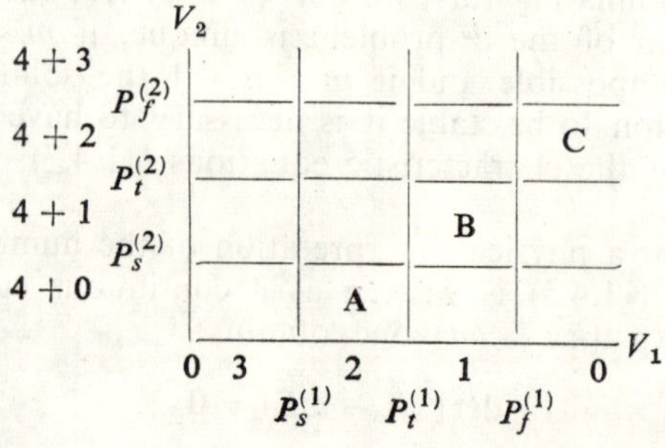

Now studying the velocity sign for each square one writes down horizontally the number of the divergent waves in D_1 (those having negative velocities) and vertically the number of the divergent waves in D_2 (those having positive velocities). One also writes down vertically the four divergent waves of the first category. It can be easily found that the number of divergent waves is equal to 6 in the squares A, B, C, To conclude, the stability condition is satisfied if:

$$A: P_s^{(1)} < V_1 < P_t^{(1)}, V_2 < P_s^{(2)}$$

$$B: P_t^{(1)} < V_1 < P_f^{(1)}, P_s^{(2)} < V_2 < P_t^{(2)}.$$

$$C: P_f^{(1)} < V_1, P_t^{(2)} < V_2 < P_f^{(2)}.$$

Particularly if the magnetic field is parallel to the wave front then $P_t^{(1)} = P_f^{(2)} = P_s^{(1)} = P_s^{(2)} = 0$. The stability condition is fulfilled if:

$$V_1 > \sqrt{a_1^2 + b_1^2}, V_2 < \sqrt{a_2^2 + b_2^2}$$

If the specific mass, pressure and magnetic field are known quantities then from the expression of the propagation velocities one determines $P_t^{(1)}$, $P_f^{(1)}$ and $P_s^{(1)}$. Once V_1 is also known one determines the parameters of the motion in D_2 from the shock equations and therefore one obtains V_2, $P_t^{(2)}$, $P_f^{(2)}$ and $P_s^{(2)}$.

However, $m = n - 1$ does not represent a sufficient condition for the $\mathscr{P}$—problem to have a unique solution since it is possible that the shock equations be separated into groups with respect to the same unknowns. In such a case the evolutionary condition (i.e. the condition that the number of divergent waves is equal to the number of independent shock conditions) should be fulfilled not only for all the unknowns as a whole but also for any individual group. Such a problem appears when the waves propagate perpendicularly to the discontinuity surface.

In the reference system ($\boldsymbol{n}, \boldsymbol{t}, \boldsymbol{s}$) defined in the previous section we have:

$$\boldsymbol{V} = V\boldsymbol{n}, \quad \boldsymbol{V}' = V'\boldsymbol{n} + v_t'\,\boldsymbol{t} + v_s'\boldsymbol{s}$$

$$\boldsymbol{H} = H_n\boldsymbol{n} + h\boldsymbol{t}, \quad \boldsymbol{H}' = H_n'\boldsymbol{n} + h_t'\boldsymbol{t} + h_s'\boldsymbol{s}$$

Linearizing the shock equations (11.1.5), (11.1.6) and (11.1.9) and taking the projections on the above reference system we obtain:

$$\left.\begin{aligned} &[\rho' V + \rho(V' - d')] = 0 \\ &[\rho' V^2 + \rho V V' + \rho V(V' - d') + p' + \mu h h_t'] = 0 \\ &[\rho V v_t' - B_n h_t'] = 0 \\ &[-H_n v_t' + V h_t' + h(V' - d')] = 0 \end{aligned}\right\} \qquad (11.4.12)$$

$$\left.\begin{aligned} &[\rho V v_s' - B_n h_s'] = 0 \\ &[H_n v_s' - V h_s'] = 0 \end{aligned}\right\} \qquad (11.4.13)$$

For instance the linearized equation (11.1.9) has the form:

$$[V'H_n + VH'_n - H'V - H(V'_n - d')] = 0$$

and its projection yields the last equations of (11.4.12) and (11.4.13). As already pointed out for the unperturbed motion $d = 0$ since the reference system is bound to the shock.

Another relation provided by the energy equation (11.1.7) should be added to the system (11.4.12).

It follows that the system of shock equations is divided into two subsystems. We have two equations (11.4.13) independent of the shock velocity for the transverse discontinuities such that two of the waves drifting with one of the velocities $V_1 \mp P_t^{(1)}$ or $V_2 \mp P_t^{(2)}$ should be divergent. Since the wave drifting with velocity $V_1 + P_t^{(1)}$ is convergent and the one drifting with $V_2 + P_t^{(2)}$ is divergent it follows that one of the waves drifting with the velocities $V_1 - P_t^{(1)}$, $V_2 - P_t^{(2)}$ should be divergent. This occurs for one of the two cases:

$$1°.\ V_1 < P_t^{(1)} \text{ and } V_2 < P_t^{(2)}$$

$$2°.\ V_1 > P_t^{(1)} \text{ and } V_2 > P_t^{(2)}$$

It follows that of the conditions A, B, C only A and C are sufficient for the stability of transverse waves.

We have five shock conditions for the magneto-acoustic and entropy waves but they are not independent since they also contain the perturbation of the drift velocity d'. After this unknown is eliminated we are left with four equations. It follows that of the waves drifting with one of the velocities

$$V_1, V_1 \mp P_s^{(1)}, V_1 \mp P_f^{(1)}$$

$$V_2, V_2 \mp P_s^{(2)}, V_2 \mp P_f^{(2)}$$

four should be divergent. The waves drifting with the velocities $V_1, V_1 + P_s^{(1)}$, $V_1 + P_f^{(1)}$ are convergent while those drifting with the velocities $V_2, V_2 + P_s^{(2)}$, $V_2 + P_f^{(2)}$ are divergent. The stability condition is fulfilled when one of the remaining four waves is divergent. This happens in one of the two cases:

$$1°.\ P_s^{(1)} < V_1 < P_f^{(1)} \text{ and } V_2 < P_s^{(2)}$$

$$2°.\ P_f^{(1)} < V_1 \text{ and } P_s^{(2)} < V_2 < P_f^{(2)}$$

It follows that in this case again only A and C are stability domains.

The table below contains a recapitulation of the results. The number of transverse divergent waves (the first number) and that of the entropy or magnetoacoustic waves (the second number) are written inside each rectangle. One notices that the rectangle previously denoted by B represents an instability region since although the stability condition is satisfied overall the $\mathscr{P}$-problem is undetermined for the Alfvén waves and impossible for the entropy and magnetoacoustic waves.

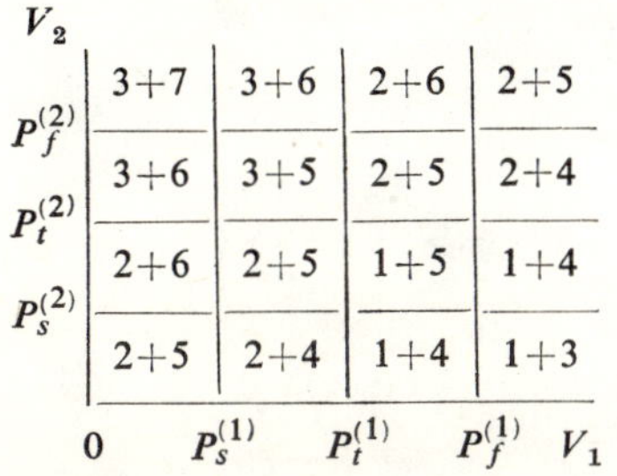

For various other consequences one should refer to [9] and [G. 23].

P. Germain [4] has proposed a definition of the *dissipation stability* (another stability). In general this consists in the following: if all the dissipation coefficients appearing in the problem are non-zero and simultaneously tend to zero the shock is "stable" if it exists and "unstable" if it does not. A first remark due to Germain is that the dissipation stability implies the fulfillment of the evolutionary condition. The reverse is not always true; more precisely the slow shocks are not always stable to dissipation.

Anderson [3] and B. P. Leonard [10] have shown that the dissipation stability condition can be broader (including the intermediate shock) but also narrower (since it allows too little to be said about the slow shock) than the evolutionary condition. Anderson and Leonard also considered the influence of the Hall effect and proved that errors could turn up whenever this is neglected.

Part 2

The microscopic theory

12

The electromagnetic field theory

12.1 Particle motion in an electromagnetic field

12.1.1 Hamilton's principle

Let us consider a particle of charge q in an electromagnetic field in the free space. The equation of motion of the particle and the electromagnetic field equations can be derived by considering the particle-field system as a whole and using Hamilton's principle [1]. According to this principle *any natural system is characterized by a function* $L\ (t,\ \boldsymbol{x},\ \dot{\boldsymbol{x}})$, *called the Lagrangian function; the evolution of the system during the time interval* $(t_0,\ t_1)$ *is determined by the condition that the functional* (called the action integral)

$$S = \int_{t_0}^{t_1} L dt \tag{12.1.1}$$

is stationary (i.e., $\delta S = 0$). In L the quantity $\boldsymbol{x}$ is a vector in the Lagrangian space, which, for a system without constraints like the present one coincides with the physical space.

From both the variational calculus and the classical mechanics [2] we know that the condition $\delta S = 0$ is equivalent to the Euler-Lagrange equations

$$\frac{\mathrm{d}}{\mathrm{d}t}\left(\frac{\partial L}{\partial \boldsymbol{v}}\right) - \frac{\partial L}{\partial \boldsymbol{x}} = 0, \quad \frac{\partial L}{\partial \boldsymbol{v}} \stackrel{\text{def}}{=} \left(\frac{\partial L}{\partial \dot{x}_i}\right), \quad \frac{\partial L}{\partial \boldsymbol{x}} \stackrel{\text{def}}{=} \left(\frac{\partial L}{\partial x_i}\right) \tag{12.1.2}$$

and that the Lagrangian function L is determined up to the total time derivative of some function.

12.1.2 The Lorentz equation

In order to apply this formalism to the particle-field system we should postulate the expression of L. In this respect we shall write

$$S = S_p + S_{pf} + S_f \tag{12.1.3}$$

where S_p represents the functional characterizing the inertial motion of the particle (in the absence of the electromagnetic field); S_{pf} the functional characterizing the particle-field interaction and S_f the functional characterizing the electromagnetic field only. In a similar way we write

$$L = L_p + L_{pf} + L_f \tag{12.1.3'}$$

We shall derive the expression for L from the condition that the resulting equations for the particle-field system will be *invariant* with respect to the *Lorentz group* as required by the *first principle of the theory of relativity*.

In order to derive the expression of L_p we shall assume (following the laws of mechanics) that the particle equations of motion should be second-order differential equations. Since the stationary condition (12.1.2) increases the differentiation order by one unit it follows that L_p may contain only first-order derivatives of the particle coordinates (x_1, x_2, x_3). It follows [1], [3] that the only invariant satisfying the above requirements is $k\mathrm{d}s$ ($\mathrm{d}s$ being the Minkowski line element (1.2.3) and k an unknown constant) so that

$$S_p = k\int_{M_0}^{M_1} \mathrm{d}s = ck\int_{t_0}^{t_1} \sqrt{1 - \frac{v^2}{c^2}}\,\mathrm{d}t \tag{12.1.4}$$

where M_0 is the event corresponding to t_0 and M_1 to t_1. In order to determine the constant k we shall use the correspondence principle according to which, for velocities small as compared with the velocity of light ($v^2/c^2 = 0$), the relativistic mechanics structures coincide with the Newtonian mechanics ones. The Lagrangian function for a free (point) particle has according to the Newtonian mechanics, the expression $m_0v^2/2$, m_0 being the particle mass taken as constant, and v the velocity. Expanding L_p in series we have

$$L_p = ck\sqrt{1 - \frac{v^2}{c^2}} = ck - \frac{1}{2}\frac{k}{c}v^2 + O\left(\frac{v^2}{c^2}\right)$$

Neglecting the constant ck (considered as a total derivative with respect to time) and the term of order v^2/c^2 we find that $k = m_0c$. Therefore

$$L_p = -\, m_0c^2\sqrt{1 - \frac{v^2}{c^2}} \tag{12.1.5}$$

In order to find L_{pf} we have to use again Lorentz group invariant this time built up of both field describing variables and particle characterizing quantities. A scalar product $A_\alpha\,\mathrm{d}x_\alpha$ ($\alpha = 1, 2, 3, 4$) imposes itself, A_α being also a quadrivector. The way this quadrivector will characterize the field will follow from the equations it will have to satisfy. It should only be specified that since the action should be real the first three components of A_α should be real and the fourth imaginary since

dx_4 (= ic dt) is imaginary. Therefore we shall write $A_4 = i\varphi/c$ and shall call the scalar φ the *scalar potential of the field* and the vector $\boldsymbol{A} = (A_1, A_2, A_3)$ the *vector potential*. We have, obviously: $\varphi = \varphi(t, \boldsymbol{x})$, $\boldsymbol{A} = \boldsymbol{A}(t, \boldsymbol{x})$.

It is experimentally established [1] that the particle properties in the interaction with the field are determined by one parameter q (which we shall call the electric charge) and that this parameter is an invariant. Therefore

$$S_{pf} = q\mu_0 \int_{M_0}^{M_1} A_\alpha \, dx_\alpha = q\mu_0 \int_{t_0}^{t_1} (\boldsymbol{A} \cdot \boldsymbol{v} - \varphi) \, dt \tag{12.1.6}$$

μ_0 being a constant depending on the choice of the system of units. It seems that the equation

$$L_{pf} = q\mu_0(\boldsymbol{A} \cdot \boldsymbol{v} - \varphi) \tag{12.1.7}$$

was first proposed by Schwarzchild in 1903 [1.3].

Since the particle presence manifests itself only in the first two terms of (12.1.2(3), the equations (12.1.5) and (12.1.7) are sufficient to determine its motion. We have:

$$\frac{\partial}{\partial \boldsymbol{v}}(L_p + L_{pf}) = \boldsymbol{p} + q\mu_0 \boldsymbol{A}$$

$\boldsymbol{p}$ being the particle momentum

$$\boldsymbol{p} = m\boldsymbol{v}, \; m = \frac{m_0}{\sqrt{1 - \dfrac{v^2}{c^2}}} \tag{12.1.8}$$

Using equation (A. 26) it also follows that:

$$\frac{\partial}{\partial \boldsymbol{x}}(L_p + L_{pf}) = q\mu_0\{\operatorname{grad}(\boldsymbol{A} \cdot \boldsymbol{v}) - \operatorname{grad} \varphi\}$$

$$= q\mu_0\{(\boldsymbol{v} \cdot \nabla)\boldsymbol{A} + \boldsymbol{v} \times \operatorname{curl} \boldsymbol{A} - \operatorname{grad} \varphi\}$$

Taking into account the fact that $\boldsymbol{A}$ also depends on t through the particle coordinates x_1, we have:

$$\frac{d\boldsymbol{A}}{dt} = \frac{\partial \boldsymbol{A}}{\partial t} + (\boldsymbol{v} \cdot \nabla)\boldsymbol{A}$$

Thus the condition (12.1.3) yields

$$\frac{d\boldsymbol{p}}{dt} = q(\boldsymbol{E} + \boldsymbol{v} \times \boldsymbol{B}) \tag{12.1.9}$$

where

$$E = -\mu_0(\partial_t A + \text{grad}\,\varphi)$$
$$B = \mu_0\,\text{curl}\,A, \quad B \overset{\text{def}}{=} \mu_0 H \tag{12.1.10}$$

The equation (12.1.9) represents the particle equation of motion: the LHS term is the time derivative of the momentum and the RHS is a force which we shall call the *Lorentz force*. In the non-relativistic case this equation reduces to

$$\frac{\mathrm{d}}{\mathrm{d}t} m_0 v = q(E + v \times B) \tag{12.1.9'}$$

The vector E defined by (12.1.10) will be called the electric field vector, B the magnetic induction vector and H the magnetic field vector. From equation (12.1.10) it follows:

$$\text{curl}\,E = -\partial_t B, \quad \text{div}\,B = 0 \tag{12.1.11}$$

Here we recognize the first two Maxwell equations, with the remark that this time E and B are microscopic quantities (free space quantities). In order to determine completely the field equations the expression for S_f should be taken into account. This will be done in the next section.

Taking the scalar product of equation (12.1.9) and v we get the energy equation:

$$\frac{\mathrm{d}}{\mathrm{d}t} mc^2 = qE \cdot v \tag{12.1.12}$$

Indeed, we know from classical mechanics that in the case of a schleronomous conservative system (like the one considered above) the total energy is given by the Hamiltonian function:

$$\mathcal{H} \overset{\text{def}}{=} p \cdot v - (L_p + L_{pf}) = mc^2 \tag{12.1.12'}$$

In the non-relativistic case equation (12.1.12) becomes:

$$\frac{\mathrm{d}}{\mathrm{d}t} \frac{1}{2} m_0 v^2 = qE \cdot v$$

We notice that the energy charge is determined only by the mechanical work performed by the electric field. If the electric field is zero the energy of the particle remains constant for any magnetic field value.

As a particular case we shall consider the motion of a particle in a constant and homogeneous magnetic induction $\boldsymbol{B}_0$. According to the above results in this case the particle energy remains constant, equal to $\mathscr{H}_0(m = \mathscr{H}_0/c^2)$. Let us choose the reference system so that the Ox–axis is along the direction of $\boldsymbol{B}_0$. Then equation (12.1.9) gives

$$\dot{v}_1 = 0, \quad \dot{v}_2 = \omega v_3, \quad \dot{v}_3 = -\omega v_2$$

where

$$\omega = \frac{c^2 q B_0}{\mathscr{H}_0} \left(\lim_{v^2/c^2 \to 0} \omega = \frac{qB_0}{m_0} \right)$$

Thus it follows that

$$\frac{\mathrm{d}}{\mathrm{d}t}(v_2 + \mathrm{i}v_3) = -\mathrm{i}\omega(v_2 + \mathrm{i}v_3) \Rightarrow v_2 + \mathrm{i}v_3 = a\mathrm{e}^{-\mathrm{i}\omega t}$$

a being a complex constant determined by the initial conditions for v_2 and v_3:

$$a = v_2^0 + \mathrm{i}v_3^0 \stackrel{\text{def}}{=} w\mathrm{e}^{-\mathrm{i}\alpha} (w = \sqrt{(v_2^0)^2 + (v_3^0)^2})$$

In this way we find

$$v_2 = w \cos(\omega t + a)$$

$$v_3 = -w \sin(\omega t + a)$$

Integrating we have

$$x_2 = \left(x_2^0 + \frac{v_3^0}{\omega} \right) + \frac{w}{\omega} \sin(\omega t + a),$$

$$x_3 = \left(x_3^0 - \frac{v_2^0}{\omega} \right) + \frac{w}{\omega} \cos(\omega t + a)$$

$$x_1 = x_1^0 + v_1^0 t$$

The above solution means that the particle moves along a spiral whose axis is along the magnetic induction and whose radius is:

$$\frac{w}{\omega} = \frac{w\mathscr{H}_0}{c^2 q B_0} \frac{p_t}{qB_0} \left(\lim_{v^2/c^2 \to 0} \frac{w}{\omega} = \frac{w m_0}{qB_0} \right)$$

where p_t is the projection of momentum on the x_2Ox_3 plane. The particle velocity is constant in magnitude. In the case $v_1^0 = 0$ the particle moves along a circle in the plane perpendicular to $\boldsymbol{B}_0$. From the above solution it follows that ω is the *particle gyration frequency*. This quantity will also turn up in the last chapter.

For the other considerations on the integration of equation one may use [4], [5], for instance.

12.1.3 The quadridimensional form of the equation of motion

Following Landau and Lifschitz we shall write directly the stationary condition $\delta(S_p + S_{pf}) = 0$. Thus:

$$\delta\int_{M_1M_2} (-m_0c\,\mathrm{d}s + q\mu_0A_\alpha\,\mathrm{d}x_\alpha) = 0 \tag{12.1.13}$$

But since from (1.2.3) and (1.2.6) it follows that

$$\delta\,\mathrm{d}s = -\frac{\mathrm{d}x_\alpha\,\delta\,\mathrm{d}x_\alpha}{\mathrm{d}s} = -U_\alpha\frac{\mathrm{d}\,\delta x_\alpha}{c}$$

the condition (12.1.13) becomes:

$$\int_{M_1M_2} (m_0U_\alpha\,\mathrm{d}\,\delta x_\alpha + q\mu_0\delta A_\alpha\,\mathrm{d}x_\alpha + q\mu_0A_\alpha\,\mathrm{d}\delta x_\alpha) = 0$$

Integrating by parts and using the relationship $\delta x_\alpha|_{M_1} = \delta x_\alpha|_{M_2} = 0$ we obtain

$$\int_{M_1M_2}\left\{-m_0\frac{\mathrm{d}U_\alpha}{\mathrm{d}\tau} + q\mu_0\left(\frac{\partial A_\beta}{\partial x_\alpha} - \frac{\partial A_\alpha}{\partial x_\beta}\right)\frac{\mathrm{d}x_\beta}{\mathrm{d}\tau}\right\}\mathrm{d}\tau\,\delta x_\alpha = 0$$

The variations δx_α being arbitrary class C^1 functions, according to the fundamental lemma of the variational calculus it follows that

$$m_0\frac{\mathrm{d}U_\alpha}{\mathrm{d}\tau} = qU_\beta f_{\beta\alpha},\ \alpha = 1, 2, 3, 4 \tag{12.1.14}$$

where

$$f_{\beta\alpha} = \mu_0\left(\frac{\partial A_\beta}{\partial x_\alpha} - \frac{\partial A_\alpha}{\partial x_\beta}\right) \tag{12.1.14'}$$

The equations (12.1.14) represent the quadridimensional form of the vector equations (12.1.9). We should notice that the four equations (12.1.14) are not independent. Indeed, multiplying (12.1.14) by U_α and summing we get zero on both sides: on the right-hand side due to the symmetry of the tensor $f_{\beta\alpha}$; on the left-hand side due to equation (1.2.6′). As a matter of fact the fourth equation (12.1.14) (the equation for $\alpha = 4$) is just the energy equation which is a consequence of the first three. It should be mentioned that the quadridimensional form of the equations of motion contains both the momentum equation and the energy equation.

Let us now return to the definition (12.1.14′) and write explicitly the form of $f_{\beta\alpha}$ taking into account (12.1.10). We have

$$(f_{\beta\alpha}) = \begin{pmatrix} 0 & -B_3 & B_2 & \frac{i}{c}E_1 \\ B_3 & 0 & -B_1 & \frac{i}{c}E_2 \\ -B_2 & B_1 & 0 & \frac{i}{c}E_3 \\ -\frac{i}{c}E_1 & -\frac{i}{c}E_2 & -\frac{i}{c}E_3 & 0 \end{pmatrix}$$

We recognize the *microscopic form* of the tensor introduced in Section 1.2. It is the electromagnetic field tensor. As in Section 1.2, the Maxwell equations (12.1.11) may be written in the covariant form using this tensor.

This tensor has the following two basic invariants:

$$I_1 = f_{\beta\alpha}f_{\beta\alpha} = 2(|\boldsymbol{B}|^2 - \frac{1}{c^2}|\boldsymbol{E}|^2)$$

$$I_2 = f_{\beta\alpha}f^*_{\beta\alpha} = -\frac{4i}{c}\boldsymbol{B}\cdot\boldsymbol{E} \tag{12.1.15}$$

where $f^*_{\beta\alpha}$ is the dual tensor defined in Section 1.2. As a matter of fact I_2 is not a genuine scalar but a pseudo-scalar since $\delta_{\alpha\beta\gamma\delta}$ is a pseudo-tensor (as shown in the Appendix).

12.2 The electromagnetic field equations

In order to find the electromagnetic field equations one should postulate the expression of S_f from (12.1.3) i.e., that part of the action which depends only on the field properties and not on the properties of the particle. Thus L_f will depend only on the dynamical parameters of the $\boldsymbol{E}$ and $\boldsymbol{B}$ field (or $f_{\beta\alpha}$ in the Universe).

The following properties [1] establish the form of L_f: (1) L_f (like L_p and L_{pf}) should be an invariant of the Lorentz group (the field equations should be invariant); (2) L_f should not contain any derivative of the tensor $f_{\beta\alpha}$ since this would mean second (or higher) order derivatives of the potentials $\boldsymbol{A}$ and φ (in the present case the role of the coordinates in the Lagrangian function will be taken up by the potentials $\boldsymbol{A}$ and φ).

(3) It is experimentally* established that two superposed fields produce the same effect as though each acted separately (the superposition principle). This means that the field equations should be linear equations (the sum of two solutions should be a solution as well). Hence it follows that L_f should be a quadratic expression of $f_{\beta\alpha}$ since the field equations are derived applying the Eulerian differential.

Since the only invariant obeying these requirements is I_1 it follows that:

$$S_f = \varkappa \int_{\Omega} f_{\beta\alpha} f_{\beta\alpha} \, \mathrm{d}\omega \tag{12.2.1}$$

$\varkappa$ being a constant depending (like in the case of S_{pf}) on the system of units and $\mathrm{d}\omega$ the volume element in the Universe (this element also being an invariant of the Lorentz group (1.2.2). One should note that the first two requirements are not sufficient to determine L_f. This simply follows from the fact that any function of invariants is an invariant itself. Thus, the linearity condition is an essential one. Without this condition one obtains another expression for L_f and hence another (non-linear) equations for the field [6].

In order to determine the field we shall therefore use the functional:

$$S_{pf} + S_f = q\mu_0 \int_{M_1 M_2} A_\alpha \, \mathrm{d}x_\alpha + \varkappa \int_{\Omega} f_{\beta\alpha} f_{\beta\alpha} \, \mathrm{d}\omega \tag{12.2.2}$$

Ω being the Universe of the field action.

To write the stationary condition one should extend the first integral of (12.2.2) to the region Ω. In this respect the point charge q is written as a continuous distribution of charge

$$\rho^{(q)} \stackrel{\mathrm{def}}{=} q\delta(\boldsymbol{x} - \boldsymbol{x}_0) \tag{12.2.3}$$

δ being the Dirac distribution and $\boldsymbol{x}_0$ the position vector of q. The equivalence of the two distributions (the point and continuous ones) follows imediately if we integrate (12.2.3) over the whole region of influence E_3. Indeed we have:

$$\int_{E_3} \rho^{(q)} \, \mathrm{d}v = q \tag{12.2.3'}$$

In the case of more point charges q_i we have

$$\rho^{(q)} = \Sigma \, q_i \delta(\boldsymbol{x} - \boldsymbol{x}_i) \tag{12.2.4}$$

* The electric field created by two particles at some point P is the vector sum of the fields created by each particle at that point.

Let use now define

$$J_\alpha = \rho^{(q)} \frac{dx_\alpha}{dt}, \quad \alpha = 1, 2, 3, 4 \tag{12.2.5}$$

Noticing that from equation (12.2.3) it follows that

$$\rho^{(q)} \, dv = dq \tag{12.2.3''}$$

we find

$$dq \, dx_\alpha = J_\alpha \, dv \, dt = \frac{1}{ic} J_\alpha \, d\omega, \quad \alpha = 1, 2, 3, 4 \tag{12.2.6}$$

Since dq is an invariant of the Lorentz group, and dx_α is a quadrivector it follows that J_α is also a quadrivector. We shall call it the *current-charge density quadrivector.* With these specifications we have

$$S_{pf} = \mu_0 \int_{t_1}^{t_2} \int_{E_3} \rho^{(q)} \frac{dx_\alpha}{dt} A_\alpha \, dv \, dt = \frac{\mu_0}{ic} \int_\Omega J_\alpha A_\alpha \, d\omega \tag{12.2.7}$$

Applying the stationary condition to the functional (12.2.2) we should take into account the fact that A_α are the variables defining the field and so these are to undergo variations. Moreover, we shall assume that the variables vanish at large distance. Finally:

$$\begin{aligned} \delta(S_{pf} + S_f) = 0 &\Rightarrow \int_\Omega \left(\frac{\mu_0}{ic} J_\alpha \delta A_\alpha + 2\varkappa f_{\beta\alpha} \delta f_{\beta\alpha} \right) d\omega = 0 \\ &\Rightarrow \int_\Omega \left(J_\alpha \delta A_\alpha - 4ic\varkappa f_{\beta\alpha} \frac{\partial}{\partial x_\beta} \delta A_\alpha \right) d\omega = 0 \\ &\Rightarrow \int_\Omega (J_\alpha + 4ic\varkappa f_{\beta\alpha,\,\beta}) \delta A_\alpha \, d\omega = 0 \end{aligned} \tag{12.2.8}$$

where we used (12.1.4′) and the fact that

$$\int_\Omega f_{\beta\alpha} \frac{\partial}{\partial x_\beta} \delta A_\alpha \, d\omega = \int_\Omega \frac{\partial}{\partial x_\beta} (f_{\beta\alpha} \delta A_\alpha) \, d\omega - \int_\Omega f_{\beta\alpha,\,\beta} \delta A_\alpha \, d\omega$$

$$\int_\Omega \frac{\partial}{\partial x_\beta} (f_{\beta\alpha} \delta A_\alpha) \, d\omega = \int_\Sigma f_{\beta\alpha} \delta A_\alpha n_\beta \, d\Sigma = 0$$

since the variations $\delta A_\alpha = 0$ on a sphere Σ of very large radius.

Since δA_α are arbitrary variations it follows that

$$J_\alpha + 4ic\varkappa f_{\beta\alpha,\,\beta} = 0, \quad \alpha = 1, 2, 3, 4 \tag{12.2.9}$$

Introducing now the constant ε_0 by the equation:

$$\varepsilon_0\mu_0 c^2 = 1 \tag{12.2.10}$$

and choosing for $\varkappa$ the value given by $4ic\mu_0\varkappa = -1$, equation (12.2.7) becomes

$$f_{\beta\alpha,\beta} = \mu_0 J_\alpha, \quad \alpha = 1, 2, 3, 4 \tag{12.2.11}$$

From equation (12.2.11) we find (for $\alpha = 1, 2, 3$):

$$\operatorname{curl} \boldsymbol{H} = \boldsymbol{J} + \partial_t \boldsymbol{D}, \quad \boldsymbol{D} \overset{\text{def}}{=} \varepsilon_0 \boldsymbol{E} \tag{12.2.12}$$

and for $\alpha = 4$

$$\operatorname{div} \boldsymbol{D} = \rho^{(q)} \tag{12.2.13}$$

The equations (12.2.12) and (12.2.13) represent the second pair of equations for the determination of the field. Their correspondence with the macroscopic equations of Section 1.1 is obvious. We mention once more that these are microscopic quantities (in the absence of material media).

The complete expression of the Lagrangian density $L_p + L_{pf} + L_f$ was for the first time derived by M. Born and H. Weill in 1909 [1.3].

Finally we notice that in the absence of charges the field equations reduce to

$$\operatorname{curl} \boldsymbol{E} = -\mu_0 \partial_t \boldsymbol{H}, \quad \operatorname{curl} \boldsymbol{H} = \varepsilon_0 \partial_t \boldsymbol{E} \tag{12.2.14}$$

the other two following from these. Applying the *curl* operator to equations (12.2.14) we get

$$\left(\Delta - \frac{1}{c^2}\frac{\partial^2}{\partial t^2}\right)(\boldsymbol{E}, \boldsymbol{H}) = 0$$

hence we conclude that c represents the velocity of propagation of electromagnetic waves in vacuum.

12.3 Macroscopic quantities. The field equations for material media

12.3.1 Macroscopic quantities

Let us now consider a material medium in the presence of an electromagnetic field. As we know any material medium is made up of neutral entities (molecules and atoms) which in turn are made up of charged particles (electrons and protons). If for the determination of the charged-particle motion we use the results of the first section we shall encounter at least two major difficulties. The first will be the large number of these particles (and the imposibility of knowing their initial states) and the second the fact that these particles are not free but bound* in entities. For a given intensity of the field only some particles may be pulled out of their entities and carried away by the field as free charges the others remaining within the bound entity and undergoing only small variations from their usual course (in the entity field they have a determined motion).

Therefore in the case of material media one should distinguish between *free* charges which determine the conduction current $\boldsymbol{J}$ (and density $\rho^{(q)}$) and *bound* charges (which remain within the entity). The latter altering their motion under the action of the external field transforms the entity into a *dipole*: under the action of the electric field the positive charges will polarize at one end of the entity (in the direction of the field) and the negative ones at the opposite and (electric dipole); under the magnetic field action the electron trajectory planes will be orientated so that the molecular currents due to the electron motion will give a non-vanishing magnetic field (magnetic dipole). This phenomenon is called *polarization.* The electric and magnetic fields due to the polarization will combine with the external field giving rise to inhomogeneous and variable in time fields. These (time and space) non-uniformities will have atomic dimensions. In the macroscopic theory we shall not be interested in the inhomogeneities but in the resultant field only. Therefore we shall define the *macroscopic* field by mean values.

We define the mean value $\langle f(t, \boldsymbol{x})\rangle$ of some quantity $f(t, \boldsymbol{x})$ by the equation:

$$\langle f(t, \boldsymbol{x})\rangle = \frac{1}{\Delta v \Delta t} \int_{\Delta v} \int_{\Delta t} f(t + t', \boldsymbol{x} + \boldsymbol{x}')\, \mathrm{d}t'\, \mathrm{d}\boldsymbol{x}' \qquad (12.3.1)$$

Δv being a vicinity of the point P of position vector $\boldsymbol{x}$ and Δt a vicinity of time t. The vector $\boldsymbol{x}'$ having its origin at P belongs to the vicinity Δv and t' to the vicinity Δt. In (12.3.1) Δv and Δt are from the physical point of view infinitesimal quantities: sufficiently large compared with the atomic dimensions** (thus, compared with the microscopic inhomogeneities) and sufficiently small compared with the macroscopic inhomogeneities of the field and medium. So, the mean values differ very little from one point to another and thus they may be considered continuous and differentiable functions. One also proves [7] that *the derivatives of the mean values are equal to the mean values of the derivatives.* Obviously, as uniquely determined quantities, the mean values depend on the shape of the vicinities Δv and Δt.

To conclude, by a macroscopic quantity characterizing the field we mean the mean value of the corresponding microscopic quantity

$$f_{\text{macro}} = \langle f_{\text{micro}}\rangle \qquad (12.3.2)$$

in the sense specified above.

---

* The strength of the binding varies from one substance to the other. According to the nature of this binding the substances are of two classes: conductors and dielectrics. In a conductor any external field produces a motion of the charges thus giving rise to a current. As we mentioned in Section 1.1 this division is a relative one since there exist neither ideal conductors nor ideal dielectrics.

** If for instance Δv is a cube of side 10^{-2} cm it will contain approximately 10^{18} atoms (the volume of the atom being of the order of 10^{-24}cm^3).

12.3.2 The Maxwell equations

The problem of averaging the field equations is a very complex one. In this respect we shall give the solution of this problem due to de Groot and Vilieyer [8].

Let us consider the equations of the microscopic field in the presence of a system of particles P_i of position vectors $\boldsymbol{x}_i$ and charges q_i.

$$\operatorname{curl} \boldsymbol{E} = -\partial_t \boldsymbol{B}, \quad \operatorname{div} \boldsymbol{B} = 0 \tag{12.3.3}$$

$$\operatorname{curl} \boldsymbol{H} = \Sigma\, q_i \delta(\boldsymbol{x} - \boldsymbol{x}_i(t))\dot{\boldsymbol{x}}_i(t) + \partial_t \boldsymbol{D} = \boldsymbol{J}_{\text{tot}} + \partial_t \boldsymbol{D} \tag{12.3.4}$$

$$\operatorname{div} \boldsymbol{D} = \Sigma q_i \delta(\boldsymbol{x} - \boldsymbol{x}_i(t)) = \rho_{\text{tot}}^{(q)} \tag{12.3.5}$$

These equations have been established in the first two sections of this chapter. Let us assume the existence of stable groups of particles. We shall call these atoms and denote them by s. Their constituents (electrons and nuclei) will be denoted by l. Substituting in (12.3.4) and (12.3.5) i by sl we have

$$\boldsymbol{x}_{sl}(t) = \boldsymbol{x}_s(t) + \boldsymbol{x}'_{sl}(t)$$

$$\dot{\boldsymbol{x}}_{sl}(t) = \dot{\boldsymbol{x}}_s(t) + \dot{\boldsymbol{x}}'_{sl}(t)$$

$\boldsymbol{x}_s$ being the position vector of some point A_s within the atom and $\boldsymbol{x}'_{sl}$ the position vector of particles P_{sl} with respect to A_s. Since we are interested in the field produced by the atomic entity at some external point $\boldsymbol{x}$ we shall have (Figure 53)

$$|\boldsymbol{x}'_{sl}| < |\boldsymbol{x} - \boldsymbol{x}_s|$$

so that we may use the following series expansion

$$\delta(\boldsymbol{x} - \boldsymbol{x}_s - \boldsymbol{x}'_{sl}) =$$

$$\delta(\boldsymbol{x} - \boldsymbol{x}_s) - (\boldsymbol{x}'_{sl} \cdot \nabla_{\boldsymbol{x}})\, \delta(\boldsymbol{x} - \boldsymbol{x}_s) + \frac{1}{2}(\boldsymbol{x}'_{sl} \cdot \nabla_{\boldsymbol{x}})(\boldsymbol{x}'_{sl} \cdot \nabla_{\boldsymbol{x}})\, \delta(\boldsymbol{x} - \boldsymbol{x}_s) + \cdots$$

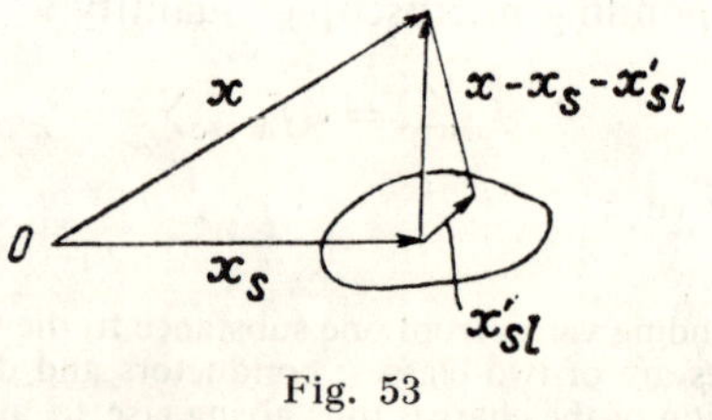

Fig. 53

With this we have

$$\rho_{\text{tot}}^{(q)} = \rho_0^{(q)} + \rho_1^{(q)} + \rho_2^{(q)} \tag{12.3.6}$$

where

$$\rho_0^{(q)} = \sum_s q_s\, \delta(\boldsymbol{x} - \boldsymbol{x}_s),\ q_s = \sum_l q_{sl}$$

$$\rho_1^{(q)} = -\operatorname{div}_{\boldsymbol{x}} \{\sum_s \boldsymbol{\mu}_s \delta(\boldsymbol{x} - \boldsymbol{x}_s)\},\ \boldsymbol{\mu}_s = \sum_l q_{sl}\boldsymbol{x}'_{sl} \tag{12.3.6'}$$

$$\rho_2^{(q)} = \operatorname{div}_{\boldsymbol{x}} \{\sum_s (\underline{\boldsymbol{\mu}}_s \cdot \nabla_{\boldsymbol{x}})\, \delta(\boldsymbol{x} - \boldsymbol{x}_s)\},\ \underline{\boldsymbol{\mu}}_s = \frac{1}{2} \sum_l q_{sl}\boldsymbol{x}'_{sl}\boldsymbol{x}'_{sl}$$

q_s (scalar) being the total charge of the atom (also called the electric moment of order zero), $\boldsymbol{\mu}_s$ (vector) the first order moment and $\underline{\boldsymbol{\mu}}_s$ (tensor) the second order moment.

So equation is written

$$\operatorname{div}(\boldsymbol{D} + \boldsymbol{P}) = \rho_0^{(q)} \tag{12.3.7}$$

$$\boldsymbol{P} \overset{\text{def}}{=} \sum_s (\boldsymbol{\mu}_s - \underline{\boldsymbol{\mu}}_s \cdot \nabla_{\boldsymbol{x}})\, \delta(\boldsymbol{x} - \boldsymbol{x}_s)$$

In a similar way, using notations

$$\boldsymbol{J} = \sum_s q_s\, \delta(\boldsymbol{x} - \boldsymbol{x}_s)\dot{\boldsymbol{x}}_s$$

$$\boldsymbol{M} = \sum_s \{\boldsymbol{\nu}_s - \dot{\boldsymbol{x}}_s \times (\boldsymbol{\mu}_s - \underline{\boldsymbol{\mu}}_s \cdot \nabla_{\boldsymbol{x}})\}\, \delta(\boldsymbol{x} - \boldsymbol{x}_s),\ \boldsymbol{\nu}_s = \frac{1}{2} \sum_l q_{sl}\boldsymbol{x}'_{sl} \times \dot{\boldsymbol{x}}'_{sl}$$

one finds

$$\boldsymbol{J}_{\text{tot}} = \boldsymbol{J} + \partial_t \boldsymbol{P} + \operatorname{curl}_{\boldsymbol{x}} \boldsymbol{M} + \frac{1}{2} \sum_{s,l} q_{sl}\dot{\boldsymbol{x}}'_{sl}(\boldsymbol{x}'_{sl} \cdot \nabla_{\boldsymbol{x}})(\boldsymbol{x}'_{sl} \cdot \nabla_{\boldsymbol{x}})\, \delta(\boldsymbol{x} - \boldsymbol{x}_s) + \ldots$$

Assuming that the products of the form $(\dot{\boldsymbol{x}}'_{sl})_i\ (\boldsymbol{x}'_{sl})_j\ (\boldsymbol{x}'_{sl})_k$ (i, j, k, being coordinates indices) which appear in the last term are small compared with the products appearing in the expression of $\boldsymbol{M}$ (in the term $\dot{\boldsymbol{x}}_s \times \underline{\boldsymbol{\mu}}_s \cdot \nabla_{\boldsymbol{x}}$) of the form $(\dot{\boldsymbol{x}}_s)_i\ (\boldsymbol{x}'_{sl})_j\ (\boldsymbol{x}'_{sl})_k$, the last term will be neglected. So, the equation (12.3.4) becomes

$$\operatorname{curl}(\boldsymbol{H} - \boldsymbol{M}) = \boldsymbol{J} + \partial_t(\boldsymbol{D} + \boldsymbol{P}) \tag{12.3.8}$$

The equations (12.3.3), (13.3.7) and (12.3.8) represent the electromagnetic field equations at the atomic level.

In order to derive the macroscopic equations we should average the microscopic equations (12.3.3), (12.3.7) and (12.3.8). Taking into account the fact that the averaging operation may be permuted with that of differentiation we get:

$$\operatorname{curl}\langle \boldsymbol{E}\rangle + \partial_t\langle \boldsymbol{B}\rangle = 0, \ \operatorname{div}\langle \boldsymbol{B}\rangle = 0 \tag{12.3.9}$$

$$\operatorname{div}(\langle \boldsymbol{D}\rangle + \langle \boldsymbol{P}\rangle) = \langle \rho_0^{(q)}\rangle \tag{12.3.10}$$

$$\operatorname{curl}(\langle \boldsymbol{H}\rangle - \langle \boldsymbol{M}\rangle) = \langle \boldsymbol{J}\rangle + \partial_t(\langle \boldsymbol{D}\rangle + \langle \boldsymbol{P}\rangle) \tag{12.3.11}$$

With the notations

$$\begin{aligned}
&\langle \boldsymbol{E}\rangle = \boldsymbol{E}(t, \boldsymbol{x}), \ \langle \boldsymbol{B}\rangle = \boldsymbol{B}(t, \boldsymbol{x})\\
&\mu_0\langle \boldsymbol{M}\rangle = \boldsymbol{M}(t, \boldsymbol{x}), \ \langle \boldsymbol{P}\rangle = \boldsymbol{P}(t, \boldsymbol{x})\\
&\langle \rho_0^{(q)}\rangle = \rho^{(q)}(t, \boldsymbol{x}), \ \langle \boldsymbol{J}\rangle = \boldsymbol{J}(t, \boldsymbol{x})\\
&\langle \boldsymbol{D}\rangle + \langle \boldsymbol{P}\rangle = \varepsilon_0\boldsymbol{E} + \boldsymbol{P} = \boldsymbol{D}(t, \boldsymbol{x})\\
&\mu_0\langle \boldsymbol{H}\rangle - \mu_0\langle \boldsymbol{M}\rangle = \boldsymbol{B} - \boldsymbol{M} = \mu_0\boldsymbol{H}(t, \boldsymbol{x})
\end{aligned} \tag{12.3.12}$$

equations (12.3.9) — (12.3.11) become

$$\begin{aligned}
&\operatorname{curl}\boldsymbol{E} + \partial_t\boldsymbol{B} = 0, \ \operatorname{div}\boldsymbol{B} = 0\\
&\operatorname{div}\boldsymbol{D} = \rho^{(q)}, \ \operatorname{curl}\boldsymbol{H} = \boldsymbol{J} + \partial_t\boldsymbol{D}
\end{aligned} \tag{12.3.13}$$

These are the Maxwell equations postulated in Section 1.1. While equations (12.3.3)—(12.3.5) refer to the microscopic quantities $\boldsymbol{E}, \boldsymbol{B}, \boldsymbol{D}, \boldsymbol{H}, \boldsymbol{J}$ and $\rho^{(q)}$, equations (12.3.13) refer to the macroscopic quantities $\boldsymbol{E}, \boldsymbol{B}, \boldsymbol{D}, \boldsymbol{H}, \boldsymbol{J}$ and $\rho^{(q)}$ defined by (12.2.2). From equation (12.3.12) one can see that the polarization vector $\boldsymbol{P}$ appears as the difference between the electric induction of the medium (dielectric) and the vacuum induction, while the magnetization vector $\boldsymbol{M}$ appears as the difference between the magnetic induction of the medium and the vacuum magnetic induction. In vacuum $\boldsymbol{P}$ and $\boldsymbol{M}$ vanish.

13

The kinetic theory of homogeneous gases

13.1 The Liouville equation. The BBGKY chain

13.1.1 Phase space

Let us consider some gas made up of particles (molecules, atoms, ions, electrons, etc.) reduced to their centre of inertia P_r, of position vectors $\boldsymbol{x}_r$ and velocities $\boldsymbol{v}_r$, $(r = 1, 2, \ldots, N)$. The representation of material particles by material points is justified by the theorem of the centre of inertia motion of classical mechanics. This means that the particle rotation with respect to their centre as well as their vibration are neglected. Otherwise the position and velocity of a particle could not be each described by three coordinates. For instance, six position parameters (three for the centre of inertia and three for orientation) and six velocity parameters (three for velocities of the centre and three for rotation) would be required for the case of a rigid particle. The above hypothesis is plausible enough for the case of a smooth particle of spherical symmetry. For this case the orientation and angular velocities are of no interest.

We suppose that the inter-particle distance is large compared with the particle dimensions so that their interaction is weak for most of the time. The strong interaction of the two particles will be called a collision.

In this chapter we shall suppose that all the particles have the same mass (m). A multiple-species (electrons, ions, atoms, etc.) gas will be presented in the next chapter.

For a single-species gas its evolution will be described by Newton's law:

$$\boldsymbol{x}_r = \boldsymbol{v}_r, \; m\boldsymbol{v}_r = \boldsymbol{X}_r + \sum_s \boldsymbol{X}_{rs}, \; r = 1, \ldots N \tag{13.1.1}$$

where $\boldsymbol{X}_r$ represents the external force resultant acting on the particle P_r and $\boldsymbol{X}_{rs}$ the action of particle P_s on P_r.

We assume that the interactions depend on the particle positions only:

$$\boldsymbol{X}_{rs} = \boldsymbol{X}_{rs}(\boldsymbol{x}_r, \boldsymbol{x}_s), \; \boldsymbol{X}_{rs} = \boldsymbol{X}_{sr}$$

We shall denote by Γ the $6N$ — dimensional phase space of representative point $Z = (\boldsymbol{x}_1, \ldots, \boldsymbol{x}_N;\ \boldsymbol{v}_1, \ldots, \boldsymbol{v}_N)$. The dimensional phase space of representative point $z_r = (\boldsymbol{x}_r, \boldsymbol{v}_r)$ will be denoted by γ_r or simply γ when the particle need not be specified.

With the notations:

$$(x_1, y_1, z_1, \ldots x_N, y_N, z_N, \dot{x}_1, \dot{y}_1, \dot{z}_1, \ldots \dot{x}_N, \dot{y}_N, \dot{z}_N)$$

$$= (Z_1\, Z_2\, Z_3 \ldots Z_{3N-2}, Z_{3N-1}, Z_{3N}, Z_{3N+1}, Z_{3N+2}, Z_{3N+3}, \ldots Z_{6N-2}, Z_{6N-1}\, Z_{6N}) \tag{13.1.2}$$

$$\left(\dot{x}_1, \dot{y}_1, \dot{z}_1, \ldots \dot{x}_N, \dot{y}_N, \dot{z}_N, \frac{X_1 + \Sigma X_{1s}}{m}, \frac{Y_1 + \Sigma\, Y_{1s}}{m}, \frac{Z_1 + \Sigma Z_{1s}}{m}, \ldots, \frac{Z_N + \Sigma Z_{Ns}}{m}\right)$$

$$= (W_1, W_2, W_3, \ldots, W_{3N-2}, W_{3N-1}, W_{3N}, W_{3N+1}, W_{3N+2}, W_{3N+3}, \ldots W_{6N})$$

system (13.1.1) becomes:

$$\dot{Z}_i = W_i,\ i = 1, \ldots, 6N \tag{13.1.3}$$

and has a solution of the form:

$$Z_i = Z_i(t, Z_1^0, \ldots, Z_{6N}^0),\ \mathrm{i} = 1 \ldots 6N \tag{13.1.3'}$$

(Z_i^0) representing the initial state of the system. In the space Γ equation (13.1.3′) defines a class C^1 transformation of the initial-states domain D_0 into the domain of time-t-states D_t.

The system (13.1.3) has the following fundamental property:

$$\frac{\partial W_i}{\partial Z_i} = 0,\ i = 1, \ldots, 6N \tag{13.1.4}$$

Indeed, the above property is obvious if the external forces do not depend on the velocity. Property (13.1.4) is also valid for Lorentzian-type external forces:

$$\boldsymbol{X}_r = q_r(\boldsymbol{v}_r \times \boldsymbol{B}),\ \boldsymbol{B} = \boldsymbol{B}(t, \boldsymbol{x}_r) \tag{13.1.5}$$

since a certain force coordinate does not depend on the velocity coordinate of the same kind. For instance

$$X_1 = q_1(\dot{y}_1 B_z - \dot{z}_1 B_y) \tag{13.1.6}$$

does not depend on $\dot{x}_1$.

13.1.2 The Liouville theorem

If we denote by J the Iacobian of transformation (13.1.3′), i.e.,

$$J = \frac{\partial(Z_1, \ldots, Z_{6N})}{\partial(Z_1^0, \ldots, Z_{6N}^0)} \tag{13.1.7}$$

we find, using equations (13.1.3) and (13.1.4):

$$\frac{\mathrm{d}J}{\mathrm{d}t} = J \sum_i \frac{\partial W_i}{\partial Z_i} = 0 \tag{13.1.8}$$

Indeed, the derivative of any determinant is a sum of determinants each with a differentiated line

$$\frac{\mathrm{d}J}{\mathrm{d}t} = \Sigma\, J_i, \; J_i = \frac{\partial(Z_1, \ldots, Z_{i-1}\dot{Z}_i, Z_{i+1}, \ldots Z_{6N})}{\partial(Z_1^0, \ldots \qquad , Z_{6N}^0)}$$

Using equation (13.1.3) and noticing that

$$\frac{\partial W_i}{\partial Z_j^0} = \sum_{k=0}^{6N} \frac{\partial W_i}{\partial Z_k} \frac{\partial Z_k}{\partial Z_j^0}$$

it follows that each determinant J_i breaks up into a sum of $6N$ determinants

$$J_i = \sum_k \frac{\partial W_i}{\partial Z_k} \frac{\partial(Z_1, \ldots, Z_{i-1}, Z_k, Z_{i+1}, \ldots Z_{6N})}{\partial(Z_1^0, \ldots \qquad Z_{6N}^0)}$$

Of these determinants only those corresponding to $k=i$ are different from zero and equal to J (the others have two identical lines). This proves equation (13.1.8).

From equation (13.1.8) it follows that $J = 1$. Indeed, J is a continuous function of t (transformation (13.1.3′) being of class C^1), its derivative vanishes and has the initial value of 1. Thus it follows that *the volume element of space Γ is conserved during the motion* (Liouville's theorem):

$$\Delta Z = \Delta Z^0$$

More generally, the volume of some continuum of representative points is conserved:

$$\frac{\mathrm{d}}{\mathrm{d}t} \int_{D_t} \mathrm{d}Z = \int_{D_0} \frac{\mathrm{d}J}{\mathrm{d}t}\, \mathrm{d}Z^0 = 0 \tag{13.1.9}$$

13.1.3 The probability density

In order to study the motion of the gas considered as an assembly of N material particles we have to integrate the system of equations (13.1.3). In doing this two major difficulties arise. The first is due to the large number of equations and the second to the practical and theoretical (from the quantum mechanical point of view) imposibility of knowing the initial state of the system (the position and velocity of each particle). Fortunately, from the macroscopic point of view, the individual motion of each particle is of no interest. Only their overall motion matters. Indeed, many microscopic motions consistent with the same macroscopic motion may exist. By *macroscopic motion we mean the one that may be experimentally determined*, i.e. the motion characterized by mean values.

We established that a microscopic state of the system is represented by a point in the phase space. The knowledge of the distribution of such points can provide information about the actual state of the system. It is very likely that the actual state will be found where more representative points of the microscopic states appear at time t. Thus we have to use statistical methods.

We consider a point $P(Z)$ in Γ and a volume element ΔZ about P. It is natural to assume that the probability of finding the representative point of the state of the system at time t in ΔZ is proportional to this volume. We denote this probability by $F_N(t, Z)\,\Delta Z$, $F_N(t, Z)$ being the probability density. It is obvious that the representative point of the material system is in ΔZ when particle 1 is in the vicinity Δz_1 about z_1 (it has the position in $\Delta \boldsymbol{x}_1$, about $\boldsymbol{x}_1$, and the velocity in $\Delta \boldsymbol{v}_1$, about $\boldsymbol{v}_1$), particle 2 is in the vicinity Δz_2 about $z_2, \ldots$, particle N is in the vicinity Δz_N about z_N. We shall thus write:

$$F_N(t, Z)\Delta Z = F_N(t, z_1, \ldots, z_N)\Delta z_1 \ldots \Delta z_N$$

$$= F_N(t, \boldsymbol{x}_1, \boldsymbol{v}_1, \ldots, \boldsymbol{x}_N, \boldsymbol{v}_N)\,\Delta \boldsymbol{x}_1\,\Delta \boldsymbol{v}_1 \ldots \Delta \boldsymbol{x}_N\,\Delta \boldsymbol{v}_N$$

F_N being a probability we have $F_N \geqslant 0$ and

$$\int_\Gamma F_N(t, Z)\,\mathrm{d}Z = 1 \qquad (13.1.10)$$

for any value of t since the representative point of the system certainly lies somewhere in the phase space. Also, we have that function F_N is symmetrical with respect to $z_1, \ldots, z_N$ since the particles of the system are identical.

It is obvious that the probability of finding the representative point of the system at time t in ΔZ is equal to the probability of finding the point at the initial time in ΔZ^0. Therefore the probability $F_N\Delta Z$ is conserved during the motion of the system. Since ΔZ is also conserved (Liouville's theorem) it follows that F_N itself is conserved. This means that F_N is an integral of motion (a first integral of the equations of motion (13.1.1)) and consequently its total time derivative is zero if the variables satisfy the equations of motion. Finally we have

$$\frac{\mathrm{d}F_N}{\mathrm{d}t} = \frac{\partial F_N}{\mathrm{d}t} + \sum_r \left(\frac{\partial F_N}{\partial \boldsymbol{x}_r} \cdot \boldsymbol{v}_r + \frac{\partial F_N}{\partial \boldsymbol{v}_r} \cdot \frac{\boldsymbol{X}_r + \sum_s \boldsymbol{X}_{rs}}{m} \right) = 0 \qquad (13.1.11)$$

This is *Liouville's equation.*

13.1.4 Distribution functions

The problem of finding the function F_N from the Liouville equation would lead us (according to the classical theory) to the integration of the characteristic system which is just the system of equations (13.1.1). Hence we shall not be concerned with the function F_N in what follows but with other functions obtained from F_N in the following way:

$$F_1(t, z_1) = \int F_N(t, z_1, \ldots z_N) \, \mathrm{d}z_2 \ldots \mathrm{d}z_N$$

$$F_2(t, z_1, z_2) = \int F_N(t, z_1, \ldots, z_N) \, \mathrm{d}z_3 \ldots \mathrm{d}z_N \qquad (13.1.12)$$

$$\cdots\cdots\cdots\cdots\cdots\cdots\cdots\cdots$$

$$F_r(t, z_1, \ldots z_r) = \int F_N(t, z_1, \ldots, z_N) \, \mathrm{d}z_{r+1} \ldots \mathrm{d}z_N$$

$$\cdots\cdots\cdots\cdots\cdots\cdots\cdots\cdots$$

We shall call F_1 a distribution function of order 1, F_2 a distribution function of order 2, F_r a distribution function of order r. In this case F_N is an N order distribution function.

There exist N distribution functions of order 1 defined as the integral of F_N over the positions and velocities of all points but one:

$$F_1(t, z_i) = \int F_N(t, z_1, \ldots, z_N) \, \mathrm{d}z_1 \ldots \mathrm{d}z_{i-1} \, \mathrm{d}z_{i+1} \ldots \mathrm{d}z_N$$

Since F_N is symmetrical with respect to $z_1 \ldots z_N$ it follows that all these functions are equal. Therefore F_1 does not depend on the chosen particle.

Taking into account the meaning of F_N it follows that $F_1(t, z_1) \, \Delta z_1$ is the probability of finding particle 1 at time t in the volume element Δz_1 about z_1; $F_2(t, z_1, z_2) \, \Delta z_1 \, \Delta z_2$ the probability of finding particle 1 in Δz_1 and particle 2 in Δz_2, etc. Particularly if the event of finding particle 1 within the unit volume about z_1 (at time t) is independent of the event of finding particle 2 within the unit volume about z_2 (at the same time t), then, according to the theorem of the probability of independent events we shall have

$$F_2(t, z_1, z_2) = F_1(t, z_1) \, F_1(t, z_2) \qquad (13.1.13)$$

This hypothesis (the molecular chaos hypothesis) is basic for the derivation of the Boltzmann equation.

We also note here that, from the interpretation of the function F_1, it follows that the number of particles which lie within the unit volume element about some point $z = (\boldsymbol{x}, \boldsymbol{v})$ at time t is given by the function

$$f(t, \boldsymbol{x}, \boldsymbol{v}) = NF_1(t, \boldsymbol{x}, \boldsymbol{v}), \tag{13.1.14}$$

This is a fundamental function of the kinetic theory. If instead of the particles which have their velocity within the unit volume element about $\boldsymbol{v}$ we consider all the particles (of any velocity) which at time t lie within the unit volume element about $\boldsymbol{x}$, then their number $n(t, \boldsymbol{x})$ (the particle density) is obtained from equation (13.1.14) integrating over the whole velocity space:

$$n(t, \boldsymbol{x}) = \int f(t, \boldsymbol{x}, \boldsymbol{v})\, \mathrm{d}\boldsymbol{v}, \quad \mathrm{d}\boldsymbol{v} = \mathrm{d}v_x \mathrm{d}v_y \mathrm{d}v_z \tag{13.1.15}$$

In Section 13.3 all the macroscopic quantities will be obtained using function f.

13.1.5 The BBGKY chain

We shall now attempt to find the equations obeyed by the distribution functions (13.1.12). In this respect we note that for both external forces independent of the velocity and Lorentzian forces the Liouville equation may be written as follows:

$$\frac{\partial F_N}{\partial t} + \sum_r \frac{\partial}{\partial \boldsymbol{x}_r} \cdot (F_N \boldsymbol{v}_r) + \frac{1}{m} \sum_r \frac{\partial}{\partial \boldsymbol{v}_r} \cdot \{F_N(\boldsymbol{X}_r + \sum_s \boldsymbol{X}_{rs})\} = 0 \tag{13.1.11'}$$

Concerning the function F_N, we shall assume that this vanishes at least for one space or velocity variable going to infinity. From the physical point of view this means that the gas particles lie within a finite volume and have finite velocities.

Let us now multiply equation (13.1.11') by $\mathrm{d}z_2 \ldots \mathrm{d}z_N$ and integrate with respect to these variables over the whole phase space. We note that the derivatives with respect to t, $\boldsymbol{x}_1$ and $\boldsymbol{v}_1$ may be permutted with the integration operator since the latter does not depend on these variables. More, the terms containing derivatives with respect to $\boldsymbol{x}_2, \ldots, \boldsymbol{x}_N, \boldsymbol{v}_2, \ldots, \boldsymbol{v}_N$ will be once integrated and then vanish due to the above hypothesis concerning F_N. Thus equation (13.1.11') becomes:

$$\frac{\partial F_1}{\partial t} + \boldsymbol{v}_1 \cdot \frac{\partial F_1}{\partial \boldsymbol{x}_1} + \frac{1}{m} \frac{\partial}{\partial \boldsymbol{v}_1} \cdot \int (\boldsymbol{X}_1 + \sum_s \boldsymbol{X}_{1s}) F_N\, \mathrm{d}z_2 \ldots \mathrm{d}z_N = 0 \tag{13.1.11''}$$

One should further note that the factor $\boldsymbol{X}_1$ does not depend on the integration variables: neither in the case when the external force is a Lorentzian one, nor in the case this force is a positional one. The factor X_1 may be taken outside the integral. Moreover, due to the symmetry of F_N the integrals with respect to the last term are equal to each other for various values of s :

$$\int X_{12}(x_1, x_2)\, F_N\, dz_2 \ldots dz_N = \int X_{13}(x_1, x_3)\, F_N\, dz_2 \ldots dz_N$$

$$= \int X_{12}(x_1, x_2)\left\{\int F_N\, dz_3 \ldots dz_N\right\} dz_2 = \int X_{12}\, F_2(t, z_1, z_2)\, dz_2$$

Thus equation (13.1.11″) becomes

$$\frac{\partial F_1}{\partial t} + v_1 \cdot \frac{\partial F_1}{\partial x_1} + \frac{X_1}{m} \cdot \frac{\partial F_1}{\partial v_1} + \frac{N-1}{m} \frac{\partial}{\partial v_1} \cdot \int X_{12} F_2 dz_2 = 0 \qquad (13.1.16)$$

One can similarly obtain equations for higher order functions:

$$\frac{\partial F_2}{\partial t} + \sum_{s=1,2} \left(v_s \cdot \frac{\partial F_2}{\partial x_s} + \frac{X_s + \sum\limits_{q=1,2} X_{sq}}{m} \cdot \frac{\partial F_2}{\partial v_s} \right)$$
$$+ \frac{N-2}{m} \sum_s \frac{\partial}{\partial v_s} \cdot \int X_{s3} F_3\, dz_3 = 0 \qquad (13.1.17)$$

$$\frac{\partial F_r}{\partial t} + \sum_{s=1}^{r} \left(v_s \cdot \frac{\partial F_r}{\partial x_s} + \frac{X_s + \sum\limits_{q=1}^{r} X_{sq}}{m} \cdot \frac{\partial F_r}{\partial v_s} \right)$$
$$+ \frac{N-r}{m} \sum_{s=1}^{r} \frac{\partial}{\partial v_s} \cdot \int X_{s,r+1} F_{r+1} dz_{r+1} = 0 \qquad (13.1.18)$$

X_{sq} (x_s, x_q) may be taken outside the integral for $q = 1, \ldots r$, since they do not depend on the integration variables $z_{r+1}, \ldots z_N$.

In this manner one obtains a chain of equations for the consecutive distribution functions, the equation determining the function F_r containing the higher-order distribution function F_{r+1} under the integral. The chain has N equations the last one being just Liouville's equation. This chain of equations is called the BBGKY (Bogoliubov, Born, Green, Kirkwood, Yvon) chain.

In order to be able to find a distribution function one should introduce some hypothesis to break up the BBGKY chain. Various degrees of approximation of the information will be obtained corresponding to the different hypotheses.

The roughest hypothesis will obviously be to break the chain at the first equation. This simply means that the particle interactions are neglected (this is a plausible enough hypothesis for a very dilute gas). In this case we get the equation:

$$\frac{\partial f}{\partial t} + \boldsymbol{v} \cdot \frac{\partial f}{\partial \boldsymbol{x}} + \frac{\boldsymbol{X}}{m} \cdot \frac{\partial f}{\partial \boldsymbol{v}} = 0 \tag{13.1.19}$$

which is the *collisionless Boltzmann equation*. This is identical to the Liouville equation for a single particle. Equation (13.1.19) may be used to describe the evolution of a charged particle gas in an external electromagnetic field (if the particle density is small enough in order not to modify the field).

If the particle density is such that their interactions cannot be neglected then the simplest hypothesis consists in neglecting the particle correlations. So, taking into account equation (13.1.16) and condition (13.1.13) and noting that $F_1(z_1)$ may be taken outside the integral we find:

$$\frac{\partial f}{\partial t} + \boldsymbol{v} \cdot \frac{\partial f}{\partial \boldsymbol{x}} + \frac{\boldsymbol{X} + \boldsymbol{X}'}{m} \cdot \frac{\partial f}{\partial \boldsymbol{v}} = 0$$

where

$$\boldsymbol{X}' = (N-1) \int \boldsymbol{X}_{12} F_1(z_2) \mathrm{d}z_2 = \frac{N-1}{N} \int \boldsymbol{X}_{12} n_2(t, \boldsymbol{x}_2)\, \mathrm{d}\boldsymbol{x}_2$$

is interpreted as a space charge field. This is the *Vlasov equation* which is identical to the collisionless Boltzmann equation provided that the term $\boldsymbol{X}'$ is included in the external forces.

Other, less restricting hypotheses lead us to more general equations (Boltzmann, Fokker-Planck, etc.) of which the Boltzmann equation is the most important.

13.2 The Boltzmann equation

The Boltzmann equation is based on the following classical hypotheses [1], [2]:
(1) the binary collisions hypothesis;
(2) the truncated action hypothesis;
(3) the molecular chaos hypothesis;
(4) the slow variation of F_1 hypothesis;

On account of the first hypothesis we assume that any particle is acted upon by a single particle only*. Hence, if $\boldsymbol{X}_{12}$ of (13.1.16) is non-vanishing, then both $\boldsymbol{X}_{13}$ and $\boldsymbol{X}_{23}$

* This hypothesis is rigourously valid only for a gas made up of spherical, inert molecules since their only interaction is expressed through a collision. However, since the inter-molecular distances are quite large we may quite well assume that the hypothesis is generally valid.

of (13.1.17) vanish so that the BBGKY chain breaks up at the following equation:

$$\frac{\partial F_2}{\partial t} + \boldsymbol{v}_1 \cdot \frac{\partial F_2}{\partial \boldsymbol{x}_1} + \boldsymbol{v}_2 \cdot \frac{\partial F_2}{\partial \boldsymbol{x}_2} + \frac{\boldsymbol{X}_1 + \boldsymbol{X}_{12}}{m} \cdot \frac{\partial F_2}{\partial \boldsymbol{v}_1} + \frac{\boldsymbol{X}_2 + \boldsymbol{X}_{21}}{m} \cdot \frac{\partial F_2}{\partial \boldsymbol{v}_2} = 0 \quad (13.2.1)$$

We may now use this equation to eliminate F_2 from (13.1.16).

According to the second hypothesis we shall assume that the internal forces are finite-range (range σ) forces. Then considering a sphere S_σ of centre $\boldsymbol{x}_1$ and radius σ we suppose that $X_{12} = 0$ for $|x_2 - x_1| \geqslant \sigma$ so that the integral of (13.1.16) is limited to the volume of the sphere S_σ only. With the following substitution

$$\boldsymbol{x} = \boldsymbol{x}_2 - \boldsymbol{x}_1 \quad (13.2.2)$$

equation (13) becomes:

$$\frac{\partial F_2}{\partial t} + \boldsymbol{v}_1 \cdot \frac{\partial F_2}{\partial \boldsymbol{x}_1} + \boldsymbol{w} \cdot \frac{\partial F_2}{\partial \boldsymbol{x}} + \frac{\boldsymbol{X}_1 + \boldsymbol{X}_{12}}{m} \cdot \frac{\partial F_2}{\partial \boldsymbol{v}_1} + \frac{\boldsymbol{X}_2 + \boldsymbol{X}_{21}}{m} \cdot \frac{\partial F_2}{\partial \boldsymbol{v}_3} = 0 \quad (13.2.3)$$

$$\boldsymbol{w} \overset{\text{def}}{=} \boldsymbol{v}_2 - \boldsymbol{v}_1$$

Multiplying equation (13.3) by $\mathrm{d}\boldsymbol{x}\, \mathrm{d}\boldsymbol{v}_2$ and integrating with respect to $\boldsymbol{x}$ over the volume S_σ and with respect to $\boldsymbol{v}_2$ over the whole velocity space, we get the last term of (13.1.16):

$$J \overset{\text{def}}{=} -\frac{1}{m}\int \boldsymbol{X}_{12} \cdot \frac{\partial F_2}{\partial \boldsymbol{v}_1} \mathrm{d}\boldsymbol{x}\, \mathrm{d}\boldsymbol{v}_2 = \int \boldsymbol{w} \cdot \frac{\partial F_2}{\partial \boldsymbol{x}} \mathrm{d}\boldsymbol{x}\, \mathrm{d}\boldsymbol{v}_2 \quad (13.2.4)$$

In order to get this result we assumed that the time and space changes of F_2 within the interaction region S_σ are negligible (this hypothesis may be justified considering the significance of F_2 [3], [4] and [5]) and that, within the same region, the external force effect is negligible compared with that of the collision (i.e., $\boldsymbol{X}_1 \ll \boldsymbol{X}_{12}$).Grad [1] clears up this approximation by introducing the concept of truncated function. The last term of (13.2.3) is rigourously zero since we have

$$\int (\boldsymbol{X}_2 + \boldsymbol{X}_{21}) \cdot \frac{\partial F_2}{\partial \boldsymbol{v}_2} \mathrm{d}\boldsymbol{x}\, \mathrm{d}\boldsymbol{v}_2 = \int \frac{\partial}{\partial \boldsymbol{v}_2} \cdot (\boldsymbol{X}_2 + \boldsymbol{X}_{21}) F_2 \mathrm{d}\boldsymbol{x}\, \mathrm{d}\boldsymbol{v}_2 = 0$$

Us'ng Gauss' theorem in (13.2.4) we may change to a surface integral, so that

$$J = \int \frac{\partial}{\partial \boldsymbol{x}} \cdot (\boldsymbol{w} F_2)\, \mathrm{d}\boldsymbol{x}\, \mathrm{d}\boldsymbol{v}_2 = \int_\Sigma \int_{(\boldsymbol{v}_2)} \boldsymbol{w} F_2 \cdot \boldsymbol{n}\, \mathrm{d}\Sigma\, \mathrm{d}\boldsymbol{v}_2 \quad (13.2.5)$$

where $\boldsymbol{\Sigma}$ is the surface of sphere S_σ and $\boldsymbol{n}$ its external normal.

According to the molecular chaos hypothesis we assume that if particle 2 lies on Σ (or outside) particles 1 and 2 are uncorrelated. Hence equation (13.1.13) is valid and (13.1.5) becomes:

$$J = \int_{\Sigma}\int_{(\boldsymbol{v}_2)} \boldsymbol{w} F_1(t, \boldsymbol{x}_1, \boldsymbol{v}_1) F_1(t, \boldsymbol{x}_2, \boldsymbol{v}_2) \cdot \boldsymbol{n} \mathrm{d}\Sigma \, \mathrm{d}\boldsymbol{v}_2 \qquad (13.2.6)$$

In order to calculate J we shall first hold $\boldsymbol{v}_2$ constant (and thus $\boldsymbol{w}$) and integrate over Σ. In this respect, let Π be the plane containing particle 1, perpendicular to $\boldsymbol{w}$. This plane divides surface Σ into two hemispheres: Σ^+ for which $\boldsymbol{w} . \boldsymbol{n} > 0$ and Σ^- for which $\boldsymbol{w} . \boldsymbol{n} < 0$. When particle 2 is on Σ^+ the collision is terminated and when particle 2 is on Σ^- the collision just starts. This will be indicated by substituting $\boldsymbol{x}_2^+$ for $\boldsymbol{x}_2$ on Σ^+ and $\boldsymbol{x}_2^-$ for $\boldsymbol{x}_2$ on Σ^-. With the notation $\omega = S_\sigma \cap \Pi$ we note that Σ^+ is projected on the upper part and Σ^- on the lower part of the disc ω. Hence we have:

$$\boldsymbol{w} \cdot \boldsymbol{n} \, \mathrm{d}\Sigma = w \, \mathrm{d}\omega, \text{ on } \Sigma^+$$

$$\boldsymbol{w} \cdot \boldsymbol{n} \, \mathrm{d}\Sigma = -w \, \mathrm{d}\omega, \text{ on } \Sigma^- \qquad (13.2.7)$$

$\mathrm{d}\omega$ being the area element on ω and $w = |\boldsymbol{w}|$. With b and ε for the polar coordinates of the projection of point $\boldsymbol{x}_2$ (when this lies on Σ) on ω, we find that $\mathrm{d}\omega = b\mathrm{d}b \, \mathrm{d}\varepsilon$ so that

$$J = \int_{(\boldsymbol{v}_2)}\int_{\omega} \{F_1(t, \boldsymbol{x}_1, \boldsymbol{v}_1) F_1(t, \boldsymbol{x}_2^+, \boldsymbol{v}_2) - F_1(t, \boldsymbol{x}_1, \boldsymbol{v}_1) F_1(t, \boldsymbol{x}_2^-, \boldsymbol{v}_2)\} w \mathrm{d}\omega \, \mathrm{d}\boldsymbol{v}_2$$

$$= \int_{(\boldsymbol{v}_2)}\int_0^{\sigma}\int_0^{2\pi} \{F_1(t, \boldsymbol{x}_1, \boldsymbol{v}_1) F_1(t, \boldsymbol{x}_2^+, \boldsymbol{v}_2) - F_1(t, \boldsymbol{x}_1, \boldsymbol{v}_1) F_1(t, \boldsymbol{x}_2^-, \boldsymbol{v}_2)\} w b \, \mathrm{d}b \, \mathrm{d}\varepsilon \, \mathrm{d}\boldsymbol{v}_2 \qquad (13.2.8)$$

Since outside the sphere S_σ particle 2 moves freely (both the incident and the emergent trajectories being straight lines) it follows that b and ε are just the impact parameters of the two-body problem (to be defined in the following subsection).

Finally, the hypothesis of slow variation of F_1 amounts to the assumption that F_1 as a function of t and $\boldsymbol{x}$ changes little during a collision. Denoting by t' the time, particle 2 reaches Σ^+ (thus, the time of collision termination) and by $\boldsymbol{x}_1'$, $\boldsymbol{v}_1'$, $\boldsymbol{x}_2'$, $\boldsymbol{v}_2'$ the parameters of the particles at this time, the last hypothesis amounts to the following:

$$F_1(t, \boldsymbol{x}_1, \boldsymbol{v}_1) F_1(t, \boldsymbol{x}_2^+, \boldsymbol{v}_2) = F_1(t', \boldsymbol{x}_1', \boldsymbol{v}_1') F_1(t', \boldsymbol{x}_2', \boldsymbol{v}_2')$$

$$\simeq F_1(t, \boldsymbol{x}_1, \boldsymbol{v}_1') F_1(t, \boldsymbol{x}_1, \boldsymbol{v}_2') \stackrel{\text{def}}{=} F_1(\boldsymbol{v}_1') F_1(\boldsymbol{v}_2')$$

$$F_1(t, \boldsymbol{x}_1, \boldsymbol{v}_1) F_1(t, \boldsymbol{x}_2^-, \boldsymbol{v}_2) \simeq F_1(t, \boldsymbol{x}_1, \boldsymbol{v}_1) F_1(t, \boldsymbol{x}_1, \boldsymbol{v}_2) \stackrel{\text{def}}{=} F_1(\boldsymbol{v}_1) F_1(\boldsymbol{v}_2)$$

The un-primed variables represent quantities prior to the collision. Thus J has the following final form:

$$J = \int \{F_1(v_1') F_1(v_2') - F_1(v_1) F_1(v_2)\} w \mathrm{d}\omega \, \mathrm{d}v_2 \tag{13.2.9}$$

The Boltzmann equation is obtained from equations (13.1.16) and (13.2.9) using the notation (13.1.14) and substituting N for $N-1$. One obtains:

$$\frac{\partial f}{\partial t} + v \cdot \frac{\partial f}{\partial x} + \frac{X}{m} \cdot \frac{\partial f}{\partial v} = \int (f' f_1' - f f_1) w b \, \mathrm{d}b \, \mathrm{d}\varepsilon \, \mathrm{d}v_2 \tag{13.2.10}$$

where the subscript 1 being left out (the test particle may be anyone) and the subscript 2, designating the particle interacting with the test particle, has been replaced by subscript 1. The following notations have also been used:

$$f = f(t, x, v), \; f_1 = f(t, x, v_1)$$

$$f' = f(t, x, v'), \; f_1' = f(t, x, v_1') \tag{13.2.10'}$$

The Boltzmann equation may also be written as

$$\mathscr{D} f = J(f f_1) \tag{13.2.10''}$$

with obvious notations.

As a matter of fact, one does not use in the kinetic theory the function f but its average $\langle f \rangle$ over a time interval centered about t:

$$\langle f \rangle = \frac{1}{\Delta t} \int_{\Delta t} f(t + \tau, x, v) \, \mathrm{d}\tau \tag{13.2.11}$$

But one can show [4] that this mean satisfies the same equation (13.2.10). The motivation of the use of $\langle f \rangle$ consists in the fact that the macroscopic values obtained from f should also be time averages since any practical measurement is performed over some time interval (the macroscopic values are experimentally measurable). Since the whole theory is based on the equation (13.2.10) it follows that its conclusions are also valid for the substitution of f with $\langle f \rangle$. This will be the procedure in defining the macroscopic quantities in Section 13.3.

13.2.1 The two-body problem

Let us now consider more thouroughly the problem of the interaction of two particle P_1 and P_2. Let X_{12} be the action performed by P_2 on P_1 and X_{21} the action of P_1 on

P_2. According to the principle of action and reaction $X_{12} = -X_{21}$ so that the equations of motion of the particles P_1 and P_2 are

$$m_1\ddot{x}_1 = -X_{21},\ m_2 x_2 = X_{21} \tag{13.2.12}$$

Multiplying the first equation by m_2 and the second by m_1 and subtracting one from the other one has

$$\mu\ddot{x} = X_{21}, \tag{13.2.13}$$

with the notations

$$\mu = \frac{m_1 m_2}{m_1 + m_2},\ x = x_2 - x_1$$

It follows that the second particle moves with respect to the first according to Newton's equation μ being the reduced mass. In the above considered case in which the interactions depend on the distance only we have:

$$X_{21} = X(r)\frac{x}{r},\ r = |x| \tag{13.2.14}$$

X being positive for the case of repulsive forces and negative for attractive ones. Such forces are conservative ones:

$$X_{21} = -\text{grad } V \tag{13.2.15}$$

with

$$V = \int_r^{\infty} X(r)\,\mathrm{d}r$$

In this case the energy equation (obtained by scalar multiplication of equation (13.2.13) by dx) gives:

$$\mathrm{d}\left(\frac{1}{2}\mu\dot{x}^2\right) = -\mathrm{d}V \tag{13.2.16}$$

and hence

$$\frac{1}{2}\mu\dot{x}^2 = -V + \mu h$$

h being a constant.

It is well known that the trajectory of a point subject to a force of type (13.2.14) is a planar trajectory and its areolar velocity is constant. These results are obtained from the equation $\boldsymbol{x} \times \dot{\boldsymbol{x}} = \boldsymbol{x}^0 \times \dot{\boldsymbol{x}}_0$ obtained by taking the vector product of equation (13.2.13) with $\boldsymbol{x}$. Hence we may use polar coordinates (r, θ) and write the following integrals of motion

$$|\boldsymbol{x} \times \dot{\boldsymbol{x}}| = r^2\dot{\theta} = C \tag{13.2.17}$$

(the areolar velocity integral)

$$\dot{r}^2 + r^2\dot{\theta} = -\frac{2}{\mu}V + 2h \tag{13.2.18}$$

(the energy integral).
Eliminating $\dot{\theta}$ from (13.2.18) we find

$$\frac{du}{d\theta} = \pm\sqrt{G(u)} \tag{13.2.19}$$

with the following notations:

$$u = \frac{1}{r}, \quad G(u) = \frac{2h}{C^2} - \frac{2}{\mu C^2}V(u) - u^2$$

The equation (13.2.19) determines the trajectory of particle P_2 in polar coordinates and then from (13.2.17) it follows that $\theta = \theta(t)$. The constants h and C are determined from the initial conditions.

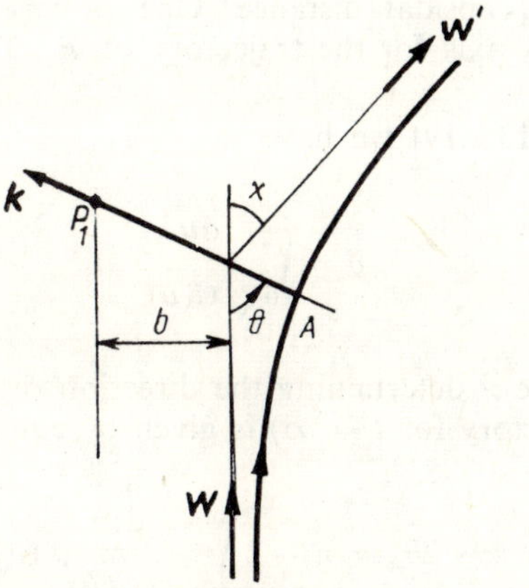

Fig. 54

Let us consider the time prior to the collision and corresponding to $r = \infty$ $(u = 0)$ as the initial time. Let us denote by b the distance between P_1 and the relative velocity $\boldsymbol{w}$ $(= \boldsymbol{v}_2 - \boldsymbol{v}_1)$ axis at the initial time.
Then from (13.2.17) it follows that $C = bw$, and from (13.2.16) that $2h = w^2$. The time integral during which P_2 goes from infinity to some position u is found from

(13.2.17) and (13.2.19):

$$t = \frac{1}{bw}\int_0 \frac{du}{u^2\sqrt{G(u)}} \tag{13.2.20}$$

the sign of the square root being positive since u increases with increasing t.

To have a real motion one should have $G(0) > 0$. If $G(u) > 0$ for any u, u increases infinitely (and $r \to 0$) with increasing t. Particle P_2 is said to be *trapped*. If equation $G(u) = 0$ has a solution u_A then one may write:

$$G(u) = (u - u_A)^\lambda\, g(u),\ g(u_A) \neq 0$$

If $\lambda \geqslant 2$ the integral of (13.2.20) diverges, so that u tends asymptotically to u_A. This is a trapping phenomenon as well. If $\lambda < 2$ then the integral of (13.2.20) converges. This means that there exists a finite time t_A for which $u = u_A$, u_A being the maximum value of u (and r_A the minimum value of r). At this position we have:

$$\frac{du}{d\theta} = 0,\ r\left(= \frac{c}{r^2}\frac{dr}{d\theta}\right) = 0$$

so that velocity $\boldsymbol{w}$ is perpendicular to the position vector (in polar coordinates the velocity is just $\dot{\theta}$ at this position). The position A corresponding to $u = u_A$ is called *apside* and r_A apsidal distance. One proves [4], [6] that the apsidal axis P_1A is a symmetry axis for the trajectory of P_2. This phenomenon is called *diffraction.*

Noting that from (13.2.19) we have

$$\theta = \int_0^u \frac{du}{\sqrt{G(u)}} \tag{13.2.21}$$

it follows that the angle $\varkappa$ determining the direction of diffraction (the direction asymptotic to the trajectory for $t \to \infty$) is given by equation

$$\varkappa = \pi - 2\theta_A = \pi - 2\int_0^{uA} \frac{du}{\sqrt{G(u)}},\quad G(u_A) = 0 \tag{13.2.22}$$

If, for instance, the particles P_1 and P_2 are two inert elastic spheres ($V = 0$) of radii R_1 and R_2, their interaction reduces to a collision so that the apsidal distance is just the distance between particle centres at the time of collision:

$$r_A = R_1 + R_2$$

In conclusion one has

$$\varkappa = \pi - 2\int_0^{u_A} \frac{du}{\sqrt{b^2 - u^2}} = \pi - 2 \arcsin \frac{b}{R_1 + R_2} \tag{13.2.23}$$

More generally, for an interaction of the form

$$X = \frac{k}{r^s} \left(V = \frac{k}{s-1} u^{s-1} \right) \tag{13.2.24}$$

the apsidal distance is given by the following equation

$$1 - v^2 - \frac{2}{s-1} \left(\frac{v}{\rho} \right)^{s-1} = 0 \tag{13.2.25}$$

where

$$v = bu, \; \rho = b \left(\frac{\mu w^2}{k} \right)^{\frac{1}{s-1}} \tag{13.2.25$'$}$$

v varying from 0 to ∞ when r varies from ∞ to 0. If $s > 1$ equation (13.2.25) has only one real positive solution v_A (except for $s = 3$ [4], [7]). In this case the diffraction phenomenon is produced:

$$\theta_A = \int_0^{v_A} \left\{ 1 - v^2 - \frac{2}{s-1} \left(\frac{v}{\rho} \right)^{s-1} \right\}^{-1/2} dv \tag{13.2.26}$$

or

$$\theta_A = \int_{r_0}^{\infty} \left\{ \frac{r^4}{b^2} - r^2 - \frac{2kr^{5-s}}{b^2 w^2 \mu (s-1)} \right\}^{-1/2} dr \tag{13.2.26$'$}$$

We see that the case $s = 5$ (Maxwellian interaction) plays a special role.

Since from (13.2.25′) we have

$$b^2 = \rho^2 \left(\frac{\mu w^2}{k} \right)^{\frac{2}{1-s}}$$

it follows that

$$wb\, db = \left(\frac{\mu w^2}{k} \right)^{\frac{2}{1-s}} w\rho\, d\rho = \left(\frac{k}{\mu} \right)^{\frac{2}{s-1}} w^{\frac{s-5}{s-1}} \rho\, d\rho \tag{13.2.27}$$

which expression is simplified for the case of Maxwellian interaction.

The calculation of the diffraction angle for various interactions may be found for instance in [6].

Integrating the energy equation along the trajectory and noting that V vanishes for $r \to \infty$ we get

$$w = w' \tag{13.2.28}$$

From the relationship $C = bw$ it follows that $b = b'$ since C has the same value along the trajectory. From (13.2.25) it follows that the relative velocity remains unchanged during the collision; only its direction is changed according to (13.2.22).

13.2.2 The binary collisions

The following quantities remain unchanged during the collision

$$m_1 + m_2,\ m_1\boldsymbol{v}_1 + m_2\boldsymbol{v}_2,\ m_1\boldsymbol{v}_1^2 + m_2\boldsymbol{v}_2^2 \tag{13.2.29}$$

that is, the mass, momentum and total energy. We may thus write:

$$m_1\boldsymbol{v}_1 + m_2\boldsymbol{v}_2 = m_1\boldsymbol{v}_1' + m_2\boldsymbol{v}_2' \tag{13.2.30}$$

$$m_1 v_1^2 + m_2 v_2^2 = m_1 v_1'^2 + m_2 v_2'^2 \tag{13.2.31}$$

We introduce the unit vector $\boldsymbol{k}$ having the direction along the direction of the change of velocity $\boldsymbol{v}_1$, i.e.,

$$\alpha\boldsymbol{k} = m_1(\boldsymbol{v}_1' - \boldsymbol{v}_1) \tag{13.2.32}$$

α being a scalar. From equation (13.2.30) it follows that $m_2(\boldsymbol{v}_2' - \boldsymbol{v}_2) = -\alpha\boldsymbol{k}$ and from (13.2.31), that $\alpha = 2\mu(\boldsymbol{w}.\boldsymbol{k})$.
Finally we have:

$$\begin{aligned} \boldsymbol{v}_1' &= \boldsymbol{v}_1 + \mu_2(\boldsymbol{w}\cdot\boldsymbol{k})\boldsymbol{k},\ \mu_2 \overset{\text{def}}{=} m_2/(m_1 + m_2) \\ \boldsymbol{v}_2' &= \boldsymbol{v}_2 - \mu_1(\boldsymbol{w}\cdot\boldsymbol{k})\boldsymbol{k},\ \mu_1 \overset{\text{def}}{=} m_1/(m_1 + m_2) \end{aligned} \tag{13.2.33}$$

Thus the velocities after collision which appear in f' and f_1' are expressed in terms of the velocities before the collision through parameters introduced by $\boldsymbol{k}$.

The interpretation of $\boldsymbol{k}$ follows imediately if we note that

$$\boldsymbol{w}' \overset{\text{def}}{=} \boldsymbol{v}_2' - \boldsymbol{v}_1' = \boldsymbol{w} - 2(\boldsymbol{w}\cdot\boldsymbol{k})\boldsymbol{k} \tag{13.2.34}$$

and thus

$$w' \cdot k = -(w \cdot k), \; w' \times k = w \times k$$

From the last equations if follows (taking into account (13.2.28)), that the projections of vectors w and w' on k are equal in magnitude and of opposite sign, while their projections on the perpendicular to k are equal in both magnitude and sign. It should also be noticed that the substitution $-k$ to k leaves the correspondence $(v_1, \; v_2) \rightarrow (v_1', \; v_2')$ unchanged. In order to make this a one-to-one correspondence we must fix the sign of α. For positive α ($w.\; k > 0$) it follows that the unit vector k is orientated along the direction of AP_1 (Figure 54). In conclusion $w.k$ is expressed according to the interaction law:

$$w \cdot k = w \cos \theta_A = w \sin (\varkappa/2) \tag{13.2.35}$$

From (13.2.33) and (13.2.35) it also follows that

$$v_1 = v_1' + 2\mu_2(w' \cdot k)k$$
$$v_2 = v_2' - 2\mu_1(w' \cdot k)k \tag{13.2.36}$$

and also that

$$\frac{\partial(v_1', \; v_2')}{\partial(v_1, \; v_2)} = 1 \tag{13.2.37}$$

13.3 Macroscopic quantities

13.3.1 Mean values. Transport of molecular quantities

The mean number of the particles lying in the interval Δt about t and the interval Δx about x will be denoted by $n(t, \; x)\Delta x$, $n(t, \; x)$ being the particle number density. From the interpretation of function f it follows:

$$n(t, x) = \int f(t, x, v) \, \mathrm{d}v, \tag{13.3.1}$$

the integration being performed over the whole velocity space. If m is the particle mass then the mass density (the mass of particles within in the unit volume) is $\rho = nm$.

Let $\Phi(t, \; x, \; v)$ be a quantity concerning the particle of velocity v, which lies within Δx during the time interval Δt. Since each of the $f\Delta v$ particles of velocity in $\Delta \boldsymbol{v}$ (situated in the time interval Δt about t within the unit volume about $\boldsymbol{x}$) provide the contribution Φ to the mean value $\langle \Phi \rangle$ it follows that the contribution of all these particles will be $\Phi f \Delta \boldsymbol{v}$. Thus the average value $\langle \Phi \rangle$ of quantity Φ will be defined by the equation

$$\langle \Phi \rangle = \frac{1}{n} \int \Phi f \, d\boldsymbol{v} \tag{13.3.2}$$

Particularly the mean value V will have the following expression:

$$V(t, x) = \frac{1}{n} \int \boldsymbol{v} f \, d\boldsymbol{v} \tag{13.3.3}$$

The difference $\boldsymbol{v}' = \boldsymbol{v} - V$ will be called the random velocity. It is obvious that $\langle \boldsymbol{v}' \rangle = 0$.

Let us now define *the transport of molecular quantities.* In this respect let ΔS be an orientated element and $\boldsymbol{\nu}$ its positive normal. We assume that this moves with velocity $V_0(t, \boldsymbol{x})$ and let $W = \boldsymbol{v} - V_0$. Those molecules whose velocities lie within $\Delta \boldsymbol{v}$ and which in the time interval Δt about t cross the surface element ΔS are those which at the beginning of the interval were situated within a cylinder of base ΔS and generatrix $W \Delta t$. The number of these particles is $f \Delta \boldsymbol{v} \Delta S W \cdot \boldsymbol{\nu} \Delta t$. As each particle carries Φ it follows that these particles all carry the quantity $\Phi f \Delta \boldsymbol{v} \Delta S W \cdot \boldsymbol{\nu} \Delta t$. Integrating over the whole velocity space we get the total flux of the quantity Φ through ΔS in the time interval Δt:

$$\Delta S \, \Delta t \int \Phi f (W \cdot \boldsymbol{\nu}) \, d\boldsymbol{v} = \Delta S \, \Delta t \, n \langle \Phi (W \cdot \boldsymbol{\nu}) \rangle$$

the flux per unit time and unit area being:

$$n \langle \Phi (W \cdot \boldsymbol{\nu}) \rangle = n \langle \Phi W \rangle \cdot \boldsymbol{\nu}$$

Particularly, if $\Phi = 1$ one gets the number of particles that cross the unit area of normal $\boldsymbol{\nu}$ in unit time. This number is

$$n \langle W \rangle \cdot \boldsymbol{\nu} = n (V - V_0) \cdot \boldsymbol{\nu}$$

If $V_0 \equiv V$ this vanishes. We note that the particle flux is given by the mean relative velocity as in the macroscopic theory.

The above definitions are completely valid for the case of a dilute gas (taking also into account the collisions what would make the problem very difficult).

13.3.2 The pressure tensor

The case $\Phi = m\boldsymbol{v}$ is of particular importance since the transfer of momentum produces the pressure. Indeed, each fluid particle colliding with a wall gives up to the latter some part of its momentum. For a sufficiently large number of collisions the momentum given up by the particles will give rise to a force acting on the wall and which we call *pressure* (the force per unit area). Therefore the pressure is a vector whose direction does not necessarly coincide with the normal at the surface under consideration. Quantitatively the momentum transferred to the wall is given by the formula:

$$\Delta S\,\Delta t \int m\boldsymbol{v}f(\boldsymbol{W}\cdot\boldsymbol{\nu})\,\mathrm{d}\boldsymbol{v} = \Delta S\,\Delta t\; nm\langle \boldsymbol{v}(\boldsymbol{W}\cdot\boldsymbol{\nu})\rangle$$

We have taken into account the fact that in the momentum of the particles incident on the wall $\boldsymbol{W}\cdot\boldsymbol{\nu} > 0$ while in the momentum of the particles reflected by the wall $\boldsymbol{W}\cdot\boldsymbol{\nu} < 0$, the momentum given up being the difference between the momentum of the particle before and after the collision, i.e. exactly what the above integral contains. Thus the pressure vector will have the following expression with respect to the orientation element $\boldsymbol{\nu}$

$$\boldsymbol{p}_\nu = nm\langle \boldsymbol{v}(\boldsymbol{W}\cdot\boldsymbol{\nu})\rangle = nm\langle \boldsymbol{v}\boldsymbol{W}\rangle\cdot\boldsymbol{\nu} \tag{13.3.4}$$

$\boldsymbol{v}\boldsymbol{W}$ being the dyadic product of the vectors $\boldsymbol{v}$ and $\boldsymbol{W}$. Within the fluid the pressure vector acting on ΔS in the positive sense represents the momentum flux in the positive sense through the unit area in unit time and is expressed by the same equation (13.3.4).

If the gas does not condense on the test wall then the number of incident particles is equal to the number of reflected particles. Therefore:

$$\int f(\boldsymbol{W}\cdot\boldsymbol{\nu})\,\mathrm{d}\boldsymbol{v} = 0 \Rightarrow \langle \boldsymbol{W}\rangle\cdot\boldsymbol{\nu} = 0 \tag{13.3.5}$$

From this conservation equation it follows that:

$$\langle \boldsymbol{v}\boldsymbol{W}\rangle\cdot\boldsymbol{\nu} = \langle(\boldsymbol{V}+\boldsymbol{v}')\boldsymbol{W}\rangle\cdot\boldsymbol{\nu} = \langle \boldsymbol{v}'\boldsymbol{W}\rangle\cdot\boldsymbol{\nu}$$

$$= \langle \boldsymbol{v}'(\boldsymbol{v}' + \boldsymbol{V} - \boldsymbol{V}_0)\rangle\cdot\boldsymbol{\nu} = \langle \boldsymbol{v}'\boldsymbol{v}'\rangle\cdot\boldsymbol{\nu}$$

so that the pressure vector becomes

$$\boldsymbol{p}_\nu = \rho\langle \boldsymbol{v}'\boldsymbol{v}'\rangle\cdot\boldsymbol{\nu} = \underline{\boldsymbol{P}}\cdot\boldsymbol{\nu} = \boldsymbol{\nu}\cdot\underline{\boldsymbol{P}} \tag{13.3.6}$$

$$\underline{\boldsymbol{P}} \overset{\text{def}}{=} \rho\langle \boldsymbol{v}'\boldsymbol{v}'\rangle$$

$\underline{P}$ being the *pressure tensor* (a symmetrical tensor). This is different from the stress tensor introduced in the macroscopic theory only through its sign. Obviously

$$P_{ij} = \int m(v_i - V_i)(v_j - V_j) f \, d\boldsymbol{v} \tag{13.3.7}$$

In order to define the pressure at a point P within the fluid a surface element moving with the fluid ($V = V_0$) is considered at P. Then equations (13.3.5) and (13.3.6) remain valid. The orientation of the normal is arbitrary.

The following equation

$$\boldsymbol{p}_\nu \cdot \boldsymbol{\nu} = \rho\langle v_i' v_j' \rangle \nu_j \nu_i = \rho\langle(\boldsymbol{\nu} \cdot \boldsymbol{v}')(\boldsymbol{\nu} \cdot \boldsymbol{v}')\rangle = \rho\langle(\boldsymbol{\nu} \cdot \boldsymbol{v}')^2\rangle 0$$

shows us that the perpendicular force exerted by the gas on a wall is always a pressure and not a traction.

The sum of the pressures (at point P) normal to three planes parallel to the (orthogonal) coordinate planes is called *the hydrostatic pressure*. This is given by the following equation:

$$\sum_{i=1,2,3} \boldsymbol{p}_i \cdot \boldsymbol{i}_i = \rho\langle v_1'^2 + v_2'^2 + v_3'^2 \rangle = \rho\langle v'^2 \rangle \tag{13.3.8}$$

The average hydrostatic pressure defines *the pressure* at point P. It has the expression

$$p = \frac{1}{3}\rho\langle v'^2 \rangle = \frac{1}{3}\underline{P} : \underline{U} \tag{13.3.9}$$

where $\underline{U}$ is the unit second order tensor.

If the non-diagonal components of $\underline{P}$ are zero and the diagonal ones are equal to each other, then $P_{ii} = p$ and

$$\underline{P} = p\underline{U} \tag{13.3.10}$$

whence

$$\boldsymbol{p}_\nu = p\underline{U} \cdot \boldsymbol{\nu} = p\boldsymbol{\nu} \tag{13.3.10'}$$

This shows that the pressure on a surface element is normal to this element. Its value does not depend on the orientation of the surface element and is equal to the hydrostatic pressure. Such systems will be called *hydrostatic* systems (in hydrostatics the pressure has all these properties). The ideal fluid is a hydrostatic system. It can be shown that for a system to be hydrostatic it is sufficient for the distribution function to be an even function of $\boldsymbol{v}$.

13.3.3 Kinetic energy. Heat. Temperature

The kinetic energy of a particle is $\frac{1}{2}mv^2$. The sum of the kinetic energy of the particles lying within the unit volume about x during Δt is given by:

$$n\left\langle \frac{1}{2}mv^2 \right\rangle = \frac{1}{2}\rho\langle(\boldsymbol{V}+\boldsymbol{v}')^2\rangle = \frac{1}{2}\rho V^2 + \frac{1}{2}\rho\langle v'^2\rangle \qquad (13.3.11)$$

The term $\frac{1}{2}\rho V^2$ gives us what is macroscopically called the kinetic energy density. The term $\frac{1}{2}\rho\langle v'^2\rangle$ is due to the deviation (caused by the random motion of the particles) of the particle velocities from the mean velocity. Macroscopically this term is identified as the thermal energy density or the heat density. The thermal energy corresponding to one particle is $\frac{1}{2}mv'^2$, so that the heat flux through the unit area per unit time is $n\left\langle \frac{1}{2}mv'^2\boldsymbol{W}\right\rangle\cdot\boldsymbol{\nu}$. As for the case of pressure we consider the velocity of the surface element to be equal to the mean velocity of the fluid (there is no particle transport through the surface). Then we have $\boldsymbol{W} = \boldsymbol{v}'$ and *the heat-flow vector* $\boldsymbol{q}$ is given by

$$\boldsymbol{q} = \frac{1}{2}\rho\langle v'^2\boldsymbol{v}'\rangle \qquad (13.3.12)$$

The absolute temperature T is defined through the equation

$$\frac{3}{2}kT = \frac{1}{2}m\langle v'^2\rangle \stackrel{\text{def}}{=} mU \qquad (13.3.13)$$

k being a universal constant (valid for all gases) and U the *internal energy per unit mass.*

The following relationship between pressure and temperature follows from equations (13.3.9) and (13.3.13):

$$p = nkT \qquad (13.3.14)$$

This equation contains Avogadro's hypothesis: equal volumes of different gases at the same pressure and temperature contain the same number of molecules.

If we denote by M the mass of gas contained within a volume $\mathscr{V}$ then the number of molecules per unit volume will be $(M/m)/\mathscr{V} = M/m\mathscr{V}$. Thus (13.3.14) becomes (the Boyle-Mariotte law)

$$p\mathscr{V} = k\left(\frac{M}{m}\right)T \qquad (13.3.14')$$

Finally, taking $M = 1$ and with the rotation $(k/m) = R$ equation (13.3.4') becomes

$$p = \rho RT \tag{13.3.15}$$

which is *the equation of state* of an (ideal) gas. Equation (13.3.15) is only an approximation of the equation of state of real gases (in its derivation we neglected the molecular dimensions, the long-range interactions, etc.) and is not valid, e.g. for high pressures and low temperatures.

13.4 The conservation equations

13.4.1 The transport equation. Additive invariants

Let us multiply the Boltzmann equation (13.2.10) by a velocity function $\Phi(\boldsymbol{v})$ and integrate over the whole velocity space $(\boldsymbol{v})$. One obtains

$$\int \Phi\, \mathscr{D} f \,\mathrm{d}\boldsymbol{v} = \int \Phi J(ff_1)\,\mathrm{d}\boldsymbol{v} \tag{13.4.1}$$

where

$$\int \Phi \frac{\partial f}{\partial t}\,\mathrm{d}\boldsymbol{v} = \frac{\partial}{\partial t}\int \Phi f\,\mathrm{d}\boldsymbol{v} - \int f \frac{\partial \Phi}{\partial t}\,\mathrm{d}\boldsymbol{v} = \frac{\partial}{\partial t} n\langle \Phi\rangle - n\left\langle \frac{\partial \Phi}{\partial t}\right\rangle$$

$$\int \Phi v_i \frac{\partial f}{\partial x_i}\,\mathrm{d}\boldsymbol{v} = \frac{\partial}{\partial x_i}\int \Phi v_i f\,\mathrm{d}\boldsymbol{v} - \int f v_i \frac{\partial \Phi}{\partial x_i}\,\mathrm{d}\boldsymbol{v} = \frac{\partial}{\partial x_i} n\langle \Phi v_i\rangle - n\left\langle v_i \frac{\partial \Phi}{\partial x_i}\right\rangle \tag{13.4.1'}$$

$$\int \Phi \frac{X_i}{m}\frac{\partial f}{\partial v_i}\,\mathrm{d}\boldsymbol{v} \overset{*}{=} \frac{1}{m}\int \Phi \frac{\partial}{\partial v_i}(X_i f)\,\mathrm{d}\boldsymbol{v}$$

$$\overset{**}{=} -\frac{1}{m}\int X_i f \frac{\partial \Phi}{\partial v_i}\,\mathrm{d}\boldsymbol{v} = -\frac{n}{m}\left\langle X_i \frac{\partial \Phi}{\partial v_i}\right\rangle \overset{***}{=} -\frac{n}{m} X_i \left\langle \frac{\partial \Phi}{\partial v_i}\right\rangle$$

The equality denoted by * is valid both for the case of $\boldsymbol{X}$ independent of $\boldsymbol{v}$ and for $\boldsymbol{X}$ being a Lorentzian force; the equality denoted by ** is obtained integrating by parts with respect to $\boldsymbol{v}_i$; and equality *** is valid only for the case when $\boldsymbol{X}$ does not depend on $\boldsymbol{v}$.

Let us calculate the term

$$I_\Phi \overset{\text{def}}{=} \int \Phi J(ff_1)\,\mathrm{d}\boldsymbol{v} = \int \Phi(f'f_1' - ff_1) wb\,\mathrm{d}b\,\mathrm{d}\varepsilon\,\mathrm{d}\boldsymbol{v}_1\,\mathrm{d}\boldsymbol{v} \tag{13.4.2}$$

Noting that the collision equations (13.2.30) and (13.2.31) are symmetrical with respect to $\boldsymbol{v}$, $\boldsymbol{v}_1$ and $\boldsymbol{v}'$, $\boldsymbol{v}_1'$ in the sense that if $\boldsymbol{v}'$, $\boldsymbol{v}_1'$ are velocities before the collision then $\boldsymbol{v}$, $\boldsymbol{v}_1$ are velocities after collision, we shall substitute $(\boldsymbol{v}, \boldsymbol{v}_1) \to (\boldsymbol{v}', \boldsymbol{v}_1')$ in I_Φ. Since $w' = w$ and $b' = b$ it follows that $w'b'\,db'\,d\varepsilon' = wb\,db\,d\varepsilon$. Taking into account (13.2.37) it also follows that $d\boldsymbol{v}'\,d\boldsymbol{v}_1' = d\boldsymbol{v}\,d\boldsymbol{v}_1$ so that

$$I_\Phi = -\int \Phi(\boldsymbol{v}')(f'f_1' - ff_1)w\,b\,db\,d\varepsilon\,d\boldsymbol{v}_1\,d\boldsymbol{v} \tag{13.4.3}$$

On the other hand a change of $\boldsymbol{v}$ with $\boldsymbol{v}_1$ implies a change of $\boldsymbol{v}'$ with $\boldsymbol{v}_1'$ (the particles having the same mass). Performing this change in (13.4.2) and (13.4.3) we have:

$$I_\Phi = \int \Phi(\boldsymbol{v}_1)(f'f_1' - ff_1)wb\,db\,d\varepsilon\,d\boldsymbol{v}_1\,d\boldsymbol{v} \tag{13.4.4}$$

$$I_\Phi = -\int \Phi(\boldsymbol{v}_1')(f'f_1' - ff_1)wb\,db\,d\varepsilon\,d\boldsymbol{v}_1\,d\boldsymbol{v} \tag{13.4.5}$$

From equations (13.4.2)—(13.4.5) it follows

$$I_\Phi = \frac{1}{4}\int (\Phi + \Phi_1 - \Phi' - \Phi_1')(f'f_1' - ff_1)wb\,db\,d\varepsilon\,d\boldsymbol{v}_1\,d\boldsymbol{v} \tag{13.4.6}$$

Equation (13.4.6) shows us that if

$$\Phi + \Phi_1 = \Phi' + \Phi_1' \tag{13.4.7}$$

then $I_\Phi = 0$. The functions obeying equation (13.4.7) are called *additive invariants*. The mass, momentum and kinetic energy

$$\Phi = m,\ mv_i,\ \frac{1}{2}mv^2 \tag{13.4.8}$$

are fundamental additive invariants.
The following theorem [8], [9] points out their importance:

"*any additive invariant is expressed as a linear combination of the invariants* [8]".

13.4.2 The conservation equations

For $\Phi = m$ equation (13.4.1) together with definitions of Section 13.3 gives the following equation

$$\frac{\partial \rho}{\partial t} + \frac{\partial}{\partial x_i}(\rho V_i) = 0 \tag{13.4.9}$$

which is the continuity equation (*mass conservation*).

In a similar way for $\Phi = mv_j$ ($j = 1, 2, 3$) we get

$$\frac{\partial}{\partial t}(\rho V_j) + \frac{\partial}{\partial x_i}\rho\langle v_i v_j\rangle - \frac{X_j}{m}\rho = 0,\ j = 1, 2, 3 \qquad (13.4.10)$$

But

$$\rho\langle v_i v_j\rangle = \rho\langle (V_i + v_i')(V_j + v_j')\rangle = \rho V_i V_j + P_{ij}$$

so that equation (13.4.10) becomes (using equation (13.4.9) as well):

$$\left(\frac{\partial}{\partial t} + V_i\frac{\partial}{\partial x_i}\right)V_j = \frac{1}{m}X_j - \frac{1}{P}\frac{\partial P_{ij}}{\partial x_i},\ j = 1, 2, 3 \qquad (13.4.10')$$

which is the equation of *momentum conservation*. If $\boldsymbol{F}$ represents the force per unit mass (as in the macroscopic theory) then we shall put $\boldsymbol{X} = m\boldsymbol{F}$ in (13.4.10')

Finally, for $\Phi = \dfrac{1}{2}mv^2$ we find from (13.4.1):

$$\frac{1}{2}\frac{\partial}{\partial t}\rho\langle v^2\rangle + \frac{1}{2}\frac{\partial}{\partial x_i}\rho\langle v^2 v_i\rangle - \frac{\rho}{m}\boldsymbol{X}\cdot\boldsymbol{V} = 0 \qquad (13.4.11)$$

But

$$\frac{1}{2}\rho\langle v^2\rangle = \frac{1}{2}\rho\langle(\boldsymbol{V} + \boldsymbol{v}')^2\rangle = \frac{1}{2}\rho V^2 + \frac{1}{2}\rho\langle v'^2\rangle = \frac{1}{2}\rho V^2 + \rho U$$

$$\frac{1}{2}\rho\langle v^2 v_i\rangle = \frac{1}{2}\rho\langle(\boldsymbol{V} + \boldsymbol{v}')^2(V_i + v_i')\rangle = \left(\frac{1}{2}V^2 + U\right)\rho V_i + V_j P_{ji} + q_i$$

Thus equation (13.4.11) becomes

$$\frac{\partial}{\partial t}\rho\left(\frac{1}{2}V^2 + U\right) + \frac{\partial}{\partial x_i}\left\{\left(\frac{1}{2}V^2 + U\right)\rho V_i + V_j P_{ji} + q_i\right\} = \frac{\rho}{m}\boldsymbol{X}\cdot\boldsymbol{V} \qquad (13.4.11')$$

which has obviously the form of a conservation equation.

Multiplying (13.4.10') by V_j and summing we get

$$\rho\frac{\partial}{\partial t}\left(\frac{1}{2}V^2\right) + \rho V_i\frac{\partial}{\partial x_i}\left(\frac{1}{2}V^2\right) = \frac{\rho}{m}\boldsymbol{X}\cdot\boldsymbol{V} - V_j\frac{\partial P_{ij}}{\partial x_i}$$

or, using equation (13.4.9) as well,

$$\frac{\partial}{\partial t}\left(\frac{1}{2}\rho V^2\right) + \frac{\partial}{\partial x_i}\rho V_i\left(\frac{1}{2}V^2\right) = \frac{\rho}{m}\boldsymbol{X}\cdot\boldsymbol{V} - V_j\frac{\partial P_{ij}}{\partial x_i} \tag{13.4.12}$$

Taking into account equation (13.4.12) and the continuity equation (13.4.9), equation (13.4.11′) becomes

$$\rho\left(\frac{\partial}{\partial t} + V_i\frac{\partial}{\partial x_i}\right)U = -\frac{\partial q_i}{\partial x_i} - P_{ji}\frac{\partial V_j}{\partial x_i},\quad U = \frac{3}{2}RT \tag{13.4.13}$$

This is the *energy equation.*

13.5 The Maxwellian distribution. The H-theorem

13.5.1 The Maxwellian distribution

Let us consider the collision term $J(f, f_1)$ of the Boltzmann equation and show that the necessary and sufficient condition for this term to vanish is

$$f'f_1' = ff_1 \tag{13.5.1}$$

The sufficiency of condition (13.5.1) follows immediately from the expression of $J(ff_1)$. In order to show the necessity we multiply this term by $(1 + \ln f)$, integrate with respect to $\boldsymbol{v}$ and take into account (13.4.6). We obtain:

$$I_{(1+\ln f)} = -\frac{1}{4}\int \ln\frac{f'f_1'}{ff_1}(f'f_1' - ff_1)w\mathrm{d}\omega\,\mathrm{d}\boldsymbol{v}_1\,\mathrm{d}\boldsymbol{v} \overset{\text{def}}{=} -G \tag{13.5.2}$$

Since $\ln(f'f_1'/ff_1)$ and $f'f_1' - ff_1$ have the same sign for any of the variables upon which they depend it follows that, in order to have $G = 0$ the condition (13.5.1) should be fulfilled.

Condition (13.5.1) may be written

$$\ln f + \ln f_1 = \ln f' + \ln f_1' \tag{13.5.1′}$$

which shows that the function $\ln f$ is an additive invariant. But, according to a theorem given above any additive invariant is a linear combination of the five fundamental invariants. Therefore, it follows:

$$\ln f = -\alpha(\boldsymbol{v} - \boldsymbol{\beta})^2 + \ln\gamma \tag{13.5.3}$$

or

$$f = \gamma\,\mathrm{e}^{-\alpha(\boldsymbol{v}-\boldsymbol{\beta})^2} \tag{13.5.3′}$$

with the coefficients α, $\boldsymbol{\beta}$, γ possible functions of t and $\boldsymbol{x}$. Using the definitions of the macroscopic quantities of Section 13.33 we shall express these coefficients in terms of n, V_i and T. Indeed, using polar coordinates (in the velocity space) $\rho = |\boldsymbol{v} - \boldsymbol{\beta}|$ θ and φ we find:

$$n = \gamma \int e^{-\alpha(\boldsymbol{v}-\boldsymbol{\beta})^2} \mathrm{d}(\boldsymbol{v} - \boldsymbol{\beta}) = \gamma \int_0^\infty \rho^2 e^{-\alpha\rho^2} \mathrm{d}\rho \int_0^\pi \sin\theta \, \mathrm{d}\theta \int_0^{2\pi} \mathrm{d}\varphi = \gamma \left(\frac{\pi}{\alpha}\right)^{3/2}$$

whence

$$\alpha > 0, \; \gamma = \left(\frac{\alpha}{\pi}\right)^{3/2} n \tag{13.5.4}$$

We also obtain

$$\boldsymbol{V} = \frac{\gamma}{n} \int \boldsymbol{v} \, e^{-\alpha(\boldsymbol{v}-\boldsymbol{\beta})^2} \mathrm{d}\boldsymbol{v} = \frac{\gamma}{n} \int (\boldsymbol{\rho} + \boldsymbol{\beta}) e^{-\alpha\rho^2} \mathrm{d}\boldsymbol{\rho} = \boldsymbol{\beta} \frac{\gamma}{n} \left(\frac{\pi}{\alpha}\right)^{3/2} = \boldsymbol{\beta}$$

since for r, an integer, [1] one has

$$\int_0^\infty e^{-\alpha\rho^2} \rho^r \mathrm{d}\rho = \begin{cases} \dfrac{\sqrt{\pi}}{2} \dfrac{1}{2} \dfrac{3}{2} \cdots \dfrac{r-1}{2} \alpha^{-\frac{r+1}{2}}, & r \text{ even} \\[2ex] \dfrac{1}{2} \alpha^{-\frac{r+1}{2}} \left(\dfrac{r-1}{2}\right)!, & r \text{ odd} \end{cases} \tag{13.5.5}$$

the calculations for $r = 0$ and 1 being very simple. Finally, using equation (13.3.13) as well, it follows:

$$3kT = \gamma \frac{m}{n} \int v'^2 e^{-\alpha v'^2} \mathrm{d}\boldsymbol{v}' = \left(\frac{\alpha}{\pi}\right)^{3/2} 4\pi m \int_0^\infty v'^4 e^{-\alpha v'^2} \mathrm{d}v' = \frac{3m}{2\alpha}$$

which gives $\alpha = m/2kT = 1/2RT$.

Thus it follows that the solution of equation (13.5.1) is $f = f_0$ where

$$f_0 = n \left(\frac{1}{2\pi RT}\right)^{3/2} e^{-\frac{1}{2RT}(\boldsymbol{v}-\boldsymbol{V})^2} \tag{13.5.6}$$

This is the *Maxwell-Boltzmann distribution*. It obeys equation $\mathscr{D} f_0 = 0$.

If n, $\boldsymbol{V}$ and T are constants the function f_0 represents the *equilibrium* state of a gas. Indeed, f_0 identically satisfies Boltzmann's equation with $\boldsymbol{X} \equiv 0$. In this case the Maxwell-Boltzmann distribution is said to be *absolute*.

13.5.2 Hydrostatic systems

If n, V and T depend on t and $\boldsymbol{x}$ then the gas is said to be *locally Maxwellian* (the five quantities, n, V_i and T, may be considered, in the vicinity of any point and at any moment, to be constant and equal to their values at the considered point and time). In this case the Boltzmann equation is only locally obeyed.

Let us now show that a gas characterized by f_0 is a hydrostatic system. To do this we have to calculate the pressure tensor:

$$P_{ij}^0 = m\gamma \int v_i' v_j' \,\mathrm{e}^{-\alpha v'^2} \,\mathrm{d}\boldsymbol{v}' \tag{13.5.7}$$

For $i = j$ we have

$$P_{11}^0 = P_{22}^0 = P_{33}^0 = \frac{4}{3}\pi m\gamma \int_0^\infty v'^4 \mathrm{e}^{-\alpha v'^2} \,\mathrm{d}v' = nkT = p$$

using spherical coordinates and equation (13.5.5). For $i \neq j$ the integral of (13.5.7) vanishes since the integrand is an odd function of v_i' and v_j'. Finally we get:

$$P_{ij}^0 = p\delta_{ij} \tag{13.5.8}$$

which proves the above statement.

Since the integrand is an odd function we also have

$$q_i^0 = \frac{1}{2} m\gamma \int v'^2 v_i' \,\mathrm{e}^{-\alpha v'^2} \,\mathrm{d}\boldsymbol{v}' = 0$$

which shows the conservation equations (13.4.9), (13.4.10) and (13.4.13) reduce to the Euler equations (for an ideal fluid)

$$\begin{gathered} \frac{\mathrm{d}\rho}{\mathrm{d}t} + \rho \operatorname{div} \boldsymbol{V} = 0, \quad \frac{\mathrm{d}}{\mathrm{d}t} = \frac{\partial}{\partial t} + V_i \frac{\partial}{\partial x_i} \\ \frac{\mathrm{d}\boldsymbol{V}}{\mathrm{d}t} = \frac{1}{m}\boldsymbol{X} - \frac{1}{\rho}\nabla p, \quad \frac{\mathrm{d}}{\mathrm{d}t}\frac{3}{2}RT = -\frac{p}{\rho}\operatorname{div} \boldsymbol{V} \end{gathered} \tag{13.5.9}$$

13.5.3 The H-theorem

In Section 13.1 we proved that F_N is an invariant of the equations of motion (13.1.1). The distribution function f also has this property, as well as any function of f. Therefore

$$\mathscr{H}(t, x) \overset{\text{def}}{=} \int f \ln f \,\mathrm{d}\boldsymbol{v} \tag{13.5.10}$$

is a invariant of the equations of motion. The function $\mathscr{H}$ has an interesting property which can be pointed out by multiplying the Boltzmann equation by $(1 + \ln f)$ and integrating with respect to $\boldsymbol{v}$. We note that:

$$(1 + \ln f)\frac{\partial f}{\partial t} = \frac{\partial}{\partial t}(f \ln f), \; (1 + \ln f)v_i\frac{\partial f}{\partial x_i} = \frac{\partial}{\partial x_i}(v_i f \ln f)$$

$$(1 + \ln f)\frac{X_i}{m}\frac{\partial f}{\partial v_i} = \frac{1}{m}\frac{\partial}{\partial v_i}(X_i f \ln f)$$

(the last equation being also valid for Lorentzian forces). Then we have:

$$\frac{\partial \mathscr{H}}{\partial t} + \frac{\partial \mathscr{H}_i}{\partial x_i} = -G \leqslant 0, \; \mathscr{H}_i \overset{\text{def}}{=} \int v_i f \ln f \, d\boldsymbol{v} \qquad (13.5.11)$$

Integrating this relationship over a domain D of the Euclidean space x and supposing that there is no $\mathscr{H}$-flux through ∂D we get *Boltzmann's H-theorem* [1], [2], [10]:

$$\frac{dH}{dt} \leqslant 0, \; H \overset{\text{def}}{=} \int_D \mathscr{H} d\boldsymbol{x} \qquad (13.5.12)$$

Therefore, for such a container D the function H does not increase. Particularly, if the system is in equilibrium ($G = 0$) then H is a constant. In this case we have:

$$\mathscr{H}_0 = n\langle \ln f_0 \rangle = n\left\{\ln n\left(\frac{1}{2\pi RT}\right)^{3/2} - \frac{1}{2RT}\langle v'^2\rangle\right\} = -\frac{nm}{k}S$$

$$S \overset{\text{def}}{=} -\frac{k}{m}\left\{\ln n\left(\frac{1}{2\pi RT}\right)^{3/2} - \frac{3}{2}\right\} = -R\ln\frac{\rho}{T^{3/2}} + S_0 \qquad (13.5.13)$$

S being the *macroscopic thermodynamic entropy* ($c_v/R = 3/2$) and S_0 a constant. For absolute equilibrium $\mathscr{H}_0$ does not depend on $\boldsymbol{x}$ so that

$$H_0 = \int_D \mathscr{H}_0 \, d\boldsymbol{x} = \mathscr{H}_0\frac{M}{nm} = -\frac{M}{k}S \qquad (13.5.14)$$

From the H-theorem it follows that *the motion is isentropic*. The same result can also be obtained from (13.5.9) eliminating $\operatorname{div} V$ from the energy equation (using the continuity equation) and taking into account equation (13.3.15). We have

$$\frac{3}{2}\frac{1}{T}\frac{dT}{dt} = \frac{1}{\rho}\frac{d\rho}{dt} \Rightarrow \frac{dS}{dt} = 0.$$

14

The theory of mixtures. The plasma

14.1 The Boltzmann equation

Let us consider a mixture made up of some particle species denoted by s. The plasma is such a mixture whose components are *neutral particles* (molecules and atoms) *ions* and *electrons* indicated by a, i and e, respectively. Since some of the particles carry a non-zero electric charge they will interact with each other through Coulomb forces. Moreover, in the presence of an electromagnetic field, the Lorentz force (12.1.9) should be included in the expression of $\boldsymbol{X}$. The concept of plasma was introduced by Langmuir in 1923 to designate an ionized gas, globally neutral or quasi-neutral.

We shall denote by $n_s(t, \boldsymbol{x})$ the number of particles of species s which, during the time interval Δt about t, lie inside the unit volume about $\boldsymbol{x}$ and by f_s the corresponding distribution function. We have:

$$n_s(t, \boldsymbol{x}) = \int f_s \, \mathrm{d}\boldsymbol{v}_s \tag{14.1.1}$$

The gas evolution will be determined by the set of distribution functions $\{f_s\}$.

A certain distribution function f_s will be determined by the Boltzmann equation, derived as in the last chapter, with two specifications:

(1) If the mixture component has an electrical charge then a Lorentz-force term will be introduced into the expression of the external force. Thus $\boldsymbol{X}$ will be replaced by the following:

$$m_s\boldsymbol{X} + q_s(\boldsymbol{E} + \boldsymbol{v}_s \times \boldsymbol{B}) \tag{14.1.2}$$

where $\boldsymbol{X}$ represents the mechanical force (per unit mass), $\boldsymbol{E}$ the applied electric field and $\boldsymbol{B}$ the applied magnetic induction. We have:

$$\boldsymbol{X} = \boldsymbol{X}(t, \boldsymbol{x}_s), \ \boldsymbol{E} = \boldsymbol{E}(t, \boldsymbol{x}_s), \ \boldsymbol{B} = \boldsymbol{B}(t, \boldsymbol{x}_s)$$

At various stages in the derivation of the Boltzmann equation we have pointed out that the operations were also valid for Lorentz-type forces.

(2) We have to include in the collision term both the collision of the test particle s with a particle of the species, but of velocity $\boldsymbol{v}_{s1}$, and the collision of the test particle s with a particle of another species r. Therefore the Boltzmann equation becomes:

$$\mathscr{D}_s f_s = J(f_s f_{s1}) + \sum_{r \neq s} J(f_s f_r) \tag{14.1.3}$$

where

$$\mathscr{D}_s = \frac{\partial}{\partial t} + \boldsymbol{v}_s \cdot \frac{\partial}{\partial \boldsymbol{x}} + \left\{ \boldsymbol{X} + \frac{q_s}{m_s} (\boldsymbol{E} + \boldsymbol{v}_s \times \boldsymbol{B}) \right\} \cdot \frac{\partial}{\partial \boldsymbol{v}_s}$$

$$J(f_s f_{s1}) = \int (f_s' f_{s1}' - f_s f_{s1}) |\boldsymbol{v}_s - \boldsymbol{v}_{s1}| \, d\omega \, d\boldsymbol{v}_{s1}$$

$$J(f_s f_r) = \int (f_s' f_r' - f_s f_r) |\boldsymbol{v}_s - \boldsymbol{v}_r| \, d\omega \, d\boldsymbol{v}_r$$

The after-collision velocities which appear in f' are obtained using equation (13.2.33) taking into account the fact that for $I(f_s f_{s1})$ the masses are equal to each other (m_s) while for $J(f_s f_r)$ the masses are different (m_s, m_r).

It is obvious that in (14.1.3) we have a system of integral-differential equations. For instance, for a plasma made up of electrons (e), ions (i) and neutral particles (a), system (14.1.3) becomes:

$$\begin{aligned} \mathscr{D}_e f_e &= J(f_e f_{e1}) + J(f_e f_i) + J(f_e f_a) \\ \mathscr{D}_i f_i &= J(f_i f_{i1}) + J(f_i f_e) + J(f_i f_a) \\ \mathscr{D}_a f_a &= J(f_a f_{a1}) + J(f_a f_e) + J(f_a f_i) \end{aligned} \tag{14.1.4}$$

The systems become simpler only in certain cases. For instance, if the gas is *weakly ionized (Lorentz plasma)* so that the charge - carrier densities (n_e and n_i) are negligible compared with the density (n_a) of the neutral particles, then the interactions (ee) and (ei) may be neglected compared with the (ea) interaction, the (ii) and (ie) interactions compared with (ia) and (ae) and (ai) compared with the (aa) interaction. As a matter of fact one can prove [1] that the term $I(f_s f_r)$ is proportional to $n_s n_r$. System (14.1.4) becomes in this case:

$$\mathscr{D}_e f_e = J(f_e f_a), \; \mathscr{D}_i f_i = J(f_i f_a), \; \mathscr{D}_a f_a = J(f_a f_{a1}) \tag{14.1.4'}$$

No Lorentz force appears in the last equation since the particles are neutral. One may thus assume that f_a is exactly the Maxwellian distribution. Then the first two equations are decoupled and can be linearized. This way the ion and electron evolution may be distinctly treated [2].

Another simpler case is that of the *fully ionized gas* for which the density n_a is negligible. In this case the (ea) and (aa) interactions are neglected. Considering also that the electron mass is negligible compared with the ion one we find that the (ie) interaction may be neglected. Thus system (14.1.4) is reduced to

$$\mathscr{D}_e f_e = J(f_e f_{e1}), \; \mathscr{D}_i f_i = J(f_i f_{i1}), \; \mathscr{D}_a f_a = 0 \tag{14.1.4''}$$

The above system may be further simplified in the case of a very dilute gas for which the interactions are practically zero. In this case we have:

$$\mathscr{D}_e f_e = 0, \ \mathscr{D}_i f_i = 0, \ \mathscr{D}_a f_a = 0 \tag{14.1.4'''}$$

14.2 Conservation equations

14.2.1 Mean values. Macroscopic quantities

The knowledge of the distribution functions makes it possible to determine the macroscopic properties of the mixture as well as its equations of evolution. Let us first define the macroscopic characteristics. Knowing $\{f_s\}$, $\{n_s\}$ follows and thus the density n of the mixture

$$n = \Sigma\, n_s \tag{14.2.1}$$

If m_s is the mass of the particle of species s it follows that $\rho_s = m_s n_s$ is the mass density of the s species and

$$\rho = \Sigma\, \rho_s = \Sigma\, n_s m_s \tag{14.2.2}$$

is the mass density of the mixture.

The mean value $\langle \Phi_s \rangle$ of a quantity Φ related to the species particle is defined by

$$\langle \Phi_s \rangle = \frac{1}{n_s} \int \Phi_s f_s \, \mathrm{d}\boldsymbol{v}_s \tag{14.2.3}$$

and the mean value $\langle \Phi \rangle$ of the whole assembly

$$\langle \Phi \rangle = \frac{1}{n}\, \Sigma\, n_s \langle \Phi_s \rangle = \frac{1}{n}\, \Sigma \int \Phi_s f_s \, \mathrm{d}\boldsymbol{v}_s \tag{14.2.4}$$

We also have the notation $\boldsymbol{V}_s = \langle \boldsymbol{v}_s \rangle$ and

$$\boldsymbol{V} = \frac{1}{\rho}\, \Sigma\, \rho_s \boldsymbol{V}_s = \frac{1}{\rho}\, \Sigma \int m_s \boldsymbol{v}_s f_s \mathrm{d}\boldsymbol{v}_s \tag{14.2.5}$$

$\boldsymbol{V}$ being the mean value of the (mass) velocity of the mixture at point $\boldsymbol{x}$ $\Big($this should not be mistaken for the mean velocity calculated according to (14.2.4): $\langle \boldsymbol{v} \rangle = \frac{1}{n}\, \Sigma n_s \boldsymbol{V}_s \Big)$.

During the calculations to follow we shall also use the following notations:

$$v_s' = v_s - V_s,\ v_s'' = v_s - V \tag{14.2.6}$$

Obviously

$$\langle v_s' \rangle = 0,\ \Sigma \rho_s \langle v_s'' \rangle = 0 \tag{14.2.6$'$}$$

The diffusion velocity of species s is defined by the flow of the s particles with respect to the mean velocity of the assembly, that is, by the equation

$$\frac{1}{n_s}\int (v_s - V) f_s \mathrm{d}v_s = \langle v_s \rangle - V = \langle v_s'' \rangle \tag{14.2.6$''$}$$

The mixture temperature is defined (like for the case of a homogeneous gas) by the relationship

$$\frac{3}{2} kT = \frac{1}{2} \langle mv''^2 \rangle = \frac{1}{n} \Sigma \frac{1}{2} \int m_s v_s''^2 f_s \mathrm{d}v_s \tag{14.2.7}$$

The total gas pressure is the sum of the partial pressures of the gas components Thus the pressure tensor will be written as

$$\underline{P} = \Sigma\, \underline{P}_s = \Sigma\, n_s m_s \langle v_s'' v_s'' \rangle = n \langle m v'' v'' \rangle \tag{14.2.8}$$

Particularly, the mean hydrostatic pressure of the gas, $\frac{1}{3}\underline{P} : \underline{U}$, has, according to equations (14.2.7) and (14.2.8), the following form

$$p = \frac{1}{3} n \langle mv''^2 \rangle = nkT = \Sigma (n_s kT) \tag{14.2.9}$$

The equality of the first and last terms in equation (14.2.9) gives Dalton's law: "*the hydrostatic pressure of a mixture is equal to the sum of the hydrostatic pressure p_s exerted by the gas components as through each component would separately fill the same volume at the same temperature*".

Similarly, the heat flow vector has the form

$$q = \Sigma q_s = \frac{1}{2} \Sigma \rho_s \langle v_s''^2 v_s'' \rangle \tag{14.2.10}$$

Finally, with the notation $q_s = Z_s e$ (not to be mistaken for the heat flow vector) for the electric charge of particle s (e being the electron charge, $Z_e = -1$), then we have the following expression for the electric charge of the mixture:

$$\rho^{(q)} = \Sigma n_s q_s \tag{14.2.11}$$

and for the total electric current density:

$$J = \Sigma\, n_s q_s \langle v_s \rangle \tag{14.2.12}$$

Using equation (14.2.6) we have:

$$J = \Sigma\, n_s q_s (V + \langle v_s'' \rangle) = \rho^{(q)} V + j \tag{14.2.13}$$

$$j \overset{\text{def}}{=} \Sigma\, n_s q_s \langle v_s'' \rangle$$

The term $\rho^{(q)} V$ represents the convection current density (due to the overall motion of the gas) and j the current density due to the relative motion (with respect to the overall motion) of the various components.

14.2.2 The transport equation

In order to derive the conservation equations we first need the (Maxwell) transport equation and this is derived by multiplying the Boltzmann equation with $\Phi_s = \Phi_s(t, x, v_s)$ and integrating over the velocity space of species s. One obtains:

$$\int \Phi_s \mathscr{D}_s f_s \, dv_s = I_{ss}(\Phi) + \sum_{r \neq s} I_{sr}(\Phi) \tag{14.2.14}$$

with the notations:

$$I_{ss}(\Phi) = \int \Phi_s J(f_s f_{s1}) \, dv_s, \quad I_{sr}(\Phi) = \int \Phi_s J(f_s f_r) \, dv_s$$

Denoting by upper indices (1, 2, 3) the coordinates, integrating by parts and taking into account the fact that for $v_s^i \to \infty$, $f_s \to 0$ we find:

$$\int \Phi_s (v_s \times B)^i \frac{\partial f_s}{\partial v_s^i} \, dv_s = -\int (v_s \times B) \cdot \frac{\partial \Phi_s}{\partial v_s} f_s \, dv_s = -n_s \left\langle (v_s \times B) \frac{\partial \Phi_s}{\partial v_s} \right\rangle$$

With this result and those of (13.4.1′) equation (14.1.14) reduces to

$$\frac{\partial}{\partial t} n_s \langle \Phi_s \rangle + \frac{\partial}{\partial x} \cdot n_s \langle \Phi_s v_s \rangle = \tag{14.2.15}$$

$$n_s \left\{ \left\langle \frac{\partial_s \Phi}{\partial t} \right\rangle + \left\langle v_s \cdot \frac{\partial \Phi_s}{\partial x} \right\rangle + \left(X + \frac{q_s}{m_s} E \right) \cdot \left\langle \frac{\partial \Phi_s}{\partial v_s} \right\rangle + \frac{q_s}{m_s} \left\langle (v_s + B) \cdot \frac{\partial \Phi_s}{\partial v_s} \right\rangle \right\}$$

$$+ I_{ss}(\Phi) + \sum_{r \neq s} I_{rs}(\Phi)$$

According to the arguments of Section 13.4 we have:

$$I_{ss}(\Phi)=\frac{1}{4}\int(\Phi_s+\Phi_{s1}-\Phi_s'-\Phi_{s1}')\,J(f_sf_{s1})\,\mathrm{d}\boldsymbol{v}_s \tag{14.2.16}$$

and this term is identically zero for

$$\Phi_s=m_s,\ m_s\boldsymbol{v}_s,\ m_s\boldsymbol{v}_s^2 \tag{14.2.17}$$

The term I_{sr} cannot be written in a form similar to (14.1.16) since particles s and r which participate in a collision are no longer identical. Thus the equations corresponding to (13.4.4) and (13.4.5) are no longer valid. Only the equation corresponding to (13.4.3) is still valid:

$$I_{sr}(\Phi)=\frac{1}{2}\int(\Phi_s-\Phi_s')\,J(f_sf_s)\,\mathrm{d}\boldsymbol{v}_s \tag{14.2.18}$$

which vanishes (only) for $\Phi_s=m_s$.

But summing over r and s one obtains an expression involving all the collisions between different species taken two by two. One may thus interchange r and s in this sum and get

$$\sum_s\sum_{r\neq s}I_{sr}(\Phi)=\frac{1}{2}\sum_s\sum_{r\neq s}\int(\Phi_r-\Phi_r')\,J(f_sf_r)\,\mathrm{d}\boldsymbol{v}_s$$

Finally we have [2]

$$\sum_s\sum_{r\neq s}I_{sr}(\Phi)=\frac{1}{2}\sum_s\sum_{r\neq s}\int(\Phi_s+\Phi_r-\Phi_s'-\Phi_r')\,J(f_sf_r)\,\mathrm{d}\boldsymbol{v}_s \tag{14.2.19}$$

Since the total momentum and total kinetic energy of particles s and r are conserved during a collision it follows:

$$\sum_s\sum_{r\neq s}I_{sr}(m\boldsymbol{v})=0,\quad \sum_s\sum_{r\neq s}I_{sr}(mv^2)=0 \tag{14.2.20}$$

For instance, for a binary mixture (electrons plus ions) we have:

$$I_{ei}(m\boldsymbol{v})=\int m_e\boldsymbol{v}_eJ(f_ef_i)\mathrm{d}\boldsymbol{v}_e=\frac{1}{2}\int m_e(\boldsymbol{v}_e-\boldsymbol{v}_e')(f_e'f_i'-f_ef_i)|\boldsymbol{v}_e-\boldsymbol{v}_i|\,\mathrm{d}\omega\,\mathrm{d}\boldsymbol{v}_e\,\mathrm{d}\boldsymbol{v}_i$$

$$I_{ie}(m\boldsymbol{v})=\int m_i\boldsymbol{v}_iJ(f_if_e)\mathrm{d}\boldsymbol{v}_i=\frac{1}{2}\int m_i(\boldsymbol{v}_i-\boldsymbol{v}_i')(f_i'f_e'-f_if_e)\,|\boldsymbol{v}_e-\boldsymbol{v}_i|\,\mathrm{d}\omega\,\mathrm{d}\boldsymbol{v}_e\,\mathrm{d}\boldsymbol{v}_i$$

so that using the momentum conservation theorem, $m_e\boldsymbol{v}_e+m_i\boldsymbol{v}_i=m_e\boldsymbol{v}_e'+m_i\boldsymbol{v}_i'$, it follows that

$$I_{ei}(m\boldsymbol{v})+I_{ie}(m\boldsymbol{v})=0 \tag{14.2.21}$$

Similarly we get

$$I_{ei}(mv^2) + I_{ie}(mv^2) = 0 \tag{14.2.22}$$

14.2.3 Conservation equations

For $\Phi_s = m_s$ one derives the continuity equation for the s-component from (14.2.15):

$$\frac{\partial}{\partial t}\rho_s + \frac{\partial}{\partial \boldsymbol{x}}\cdot\rho_s \boldsymbol{V}_s = 0 \tag{14.2.23}$$

Summing over s in (14.2.23) we get the continuity equation of the mixture

$$\frac{\partial \rho}{\partial t} + \frac{\partial}{\partial \boldsymbol{x}}\cdot\rho \boldsymbol{V} = 0 \tag{14.2.24}$$

If the degree of ionization is not constant in time instead of equation (14.2.23) one uses [G.17]:

$$\frac{\partial}{\partial t}\rho_s + \frac{\partial}{\partial \boldsymbol{x}}\cdot\rho_s \boldsymbol{V}_s = \dot{\rho}_s \tag{14.2.23'}$$

the equation for the whole mixture, (14.2.24), remaining valid.
For $\Phi_s = m_s\boldsymbol{v}_s$ equation (14.2.15) reduces to

$$\frac{\partial}{\partial t}\rho_s\boldsymbol{V}_s + \frac{\partial}{\partial \boldsymbol{x}}\cdot\rho_s\langle\boldsymbol{v}_s\boldsymbol{v}_s\rangle = n_s\{m_s\boldsymbol{X} + q_s(\boldsymbol{E} + \boldsymbol{V}_s\times\boldsymbol{B})\} + \sum_{r\neq s} I_{sr}(m\boldsymbol{v}) \tag{14.2.25}$$

Now using (14.2.6) and (14.2.6′) we have

$$\rho_s\langle\boldsymbol{v}_s\boldsymbol{v}_s\rangle = \rho_s\langle(\boldsymbol{v}_s' + \boldsymbol{V}_s)(\boldsymbol{v}_s' + \boldsymbol{V}_s)\rangle = \underline{\boldsymbol{P}}_s' + \rho_s\boldsymbol{V}_s\boldsymbol{V}_s \tag{14.2.26}$$

with the notations

$$\underline{\boldsymbol{P}}_s' = \rho_s\langle\boldsymbol{v}_s'\boldsymbol{v}_s'\rangle \tag{14.2.27}$$

and

$$\underline{\boldsymbol{P}}_s = \underline{\boldsymbol{P}}_s' + \rho_s(\boldsymbol{V}_s - \boldsymbol{V})(\boldsymbol{V}_s - \boldsymbol{V}) \tag{14.2.27'}$$

Since we also have

$$\rho_s\boldsymbol{V}_s\boldsymbol{V}_s = \boldsymbol{V}_s\left(\frac{\partial}{\partial \boldsymbol{x}}\cdot\rho_s\boldsymbol{V}_s\right) + \rho_s\left(\boldsymbol{V}_s\cdot\frac{\partial}{\partial \boldsymbol{x}}\right)\boldsymbol{V}_s$$

equation (14.2.25) reduces to

$$\rho_s \frac{d_s}{dt} V_s = -\frac{\partial}{\partial x} \cdot \underline{P}'_s + \rho_s X + n_s q_s (\mathbf{E} + V_s \times B) + \sum_{r \neq s} I_{sr}(mv) \tag{14.2.28}$$

taking into account equation (14.2.23) and with the notation

$$\frac{d_s}{dt} = \frac{\partial}{\partial t} + V_s \cdot \frac{\partial}{\partial x}$$

This is the momentum conservation equation for the s-component.

Summing (14.2.28) over s and noticing that

$$\Sigma \rho_s \langle v_s v_s \rangle = \underline{P} + \rho VV$$

one obtains the equation for the whole assembly. Taking into account the equations defining the macroscopic quantities as well as equation (14.2.24) it follows that

$$\rho \frac{d}{dt} V = -\frac{\partial}{\partial x} \cdot \underline{P} + \rho X + \rho^{(q)} E + J \times B \tag{14.2.29}$$

with the definition

$$\frac{d}{dt} = \frac{\partial}{\partial t} + V \cdot \frac{\partial}{\partial x} \tag{14.2.29$'$}$$

This is the equation used in Chapter 2.

For $\Phi_s = \frac{1}{2} m_s v_s^2$ equation (14.2.15) gives

$$\frac{1}{2} \frac{\partial}{\partial t} \rho_s \langle v_s^2 \rangle + \frac{1}{2} \frac{\partial}{\partial x} \cdot \rho_s \langle v_s^2 v_s \rangle = (\rho_s X + n_s q_s E) \cdot V_s$$

$$+ \sum_{r \neq s} I_{sr}(mv^2) \tag{14.2.30}$$

Using equations (14.2.6) and (14.2.6′) and doing some calculations similar to those of (13.4.11) equation (14.2.30) is written in the form:

$$\frac{\partial}{\partial t} \rho_s \left(\frac{1}{2} V_s^2 + U'_s \right) + \frac{\partial}{\partial x} \cdot \left\{ \rho_s V_s \left(\frac{1}{2} V_s^2 + U'_s \right) + (V_s \cdot \underline{P}'_s) + q'_s \right\}$$

$$= (\rho_s X + n_s q_s E) \cdot V_s + \sum_{r \neq s} I_{sr}(mv^2) \tag{14.2.31}$$

which represents the energy conservation equation for the s-component. Using equations (14.2.6) and (14.2.6′) we again find relationships between U'_s and $\boldsymbol{q}'_s$ and U_s and $\boldsymbol{q}_s$, respectively. For instance we have for the internal energy:

$$U_s = U'_s + \frac{1}{2}(V_s - V)^2 \tag{14.2.32}$$

In order to obtain the mixture energy equation we shall introduce velocity $\boldsymbol{V}$ summing (14.2.30) over s and using (14.2.6) and (14.2.6′). Now, since

$$\frac{1}{2}\Sigma\,\rho_s\langle v_s^2\rangle = \frac{1}{2}\Sigma\,\rho_s\langle(\boldsymbol{v}_s'' + \boldsymbol{V})^2\rangle = \rho U + \frac{1}{2}\rho V^2$$

$$\rho U \overset{\text{def}}{=} \Sigma\,\rho_s U_s$$

and

$$\frac{1}{2}\Sigma\,\rho_s\langle v_s^2\boldsymbol{v}_s\rangle = \frac{1}{2}\Sigma\,\rho_s\langle(\boldsymbol{v}_s'' + \boldsymbol{V})^2(\boldsymbol{v}_s'' + \boldsymbol{V})\rangle$$

$$= \left(U + \frac{1}{2}V^2\right)\rho\boldsymbol{V} + \boldsymbol{V}\cdot\underline{\boldsymbol{P}} + \boldsymbol{q}$$

it follows that

$$\frac{\partial}{\partial t}\rho\left(U + \frac{1}{2}V^2\right) + \frac{\partial}{\partial \boldsymbol{x}}\cdot\left\{\left(U + \frac{1}{2}V^2\right)\rho\boldsymbol{V} + \boldsymbol{V}\cdot\underline{\boldsymbol{P}} + \boldsymbol{q}\right\}$$

$$= \rho\boldsymbol{X}\cdot\boldsymbol{V} + \boldsymbol{J}\cdot\boldsymbol{E} \tag{14.2.33}$$

which is *the energy equation for the whole assembly*. This takes the usual form if one performs calculations similar to those of section 13.4 (i.e. one multiplies (14.2.29) by $\boldsymbol{V}$ and subtracts this from (14.2.33) also taking into account (14.2.24). Then we find:

$$\rho\frac{dU}{dt} = -\frac{\partial q_i}{\partial x_i} - P_{ji}\frac{\partial V_j}{\partial x_i} + (\boldsymbol{E} + \boldsymbol{V}\times\boldsymbol{B})\cdot\boldsymbol{j} \tag{14.2.34}$$

This is the form used for the energy equation in Chapter 2.

For $\Phi_s = q_s$ we find from (14.2.15):

$$\frac{\partial}{\partial t}n_s q_s + \frac{\partial}{\partial \boldsymbol{x}}\cdot n_s q_s \boldsymbol{V}_s = \begin{cases} 0 \text{ for constant degree of ionization} \\ \dot{n}_s q_s \text{ for variable degree of ionization}\end{cases} \tag{14.2.35}$$

Summing (14.2.35) over s we get the charge conservation equation for the assembly:

$$\frac{\partial}{\partial t}\rho^{(q)} + \frac{\partial}{\partial \boldsymbol{x}}\cdot \boldsymbol{J} = 0 \tag{14.2.36}$$

14.2.4 Ohm's law

Let us make the substitution $\Phi_s = (q_s/m_s)m_s\boldsymbol{v}_s$ in (14.2.15) and sum with respect to s over the whole assembly. One obtains

$$\frac{\partial}{\partial t}\boldsymbol{J} + \frac{\partial}{\partial \boldsymbol{x}}\cdot \Sigma n_s q_s \langle \boldsymbol{v}_s\boldsymbol{v}_s\rangle \tag{14.2.37}$$

$$= \sum n_s q_s\{\boldsymbol{X} + (q_s/m_s)(\boldsymbol{E} + \boldsymbol{V}_s \times \boldsymbol{B})\} + \sum_s (q_s/m_s)\sum_{r\neq s} I_{sr}(m\boldsymbol{v})$$

But

$$\Sigma\, n_s q_s \langle \boldsymbol{v}_s\boldsymbol{v}_s\rangle = \Sigma\, n_s q_s \langle(\boldsymbol{v}_s'' + \boldsymbol{V})(\boldsymbol{v}_s'' + \boldsymbol{V})\rangle = \Sigma\,(q_s/m_s)\underline{\boldsymbol{P}}_s + \boldsymbol{j}\boldsymbol{V} + \boldsymbol{V}\boldsymbol{j} + \rho^{(q)}\boldsymbol{V}\boldsymbol{V}$$

Now since

$$\frac{\partial}{\partial \boldsymbol{x}}\cdot(\boldsymbol{V}\boldsymbol{j} + \rho^{(q)}\boldsymbol{V}\boldsymbol{V}) = \frac{\partial}{\partial \boldsymbol{x}}\cdot \boldsymbol{V}\boldsymbol{J} = \left(\boldsymbol{V}\cdot\frac{\partial}{\partial \boldsymbol{x}}\right)\boldsymbol{J} + \boldsymbol{J}\left(\frac{\partial}{\partial \boldsymbol{x}}\cdot \boldsymbol{V}\right)$$

equation (14.2.37) becomes

$$\frac{\mathrm{d}}{\mathrm{d}t}\boldsymbol{J} + \boldsymbol{J}\left(\frac{\partial}{\partial \boldsymbol{x}}\cdot \boldsymbol{V}\right) + \frac{\partial}{\partial \boldsymbol{x}}\cdot \boldsymbol{j}\boldsymbol{V} \tag{14.2.38}$$

$$= -\frac{\partial}{\partial \boldsymbol{x}}\cdot \sum \frac{q_s}{m_s}\underline{\boldsymbol{P}}_s + \rho^{(q)}\boldsymbol{X} + \sum \frac{n_s q_s^2}{m_s}(\boldsymbol{E} + \boldsymbol{V}_s \times \boldsymbol{B}) + \sum_s \frac{q_s}{m_s}\sum_{r\neq s} I_{sr}(m\boldsymbol{v})$$

By writing down explicitly the first term of (14.2.38) ($\boldsymbol{J} = \rho^{(q)}\boldsymbol{V} + \boldsymbol{j}$) and noticing that on account of equation (14.2.36) we have

$$\boldsymbol{V}\frac{\mathrm{d}\rho^{(q)}}{\mathrm{d}t} + \rho^{(q)}\boldsymbol{V}\left(\frac{\partial}{\partial \boldsymbol{x}}\cdot \boldsymbol{V}\right) + \boldsymbol{V}\left(\frac{\partial}{\partial \boldsymbol{x}}\cdot \boldsymbol{j}\right)$$

$$= \boldsymbol{V}\left\{\frac{\partial}{\partial t}\rho^{(q)} + \frac{\partial}{\partial \boldsymbol{x}}\cdot(\rho^{(q)}\boldsymbol{V} + \boldsymbol{j})\right\} = 0$$

we obtain for Ohm's law (14.2.38) the following form:

$$\rho^{(q)}\frac{\mathrm{d}\boldsymbol{V}}{\mathrm{d}t} + \frac{\mathrm{d}\boldsymbol{j}}{\mathrm{d}t} + \boldsymbol{j}\left(\frac{\partial}{\partial \boldsymbol{x}}\cdot \boldsymbol{V}\right) + \left(\boldsymbol{j}\cdot\frac{\partial}{\partial \boldsymbol{x}}\right)\boldsymbol{V} \tag{14.2.39}$$

$$= -\frac{\partial}{\partial \boldsymbol{x}}\cdot \sum_s \frac{q_s}{m_s}\underline{\boldsymbol{P}}_s + \rho^{(q)}\boldsymbol{X} + \sum_s \frac{n_s q_s^2}{m_s}(\boldsymbol{E} + \boldsymbol{V}_s \times \boldsymbol{B}) + \sum_s \frac{q_s}{m_s}\sum_{r\neq s} I_{sr}(m\boldsymbol{v})$$

This is the ultimate form of *Ohm's law*.

Both the conservation equations and Ohm's law (14.2.39) are too complicated to be used in actual cases. They obviously assume the knowledge of the distribution functions f_s which appear in both the terms $\boldsymbol{I}_{sr}$ and the expressions of the pressure tensor $\underline{\boldsymbol{P}}_s$, $\underline{\boldsymbol{P}}'_s$ (and of the heat flow vectors, $\boldsymbol{q}_s$, $\boldsymbol{q}'_s$).

14.2.5 Schlüter's equations

Schlüter's equations are obtained with the following hypothesis [4], [1], [2]:

$$I_{sr}(m\boldsymbol{v}) = -\frac{\rho_s}{\tau_{sr}}(\boldsymbol{V}_s - \boldsymbol{V}_r) \tag{14.2.40}$$

Combining this hypothesis with equation

$$\boldsymbol{I}_{sr} + \boldsymbol{I}_{rs} = 0, \quad \boldsymbol{I}_{sr} \overset{\text{def}}{=} \boldsymbol{I}_{sr}(m\boldsymbol{v}) \tag{14.2.41}$$

which follows from the theorem of momentum conservation during the collision (14.2.20) we can write explicitly the momentum equations. For instance, for an electron — ion mixture (binary plasma) these become:

$$\begin{aligned} \rho_e \frac{\mathrm{d}_e}{\mathrm{d}t}\boldsymbol{V}_e &= -\frac{\partial}{\partial \boldsymbol{x}}\cdot \underline{\boldsymbol{P}}'_e + \rho_e \boldsymbol{X} - n_e e(\boldsymbol{E} + \boldsymbol{V}_e \times \boldsymbol{B}) - \frac{\rho_e}{\tau_{ei}}(\boldsymbol{V}_e - \boldsymbol{V}_i) \\ \rho_i \frac{\mathrm{d}_i}{\mathrm{d}t}\boldsymbol{V}_i &= -\frac{\partial}{\partial \boldsymbol{x}}\cdot \underline{\boldsymbol{P}}'_i + \rho_i \boldsymbol{X} + Z_i n_i e(\boldsymbol{E} + \boldsymbol{V}_i \times \boldsymbol{B}) + \frac{\rho_e}{\tau_{ei}}(\boldsymbol{V}_e - \boldsymbol{V}_i) \end{aligned} \tag{14.2.42}$$

while for a mixture made up of electrons, ions and neutral particles (ternary plasma) we have:

$$\begin{aligned} \rho_e \frac{\mathrm{d}_e}{\mathrm{d}t}\boldsymbol{V}_e &= -\frac{\partial}{\partial \boldsymbol{x}}\cdot \underline{\boldsymbol{P}}'_e + \rho_e \boldsymbol{X} - n_e e(\boldsymbol{E} + \boldsymbol{V}_e \times \boldsymbol{B}) - \frac{\rho_e}{\tau_{ei}}(\boldsymbol{V}_e - \boldsymbol{V}_i) - \frac{\rho_e}{\tau_{ea}}(\boldsymbol{V}_e - \boldsymbol{V}_a) \\ \rho_i \frac{\mathrm{d}_i}{\mathrm{d}t}\boldsymbol{V}_i &= -\frac{\partial}{\partial \boldsymbol{x}}\cdot \underline{\boldsymbol{P}}'_i + \rho_i \boldsymbol{X} + Z_i n_i e(\boldsymbol{E} + \boldsymbol{V}_i \times \boldsymbol{B}) \\ &\quad + \frac{\rho_e}{\tau_{ei}}(\boldsymbol{V}_e - \boldsymbol{V}_i) - \frac{\rho_i}{\tau_{ia}}(\boldsymbol{V}_i - \boldsymbol{V}_a) \\ \rho_a \frac{\mathrm{d}_a}{\mathrm{d}t}\boldsymbol{V}_a &= -\frac{\partial}{\partial \boldsymbol{x}}\cdot \underline{\boldsymbol{P}}'_a + \rho_a \boldsymbol{X} + \frac{\rho_e}{\tau_{ea}}(\boldsymbol{V}_e - \boldsymbol{V}_a) + \frac{\rho_i}{\tau_{ia}}(\boldsymbol{V}_i - \boldsymbol{V}_a) \end{aligned} \tag{14.2.43}$$

Equations (14.2.42) and (14.2.43) can be used practically only if the expression of $\underline{P}'$ is given. If the gas viscosity is neglected then

$$\underline{P}_s = p_s\underline{U}, \quad \left(\frac{\partial}{\partial \boldsymbol{x}} \cdot \underline{P}_s = \frac{\partial}{\partial \boldsymbol{x}} p_s\right) \tag{14.2.44}$$

where $\underline{U}$ is the unit tensor. In this case one may use to a good approximation the following equation:

$$\frac{\partial}{\partial \boldsymbol{x}} \cdot \underline{P}'_s = \frac{\partial}{\partial \boldsymbol{x}} p_s \tag{14.2.45}$$

As it follows from (14.2.27′) equation (14.2.45) would be strictly valid in a linearized theory.

14.3 The plasma equations

14.3.1 Hypothesis and approximations

Let us consider a mixture made up of electrons, ions and neutral particles. For simplicity we assume that the ions are singly charges (i.e. $Z_i = +1$). For a plasma one may also introduce the following two simplifying assumptions [2], [5], [6]:

$$\rho^{(q)} \simeq 0, \ (J \simeq j) \tag{14.3.1}$$

$$\frac{m_e}{m_i} << 1 \tag{14.3.2}$$

The first equation expresses the electrical quasi-neutrality of the assembly (as a matter of fact in Chapter 2 it was shown that $\rho^{(q)}$ is of the order of the relativistic correction), while the second shows the fact that the electron mass is much smaller than that of the ion.

Let us now consider the simplifications introduced into the conservation equations by the above hypotheses. Since

$$\rho^{(q)} = n_e q_e + n_i q_i = e(n_i - n_e)$$

it follows that the quasi-neutralized hypothesis (14.3.1) is equivalent to

$$n_e = n_i \left(\overset{\text{def}}{=} n \right) \tag{14.3.3}$$

We shall further suppose that the neutral particle mass is equal to the ion mass ($m_a = m_i \stackrel{\text{def}}{=} m$) and neglect the ratio m_e/m_a. We introduce the following notation:

$$\frac{1}{f} = 1 + \frac{n}{n_a} \tag{14.3.4}$$

f giving the degree of ionization. A fully ionized gas will be characterized by $f \simeq 0$ while a weakly ionized one by $f \simeq 1$. From the mass density definition it follows:

$$\rho \simeq \rho_i\left(1 + \frac{n_a}{n}\right) = \frac{\rho_i}{1 - f} \tag{14.3.5}$$

$$\rho \simeq \rho_a\left(1 + \frac{n}{n_a}\right) = \frac{\rho_a}{f}$$

We also have:

$$\boldsymbol{J} = ne(V_i - V_e) = ne(\langle \boldsymbol{v}_i'' \rangle - \langle \boldsymbol{v}_e'' \rangle) = \boldsymbol{j} \tag{14.3.6}$$

$$\boldsymbol{j} = \boldsymbol{j}_i - \boldsymbol{j}_e$$

From the second equation it follows that

$$\rho_a \langle \boldsymbol{v}_a'' \rangle \simeq -\rho_i \langle \boldsymbol{v}_i'' \rangle \tag{14.3.7}$$

so that

$$\langle \boldsymbol{v}_a'' \rangle \simeq -\frac{n}{n_a}\langle \boldsymbol{v}_i'' \rangle = \left(1 - \frac{1}{f}\right) \langle \boldsymbol{v}_i'' \rangle \tag{14.3.7'}$$

We thus find:

$$\boldsymbol{I}_{ei} = -\frac{nm_e}{\tau_{ei}}(V_e - V_i) = \frac{m_e}{e\tau_{ei}}\boldsymbol{j} \tag{14.3.8}$$

$$\boldsymbol{I}_{ea} = -\frac{nm_e}{\tau_{ea}}(\langle \boldsymbol{v}_e'' \rangle - \langle \boldsymbol{v}_a'' \rangle) = \frac{m_e}{e\tau_{ea}}(\boldsymbol{j} - \frac{1}{f}\boldsymbol{j}_i)$$

$$\boldsymbol{I}_{ia} = -\frac{m_i}{ef\tau_{ia}}\boldsymbol{j}_i$$

Now by also introducing the electron and ion gyration frequencies

$$\omega_e = \frac{eB}{m_e}, \ \omega_i = \frac{eB}{m_i} \tag{14.3.9}$$

and the notations

$$\varkappa_{ei} = \frac{1}{\omega_e \tau_{ei}}, \ \varkappa_{ea} = \frac{1}{\omega_e \tau_{ea}}, \ \varkappa_{ia} = \frac{1}{\omega_i \tau_{ia}} \tag{14.3.10}$$

we have the following form of (14.3.8):

$$\boldsymbol{I}_{ei} = \varkappa_{ei} B\boldsymbol{j}, \ \boldsymbol{I}_{ea} = \varkappa_{ea} B\left(\boldsymbol{j} - \frac{1}{f}\boldsymbol{j}_i\right), \ \boldsymbol{I}_{ia} = -\frac{1}{f}\varkappa_{ia} B\boldsymbol{j}.$$

14.3.2 The Schlüter — Cowling equations

If we neglect ρ_e and accept hypothesis (14.2.45) equations (14.2.43) become:

$$0 = -\frac{\partial}{\partial \boldsymbol{x}} p_e - n_e(\boldsymbol{E} + \boldsymbol{V} \times \boldsymbol{B}) + (\boldsymbol{j} - \boldsymbol{j}_i) \times \boldsymbol{B}$$

$$+ (\varkappa_{ei} + \varkappa_{ea}) B\boldsymbol{j} - \frac{1}{f}\varkappa_{ea} B\boldsymbol{j}_i \tag{14.3.11}$$

$$(1-f)\rho \frac{\mathrm{d}_i}{\mathrm{d}t} \boldsymbol{V}_i = -\frac{\partial}{\partial \boldsymbol{x}} p_i + (1-f)\rho \boldsymbol{X} + n_e(\boldsymbol{E} + \boldsymbol{V} \times \boldsymbol{B})$$

$$+ \boldsymbol{j}_i \times \boldsymbol{B} - \varkappa_{ei} B\boldsymbol{j} - \frac{1}{f}\varkappa_{ia} B\boldsymbol{j}_i$$

The neutral particle equation will be replaced by the equation for the whole assembly:

$$\rho \frac{\mathrm{d}}{\mathrm{d}t} \boldsymbol{V} = -\frac{\partial}{\partial \boldsymbol{x}} p + \rho \boldsymbol{X} + \boldsymbol{j} \times \boldsymbol{B} \tag{14.3.12}$$

From equations (14.3.11) and (14.3.12) one can obtain an equation for $\boldsymbol{j}$ with the assumption:

$$\boldsymbol{V} \simeq \boldsymbol{V}_i, \ (\boldsymbol{V}_a \simeq \boldsymbol{V}_i) \tag{14.3.13}$$

This is a plausible assumption since the ion and neutral particle masses being approximately equal we have

$$V \simeq \frac{V_i + (n_a/n_i)V_a}{1 + (n_a/n_i)} = (1 - f)V_i + fV_a \tag{14.3.13'}$$

For a fully ionized gas ($f = 0$) the assumption is obviously true. Adding the two equations (14.3.11) and using (14.3.12) in (14.3.13) we find:

$$(\varkappa_{ea} + \varkappa_{ia})\frac{B}{f}j_i = -\frac{\partial}{\partial x}\{p_e + p_i - (1 - f)p\} + fj \times B + \varkappa_{ea}Bj \tag{14.3.14}$$

so that the first equation (14.3.11) gives:

$$n_e(E + V \times B) + \frac{\partial}{\partial x}\{p_e - \beta(p_e + p_i) + \beta(1 - f)p\} \tag{14.3.15}$$

$$= (\varkappa_{ei} + \beta\varkappa_{ia})Bj + (1 - 2f\beta)j \times B$$

$$+ \frac{f}{B(\varkappa_{ea} + \varkappa_{ia})}\left\{\frac{\partial}{\partial x}[p_e + p_i - (1 - f)p] + fB \times j\right\} \times B$$

where we have written

$$\beta = \frac{\varkappa_{ea}}{\varkappa_{ea} + \varkappa_{ia}} \tag{14.3.15'}$$

Due to the mass ratio we have $\varkappa_{ea} \ll \varkappa_{ia}$ so that $\beta \ll 1$.

We now assume that the gas is ideal ($p_s = n_s KT_s$) and its components have the same temperature (isothermal plasma). Taking into account (14.3.3) and (14.3.4) we find

$$\frac{p_a}{p_e} = \frac{n_a}{n} = \frac{f}{1 - f} \Rightarrow p_a = \frac{f}{1 - f}p_e \tag{14.3.16}$$

$$p = \Sigma p_s = \frac{2 - f}{1 - f}p_e \Rightarrow p_e = \frac{1 - f}{2 - f}p$$

whence

$$\begin{aligned} p_e + p_i - (1 - f)p &= fp_e \\ p_e - \beta(p_e + p_i) + \beta(1 - f)p &= (1 - f\beta)p_e \end{aligned} \tag{14.3.16'}$$

Then equation (14.3.15) becomes

$$en(\boldsymbol{E} + \boldsymbol{V} \times \boldsymbol{B}) + (1 - f\beta)\frac{\partial}{\partial \boldsymbol{x}} p_e = (\varkappa_{ei} + \beta\varkappa_{ia})B\boldsymbol{j}$$

$$+ (1 - 2f\beta)\boldsymbol{j} \times \boldsymbol{B} + \frac{f^2}{B(\varkappa_{ea} + \varkappa_{ia})}\left(\frac{\partial}{\partial \boldsymbol{x}} p_e + \boldsymbol{B} \times \boldsymbol{j}\right) \times \boldsymbol{B} \tag{14.3.17}$$

For a *fully-ionized gas* we have $f = 0$, $\varkappa_{ea} = \varkappa_{ia} = 0$ (the last equation following from (14.3.8) taking into account the fact that the *(ea)* and *(ia)* collisions are negligible). From (14.3.17) we get:

$$en(\boldsymbol{E} + \boldsymbol{V} \times \boldsymbol{B}) + \frac{1}{2}\frac{\partial}{\partial \boldsymbol{x}} p = \frac{en}{\sigma_{ei}}\boldsymbol{j} + \boldsymbol{j} \times \boldsymbol{B} \tag{14.3.18}$$

where

$$\sigma_{ei} = \frac{e^2 n\tau_{ei}}{m_e} \tag{14.3.18'}$$

For a *weakly ionized gas* we have $f_s = 1$ and $\varkappa_{ei} = 0$ (the electron — ion collisions being negligible). Since $\varkappa_{ea} \ll \varkappa_{ia}$ it follows that :

$$\beta \simeq \frac{\varkappa_{ea}}{\varkappa_{ia}} = \frac{m_e}{m_i}\frac{\tau_{ia}}{\tau_{ea}} \tag{14.3 15''}$$

With the extreme assumption $\beta \simeq 0$ from (14.3.17) it follows that ($p_e \simeq 0$):

$$en(\boldsymbol{E} + \boldsymbol{V} \times \boldsymbol{B}) = \frac{en}{\sigma_{ea}}\boldsymbol{j} + \boldsymbol{j} \times \boldsymbol{B} + \frac{\sigma_{ia}}{en}(\boldsymbol{B} \times \boldsymbol{j}) \times \boldsymbol{j} \tag{14.3.19}$$

where

$$\sigma_{ea} = \frac{e^2 n\tau_{ea}}{m_e}, \quad \sigma_{ia} = \frac{e^2 n\tau_{ia}}{m_i} \tag{14.3.19'}$$

Equations (14.3.18) and (14.3.19) give Ohm's law. Adding to Ohm's law the assembly equation (14.3.12), the continuity equation (14.2.24) and the energy equation (14.2.34) (with the assumption $P_{ji} = p\delta_{ji}$), we can determine the macroscopic motion of an inviscid plasma (we have of course to know the structure of q_i and to use the electromagnetic field equations).

14.3.3 Ohm's law

Here we shall consider equation (14.2.39) in the light of the hypotheses (14.3.1) and (14.3.2) [2]. Let us first consider a *binary plasma* (completely ionized plasma). We have

$$\Sigma \frac{q_s}{m_s} \underline{P}_s = -\frac{e}{m_e}\left(\underline{P}_e - \frac{m_e}{m_i}\underline{P}_i\right) \simeq -\frac{e}{m_e}\underline{P}_e$$

$$\Sigma \frac{n_s q_s^2}{m_s} = e^2 \frac{n}{m_e}\left(1 + \frac{m_e}{m_i}\right) \simeq \frac{e^2 n}{m_e} \tag{14.3.20}$$

$$\Sigma \frac{n_s q_s^2}{m_s} \boldsymbol{V}_s \simeq \frac{e^2 n}{m_e}\boldsymbol{V}_e = \frac{e^2 n}{m_e}\boldsymbol{V} - \frac{e}{m_e}\boldsymbol{j}$$

the last equation being obtained from (14.3.6)

$$\boldsymbol{j} \simeq en(\boldsymbol{V}_i - \boldsymbol{V}_e)$$

taking into account (14.3.13) for $f = 0$.

Since for this case

$$\Sigma \frac{q_s}{m_s} \sum_{r \neq s} I_{sr}(m\boldsymbol{v}) \simeq -\frac{e}{m_e}\boldsymbol{I}_{ei} = -\frac{\boldsymbol{j}}{\tau_{ei}} \tag{14.3.21}$$

it follows from (14.2.39) that

$$\tau_{ei}\left\{\frac{d\boldsymbol{j}}{dt} + \boldsymbol{j}\left(\frac{\partial}{\partial \boldsymbol{x}} \cdot \boldsymbol{V}\right) + \left(\boldsymbol{j} \cdot \frac{\partial}{\partial \boldsymbol{x}}\right)\boldsymbol{V}\right\}$$

$$= \sigma_{ei}\left(\boldsymbol{E} + \boldsymbol{V} \times \boldsymbol{B} + \frac{1}{en}\frac{\partial}{\partial \boldsymbol{x}} \cdot \underline{P}_e + \frac{1}{en}\boldsymbol{B} \times \boldsymbol{j}\right) - \boldsymbol{j} \tag{14.3.22}$$

For the cases in which the products of $\boldsymbol{j}$ by $\boldsymbol{V}$ may be neglected (for instance in the case of the propagation of perturbation waves in a stationary plasma) equation (14.3.22) may be used in the following form

$$\boldsymbol{j} = \sigma_{ei}\left(\boldsymbol{E} + \boldsymbol{V} \times \boldsymbol{B} + \frac{1}{en}\frac{\partial p_e}{\partial \boldsymbol{x}} + \frac{1}{en}\boldsymbol{B} \times \boldsymbol{j}\right) - \tau_{ei}\frac{\partial \boldsymbol{j}}{\partial t} \tag{14.3.22'}$$

provided that the plasma is inviscid (14.2.44). According to the last equation of (14.3.16) we should replace $2p_e$ by p in equation (14.3.22').

Adding to Ohm's law the assembly conservation equations one can solve the macroscopic problem. The evolution of the components is determined by (14.2.42). The first equation gives V_e:

$$0 = -\frac{1}{2}\frac{\partial p}{\partial \boldsymbol{x}} - en(\boldsymbol{E} + \boldsymbol{V}_e \times \boldsymbol{B}) + \frac{en}{\sigma_{ei}}\boldsymbol{j} \tag{14.3.23}$$

The corresponding ion equation added to (14.3.23) gives equation (14.3.10). As a matter of fact this is no longer necessary since we accepted the solution $\boldsymbol{V}_i = \boldsymbol{V}$, $2p_i \simeq p$, $\rho_i \simeq \rho$.

14.3.4 Ohm's law for a ternary plasma

The first two equations of (14.3.20) are also valid for a plasma made up of electrons, ions and neutral particles. But the third equation is written as follows:

$$\Sigma\frac{n_s q_s^2}{m_s}\boldsymbol{V}_s = \frac{e^2 n}{m_e}\boldsymbol{V}_e = \frac{e^2 n}{m_e}\boldsymbol{V} + \frac{e}{m_e}(\boldsymbol{j}_i - \boldsymbol{j}) \tag{14.3.24}$$

Thus the binary-plasma equation is obtained from that of a ternary plasma by putting $\boldsymbol{j}_i = 0$ in the latter. The explanation of this fact is found in equation (14.3.7) (derived from (14.2.6′)); this, in the absence of neutral particles, gives $\langle \boldsymbol{v}_i'' \rangle = 0$.

Since

$$\Sigma\frac{q_s}{m_s}\sum_{r\neq s}\boldsymbol{I}_{sr}(m\boldsymbol{v}) = \left(\frac{q_e}{m_e} - \frac{q_i}{m_i}\right)\boldsymbol{I}_{ei} + \frac{q_e}{m_e}\boldsymbol{I}_{ea} + \frac{q_i}{m_i}\boldsymbol{I}_{ia}$$

$$= -\left(\frac{1}{\tau_{ei}} + \frac{1}{\tau_{ea}}\right)\boldsymbol{j} + \frac{1}{f}\left(\frac{1}{\tau_{ea}} - \frac{1}{\tau_{ia}}\right)\boldsymbol{j}_i \tag{14.3.25}$$

Ohm's law becomes

$$\frac{m_e}{e}\left\{\frac{\mathrm{d}\boldsymbol{j}}{\mathrm{d}t} + \boldsymbol{j}\left(\frac{\partial}{\partial \boldsymbol{x}}\cdot\boldsymbol{V}\right) + \left(\boldsymbol{j}\cdot\frac{\partial}{\partial \boldsymbol{x}}\right)\boldsymbol{V}\right\} = en(\boldsymbol{E} + \boldsymbol{V}\times\boldsymbol{B})$$

$$+\frac{\partial}{\partial \boldsymbol{x}}\cdot\underline{\boldsymbol{P}}_e + (\boldsymbol{j}_i - \boldsymbol{j})\times\boldsymbol{B} - (\varkappa_{ei} + \varkappa_{ea})B\boldsymbol{j} + \left(\varkappa_{ea} - \frac{m_e\varkappa_{ia}}{m_i}\right)\frac{B}{f}\boldsymbol{j}_i \tag{14.3.26}$$

Since in this equation the current density $\boldsymbol{j}_i$ appears it cannot be used to determine the motion of the assembly without knowledge of the component motion. By accepting assumption (14.3.13) (which finally means $\langle \boldsymbol{v}_i'' \rangle \simeq 0$, a condition well satisfied for a fully ionized gas) one may use equation (14.3.14) to eliminate the ion current $\boldsymbol{j}_i$. Then also supposing that the plasma is inviscid (14.2.44) and isothermal (14.3.16′) we get from (14.3.26)

$$\frac{m_e}{e}\left\{\frac{\mathrm{d}\boldsymbol{j}}{\mathrm{d}t}+\boldsymbol{j}\left(\frac{\partial}{\partial \boldsymbol{x}}\cdot\boldsymbol{V}\right)+\left(\boldsymbol{j}\cdot\frac{\partial}{\partial \boldsymbol{x}}\right)\boldsymbol{V}\right\}=en(\boldsymbol{E}+\boldsymbol{V}\times\boldsymbol{B})+(1-f\beta)\frac{\partial}{\partial \boldsymbol{x}}p_e$$

$$+(1-2f\beta)\boldsymbol{B}\times\boldsymbol{j}-(\varkappa_{ei}+\beta\varkappa_{ia})B\boldsymbol{j}-\frac{f^2}{B(\varkappa_{ea}+\varkappa_{ia})}\left(\frac{\partial}{\partial \boldsymbol{x}}p_e+\boldsymbol{B}\times\boldsymbol{j}\right)\times\boldsymbol{B} \qquad (14.3.27)$$

by neglecting terms of the order of m_e/m_i.

For these cases in which the products of $\boldsymbol{j}$ and $\boldsymbol{V}$ may be neglected equation (14.3.27) becomes

$$\frac{m_e}{e}\frac{\partial \boldsymbol{j}}{\partial t}=en(\boldsymbol{E}+\boldsymbol{V}\times\boldsymbol{B})+(1-f\beta)\frac{\partial}{\partial \boldsymbol{x}}p_e+(1-2f\beta)\boldsymbol{B}\times\boldsymbol{j}$$

$$-(\varkappa_{ei}+\beta\varkappa_{ia})B\boldsymbol{j}-\frac{f^2}{B(\varkappa_{ea}+\varkappa_{ia})}\left(\frac{\partial}{\partial \boldsymbol{x}}p_e+\boldsymbol{B}\times\boldsymbol{j}\right)\times\boldsymbol{B} \qquad (14.3.27'$$

Equations (14.3.27) and (14.3.27′) differ from equation (14.3.17) through their left-hand side term. For stationary cases (14.3.27′) is identical with (14.3.27).

For a fully ionized gas equation (14.3.27′) reduces to (14.3.22′) while for a weakly ionized gas (using the approximation given by (14.3.15″)) it reduces to

$$\frac{m_e}{e}\frac{\partial \boldsymbol{j}}{\partial t}=en(\boldsymbol{E}+\boldsymbol{V}\times\boldsymbol{B})+\left(1-2\frac{\varkappa_{ea}}{\varkappa_{ia}}\right)\boldsymbol{B}\times\boldsymbol{j}-\varkappa_{ea}B\boldsymbol{j}$$

$$-\frac{1}{B\varkappa_{ia}}\left(\frac{\partial}{\partial \boldsymbol{x}}p_e+\boldsymbol{B}\times\boldsymbol{j}\right)\times\boldsymbol{B} \qquad (14.3.28)$$

Particularly, for

$$\frac{\partial}{\partial t}\boldsymbol{j}=0,\ \frac{\partial}{\partial \boldsymbol{x}}p_e=0,\ \boldsymbol{B}=0$$

equations (14.3.27′) and (14.3.15) give

$$\boldsymbol{E}=\left(\frac{1}{\sigma_{ei}}+\frac{1}{\sigma_{ea}}\right)\boldsymbol{j} \qquad (14.3.29)$$

which equation indicates the way the electrical conductivity of the gas enters into Ohm's law.

14.4 Ohm's law for a fully ionized gas

For a fully ionized gas we consider the following form of Ohm's law:

$$\boldsymbol{j} = \sigma(\boldsymbol{E} + \boldsymbol{E}_e + \boldsymbol{V} \times \boldsymbol{B}) + \frac{\sigma}{en}\boldsymbol{j} \times \boldsymbol{B} \tag{14.4.1}$$

$$\boldsymbol{E}_e = \frac{1}{en}\nabla p_e$$

where $\boldsymbol{E}_e$ is the electric field equivalent to the electron pressure gradient. In order to find $\boldsymbol{j}$ we shall break it up into components along the following vectors [6].

$$\mathfrak{E},\ \boldsymbol{B} \text{ and } \boldsymbol{E} \times \boldsymbol{B},\ \mathfrak{E} \overset{\text{def}}{=} \sigma(\boldsymbol{E} + \boldsymbol{E}_e + \boldsymbol{V} \times \boldsymbol{B}) \tag{14.4.2}$$

Using (14.4.1) it follows

$$\boldsymbol{j} = \alpha\mathfrak{E} + \beta\boldsymbol{B} + \gamma(\boldsymbol{E} \times \boldsymbol{B})$$

$$= \boldsymbol{E} + \frac{\alpha\sigma}{en}\boldsymbol{B} \times \boldsymbol{E} + \frac{\gamma\sigma}{en}\boldsymbol{B} \times (\mathfrak{E} \times \boldsymbol{B}) \tag{14.4.3}$$

We first assume that vectors (14.4.2) are linearly independent. Using equation (17.3) for the double vector product and identifying the components we get:

$$\alpha = 1 + \frac{\gamma\sigma}{en}B^2,\ \beta = -\frac{\gamma\sigma}{en}(\boldsymbol{B} \cdot \boldsymbol{E}),\ \gamma = -\frac{\sigma}{en}\alpha$$

or, substituting γ in α and β

$$\alpha = \frac{1}{1 + \nu^2},\ \gamma = -\frac{\nu}{B(1 + \nu^2)},\ \beta = \frac{\nu^2(\boldsymbol{B} \cdot \mathfrak{E})}{B^2(1 + \nu^2)} \tag{14.4.4}$$

$$(\nu \overset{\text{def}}{=} \omega_e \tau_{ei})$$

Finally we have:

$$\boldsymbol{j} = \sigma\sigma_1(\boldsymbol{E} + \boldsymbol{E}_e + \boldsymbol{V} \times \boldsymbol{B})$$

$$- \sigma\sigma_2(\boldsymbol{E} + \boldsymbol{E}_e + \boldsymbol{V} \times \boldsymbol{B}) \times \frac{\boldsymbol{B}}{B} + \sigma\sigma_2\nu\left\{(\boldsymbol{E} + \boldsymbol{E}_e) \cdot \frac{\boldsymbol{B}}{B}\right\}\frac{\boldsymbol{B}}{B} \tag{14.4.5}$$

If not all the vectors in (14.4.2) are linearly independent the only possibility that remains is that $\boldsymbol{E} = \lambda\sigma\boldsymbol{B}$. In this case

$$\boldsymbol{j} = \lambda\sigma\boldsymbol{B} \tag{14.4.6}$$

would be a solution of equation (14.4.1). Let us prove that there is no other solution. Supposing that such another solution existed then we would have

$$\boldsymbol{j} = \lambda\sigma\boldsymbol{B} + \boldsymbol{j}_0, \ \boldsymbol{j}_0 \neq 0$$

Substituting this into (14.4.1) we would obtain:

$$\boldsymbol{j}_0 = \frac{\sigma}{en}\boldsymbol{B} \times \boldsymbol{j}_0 \tag{14.4.7}$$

whence it follows that $\boldsymbol{j}_0 \equiv 0$.

Finally in this case we would have

$$\lambda\boldsymbol{B} = \boldsymbol{E} + \boldsymbol{E}_e + \boldsymbol{V} \times \boldsymbol{B} \tag{14.4.8}$$

For stationary fields from the Maxwell equation $\operatorname{curl} \boldsymbol{H} = \boldsymbol{j}$ it would follow that:

$$\operatorname{curl} \boldsymbol{B} = \sigma\mu\lambda\boldsymbol{B} \tag{14.4.9}$$

which equation, together with $\operatorname{div} \boldsymbol{B} = 0$, would determine the magnetic induction independent of the gas motion. Applying operator *curl* in (14.4.9) we would get:

$$\Delta\boldsymbol{B} + \sigma^2\lambda^2\mu^2\boldsymbol{B} = 0 \tag{14.4.10}$$

Such fields have been studied by Lüst and Schlüter [7]. They are called force-free fields (indeed solution (14.4.6) implies a vanishing magnetic force).

Returning to (14.4.8) we derive a condition for the electric field:

$$\boldsymbol{B}\cdot(\boldsymbol{E} + \boldsymbol{E}_e) = \lambda B^2 \tag{14.4.11}$$

The solution of such an equation is of the form

$$\boldsymbol{E} + \boldsymbol{E}_e = \lambda\boldsymbol{B} + \boldsymbol{B} \times \boldsymbol{A} \tag{14.4.12}$$

$\boldsymbol{A}$ being an arbitrary vector:

With such a representation of the electric field from (14.4.8) it follows:

$$\boldsymbol{V} \times \boldsymbol{B} = \boldsymbol{A} \times \boldsymbol{B}$$

and then

$$\boldsymbol{V} = \lambda\boldsymbol{B} + \boldsymbol{A} \tag{14.4.13}$$

We have thus represented the solution of the problem using a scalar λ and a vector $\boldsymbol{A}$, quantities that can be determined from the other equations of the field and motion.

Appendix

Cartesian vectors and tensors

A.1 Definitions

In this section we specify some of the definitions and notations used within the present work. We assume the concepts of vector and tensor to be known. The vectors are indicated by bold letters (e.g. $\boldsymbol{x}$, $\boldsymbol{v}$,...) and the tensors by underlined bold letters (e.g. $\underline{\boldsymbol{T}}$,...). The coordinates (components) are indicated by indices (e.g., x_i, v_i,..., T_{ij}) which have the values 1, 2, 3. In some particular cases one uses for the vector components the classical notation (x, y, z, v_x, v_y, v_z,...). $\boldsymbol{x}$ represents everywhere the position vector. For the notation by indices one uses the summation convention with respect to the repeating index.

Only Cartesian orthogonal reference systems are used. The sum of two vectors $\boldsymbol{u}$ (of components u_i) and $\boldsymbol{v}$ (of components v_i) is the vector $\boldsymbol{u} + \boldsymbol{v}$ of components $u_i + v_i$. In R_3 the sum is represented by the diagonal of the parallelogram constructed on the two vectors. The vector $\lambda \boldsymbol{u}$ represents the vector of components λu_i.

The *scalar product* of the vectors $\boldsymbol{u}$ and $\boldsymbol{v}$ is denoted by $\boldsymbol{u} \cdot \boldsymbol{v}$ and is defined by the formula:

$$\boldsymbol{u} \cdot \boldsymbol{v} = u_i v_i \tag{A.1}$$

The *vector product* is denoted by $\boldsymbol{u} \times \boldsymbol{v}$ and by definition is a vector of components:

$$(\boldsymbol{u} \times \boldsymbol{v})_i = \delta_{ijk} u_j v_k, \quad i = 1, 2, 3 \tag{A.2}$$

δ_{ijk} being the Levi-Civita symbol:

$$\delta_{ijk} \stackrel{\text{def}}{=} \begin{cases} 1 \text{ if } i, j, k \text{ is an even permutation of the numbers } 1, 2, 3 \\ -1 \text{ if } i, j, k \text{ is an odd permutation} \\ 0 \text{ if any index is repeated} \end{cases}$$

For the *triple vector product* one uses the identity:

$$\boldsymbol{u} \times (\boldsymbol{v} \times \boldsymbol{w}) = \boldsymbol{v}(\boldsymbol{u} \cdot \boldsymbol{w}) - \boldsymbol{w}(\boldsymbol{u} \cdot \boldsymbol{v}) \tag{A.3}$$

which can be easily verified using

A.2 Reference systems. Tensors

Let S be a Cartesian system of origin 0 and (orthogonal) unit vectors $\boldsymbol{i}_1$, $\boldsymbol{i}_2$, $\boldsymbol{i}_3$ and S' another system with the same origin and with (orthogonal) unit vectors $\boldsymbol{i}'_1$, $\boldsymbol{i}'_2$, $\boldsymbol{i}'_3$. If we use the notations

$$c_{jk} = \boldsymbol{i}'_j \cdot \boldsymbol{i}_k = \cos(\boldsymbol{i}'_j, \boldsymbol{i}_k), \quad j, k = 1, 2, 3 \tag{A.4}$$

it follows that:

$$\boldsymbol{i}'_j = c_{jk}\boldsymbol{i}_k, \; \boldsymbol{i}_j = c_{kj}\boldsymbol{i}'_k, \quad j = 1, 2, 3 \tag{A.5}$$

with summation with respect to the index k (= 1, 2, 3).

Let now $\boldsymbol{v}$ be a vector referred to the two systems:

$$v_j\boldsymbol{i}_j = v'_k\boldsymbol{i}'_k \tag{A.6}$$

Using (A.5) it follows

$$v'_k = c_{kj}v_j, \qquad k = 1, 2, 3 \tag{A.7}$$

A set of three real numbers (v_1, v_2, v_3) represents a vector if on a change of the reference system defined by (A.5) it transforms according to the formula (A.7). Similarly a set of nine elements (T_{ij}) which on a change (A.5) of the reference system is transformed according to the formula:

$$T'_{kl} = c_{ki}c_{lj}T_{ij} \tag{A.8}$$

constitutes a second order tensor. The generalization to higher order tensors and to the n-dimensional space is obvious.

Given two vectors $\boldsymbol{u}$ and $\boldsymbol{v}$ the quantity denoted by $\boldsymbol{uv}$ is a tensor (it obeys the formulae (A.8)) of components u_iv_j. It is called the *dyadics* of the two vectors.

One should notice that from the orthogonality of the unit vectors $\boldsymbol{i}'_1$, $\boldsymbol{i}'_2$, $\boldsymbol{i}'_3$ and $\boldsymbol{i}_1$, $\boldsymbol{i}_2$, $\boldsymbol{i}_3$ the orthogonality of the matrix $(C_{ij}) \stackrel{\text{def}}{=} C$ follows. Indeed we have the obvious relations:

$$\begin{aligned} \delta_{jl} &= \boldsymbol{i}'_j \cdot \boldsymbol{i}'_l = c_{jk}c_{ls}\boldsymbol{i}_k \cdot \boldsymbol{i}_s = c_{jk}c_{ls}\delta_{ks} = c_{jk}c_{lk} \\ \delta_{jl} &= \boldsymbol{i}_j \cdot \boldsymbol{i}_l = c_{kj}c_{sl}\boldsymbol{i}'_k \cdot \boldsymbol{i}'_s = c_{kj}c_{sl}\delta_{ks} = c_{kj}c_{kl} \end{aligned} \tag{A.9}$$

δ_{jl} being the Kronecker symbol (equal to 1 if $j = l$ and to 0 if $j \neq l$). The relations (A.9) define the orthogonal matrix. From (A.9) it follows:

C' being the transposed matrix and E the unit matrix. If the systems S and S' are both either positively or negatively oriented, from the last relation it follows that $\det C = 1$. Otherwise $\det C = -1$.

Another consequence of the relations (A.9) is that δ_{ks} verifies the formulae (A.8), i.e. it is a second order tensor. It is called the unit tensor.

The following criterion for the recognition of a tensor quantity has been used in this work: let S_{ij} be a quantity depending on two indices and let u_i and v_i be two vectors; if the expression $S_{ij}u_i v_j$ is invariant to the change of the reference system then S_{ij} is a second order tensor. Indeed from the equation:

$$S'_{rs}u'_r u'_s = S_{ij}u_i u_j$$

taking into account (A.7) it follows that

$$S'_{rs}c_{ri}c_{sj} = S_{ij}$$

and then using relations (A.9)

$$S'_{kl} = c_{ki}c_{lj}S_{ij} \qquad \text{q.e.d,}$$

The generalization to higher orders is straightforward.

A tensor is symmetrical with respect to two indices if it does not change on interchanging the indices and antisymmetrical if the interchange of the indices produces the change of the tensor sign. Using the transformation formulae one proves that the symmetry or antisymmetry of a tensor represent intrinsic properties.

Finally let us consider the Levi-Civita symbol defined by the relations (A.2′) for an arbitrary reference system. It is easily shown that this symbol transforms according to the formula:

$$\delta'_{lmn} = (\det C)c_{li}c_{mj}c_{nk}\delta_{ijk} \tag{A.10}$$

Indeed we have:

$$\delta'_{123} = (\det C)c_{1i}c_{2j}c_{3k}\delta_{ijk} = (\det C)^2 = 1$$

Since the symmetry and antisymmetry are intrinsic properties it is no longer necessary to check the other components. A quantity which transforms according to the formula (A.10) is called a *tensor density* of the third order. With respect to the direct orthogonal transformations ($\det C = 1$) δ_{ijk} represents a third order tensor. As a consequence it follows that the vector product is a vector only with respect to the direct orthogonal transformations. The vector product changes its sign on transformation from a positively oriented system S to a negatively oriented system S' (or *vice-versa*). For this reason it is called an *axial vector*. The magnetic induction (13.1.10) is an axial vector.

In a similar way one proves that the Levi-Civita symbol of the fourth order

$$\delta_{\alpha\beta\gamma\delta} \overset{\text{def}}{=} \begin{cases} 1 \text{ for even permutations of the numbers } 1,\ 2,\ 3,\ 4 \\ -1 \text{ for odd permutations} \\ 0 \text{ if any index is repeated} \end{cases} \tag{A.11}$$

represents a fourth order tensor density. $\delta_{\alpha\beta\gamma\delta}$ represents a fourth order tensor with respect to the Lorentz group (det $(c_{\alpha\beta}) = 1$). This result was used in chapter 1.

A.3 Tensor products

The *scalar product* $\underline{T} \cdot \boldsymbol{v}$ is a vector $\boldsymbol{V}$ of coordinates $V_i = T_{ij}v_j$. This operation may be represented by matrix notation in the following way:

$$\begin{pmatrix} V_1 \\ V_2 \\ V_3 \end{pmatrix} = \begin{pmatrix} T_{11} & T_{12} & T_{13} \\ T_{21} & T_{22} & T_{23} \\ T_{31} & T_{32} & T_{33} \end{pmatrix} \begin{pmatrix} v_1 \\ v_2 \\ v_3 \end{pmatrix} \tag{A.12}$$

The *scalar product to the left* $\boldsymbol{v} \cdot \underline{T}$ is the vector $\boldsymbol{U}$ of coordinates $U_i = v_j T_{ji}$, or, in matrix form:

$$\begin{pmatrix} U_1 \\ U_2 \\ U_3 \end{pmatrix} = (v_1\, v_2\, v_3) \begin{pmatrix} T_{11} & T_{12} & T_{13} \\ T_{21} & T_{22} & T_{23} \\ T_{31} & T_{32} & T_{33} \end{pmatrix} \tag{A.13}$$

Generally:

$$\underline{T} \cdot \boldsymbol{v} \neq \boldsymbol{v} \cdot \underline{T} \tag{A.14}$$

the equality being valid only if $\underline{T}$ is a symmetrical tensor ($T_{ij} = T_{ji}$). But we always have:

$$\underline{T} \cdot \boldsymbol{v} = \boldsymbol{v} \cdot \underline{T}' \tag{A.15}$$

$\underline{T}'$ being the conjugated tensor (the tensor represented by the transposed matrix).

The *scalar product* of two second order tensors $\underline{T} \cdot \underline{S}$ is by definition the tensor of components $T_{ij}S_{jk}$. This is represented by a matrix equal to the product of the two matrices (T_{ij}) and (S_{ij}).

The sum of two tensors of the same order is defined as the tensor of components equal to the sum of the corresponding components of the two tensors. Taking into account these definitions it follows that the tensor set constitutes a non-commutative algebra (the product of two elements is distributive, associative and non-commutative). The second order tensor algebra is isomorphic with the square matrix algebra.

The *double product* $\underline{T} : \underline{S}$ is a scalar

$$\underline{T} : \underline{S} \overset{\text{def}}{=} T_{ij}S_{ji} \tag{A.16}$$

obtained by index contraction.

Generally the product of two tensors $\underline{T}$ and $\underline{S}$ is a tensor of components $T_{ij}S_{kl}$. The products defined above represent particular cases.

A.4 Differential operators

Let us consider a scalar field $f(t, \boldsymbol{x})$ and introduce the symbol

$$\nabla \stackrel{\text{def}}{=} \boldsymbol{i}_i \frac{\partial}{\partial x_i} \tag{A.17}$$

The operator

$$\operatorname{grad} f \stackrel{\text{def}}{=} \nabla f \stackrel{\text{def}}{=} \boldsymbol{i}_i \frac{\partial f}{\partial x_i} \tag{A.18}$$

provides the possibility of expressing the derivative with respect to an arbitrary direction $\boldsymbol{n}$ by the formula:

$$\frac{\mathrm{d}f}{\mathrm{d}n} = \boldsymbol{n} \cdot \operatorname{grad} f = \boldsymbol{n} \cdot \nabla f$$

The following operators are defined for a vector field $\boldsymbol{u}(t, \boldsymbol{x})$:

$$\operatorname{div} \boldsymbol{u} \stackrel{\text{def}}{=} \nabla \cdot \boldsymbol{u} \stackrel{\text{def}}{=} \partial u_i/\partial x_i \tag{A.19}$$

$$(\operatorname{curl} \boldsymbol{u})_i \stackrel{\text{def}}{=} (\nabla \times \boldsymbol{u})_i \stackrel{\text{def}}{=} \delta_{ijk}\partial u_k/\partial x_j \tag{A.20}$$

Some identities using the above operators have been used in this work. The following are more frequent:

$$\operatorname{grad}(fg) = f \operatorname{grad} g + g \operatorname{grad} f \tag{A.21}$$

$$\operatorname{div}(f\boldsymbol{u}) = f \operatorname{div} \boldsymbol{u} + \boldsymbol{u} \cdot \operatorname{grad} f \tag{A.22}$$

$$\operatorname{curl}(f\boldsymbol{u}) = f \operatorname{curl} \boldsymbol{u} + (\operatorname{grad} f) \times \boldsymbol{u} \tag{A.23}$$

$$\operatorname{div}(\boldsymbol{u} \times \boldsymbol{v}) = \boldsymbol{v} \cdot \operatorname{curl} \boldsymbol{u} - \boldsymbol{u} \cdot \operatorname{curl} \boldsymbol{v} \tag{A.24}$$

$$\operatorname{curl}(\boldsymbol{u} \times \boldsymbol{v}) = (\boldsymbol{v} \cdot \nabla)\boldsymbol{u} - (\boldsymbol{u} \cdot \nabla)\boldsymbol{v} + \boldsymbol{u} \operatorname{div} \boldsymbol{v} - \boldsymbol{v} \operatorname{div} \boldsymbol{u} \tag{A.25}$$

$$\operatorname{grad}(\boldsymbol{u} \cdot \boldsymbol{v}) = (\boldsymbol{v} \cdot \nabla)\boldsymbol{u} + (\boldsymbol{u} \cdot \nabla)\boldsymbol{v} + \boldsymbol{v} \times \operatorname{curl} \boldsymbol{u} + \boldsymbol{u} \times \operatorname{curl} \boldsymbol{v} \tag{A.26}$$

$$\operatorname{curl} \operatorname{curl} \boldsymbol{u} = \operatorname{grad} \operatorname{div} \boldsymbol{u} - \Delta \boldsymbol{u} \tag{A.27}$$

Particularly if in (A.26) one has $\boldsymbol{u} = \boldsymbol{v}$ then it follows:

$$\operatorname{grad}(u^2/2) = (\boldsymbol{u} \cdot \nabla)\, \boldsymbol{u} + \boldsymbol{u} \times \operatorname{curl} \boldsymbol{u} \tag{A.26$'$}$$

One should notice that the *div* operator applied to a vector field produces a scalar field (the operator reduces the tensor order). It follows therefore that div $\underline{\boldsymbol{T}}$ represents a vector of components:

$$(\operatorname{div} \boldsymbol{T}_i) \overset{\text{def}}{=} \partial T_{ji}/\partial x_j \tag{A.28}$$

A.5 Integral relations

If $F(\boldsymbol{x})$ is a class C^1 function defined on the domain D of R_3 bounded by a sufficiently smooth surface ∂D then the following formula applies:

$$\int_D \partial F/\partial x_i \, \mathrm{d}v = \int_{\partial D} F n_i \, \mathrm{d}a \tag{A.29}$$

n_i being the components of the external normal, dv the volume element and da the surface element (these notations being used all over the book). Formula (A.29) could also be extended from the scalar function F to a vector and tensor function. The index i may be a fixed or a summation index.

Substituting V_i for F and adding we get the Gauss formula:

$$\int_D \operatorname{div} \boldsymbol{V} \mathrm{d}v = \int_{\partial D} \boldsymbol{n} \cdot \boldsymbol{V} \mathrm{d}a \tag{A.30}$$

Substituting $\delta_{kij} V_j$ for F from (A.29) we obtain:

$$\int_D \operatorname{curl} \boldsymbol{V} \mathrm{d}v = \int_{\partial D} \boldsymbol{n} \times \boldsymbol{V} \mathrm{d}a \tag{A.31}$$

With the substitution of F by T_{ij} it follows:

$$\int_D \operatorname{div} \underline{\boldsymbol{T}} \mathrm{d}v = \int_{\partial D} \boldsymbol{n} \cdot \underline{\boldsymbol{T}} \mathrm{d}a \tag{A.32}$$

Directly from (A.29) it also follows that:

$$\int_D \operatorname{grad} F \, \mathrm{d}v = \int_{\partial D} F \boldsymbol{n} \, \mathrm{d}a \tag{A.33}$$

Finally for a class C^1 vector $\boldsymbol{V}$ the Stokes formula is valid:

$$\int_\Gamma \boldsymbol{V} \cdot \mathrm{d}\boldsymbol{x} = \int_\Sigma \boldsymbol{n} \cdot \operatorname{curl} \boldsymbol{V} \mathrm{d}a \tag{A.34}$$

Γ being a smooth contour and Σ a smooth surface bounded by Γ and the normal $\boldsymbol{n}$ to Σ being positively oriented (the positive sense is defined in Section 1.1).

General references

Magnetofluid dynamics

1. H. Alfvén: *Cosmical electrodynamics.* Oxford University Press, (1950).
2. T. G. Cowling: *Magnetohydrodynamics*, Interscience Publishers, New York (1957).
3. L. Landau, E. Lifshitz: *Electrodynamics of continuous media*, Pergamon Press, Oxford (1960).
4. R. Landshoff (ed.): *Magnetohydrodynamics Symposium* (1957), Stanford University Press (1960).
5. D. Bershader (ed.): *The magnetohydrodynamics of conducting fluids*, Symposium (1958), Stanford University Press (1959).
6. E. Resler, W. Sears: *The prospects for magnetoaerodynamics*, J. Aeron. Sci. (1958), 235.
7. P. Germain: Introduction à l'étude de l'aéromagnétodynamique, *Cahiers de Phys.*, **103**, (1959), 98.
8. F. N. Frenkiel, W. R. Sears (eds.): *Magnetofluid dynamics*, Symposium (1960), *Rev. Mod. Phys.*, **32**, (1960).
9. V. Ferraro, C. Plumpton: *An introduction to magnetofluid mechanics*, Oxford University Press (1961).
10. A. Kulikovsky, G. Lyubimov: *Magnetohydrodynamics* (in Russian), Moscow (1962).
11. L. Napolitano, G. Contursi: *Magnetofluid dynamics*, Pergamon Press, Oxford (1962).
12. S. I. Pai: *Magnetogasdynamics and plasma dynamics*, Springer Verlag, Wien (1962).
13. H. Alfvén, C. G. Fälthammar: *Cosmical electrodynamics*, Clarendon Press, Oxford (1963).
14. I. Imai (ed.): *Progr. Theor. Phys., Suppl.*, **24**, (1963).
15. I. A. Shercliff: *A textbook of magnetohydrodynamics*, Pergamonn Press, Oxford (1965).
16. R. J. Seeger, G. Temple (eds.): *Research frontiers in fluid dynamics*, Interscience, New York (1965).
17. G. W. Sutton, A. Sherman: *Engineering magnetohydrodynamics*, McGraw-Hill, New York (1965).
18. Proceedings of symposia in applied mathematics, vol. 18: *Magnetofluid and plasma dynamics Symposium* (1965).
19. C. Agostinelli: *Magnetofluidodinamica*, Cremonese, Roma (1966).
20. W. F. Hughes, F. J. Young: *The electromagnetodynamics of fluids*, John Wiley and Sons, New York (1966).
21. A. Jeffrey: *Magnetohydrodynamics*, Oliver and Boyd, London (1966).
22. A. Jeffrey, T. Taniutti (eds.): *Magnetohydrodynamic stability and thermonuclear containment*, Academic Press, New York (1966).
23. H. Cabannes: *Magnétodynamiques des fluides*, CDU, Paris (1969).
24. L. Dragoş: *Magnetofluid dynamics* (in Romanian), Editura Academiei, Bucharest (1969).

25. La magnétohydrodynamique classique et relativiste, *Colloque CNRS*, n° 184, Lille (1969).
26. L. I. SEDOV: *Mechanics of continuous media* (in Russian), Izd. Nauka, Moscow (1970).
27. A. B. VATAJIN, G. A. LIUBIMOV, S. A. RSGHIRER: *Magnetohydrodynamic flows in channels*, (in Russian), Izd. Nauka, Moscow (1970).

Classical fluid dynamics

28. C. JACOB: *Introduction mathématique à la mécanique des fluides*, Editura Academiei-Gauthier Villars, Bucarest, Paris (1959).
29. J. SERRIN: Mathematical principles of classical fluid mechanics, *Hand. der Phys.*, VIII/1 (1959).

Mathematics

30. R. COURANT, D. HILBERT: *Methods of mathematical physics*, vol. I and II, New York (1957), (1962).
31. R. CRISTESCU, G. MARINESCU: *Applications of the theory of distributions*, Editura Academiei, Bucureşti-John Wiley & Sons, London—New York—Sydney—Toronto (1973),
32. G. M. FIKHTENGOLTZ: *Differential and integral calculus* (in Russian), vol. I—III, Gostehizdat, Moscow, (1963).
33. I. M. GHELFAND, G. F. CHILOV: *Les distributions*, Dunod, Paris, (1962).
34. N. I. MOUSKHELISHVILI: *Singular integral equations*, (in Russian), Moscow (1962).
35. W. SCHMEIDLER: *Integral Gleichungen mit Anwendungen in Physik und Technik*, Leipzig (1955).
36. L. SCHWARTZ: *Méthodes mathématiques pour les sciences physiques*, Hermann, Paris (1962).
37. N. WATSON: *A treatise on the theory of Bessel functions*, Cambridge, University Press (1922).

Special references

Chapter 1

1. J. A. STRATTON: *Electromagnetic theory*, McGraw-Hill Inc., New York (1941).
2. A. SOMMERFELD: *Electrodynamics*, Academic Press Inc., New York (1952).
3. C. MØLLER: *The theory of relativity*, Oxford University Press (1952).
4. H. MINKOWSKI: *Nachr. Ges. Wiss.* Götingen, **53** (1908).
5. A. TIMOTIN: Dynamical properties of macroscopic magnetic fields in some material media, *Ph. D. thesis*, Bucharest (1958).
6. J. D. JACKSON: *Classical electrodynamics*, John Wiley, New York (1962).
7. W. K. H. PANOFSKY, M. PHILLIPS: *Classical electricity and magnetism*, Addison-Wesley Publishers, Cambridge (1962).
8. J. L. SYNGE: *Relativity. The special theory*, Amsterdam (1956).
9. D. HOMENTCOVSCHI: On the jump conditions in electrodynamics, *Stud. Cerc. Mat.*, **21** (1969), 1041;
   Some applications of distributions in the electromagnetic field theory (to be published).
10. P. PENFIELD, JR., H. A. HAUS: *Electrodynamics of moving media*, MIT Press, Cambridge (1967).

Chapter 2

1. A. C. ERINGEN: *Non-linear theory of continuous media*, McGraw-Hill (1962).
2. W. JANNZEMIS: *Continuum mechanics*, MacMillan (1967).
3. S. LUNDQUIST: Studies in magnetohydrodynamics, *Arkiv für Physik*, **5** (1952), 297.
4. S. P. PAI: Energy equations in magnetogas dynamics, *Phys. Rev.*, **105** (1057), 1424.
5. S. GOLDSTEIN: *Lectures on fluid mechanics*, Interscience Publishers, New York (1960).
6. BOA-THEH CHY: Thermodynamics of electrically conducting fluids, *Phys. Fluids*, **2** (1959), 473.
7. L. BRAND: The pi theorem of dimensional analysis, *Arch. Rat. Mech. Anal.*, **1** (1957), 35.
8. W. M. ELSASSER: Dimensional relations in MHD, *Phys. Rev.*, **95** (1954).
9. A. BEISER, B. RAAB: *Phys. Fluids*, **4** (1961), 177.
10. M. N. KOGAN: On the conservation of vortex and currents in MHD, *Doklad Akad Nauk, SSSR*, **139** (1961).
11. W. M. ELSASSER: Induction effects in terrestrial magnetism, *Phys. Rev.*, **69** (1946), 106.
    Hydromagnetic dynamo theory, *Rev. Mod. Phys.*, **28** (1956), 135.

12. P. SMITH: The steady magnetodynamics flow of perfectly conducting fluids, *J. Math. Mech.*, **12** (1963), 505.

13. G. POWER, D. WALKER: Plane gasdynamic flow with orthogonal magnetic and velocity field distribution, *Z. Angew. Math. Phys.*, **16** (1965), 803.

14. Y. KATO, T. TANIUTTI: Hydromagnetic plane steady flow in compressible ionized gases, *Progr. Theor. Phys.*, **21** (1959), 609.

15. P. GERMAIN: Sur certains écoulements d'un fluide parfaitement conducteur, *Rech. Aéro.*, **74** (1960), 13.

16. H. GRAD: Reducible problems in magneto-fluid dynamic steady flows, *Rev. Mod. Phys.*, **32** (1960), 830.

Uniquenesss and existence theorems

17. R. NARDINI: Due teoremi di unicità nella teoria della onde magnetoidrodinamiche, *Rend. Sem. Mat. Univ. Padova* **21** (1952), 303;
Due teoremi di unicità nella magnetodinamica dei fluidi compressibili, *Bull. Un. Mat. Ital.*, **7** (1952), 403;
Lösung eines Randwertproblems der Magneto-Hydrodyn., *Z. Angew. Math. Phys.*, **33** (1953), 304.

18. P. P. KANWALL: Uniqueness of hydromagnetic flows, *J. Rat. Mech. Anal.*, **4** (1960), 335.

19. I. FERRARI: Sul teorema di unicità per equazioni della magnetodinamica di un fluido compressible in un dominio illimitato, *Atti Sem. Mat. Fiz., Univ. Modena*, **10** (1961),158.

20. D. GRAFF: Sur un théorème d'unicité pour le mouvement d'un fluide visqueux dans un domaine illimité, *C. R. Acad. Sci. Paris*, **249** (1959), 1741;
Sul teorema di unicità nella dinamica dei fluidi, *Annali di Mat.*, **50** (1960), 379;
Sul teoremà de unicità per le equazioni del moto dei fluidi compressibili in un dominio illimitato, *Atti della Acad. della Sci. Inst. Bologna*, **7** (1960), 1.

21. H. DYER, D. E. EDMUNDS: A uniqueness theorem in magnetohydrodynamics, *Arch. Rat. Mech. Anal.*, **8** (1961), 254;
On the existence of the equations of magnetohydrodynamics, *Arch. Rat. Mech. Anal.*, **9** (1962), 403.

22. D. E. EDMUNDS: Sur l'unicité des solutions des équations de la magnétohydrodynamique, *C. R. Acad. Sci. Paris*, **254** (1962), 1377;
Sur les équations différentielles de la magnétohydrodynamique, *C. R. Acad. Sci. Paris*, **254** (1962), 4248.

23. J. FÖRSTE: Ein Existenzsatz für stationäre Stromungen in der Magnetohydrodinamik, *Mon. Deut. Akad. Wiss.*, **6** (1964), 886;
Zum stationären magnetohydrodynamischen Umströmungsproblem, *Mon. Deut. Akad. Wiss, Berlin*, **7** (1965), 1; Ein Einzigkeitsatz für stationäre Strömungen in der Magnetodynamik, *Mon. Deut. Akad. Wiss., Berlin*, **9** (1967), 241.

24. M. E. SAUCHEZ-PALENCIA: Théorème d'éxistence pour certains écoulements stationnaires à nombre de Reynolds magnétique nul, *C. R. Acad. Sci. Paris*, **A263** (1966), 141;
Théorème d'existence et d'unicité pour certains écoulements non-stationnaires à nombre de Reynolds magnétique nul, *C. R. Acad. Sci. Paris*, **A264** (1967), 363; Théorème d'existence pour l'écoulement autour d'une plaque plane, *C. R. Acad. Sci. Paris*, **A265** (1967), 354;

Existence de solutions de certains problèmes aux limites et magnétohydrodynamique, *J. Méc.*, **7** (1968), 405;
Quelques résultats d'existence et d'unicité pour des écoulements magnétohydrodynamique non-stationnaires, *J. Méc.*, **8** (1969), 509.

25. G. Lassner: Über ein Rand-Anfangswertproblem der Magnetohydrodynamik, *Arch. Rat. Mech. Anal.*, **25** (1967), 388.
26. A. B. Tzinober: *Magnetohydrodynamic flows past bodies* (in Russian), Izd. Zinatne, Riga (1970).

Electrofluid dynamics

27. G. O. M. Stuetzer: Magnetohydrodynamics and electrohydrodynamics *Phys. Fluids*, **5** (1962), 534.
28. H. Hasimoto, S. Kuwabava: Electrogasdynamics, *J. Phys. Soc. Japan*, **20** (1965), 589.
29. J. R. Melcher, G. I. Taylor: Electrohydrodynamics, *Ann. Rev. Fluid Mech.* W. R. SEARS, M. VAN DYKE (editors), **1** (1969).

Supplementary references

30. W. E. Williams: Exact solutions of the magnetohydrodynamic equations, *J. Fluid. Mech.*, **8** (1960), 452.
31. M. Gourdine: Magnetohydrodynamic flow constructions with fundamental solutions, *J. Fluid. Mech.*, **10**, (1961), 439.
32. Fr. H. Clauser: Concept of field modes and the behaviour of magnetohydrodynamic field, *Phys. Fluids*, 6, (1963), 231.
33. M. S. Uberoi: Some exact solutions of magnetohydrodynamics, *Phys. Fluids*, **6**, (1963), 1379.
34. K. B. Ranger: Hydromagnetic momentum source, *Phys. Fluids*, **8** (1965), 1747.
35. M. D. Savage: A method for determining linear solutions in magnetohydrodynamics, *J. Inst. Math. Appl.*, 4 (1968), 20.
36. R. P. Kanwall: Boundary value problems in magnetohydrodynamics, Magnétohydrodynamique classique et rélativiste, Lille 1969, *Coll. Centre Nat. Rech. Sci.*, **184** (1970). 201.
37. C. Sozou: On fluid motion induced by an electric current source, *J. Fluid. Mech.*, **46** (1971), 25; On some exact solutions in magnetohydrodynamics with astrophysical applications, *Idem* **51** (1972), 33; Fluid motion induced by an electric current jet, *Phys. Fluids*, **15** (1972), 272.
38. M. L. Bruce: Velocity measurements in regions of upstream influence of a body in aligned-fields MHD flow, *J. Fluid. Mech.*, **50** (1971), 209.
39. V. E. Zakharov: Hamiltonian formalism for hydrodynamic plasma models, *Soviet Phys. JETP*, **33** (1971), 927.
40. A. Chambarel: Recherche de l'équilibre des systèmes magnétohydrostatiques, *C. R. Acad. Sci. Paris* Sér. **A274** (1972), 278: Équilibres magnétohydrostatiques stationnaires et analogie hydrodynamique, *Idem* **276** (1973), 149.
41. P. Casal: Équations à potentiels en magnétodynamique des fluids, *C. R. Acad. Sci. Paris* Sér. **A274** (1972), 806.
42. J. T. Jenkins: A theory of magnetic fluids, *Arch. Rat. Mech. Anal.* **46** (1972), 42.
43. Shin-I Liu, Ingo Müller: On the thermodynamics and thermostatics of fluids in electromagnetic fields, *Arch. Rat. Mech. Anal.*, **46** (1972), 149.

44. A. M. SOWARD: A kinematic theory of large magnetic Reynolds number dynamos, *Philos. Trans. Roy. Soc. London* Ser. **A272** (1972), 431.

45. MARIA TERESA VACCA: Alcune funzioni potenziali in magnetoidrodinamica, *Atti Acad. Naz. Lincei* VIII, Ser. Rend., Cl. Sci. Fiz. Mat. Natur., **52** (1972), 357.

46. H. K. MOFFATT: Report on the NATO advanced study institute on magnetohydrodynamic phenomena in rotating fluids, *J. Fluid Mech*., **57** (1973), 625.

47. V. I. NATH, O. P. CHANDNA: On plane viscous magnetohydrodynamic flows, *Quart. Appl. Math*., **31** (1973), 351.

Chapter 3

1. J. HARTMANN: Theory of the laminar flow of an electrically conducting liquid in a homogeneous magnetic field. *Kgl. Danske Videnskabernes Selskab, Math. Phys. Med., 15* (1937), No. 6.

2. J. HARTMANN, F. LAZARUS: Experimental investigations on the flow of mercury in a homogeneous magnetic field, *Idem* No. 7.

3. B. LEHNERT: On the behaviour of an electrically conducting liquid in a magnetic field, *Arkiv für Fys*., **5** (1952), 69.

4. J. A. SHERCLIFF: Steady motion of conducting fluids in pipes under transverse magnetic fields, *Proc. Cambr. Phil. Soc*., **49** (1953), 136.

5. E. L. RESLER JR., W. R. SEARS: Magneto-gas dynamic channel flow, *Z. Angew. Math. Phys*., **9** (1958), 509.

6. Z. O. BLEWISS: Magneto-gas dynamics of hypersonic Couette flow *J. Aero-Space Sci*., **25** (1958), 10.

7. CH. C. CHANG, TH. S. LUNDGREN: Duct flow in magneto-hydrodynamics *Z. Angew. Math. Phys*., **12** (1961) ,100.

8. G. A. GREENBERG: Steady flow of a conducting fluid in rectangular duct with two non-conducting and two conducting walls parallel to the external magnetic field, *Prikl. Math. Meh*., **25** (1961), 1024; **26** (1962), 80.

9. H. SATO: The Hall effect in the viscous flow of ionized gas between parallel plates under transverse magnetic fields, *J. Phys. Soc. Japan*, **16** (1961), 1427.

10. J. T. YEN, CH. CHANG: Magnetohydrodynamic channel flow under time-dependent pressure gradient, *Phys. Fluids*, **4** (1961), 1355; Magnetohydrodynamic Couette flow as affected by wall electrical conductances *Z. Angew. Math. Phys*., **15** (1964), 205.

11. E. CAUSSE, R. CAUSSE, Y. POIRIER: Mise en vitesse entre plans paralleles indefinis d'un liquide electroconducteur soumis à un champ magnétique transversal. *C.R. Acad. Sci. Paris*, **254** (1962), 216; **256** (1963).

12. V. I. ROSSOW: On magnetoaerodynamic boundary layers, *Z. Angew. Math. Phys*., **9**, (1958), 519.

13. MASAKAZU KATAGINI: Flow formation in Couette motion in magnetohydrodynamics, *J. Phys. Soc. Japan*, **17** (1962), 393.

14. I. TANI: Steady flow of conducting fluids in channels under transverse magnetic field with consideration of Hall effect, *J. Aero-Space Sci*., **29** (1962), 3.

15. J. A. SHERCLIFF: The flow of conducting fluids in circular pipes under transverse magnetic fields, *J. Fluid. Mech. 1* (1956) 644; Magnetohydrodynamic pipe flow. High Hartmann number, *J. Fluid. Mech*., **13** (1962), 513.

16. W. E. WILLIAMS: Magnetohydrodynamic flow in a rectangular tube at high Hartmann number, *J. Fluid Mech.*, **16** (1963).

17. PAN CHANG LU: A study of Kantorovich's variational method in MHD duct flow, *AIAA Journ.*, **5** (1967), 1519.

18. J. C. R. HUNT: Magnetohydrodynamic flow in rectangular ducts, *J. Fluid Mech.*, **21**(1965), 577; An uniqueness theorem for magnetohydrodynamic duct flow, *Proc. Cambr. Phil. Soc.* **65**(1969), 319.

19. D. CHANG, TH. LUNDGREN: Magnetohydrodynamic flow in a rectangular duct with perfectly conducting electrodes, *Z. Angew. Math. Phys.*, *18*(1967) 92; Solution of a class of singular integral equation, *Quart. Appl. Math.*, **24** (1967), 303.

20. J. D. COLE: *Perturbation methods in applied mathematics*, Blaisdell Publish. Comp., Waltham, Massachusetts (1968).

21. L. PAMELA COOK, C. S. S. LANDFORD, J. S. WALKER: Corner regions in the asymptotic solution of $\varepsilon\nabla^2 u = \partial u/\partial y$ with reference to MHD duct flow, *Proc. Camb. Phil. Soc.*, **72** (1972), 117.

22. L. P. HARRIS: *Hydromagnetic channel flow*, John Wiley, New York (1960).

23. S. D. NIGAM, S. N. SINGH: Heat transfer by laminar flow between parallel plates under the action of transverse magnetic field, *Quart. J. Mech. and Appl. Math.*, **13** (1960), 85.

24. S. A. REGHIRER: Electroconducting flow in a duct with arbitrary profiles in a magnetic field, *Prikl. Mat. Meh.*, **24** (1960), 541; An exact solution of MHD equation, **24** (1960), 383; MHD flow in ducts in the presence of a longitudinal current, **31** (1967), 356.

25. I. G. UFLIAND: Steady flow of an electroconducting fluid in rectangular duct in the presence of an external magnetic field, *J. Teh. Phys.*, **30** (1960), 10.

26. T. S. LUNDGREN, B. H. ATABEK, C. C. CHANG: Transversal magnetohydrodynamic duct flow, *Phys. Fluids*, **4** (1961), 1006.

27. E. DAHLBERG: On the one-dimensional flow of a conducting gas in crossed fields, *Quart. Appl. Math.*, **19** (1961), 177.

28. C. C. CHANG, J. T. YEN: Magneto-hydrodynamic channel flow as influenced by wall conductance, *Z. Angew. Math. Phys.*, **13** (1962), 266.

29. P. R. GOLD: Magnetohydrodynamic pipe flow I, *J. Fluid Mech.*, **13** (1962), 505.

30. F. D. HAINS, Y. A. YOLER: Axisymmetric MHD channel flow, *J. Aerosp. Sci.*, **29** (1962), 143.

31. R. K. RATHY: Hydromagnetic Couette's flow with suction and injection, *Z. Angew. Math. Phys.*, **43** (1963), 370.

32. A. E. YAKUBENKO: Some problems of electroconducting flow in a flat duct, *Prikl. Meh. i Teh. Phys.* (1963), 7.

33. ZIMIN: Flow in a flat MHD duct, *Prikl. Mech. i Tech. Phys.* (1963), 108.

34. L. F. ERIKSON, C. S. WANG, C. L. HWAND: Heat transfer to magnetic flow in a flat duct, *Z. Angew. Math. Phys.*, **15** (1964), 408.

35. P. V. SIDLOVSKI: Some non-steady flow of viscous electroconducting fluids in a magnetic field, *St. Cerc. Mec. Apl,*. **16** (1964), 1009.

36. D. SINGH, S. RIZVI: Unsteady motion of a conducting liquid between two infinite coaxial cylinders, *Phys. Fluids* **7** (1964), 760.

37. N. N. NARASHIMHAN: Transient MHD flow in an annular channel, *J. Mat. Mech.*, **14** (1965), 353.

38. R. PEYRET: Ecoulement en conduite d'un fluide compressible conducteur de l'électricité sous l'action d'un champ magnetique non uniforme, *C.R. Acad. Sci. Paris*, **261** (1965), 1591.

39. R. M. SINGER: Transient MHD flow and heat transfer, *Z. Angew. Math. Phys.*, **16** (1965), 483.

40. L. V. WOLFERSDORF: MHD flow in a duct with nonhomogeneous conductivity of walls, *St. Cerc. Mec. Apl.*, **18** (1965), 285.

41. D. G. DRAKE, A. M. ABU-SITTA: Magnetohydrodynamic flow in a rectangular channel at high Hartmann number, *Z. Angew. Math. Phys.*, **17** (1966), 519.

42. N. MARCOV: Electroconducting flow between two coaxial cylinders in transverse magnetic field, *St. Cerc. Mat.*, **18** (1966), 161.

43. D. M. SLOAN, P. SMITH: Magnetohydrodynamic flow in a rectangular pipe between conducting plates, *Z. Angew. Math. Mech.*, **46** (1966), 439.

44. L. TODD: Hartmann flow between parallel planes, *Phys. Fluids*, **9** (1966), 1602; Hartmann flow in an annular channel, *J. Fluid Mech.*, **28** (1967), 371.

45. J. C. R. HUND, S. LEIBOVICH: MHD flow in channels of variable cross-section with strong transverse magnetic fields, *J. Fluid Mech.*, **28** (1967), 241.

46. U. P. HWANG, L. T. FAN, C. L. HWANG: Compressible laminar MHD flow inside a flat duct with heat transfer, *AIAA Journ.*, **5** (1967), 2113.

47. D. A. OLIVER, M. MITCHNER: Non-uniform electrical conduction in MHD channels *AIAA J*, **5** (1967), 1424.

48. J. C. R. HUNT, W. E. WILLIAMS: Some electrically driven flows in magnetohydrodynamics, Part 1. Theory, *J. Fluid Mech.*, **31** (1968), 705.

49. D. SLOAN: MHD flow in an insulated rectangular duct with an oblique transverse magnetic field, *Z. Angew. Math. Mech.*, **47** (1967), 109.

50. J. C. R. HUNT, C. S. S. LUNDFORD: Three-dimensional MHD duct flows with strong transverse magnetic fields, *J. Fluid Mech.*, **33** (1968), 693.

51. J. C. R. HUNT, D. G MALCOLM: Some electrically driven flows in MHD, Part 2, theory and experiment, *J. Fluid Mech.*, **33** (1968), 775.

52. C. P. YU, H. K. YANG: Effect of wall conductances on convective MHD channel flow, *App. Sci. Res.*, **20** (1969), 16.

53. G. F. BUTLER: A note on MHD duct flow, *Proc. Camb. Phil. Soc.*, **66** (1969), 655.

54. K. L. ARORA, P. R. GUPTA: Magnetohydrodynamic flow between two rotating coaxial cylinders under radial magnetic field, *Phys. Fluids*, **15** (1972), 1146.

55. P. SMITH: Some extremum principles for magnetohydrodynamic flow in conducting pipes, *Proc. Cambridge Philos. Soc.*, **72** (1972), 303; Some extremum principles for pipe flow in magnetohydrodynamics, *Z. Angew. Math. Phys.*, **23** (1972), 753.

56. V. M. SOUNDALGERKAR, D. D. HALDAVNEKAR: Effects of external circuit on MHD Couette flow between conducting walls with heat transfer, *Indian J. Pure. Appl. Math.*, **3** (1972), 64.

57. J. S. WALKER, G. S. S. LUDFORD, J. C. HUNT: Three-dimensional MHD duct flows with strong transverse magnetic fields III: Variable-area rectangular ducts with insulating walls, *J. Fluid. Mech.*, **56** (1972), 121.

58. J. S. WALKER, G. S. S. LUDFORD: Three-dimensional MHD duct flows with strong transverse magnetic fields IV: Fully insulated, variable area rectangular ducts with small divergence, *J. Fluid. Mech.*, **56** (1972), 481.

59. S. T. WU: Unsteady MHD duct flow by the finite element method, *Internat. J. Number. Meth. Engng.*, **6** (1973), 3.

Chapter 4

1. M. N. SAHA: Ionization in the solar chromosphere, *Phil. Mag.*, **40** (1920), 472.
2. E. M. DEWAN: Generalized Saha's equation, *Phys. Fluids,* **4** (1961), 4.
3. R. J. ROSA: Physical principles of MHD power generation, *Phys. Fluids,* **4** (1961), 182.
4. S. C. LIN, E. L. RESLER, A. KANTOROWITZ: Electrical conductivity of highly ionized argon produced by shock waves, *J. Appl. Phys.*, **26** (1955), 96.
5. G. SUTTON, A. W. CARLSON: End effects in inviscid flow in a MHD channel, *J. Fluid Mech.*, **11** (1961), 121.
6. NGUYEN-NGOC-TRAN: Etude des effets d'extremité sur l'écoulement dans les tuyèrs MHD, *C. R. Acad. Sci. Paris,* **262A** (1966), 583.
7. L. DRAGOŞ: Sur l'influence des extrémités des électrodes sur l'écoulement MHD dans les cannaux I, *C. R. Acad. Sci. Paris,* **263A** (1966), 328; Contributions à l'étude des effets d'extrémité sur l'écoulement dans les générateurs MHD, *Rev. Roum. Math. Pures et Appl.*, **12** (1967), 1193.
8. L. DRAGOŞ: Etude de l'effet des extrémités des électrodes finies sur l'écoulement dans un générateur MHD, *C. R. Acad. Sci. Paris,* **264A** (1967), 69; Contributions à l'étude des effets des extrémités dans un générateur à électrodes finies, *Proc. Cambr. Phil. Soc.*, **64** (1968), 535.
9. NGUEN-NGOC-TRAN: Etude des effets d'extrémité et de conductivité électrique non uniforme et anisotrope sur l'écoulement dans les tuyères MHD, *C. R. Acad. Sci. Paris*, **262** A (1966), 649.
10. L. DRAGOŞ: Influence du tenseur conductivité électrique sur l'écoulement dans un générateur MHD, *C. R. Acad. Sci. Paris*, **264A** (1967), 138; Influence du tenseur conductivité éléctrique sur l'écoulement dans un générateur à électrodes finies, *Proc. Cambr. Phil. Soc.*, **64** (1968), 549.
11. NGUEN-NGOC-TRAN: Contribution à l'étude théorique des effets d'extrémités dans les tuyères MHD, *Thèse, Faculté de Sci. Paris*, 1970.
12. L. DRAGOŞ: Sur l'influence des extrémités des électrodes sur l'écoulement MHD dans les cannaux II, *C. R. Acad. Sci. Paris*, **263A** (1966), 352; Etude des effets d'extrémité et du champ magnétique induit dans un générateur MHD, *Int. J. Engng. Sci.*, **5** (1967), 919.
13. N. MAGNUS, F. OBERHETTINGER: *Formeln und Salze für die speziellen Funktionen der Math-Phys.*, Springer-Verlag, Berlin, 1948.
14. H. HURWITZ JR., R. W. KILB, G. W. SUTTON: Influence of tensor conductivity on current distributions in a MHD generator, *J. Appl. Phys.*, **32** (1961), 205.
15. L. S. DZUNG: Der MHD generator mit Hall-Effect am Kanalende, *Brown Boveri Mitt.*, **49** (1962), 211; The MHD generator in crossconnection, *Internat. Symp. MHD electrical power generation*, Paris, 1964, vol. **2**, 601; Influence of wall conductance on the performance of MHD generators with segmented electrodes, *Electricity from MHD*, vol. **2**, Vienna (1966), 169; MHD generators with ion slip and finite electrode segments, *Idem* 177.
16. R. ELCO, W. HUGHES, FR. YOUNG: Theoretical analysis of the radial field vortex magneto-gas dynamic generator, *Z. Angew. Math. Phys.*, **13** (1962), 1.

17. A. Sherman, G. W. Sutton: The combined effect of tensor conductivity and viscosity on a MHD generator with segmented electrodes, in "*Magnetohydrodynamics*" Northwest Univ. Press, Evanston (1962), 173.

18. B. Vatajin: Some two-dimensional problems on the current distribution in a conducting medium moving through a channel in a magnetic field, *Prikl. Meh. Tehn. Fiz.*, No. 2 (1963), 39, (in Russian); The definition of Joule dissipation, *idem*, No. 4 (1964), 122,(in Russian); On the velocity profile deformation in an inhomogeneous magnetic field, *Prikl. Mat. Meh.*, **31** (1967), 72, (in Russian); Braking of the conducting medium moving through a channel in an inhomogeneous magnetic field, *idem*, **32** (1968), 869, (in Russian).

19. G. W. Sutton: End losses in MHD channels with tensor electrical conductivity and segmented electrodes, *J. Appl. Phys.*, **34** (1963), 396.

20. F. Culick: Compressible magnetogasdynamic channel flow, *Z. Angew. Math. Phys.*, **15** (1964), 126.

21. E. K. Halschevnikova: The integral characteristics of the MHD generator with two pairs of finite-length electrodes, *Prikl. Meh. Teh. Fiz.*, No. 4 (1964), 16 (in Russian); The Hall-effect influence on the MHD generator characteristics, *idem*, No. 4 (1966), 74 (in Russian).

22. *Internat Sympos. MHD electrical power generation*, Paris, 1964.

23. J. T. Yen: Magnetoplasmadynamic channel flow and energy conversion with Hall currents, *Phys. Fluids*, **7** (1964), 723; Viscous, Hall and ion-slip effects on MHD channel flow and power generation, *Internat. Sympos.*, *Paris* (1964), 689; MHD power generation as affected by viscosity and non-equilibrium ionization, *Electricity from MHD* **1**, Vienna (1966), 403.

24. A. A. Barmin, A. G. Kulikovski & L. F. Lobanova: Linearized problem of supersonic flow on the entrance to the electrode region of a MHD channel, *Prikl. Mat. Meh.*, **29** (1965), 609.

25. *Electricity from MHD*, Vienna, 1966 (Symp. Salzburg 1966) and Vienna 1968 (Symp. Warsaw 1968).

26. I. P. Emetz: On the current distribution near the electrodes in the presence of the Hall effect, *Prikl. Meh. Teh. Fiz.*, No. 3 (1966), 35 (in Russian); The current distribution in a plane MHD channel for the motion of conducting fluid in a strong magnetic field, *idem*, No. 3 (1967), 4 (in Russian); On the compensation of electrode end effects in MHD conversion, *Magnit. Gidrodinamika* (1968), 103 (in Russian).

27. R. Peyret: Sur l'écoulement bidimensionnel non-stationnaire dans un accélérateur de plasma à ondes progressives, *C. R. Acad. Sci.*, *Paris*, **262A** (1966), 92; Contribution à l'étude théorique de l'écoulement dans un accélérateur de plasma à ondes progressives, *Thèse*, *Paris*, 1967.

28. E. A. Witalis: Methods for determination of currents and fields in steady two-dimensional MHD flow with tensor conductivity, *J. Nucl. Energy*, **8** (1966), 129.

29. A. B. Vatajin, E. K. Halschevnikova: The flow of a conducting anisotropic medium in the inlet region of a channel in a magnetic field, *Prikl. Meh. Teh. Fiz.* (1967), 9, No. 5.

30. A. G. Riabinin, A. I. Hojamov: The unsteady flow of an electrically-conducting fluid in a MHD generator channel, *Prikl. Meh. Teh. Fiz.*, No. 2 (1967), 31 (in Russian).

31. R. T. Waechter: Steady electrically driven flows, *Proc. Camb. Phil. Soc.*, **64** (1968), 871.

32. J. C. R. Hunt, K. Stewartson: Some electrically driven flows in magnetohydrodynamic circular electrodes, *J. Fluid Mech.*, **38** (1969), 225.

33. J. P. EMEC: The problem of electric current eddies at entry an anisotropically conducting medium to a magnetic field, *Prikl. Mat. Meh.*, **35** (1971), 229.
34. W. E. POTTER, L. M. GROSSMAN: Optimization problems in magnetohydrodynamic flow, *Z. Angew. Math. Phys.*, **22** (1971), 552.

Chapter 5

1. W. R. SEARS, E. L. RESLER JR.: Theory of thin airfoils in fluids of high electrical conductivity, *J. Fluid Mech.*, **5** (1959), 257.
2. K. STEWARTSON: Magneto-fluid dynamics of thin bodies in oblique fields, *Z. Angew. Math. Phys.*, **12** (1961), 261.
3. L. DRAGOŞ: On the motion of perfectly conducting fluids past a thin airfoil in oblique fields, *Stud. Cerc. Mat.*, **24** (1972), 159.
4. W. R. SEARS: Magnetodynamic effects in aerodynamic flows, *ARS Journ.* (1959), 397.
5. J. E. MC CUNE, E. L. RESLER JR.: Compressibility effects in magneto-aerodynamic flows past thin bodies, *J. Aero-Space Sci.*, **27** (1960), 493.
6. P. GREENBERG: Some magnetodynamics of a thin body in perpendicular fields, *Z. Angew. Math. Phys.* **18** (1967), 523.
7. L. DRAGOŞ: On the motion of a compressible, perfectly conducting fluids past a thin airfoil, *Stud. Cerc. Mat.*, **24** (1972), 173.
8. D. HOMENTCOVSCHI: On the motion of a compressible perfectly conducting fluid past a thick airfoil, *Stud. Cerc. Mat.*, **20** (1968), 191.
9. E. CUMBERBATCH, L. SARASON, H. WEITZNER: Magnetohydrodynamic flow past a thin airfoil, *AIAA Journ.*, **1** (1963), 679.
10. P. GREENBERG: Some magnetohydrodynamics of a thin body in oblique fields, *AIAA Journ.*, **5** (1967), 1047.
11. E. P. SALATHÉ and L. SIROVICH: Boundary value problems in compressible magnetohydrodynamics, *Phys. Fluids*, **10** (1967), 1477.
12. I. IMAI: Thin airfoil theory in magnetohydrodynamics, in "La magnétohydrodynamique classique et relativiste", *Colloq. CNRS, Lille* (1969), 263.
13. L. DRAGOŞ, D. HOMENTCOVSCHI: Sur l'écoulement d'un fluide incompressible conducteur limité par une paroi isolatrice infinie, *Z. Angew. Math. Phys.*, *19* (1968), 381.
14. T. TANIUTI: An example of isentropic steady flow in magnetohydrodynamics, *Progr. Theor. Phys.*, **19** (1958), 749
15. M. N. KOGAN: Two dimensional and axial-symetric magnetohydrodynamic flow of a gas with infinite electrical conductivity (in Russian), *Prikl. Mat. Meh.*, **23** (1959), 1.; Two dimensional flow of an ideal gas with infinite electrical conductivity on a magnetic field parallel to the fluid velocity (in Russian), *Idem*, **24** (1960), 1.
16. C. K. CHU, Y. M. LYNN: Steady magnetohydrodynamic flow past a nonconducting wedge, *AIAA Journ.*, **1** (1963), 1062.
17. Y. MIMURA: Magnetohydrodynamic flow past a wedge with a perpendicular magnetic field, *AIAA Journ.*, **1** (1963), 2272.
18. C. K. CHU: Linearized hyperbolic steady magnetohydrodynamic flow past nonconducting walls, *Phys. Fluids*, **7** (1964), 707.
19. C. W. SWAN: Upstream influence effects in the flow of a conducting fluid over an insulating wall, *Quart. J. Mech. Appl. Math.*, **18** (1965), 243.

20. K. Stewartson: On the motion of a non-conducting body through a perfectly conducting fluid, *J. Fluid. Mech.*, **8** (1960), 82; Motion of bodies through conducting fluids, *Rev. Mod. Phys.*, **32** (1960), 855.

21. L. E. Ring: Theory of unsteady flow about thin profiles, *J. Fluid Mech.*, **11** (Nov., 1961).

22. D. Homentcovschi: On the unsteady motion of an insulating body in electroconducting fluids *Bull. Math. de la Soc. Roum. de Math.*, **16** (1972), 31.

23. S. Leibovich, G. S. Ludford: The transient hydromagnetic flow past an airfoil for an aligned magnetic field, *J. Mec.* **4** (1965), 21.

24. G. S. Ludford, S. Leibovich: The ultimate hydromagnetic flow past an airfoil for an aligned magnetic field, *Z. Angew Math. Phys.*, **45** (1965), 113.

25. G. S. Ludford: On the controversy over an aligned magnetic field, *Proc. Symp. on Appl. Math.*, **18** (1965), 57.

26. G. S. Ludford, D. W. Yannipell: The super Alfvénic nature of Sears-Resler flow, *J. Méc.*, **6** (1967), 153.

27. D. Homentcovschi: On the motion of a perfectly conducting fluid past a thin body, *Revue Roum. Math. Pures et Appl.*, **17** (1972), 687.

28. S. Ando: General theory of electrically conducting perfect gas flow past a three-dimensional thin body, *J. Phys. Soc. Japan*, **15** (1960), 157; Some remarks on the magnetogasdynamic linearized theory, *Idem* **15** (1960), 1523; Linearized theory of conducting two dimensional gas flow in transverse magnetic field, *Ibid.*, **17** (1962), 686.

29. L. Dragoş: Mouvement en éspace d'un fluide compressible parfaitement conducteur, *C. R. Acad. Sci. Paris*, **256** (1963), 2524.

30. K. Kusukawa: On the high speed flow of a compressible conductive fluid past a slender body, *Aero-Space Sci.*, **27** (1960), 551; The flow of an inviscid compressible conducting fluid past a slender body of an arbitrary cross section, *Aeron. Res. Lab. Off. Aero-Spatial Res. U.S. Air Force, ARL*, 158.

31. T. Boggio: Sull'integrazione di alcune equazioni lineari alle derivate parziali, *Ann. di Matem. Pure et Applic.*, **8** (1903), 181.

32. P. C. T. de Boer: Most general solution of a multiple linear operator equation, *Quart. Appl. Math.*, **27** (1970), 537.

33. N. Rott, H. K. Cheng: Generalizations of the inversion formula of thin airfoil theory, *J. Rat. Mech. Anal.*, **3** (1954), 357.

34. W. E. Williams: Integral representation in two dimensional potential theory, *Z. Angew. Math. Phys.*, **14** (1963), 675.

35. D. Homentcovschi: Sur la résolution explicite du problème de Hilbert. Application au calcul de la portance d'un profile mince dans un fluide électroconducteur, *Rev. Roum. Mat. Pures et Appl.*, **14** (1969), 203; Contributions to the study of Hilbert's problem. Applications to magnetohydrodynamics, *Stud. Cerc. Mat.*, **23** (1971), 727.

36. K. Stewartson: On the Kutta-Joukovski condition in magneto-fluid dynamics, *Proc. Roy. Soc. A* **277** (1963), 107.

37. A. Busemann: Lift control in magnetohydrodynamics, *Proc. IUTAM* (Internat. Un. Theor. Appl. Mech.), *Symp.* Tbilisi, **2** (1963), 123.

38. Ken Ichi Kusukawa: On the Kutta-Joukovski condition in magnetohydrodynamics, *J. Phys. Soc. Japan*, **19** (1964), 1031.

39. G. B. Sheriazdanov: Flow of a perfectly conducting and inviscid fluid past a thin airfoil in the presence of a perpendicular magnetic field, *Magnit. Ghidrodinamika* (1970), 67.

40. M. N. Kogan: On the magnetofluid dynamics flows, *Prikl. Mat. Mekh.*, **25** (1962), 132.

41. Ko Tamada: Transonic flow of a perfectly conducting gas with aligned magnetic field, *Phys. Fluids*, **5** (1962), 871.
42. R. Peyret: Régimes de transitions elliptiques-hyperboliques dans certains écoulements de magnétodynamique des fluides, *J. Méc.*, **1** (1962), 167.
43. R. Seebass: On transcritical and hypercritical flows in magnetogasdynamics, *Quart. Appl. Math.*, **19** (1962), 231.
44. N. Geffen: Magnetogasdynamic flows with shock waves, *Phys. Fluids*, **6** (1963), 566.
45. R. Thibault: Étude d'écoulements bidimensionnels au voisinage de la vitesse d'Alfvén, *C. R. Acad. Sci. Paris*, **258** (1964), 1395.

Chapter 6

1. J C. McCune: On the motion of thin airfoils in fluids of finite electrical conductivity, *J. Fluid. Mech.*, **7** (1960), 449.
2. E C. Lary: A theory of thin airfoils and slender bodies in fluids of finite electrical conductivity with aligned fields, *J. Fluid Mech.*, **12** (1962), 209.
3. L. Dragoş: La théorie de l'aile mince en magnetohydrodynamique I, *C. R. Acad. Sci. Paris*, **255** (1962), 1251; Theory of thin airfoils in magnetohydrodynamics, *Arch. Rat. Mech. Anal.*, **13** (1963), 262.
4. K. Stewartson: Magnetofluid dynamics of thin bodies in oblique fields II, *Z. Angew. Math. Phys.*, **13** (1962), 242.
5. L. Dragoş: La théorie de l'aile mince en magnétohydrodynamique, II, *C.R. Acad. Sci. Paris*, **255** (1962), 1289; Theory of thin airfoils in magnetohydrodynamics, *Arch. Rat. Mech. Anal.*, **13** (1963), 262.; On the motion of a fluid with arbitrary electrical conductivity past thin airfoils, *Proc. IUTAM* (Internat. Un. Theor. Appl. Mech.), *Sympos.* Tbilisi, **2** (1963), 183.
6. L. Dragoş: The motion of incompressible fluids with electrical resistivity past thin airfoils in oblique fields, *Z. Angew. Math. Phys.*, **22** (1971), 96.
7. L. Dragoş: L'écoulement d'un gas conducteur en présence des profils minces, *C.R. Acad. Sci. Paris*, **256** (1963), 4158; Théorie des profils minces en magnétoaérodynamique, *J. de Méc.*, **2** (1963), 223; L'écoulement d'un fluide à conductivité électrique finie en présence d'un profil mince, *Rend. Accad. Naz. dei Lincei*, **42** (1967), 381.
8. L. Dragoş: Théorie de l'aile mince en magnétoaérodynamique, *C.R.Acad. Sci. Paris*, **256** (1963), 2772; Theory of thin airfoils in magnetoaérodynamics, *AIAA Journ.*, **2** (1964), 1223; Reply by Author to P. Greenberg, *AIAA Journ.*, **5** (1967), 1367.
9. W. R. Sears, E. L. Resler Jr.,: Magnetoaerodynamic flow past bodies, *Adv. Appl. Mech.*, **8** (1964), 1.
10. I. Y. T. Tang and R. Seebass: Finite magnetic Reynolds number effects in magnetogasdynamic flows, *Quart. Appl. Math.*, **26** (1968), 131.
11. D. N. Fan, G. S. S. Ludford: Correct formulation of airfoil problems in magnetoaerodynamics, *AIAA Journ.*, **6** (1968), 167.
12. L. Dragoş: Sur le problème des profils minces en magnétoaérodynamique, *C.R. Acad. Sci. Paris*, **268**A (1969), 831.
13. L. Dragoş: Correct formulation of airfoil problems in magnetoaerodynamics, *AIAA Journ.*, **7** (1969), 2014.

14. R. Thibault: Etude d'un écoulement bidimensionnel avec champ non aligné, *C.R. Acad. Sci. Paris*, **256** (1963), 2528; Sur une classe particulière de mouvements à champ quasi-aligné, *Idem*, **260** (1965), 798; Ecoulement d'un fluide imparfaitement conducteur avec champ quasi-aligné, *Ibid.*, **260** (1965), 1081.

15. R. Thibault: Etude d'écoulement avec champ quasi-aligné en magnétodynamique des fluides, *Thèse, Faculté de Sci. Paris*, 1965.

16. L. Dragoş, D. Homentcovschi: On the bidimensional flow of an electrically conducting fluid past thin airfoils in quasi-aligned fields, *St. Cerc. Mat.*, **19** (1967), 963.

17. D. Ivănescu: On the electrically-conducting fluid flow past thin airfoils in quasi-aligned fields, *Stud. Cerc. Mat.*, **24** (1972), 535.

18. L. Dragoş, N. Marcov: L'écoulement Alfvén des fluides dissipatifs en présence d'un profil mince, *Fluid Dynamics Transactions*, **4** (Tarda 1967), 501.

19. N. Marcov: Les transformées de Fourier de quelques distributions, *Rev. Roum. Math. Pures et Appl.*, **16** (1971), 517.

20. E. L. Resler jr., J. E. McCune: Some exact solutions in linearized magneto-aerodynamics for arbitrary magnetic Reynolds numbers, *Rev. Mod. Phys.*, **32** (1960), 848.

21. L. Dragoş, N. Marcov: Motion of incompressible fluids with electrical resistivity past an airfoil with incidence, *Bull. Math. Soc. Roum. Sci. Math.*, **14** (1970), 29.

22. O. P. Bhutani, K. D. Nanda: A general theory of thin airfoils in nonequilibrium magneto-gasdynamics, Part II: Transverse field, *AIAA Journ.*, **6** (1968), 2122.

Chapter 7

1. B. Sonnerup: Some effects of tensor conductivity in MHD., *J. Aero-Space Sci.*, **28** (1961), 612.

2. L. Dragoş: L'effet Hall dans l'écoulement des fluides en présence des profils minces, *C.R. Acad. Sci. Paris, 267A* (1968), 579.

3. L. Y. T. Tang, R. Seebas: The effect of tensor conductivity on continuum magnetogas-dynamic flows, *Quart. Applied. Math.*, **26** (1968), 311.

4. L. Dragoş: The Hall effect in the motion of incompressible fluids past thin airfoils, *Quart. Appl. Math.*, **26** (1970), 313.

5. L. Dragoş: L'étude unitaire de l'écoulement des fluides incompressibles en présence des profils minces, *Rev. Roum. Math. Pures et Appl.*, **14** (1969), 968.

6. L. Dragoş: L'écoulement des fluides incompressibles à l'effet Hall en présence d'un profil mince, *Arch. Mech. Stosow*, **22** (1970), 281.

7. L. Dragoş: Sur l'écoulement des fluides à tenseur de conductivité en présence d'un profil mince, *Coll. Internat. CNRS Magnétohydrodynamique classique et relativiste*, Lille (1969), 297.

8. L. Dragoş: L'effet Hall dans l'écoulement des fluides compressibles en présence des profils minces, *C.R. Acad. Sci. Paris*, **269A** (1969), 1021.

9. L. Dragoş: L'écoulement Alfvén des fluides compressibles avec effet Hall en présence des profils minces, *Rev. Roum. Math. Pures et Appl.*, **14** (1969), 1253.

Chapter 8

1. KO TAMADA, TOSIO MIYAGI: Laminar viscous flow past a flat plate set normal to the stream with special reference to high Reynolds number, *J. Phys. Soc. Japan*, **17** (1962), 373; On the viscous flow past a tangential flat plate, *Idem*, **20** (1965), 454.
2. T. MIYAGI: Oseen flow past a flat plate inclined to the uniform stream, *J. Phys. Soc. Japan*, **19** (1964), 1063; Numerical study of Oseen flow past a perpendicular flat plate, *Idem*, **24** (1967), 204.
3. R. SEEBASS, KO TAMADA, T. MIYAGI: Oseen flow past a finite flat plate, *Phys. Fluids*, **9** (1966), 1697.
4. H. P. GREENSPAN, G. F. CARRIER: The magnetohydrodynamic flow past a flat plate, *J. Fluid Mech.*, **6** (1959), 77.
5. G. F. CARRIER, H. GREENSPAN: The time dependent magnetohydrodynamic flow past a flat plate, *J. Fluid Mech.*, **7** (1960), 22.
6. H. P. GREENSPAN: Flat plate drag in magnetohydrodynamic flow, *Phys. Fluids*, **3** (1960), 581; On the flow of a viscous electrically conducting fluid, *Quart. Appl. Math.*, **18** (1961), 408.
7. L. DRAGOŞ, N. MARCOV: L'écoulement des fluides visqueux, électroconducteurs, en présence d'un plaque plane à incidence non nulle, *J. de Mécanique*, **7** (1968), 379.
8. N. MARCOV: Flow of viscous electrically-conducting fluids past thin bodies, *Studii Cerc. Mat.*, **20** (1968), 199 and 213 (in Romanian).
9. D. L. HECTOR: On the linearized MHD flow past a semi-infinite flat plate in the presence of a transverse magnetic field, *J. Math. Phys.*, **46** (1967), 408.
10. N. MARCOV: The motion of electrically-conducting viscous fluids past thin bodies for the special case $\sigma_e \mu_e \nu = 1$, *Studii Cerc. Mat.*, **21** (1969), 759 (in Romanian); On the motion of electrically conducting viscous fluids past a flat plate, *Idem*, **21** (1969), 1063 (in Romanian).
11. L. DRAGOŞ: Motion of viscous and electroconducting fluids past a flat plate in the case of orthogonal fields, *Int. J. Engng. Sci.*, **8** (1970), 967.
12. V. J. ROSSOW: On the flow of electrically conducting fluids over a flat plate in the presence of a transverse magnetic field, *NASA tech. note*, 3971, May 1957.
13. H. HASIMOTO: Boundary layer growth on a flat plate in the presence of a transverse magnetic fleld, *Aeron. Res. Inst., Univ. of Tokyo, Report* 388, **29** (1964), 145.
14. D. M. DIX: The MHD flow past a non-conducting flat plate in the presence of a transverse magnetic field, *J. Fluid Mech.*, **15** (1963), 449.
15. H. HASIMOTO, G. S. IANOWITZ: Hall effect in the two-dimensional flow along an insulating plane, *Phys. Fluids*, **8** (1965), 2234.
16. L. DRAGOŞ: Hall effect in the motion of viscous fluids past the flat plate, *Rev. Roum. Math. Pures et Appl.*, **15** (1970), 683.
17. S. DATTA: The effect of Hall current on unsteady slip flow over a flat plate under transverse magnetic field, *Archives of Mech.*, **24** (1972), 269.
18. L. DRAGOŞ: L'écoulement des fluides conducteurs, visqueux et compressibles en présence d'une plaque plane, *Rend. dei Lincei* **45** (1968), 507.
19. H. YOSHINOBU: A linearized theory of MHD flow past a fixed body in a parallel magnetic field, *J. Phys. Soc. Japan*, **15** (1960), 175.

20. N. Marcov: Limit theorems for the K_0 and K_1 Bessel functions, *Studii Cerc. Mat.*, **21** (1969), 405 (in Romanian).
21. N. Marcov: Viscous flow past bodies, *Studii Cerc. Mat.*, **23** (1971), 95 (in Romanian).
22. N. Marcov: Les transformées Fourier de quelques distributions, *Rev. Roum. Mat. Pures et Appl.*, **16** (1971), 517.
23. J. Vacca, Moto non stazionario di un fluido viscoso ed elettricamente conduttore attorno ad una piastra piana nel caso di campi ortogonali, *Atti Accad. Sci. Torino*, cl. Sci. fis. mat. natur, **106** (1972), 935.
24. H. Hasimoto: Viscous flow of a perfectly conducting fluid with a frozen magnetic field, *Phys. Fluids*, **2** (1959), 238; Magnetohydrodynamic wakes in a viscous conducting fluid, *Rev. Mod. Phys.*, **32** (1960), 860.
25. Tsunehiko Kabutani: Effect of transverse magnetic field on the flow due to an oscillating flat plate, II, *J. Phys., Soc. Japan*, **15** (1960), 1316.
26. M. B. Glauert: A study of the magnetohydrodynamic boundary layer on a flat plate, *J. Fluid Mech.*, **10** (1961), 276; The boundary layer on a magnetic plate, *Idem*, **12** (1962), 625.
27. F. A. Goldsworthy: Magnetohydrodynamic flows of a perfectly conducting viscous fluid, *J. Fluid Mech.*, **11** (1961), 519.
28. D. Meksyn: MHD flow past a semi-infinite plate, *J. Aero-Space Sci.*, **29** (1962), 662; Idem, *Z. Angew. Math. Phys.*, **17** (1966), 397.
29. J. Radlow, W. B. Ericson: Transverse MHD flow past a semi-infinite plate, *Phys. Fluids*, **5** (1962), 1428.
30. T. V. Davies: The MHD boundary layer in the two-dimensional steady flow past a semi-infinite flat plate, *Proc. Roy. Soc.*, **A273**, (1963), 496, 518.
31. K. R. Singh, T. G. Cowling: Thermal convection in MHD, *Quart. J. Mech. Appl. Math.*, **16** (1963), 1; Unsteady magnetohydrodynamic flow past thin bodies with inwardly diffusing magnetic field, *Phys. Fluids*, **10** (1967), 1756.
32. D. N. Fan: Aligned fields MHD wakes, *J. Fluid Mech.*, **20** (1964), 433.
33. D. Wilson: Dual solution of Greenspan-Carrier equations, *J. Fluid Mech.*, **18** (1964), 161.
34. S. A. Gupta: Hydromagnetic free convection flows from a horizontal plate, *AIAA Journ.*, **4** (1966), 1439.
35. W. Johnson: A boundary value problem from the theory of MHD flow, *Arch. Rat. Mech. Anal.*, **22** (1966), 355.
36. D. B. Ingham: The magnetogasdynamic boundary layer for a thermally conducting plate, *Quart. J. Mech. Appl. Math.*, **26** (1967), 347; Dual solutions of the magnetogasdynamic boundary layer equations, *J. Fluid Mech.*, **27** (1967), 145.
37. E. Sanchez-Palencia: Théorème d'existence pour l'écoulement autour d'une plaque plane, *C.R. Acad. Sci. Paris*, **265A** (1967), 354.
38. E. P. Salathé, E. L. Sirovich: Dissipative magnetogasdynamic flow, *J. Fluid Mech.*, **33** (1968), 361.
39. R. P. Kanwal: Motion of solids in viscous and electrically-conducting fluids, *J. Math. Mech.*, **19** (1969), 489.
40. M. Katagiri: On the separation of magnetohydrodynamic flow near the rear stagnation point, *J. Phys. Soc. Japan*, **27** (1969), 1045; Magnetohydrodynamic flow with suction or injection at the forward stagnation point, *Idem*, 1677.
41. U. N. Das: A small unsteady perturbation on the steady hydromagnetic boundary-layer flow past a semi-infinite plate, *Proc. Cambridge Philos. Soc.*, **68** (1970), 509.

42. H. Naruse: The Hall effect on the magnetogasdynamic flow past an axi-symmetric body, Part 1, Inviscid flow, *J. Phys. Soc. Japan*, 28 (1970), 238.

43. H. Naruse, S. Nishijima: The Hall effect on the magnetogasdynamic flow past an axisymmetric body, Part 2, Boundary layer flow, *Idem*, 758.

44. P. L. Rimmer: Magnetohydrodynamic flow past a yawed semi-infinite flat plate in the presence of a pressure gradient, *Z. Angew. Math. Phys.*, **21** (1970), 211.

45. R. C. Choudhary: An approximate method of calculating hydromagnetic boundary layers in the presence of transverse magnetic field, *Z. Angew. Math. Mech.*, **51** (1971), 413.

46. R. J. Cole: Aligned field MHD flow past a flat plate, *Quart. J. Mech. Appl. Math.*, **24** (1971), 187.

47. I. Pop: A note on unsteady hydromagnetic slip flow past an infinite flat plate, *Indian J. Phys.*, **45** (1971), 275.

48. K. B. Ranger: MHD flow past a flat plate in the presence of a non-aligned magnetic field, *Appl. Sci. Res.*, **25** (1972), 355.

The Rayleigh problem

49. H. Raymond, P. M. Roberts: Some elementary problems in MHD, *Advances in Appl. Mech.*, **7** (1964).

50. J. D. Jukes: On the Rayleigh-Taylor problem in magnetohydrodynamics with finite resistivity, *J. Fluid Mech.*, **16** (1963), 177.

51. P. Chang Lu: Rayleigh's problem in magnetogasdynamics, *AIAA Journ.*, **3** (1965), 2219.

52. C. Cercignani, G. Tironi: Linearized Rayleigh's problem in magnetogasdynamics, *Phys. Fluids*, **9** (1966), 343.

The flow past a sphere

53. Kanefusa Gotoh: MHD flow past a sphere, *J. Phys. Soc. Japan*, **15** (1960), 189.

54. G. Ludford, J. Murray: On the flow of a conducting fluid past a magnetized sphere, *J. Fluid Mech.*, **7** (1960), 516.

55. I. Dee Chang: Slow motion of a sphere in a compressible viscous fluid, *Z. Angew. Math. Phys.*, **16** (1965), 449.

56. G. Ludford, M. P. Singh: The hydromagnetics of an ellipsoid moving in an aligned field, *Proc. Cambr. Phil. Soc.*, **62** (1966), 95.

57. T. Levy: Ecoulement lent d'un fluide visqueux et conducteur autour d'une coque sphérique en présence d'un champ magnetique, *J. de Mécanique*, **6** (1967), 529.

58. H. Cabannes: *Magnetodynamique des fluides* [G.23]; *Ecoulements autour d'une sphère en magnétodynamique des fluides*, *Colloq. CNRS* 184, Lille (1969), 19.

59. F. L. Françoise: Écoulement d'un fluide visqueux et conducteur autour d'une sphère magnetisée, *C.R. Acad. Sci. Paris, Sèr. A* **274** (1972), 271.

60. Th. Lévy: Écoulement magnétodynamique autour d'une sphère creuse, *Ann. Inst. H. Poincaré, Sect. A* **17** (1972), 1.

Chapter 9

1. H. ALFVÉN: Existence of electromagnetic-hydrodynamic waves, *Nature*, **150** (1942); On the existence of electromagnetic hydrodynamic waves, *Arkiv Mat. Astron. Fys.*, **29B** (1943).
2. N. HERLOFSON: Magnetohydrodynamic waves in a compressible fluid conductor, *Nature*, **165** (1950), 1020.
3. H. C. VAN DE HULST: Interstellar polarization and magneto-hydrodinamic waves, *Problems of Cosmical aerodynamics, Central Air Documents Office* (1951).
4. N. S. ANDERSON: Longitudinal magneto-hydrodynamic waves, *J. Acoust. Soc. Amer.*, **25** (1953), 529.
5. A. BAÑOS: Fundamental wave equations in an unbounded magnetohydrodynamic fluid, *Phys. Rev.*, **97** (1955), 1435; Magnetohydrodynamic waves in incompressible and compressible fluids, *Proc. Roy. Soc.*, **A 233** (1955), 350.
6. G. G. LUDFORD: The propagation of small disturbances in hydromagnetics, *J. Fluid Mech.*, **5** (1969), 387.
7. F. G. FRIEDLANDER: Sound pulses in a conducting medium, *Proc. Cambr. Phil. Soc.*, **55** (1959), 341.
8. M. LIGHTHILL: Studies on magnetohydrodynamic waves and other anisotropic waves motion, *Phil. Trans. Roy. Soc. London*, A **252** (1960), 397.
9. E. G. BROADBENT: Magnetohydrodynamic wave propagation from a localized source including Hall effect, *Phil. Trans. Roy. Soc. London*, **263** (1968), 119.
10. M. IGNAT: Contributions to the theory of MHD waves, *Ph. D. Thesis*, University of Cluj, (1969) (in Romanian); Contributions to the solution of the linearized MHD equations by the Fourier method, *An. St. Univ. "A. I. Cuza", Jassy*, **16** (1970), 7, 111 (in Romanian); Sulle onde generate da una perturbazione locale in un plasma soggetto a un campo magnetico, *Atti della Accad. Sci. Torino*, **106** (1971), 69.
11. L. DRAGOŞ: Le problème de Cauchy dans le cas du fluide compressible à résistivité électrique négligeable, *C.R. Acad. Sci. Paris*, **263 A** (1966), 296; Sur la propagation des perturbations initiales dans un fluide compressible à résistivité électrique négligeable, *Revue Roum. Math. Pures et Appl.*, **12** (1967), 63.
12. L. DRAGOŞ: Propagation des petites perturbation plannes dans un fluide non limité, compressible et parfaitement conducteur, *Rev. Roum. Phys.*, **18** (1973), 735.
13. L. DRAGOŞ, N. MARCOV: Théorie de la propagation des petites perturbations dans un fluide conducteur illimité, *Bull. Math. Soc. Roum. de Math.*, **15** (1971), 277.
14. L. DRAGOŞ, L. TODOR: The propagation of small disturbances in a fluid with electrical resistivity and Hall effect (unpublished).
15. M. J. LIGHTHILL: Group velocity, *J. Inst. Math. Appl.*, **1** (1965), 1.
16. S. L. WEN: An extension of Lighthill's result on asymptotic evolution of multiple Fourier integrals, *Int. J. Engng. Sci.*, **7** (1969), 53.
17. M. FROISSART (editor): *Hyperbolic equations and waves*, Springer Verlag, Berlin (1970).
18. G. S. S. LUDFORD: On initial conditions in hydromagnetics, *Proc. Cambr. Phil. Soc.*, **55** (1959), 141.
19. G. B. WHITHAM: Some comments on wave propagation and shock wave structure with application to magnetohydrodynamics, *Comm. Pure Appl. Math.*, (1959), No. 1.
20. H. WEITZNER: Green's function for two-dimensional MHD waves I, II, *Phys. Fluids*, **4** (1961), 1238, 1246.

21. G. S. Golitzin: Propagation of an initial impulse in magnetohydrodynamics with arbitrary conductivity, *Prikl. Meh. Teh. Fiz.*, (1963), 55 (in Russian).
22. A. M. J. Davis: Small disturbances in a conducting fluid in the presence of a current- carrying conductor, *Proc. Cambr. Phil. Soc.*, **60** (1964), 325.
23. B. Granff, R. M. Lewis: Asymptotic solution of initial boundary-value problems for hyperbolic systems, *Phil. Trans.*, **A262** (1967), 387.
24. G. A. Nariboli, M. P. Rao Ranga: Wave propagation in magnetogasdynamics, *J. Mat. Phys. Sci. Madras*, **1** (1967), 302.
25. Alfred Ayoub: Quelques problèmes de perturbation en magnétohydrodynamique, *Bull. Math. Soc. Sci. Math. R.S.R.*, **14** (62) (1970), 387.
26. O. P. Bhutani, L. Raman: General three dimensional wave propagation in radiation magnetogasdynamics, *Internat. J. Engng. Sci.*, **9** (1971), 521; On plane linear radiation magnetogasdynamic waves, *Idem*, 537.
27. J. B. Helliwell: The propagation of small disturbances in radiative magnetogasdynamics, *Arch. Rat. Mech. Anal.*, **47** (1972), 380.
28. Chee-Sing Lim: Waves generated in the configuration of a magnetically confined and field permeated axisymmetric jet, *J. Fluid. Mech.*, **55** (1972), 129.

Chapter 10

1. K. O. Friedrichs: Non linear wave motion in magnetohydrodynamics, *Los Alamos Rept.* 2.105 (1954); Symmetrical hyperbolic linear differential equations, *Comm. Pure Appl. Math.*, **7** (1954), 345; Mathematical aspects of flow problems of hyperbolic type in "*General theory of high speed aerodynamics*", Oxford Univ. Press (1955), 33; *Nichtlinear differential Gleichungen*, Göttingen, 1955.
2. K. O. Friedrichs and H. Kranzer: Non-linear wave motion in magnetohydrodynamics, *Inst. Math. Sci., New York, Univ. Rept.*, M.H. 8 (1958).
3. A. Jeffrey and T. Taniuti: *Non-linear wave propagation with application to physics and magnetohydrodynamics*, Acad. Press, New York (1964).
4. I. B. Keller: Geometrical acoustics I. The theory of weak shock waves, *J. Appl. Phys.*, **25** (1954), 938.
5. N. Roth: A simple construction for the determination of the magnetohydrodynamic wave speed in a compressible conductor, *J. Aero-Space Sci.*, (1959).
6. J. Bazer and O. Fleischman: Propagation of weak hydromagnetic discontinuities, *Phys. Fluids*, **2** (1959), 366.
7. R. V. Polovin: On the theory of simple magnetohydrodynamic waves, *Zh. Eksp. Teor. Fiz.*, **39** (1960), 463 (in Russian).
8. W. R. Sears: Some remarks about flow past bodies, *Rev. Mod. Phys.*, **32** (1960), 701; Sub-Alfvénic flows in magnetoaerodynamics, *J. Aero-Space Sci.*, **28** (1961), 249.
9. L. Dinu: Studies on the shock layer theory in a plasma, *Ph. D. thesis, Bucharest University* (1974).
10. I. Bohachevsky: Simple waves and shocks in magnetohydrodynamics, *Phys. Fluids*, **5** (1962), 1456.
11. R. Courant, K. O. Friedrichs: *Supersonic flow and shock waves*, Interscience, New York (1948).

12. P. D. LAX: Hyperbolic systems of conservation laws II, *Comm. Pure Appl. Math.*, **10** (1957), 537.

13. G. LIUBARSKI, P. V. POLOVIN: Propagation of a small discontinuity in MHD, *Zh. Exp. Teor. Fiz.*, **35** (1958), 5 (in Russian).

14. A. I. AKHIEZER, G. J. LIUBARSKI, R. V. POLOVIN: Simple waves and shock waves in MHD, *Proc. 2nd. Intern. Conf. Geneva*, 1958, **31** (1959), 225; Simple waves in MHD, *Zh. Exp. Teor. Fiz.*, **36** (1959), 8 (in Russian).

15. I. M. GEL'FAND: Some problems in quasilinear equations theory, *Uspekhi Mat. Nauk*, **87** (1959), 2 (in Russian).

16. R. S. ONG: Characteristics manyfolds in three-dimensional unsteady MHD, *Phys. Fluids*, **2** (1959).

17. P. D. LAX and B. WENDROFF: Systems of conservation laws, *Comm. Pure Appl. Math.*, **13** (1960), 217.

18. E. CUMBERBATCH: MHD Mach cones, *J. Aero-Space Sci.*, **29** (1962), 1476.

19. Y. M. LYNN: Characteristic locus in MHD, *Phys. Fluids*, **5** (1962), 626.

20. A. JEFREY: The propagation of weak discontinuities in quasilinear symmetric hyperbolic systems, *Z. Angew. Math. Phys.*, **14** (1963), 301; Magnetoacoustic simple waves in a polytropic gas, *Idem*, **15** (1964), 217.

21. P. V. POLOVIN, K. P. CHERHASOV: Magnetosonic waves, *Magnit. Gidrod.* (1966), 3 (in Russian); MHD waves, *Uspekhi Fiz. Nauk*, **88** (1966), 593 (in Russian).

22. H. SAUERWEIN: The method of characteristics for the three-dimensional unsteady MFD of a multi-component medium, *J. Fluid Mech.*, **25** (1966), 17.

23. A. G. KULIKOVSKI: On discontinuity surfaces which separate two ideal fluids with different properties. Recombination waves in MHD, *Prikl. Math. Mech.*, **32** (1968), 1125 (in Russian).

24. B. L. ROZDENSTVENSKI, N. N. IANENKO: *Systems of quasilinear equations*, Moscow, Nauka (1968).

25. V. A. WYE: Wave propagating in a highly electrically-conducting radiating gas, *Quart. J. Mech. Appl. Math.*, **23** (1970), 265.

Chapter 11

1. R. LUST: Stationare magnetohydrodynamische Stosswellen beliebigen Starke, *Zeitschr. NaturForsch.*, **10** (1955), 125.

2. R. THIBAULT: Sur le théorème de Zemplen en magnétodynamique des fluides, *C.R. Acad. Sci. Paris*, **255** (1962), 834.

3. J. E. ANDERSON: *Magnetohydrodynamic shock-waves* M.I.T. Press, Cambridge, (1963).

4. R. V. POLOVIN, V. P. DEMUTSKII: The shock adiabatic in MHD, *Ukrain. Fiz. Zh.*, **5** (1960), 3 (in Russian).

5. P. GERMAIN: Contribution à la théorie des ondes de choc en magnétodynamique des fluides,, *Publication ONERA*, No. 97 (1959); Shock waves and shock structure in MFD, *Rev. Mod. Phys.*, **32** (1960), 951.

6. J. A. SHERCLIFF: One dimensional magnetogasdynamics in oblique fields, *J. Fluid. Mech.*, **9** (1960), 481.

7. A. I. Akhiezer, G. J. Liubarski, R. V. Polovin: The stability of shock waves in magnetohydrodynamics, *Zh. Eksp. Teor. Fiz.*, **35** (1958); 3.

8. S. I. Syrovatskii: Magnetohydrodynamics *Uspehi Fiz. Nauk*, **62** (1957), 247 (in Russian). The stability of shock waves in MFD, *Zh. Eksp. Teor. Fiz.*, **35** (1958), 6 (in Russian).

9. R. V. Polovin: Shock Waves in magnetohydrodynamics, *Uspehi Fiz. Nauk.*, **72** (1960), 1, 33 (in Russian).

10. B. P. Leonard: Hall currents in MHD shock waves, *Phys. Fluids*, **9** (1966), 917; Stability of transverse Alfvénic ionizing shock waves, *Idem*, **13** (1970), 833; Comments on structure of MHD ionizing shock waves, *Ibid.*, **13** (1970), 3063; Magnetic structure of ionizing shock waves, *J. Plasma Phys.*, **7** (1972), 133–185.

11. F. Hoffmann, E. Teller: Magneto-hydrodynamic shocks, *Phys. Rev.*, **80** (1950), 692.

12. N. Marshall: The structure of MHD shock waves, *Proc. Roy. Soc.*, *A* **233** (1955).

13. J. Bazer: Resolution of an initial shear-flow discontinuity in one-dimensional hydromagnetic flow, *Astrophys. J.*, **128** (1958), 686.

14. J. Bazer, W. B. Ericson: Hydromagnetic shocks, *Astrophys. J.*, **129** (1959), 758.

15. Y. Kato: Interaction of hydromagnetic waves, *Progr. Theor. Phys.*, **21** (1959), 409.

16. M. N. Kogan: Shock waves in MHD, *Prikl. Mat. Meh.*, **23** (1959), 3.

17. V. M. Kontorovich: On the interaction between small disturbances and discontinuities in MHD and on the stability of shock waves., *Zh. Eksp. Teor. Fiz.*, **8** (1959), 851 (in Russian).

18. A. G. Kulikovsky, G. Zyubimov: Observations on the structure of orthogonal MHD shock waves, *Prikl. Mat. Meh.*, **23** (1959), 6; Structure of the oblique MHD shock waves, *Idem*, **25** (1961), 1 (in Russian).

19. G. I. Liubarski, R. V. Polovin: On the break up of unstable MHD shock waves, *Zh. Eksp. Teor. Fiz.*, **36** (1959), 4, 1272 (in Russian).

20. G. S. S. Ludford: The structure of a hydromagnetic shock in steady plane motion, *J. Fluid Mech.*, **5** (1959), 67.

21. Z. O. Blewiss: A study of the structure of MHD switch-on shocks in steady plane motion, *J. Fluid Mech.*, **9** (1960).

22. H. Cabannes: Attached stationary shock wave in ionized gases, *Rev. Mod. Phys.*, **32** (1960), 973.

23. W. Ericson, J. Bazer: On certain properties of hydromagnetic shocks, *Phys. Fluids*, **3** (1960), 631.

24. M. I. Kiselev, N. I. Kolesnitzyn: Calculations for inclined shock waves in MHD, *Dokl. Akad. Nauk SSSR*, **131** (1960), 773 (in Russian).

25. V. V. Gogosov: Resolution of an arbitrary discontinuity in MHD, *J. Appl. Math. Mech.*, **25** (1961), 148; Interaction of MHD waves with contact and vortex discontinuities, *Idem*, p. 277; Interaction of MHD waves, *Ibid.*, p. 678.

26. P. Germain: Les ondes de choc dans les plasmas, *Cahiers de Phys.*, **15** (1962), 243.

27. P. Germain, Soubbaramayer: Sur un schéma d'onde de choc en MFD, *C.R. Acad. Sci.*, **255** (1962), 1856.

28. P. Helliwell: MHD shock waves in a gas with variable conductivity, *Phys. Fluids.*, **5** (1962), 738; Gas-ionizing shock waves with oblique fields, *Idem* **6** (1963), 1516.

29. A. G. Kulikovsky: On the structure of shock waves in MHD for arbitrary dissipation, *Prikl. Mat. Meh.*, **26** (1962) (in Russian).

30. K. W. MORTON: Large amplitude compression waves in an adiabatic two-fluid model of a collision free plasma, *J. Fluid Mech.*, **14** (1962), 364; Finite amplitude compression waves in a collision free plasma, *Phys. Fluids*, **7** (1964), 1800.

31. T. TANIUTI: A note on the evolutionary condition of hydromagnetic shocks, *Progr. Theor. Phys.*, **28** (1962), 756.

32. C. CHU, Y. LYNN: Steady MHD flow past a non-conducting wedge, *AIAA Journ.*, **1** (1963), 1062.

33. I. C. PANT, R. S. MISHRA: Shock waves of finite thickness in magneto-gasdynamics. *Rend. Circ. Mat. Palermo*, **12** (1963), 59.

34. H. CABANNES: Sur le problème du dièdre en magnétoaérodynamique, Etude de la stabilité, *Arch. Mech. Stosow*, **16** (1964); Sur la stabilité des ondes de choc attachées en magnétodynamique des fluides, *J. Math. Pures et Appl.* (1965), 171.

35. C. K. CHY: Dynamics of ionizing shock waves: shocks in transverse magnetic fields, *Phys. Fluids*, **7** (1964), 1349.

36. I. FERRARI: Sui fronti d'onda in magnetofluidodinamica, *Atti del Sem. Mat. et Fiz. Univ. Modena*, **13** (1964).

37. R. PEYRET: Sur la structure du choc lent dans un schèma à deux fluides nondissipatifs (avec dissipations), *C.R. Acad. Sci.*, **258** (1964), 2973 (3178).

38. L. TODD: Evolution of the trans-Alfvénic normal (oblique) shock in a gas of finite electrical conductivity, *J. Fluid. Mech.*, **18** (1964), 321 (*Idem* **21** (1965), 193).

39. E. G. BROADBENT: Plasma heating by reflected ionizing shock waves, *Phys. Fluids*, **13** (1970), 2270; Magnetohydrodynamic shocks and their stability, *Progr. Aeron. Sci.*, **6** (1965), 133.

40. Y. LYNN: MHD shoch relations in nonaligned flows, *New York Univ.*, *Rep. Nyo* — 1480—29, MF—47 (1965).

41. R. T. TAUSSING: Normal ionizing shock waves, *Phys. Fluids*, **8** (1965), 1618; Normal ionizing shock waves with equilibrium chemistry in hydrogen, *Idem* **9** (1966), 421; Comparison of oblique, normal and transverse ionizing shock waves, *Ibid.*, **10** (1967), 1162.

42. D. C. PACK, G. W. SWAN: Magnetogasdynamic flow over a wedge, *J. Fluid Mech.*, **25** (1966), 165.

43. S. URASHIMA, S. MORIOKA: MHD shock polar, *J. Phys. Soc. Japan*, **21** (1966), 1431.

44. B. G. VERMA: Reflection of a hydromagnetic shock on a rigid wall, *Nuovo Cimento*, **41** (1966), 9.

45. D. I. DEMENUNY, G. I. LIUBARSKY: On the theory of slow intensity shock waves., *Magnit. Gidrodinamika* (1967), 74 (in Russian).

46. M. EOUTANEZI, S. E. SEGRE: Strong ionising fronts in an oblique magnetic field, *Plasma Phys.*, **9** (1967), 479.

47. C. K. CHY: Some remarks on the stability of hydromagnetic shock waves, *Proc. Symp. Appl. Math.*, **18** (1967), 1 [G.18].

48. P. GERMAIN: A model of some plasma shock structures, *Idem* 17.

49. SOUBBARAMAYER: Sur les chocs dans un milieu magnéto-dynamique réactif avec application au problème du piston, *Thèse Doct. Sci. Paris*, 1967.

50. A. A. BARMIN: On the change of the velocity of a gas in a normal ionizing shock wave and the piston problem, *Prikl. Mat. Meh.*, **32** (1968), 954; The recombination fronts in arbitrary magnetic fields, *Meh. Jid. i Gaz.* (1970), 8 (in Russian).

51. A. A. BARMIN, A. G. KULICOVSKI: On the shock waves in an ionizing in an electromagnetic field, *Dokl. Akad. Nauk. S.S.S.R.*, **178** (1968), 95; Shock waves in interaction with

magnetic field, *Proc. of the 3th Symp. Theor. and Appl. Mech. Moscow*, 1968; On effect of the electric field on the ionizing shock waves in gases, *Meh. Jid. i Gaz.* (1968), 133; On the shock waves in an ionizing gas with an arbitrary magnetic field, *Hydrodynamics and Continuous Mechanical Problems*, Moscow, Nauka, 35–48 (in Russian).

52. K. V. Bryshliuski, A. I. Morozov: On the evolutionary MHD equations with Hall effect, *Prikl. Mat. Meh.* **32** (1968), 957 (in Russian).

53. M. D. Cowley, S. I. Moarin: Development of MHD waves behind an ionizing shock wave, *Phys. Fluids*, **11** (1968), 674.

54. A. Friedhoffer: A study of the magnetogasdynamics equations as applied to shock and blast waves, *Doct. diss. Univ. Dela* (1968).

55. A. Kennedy: Structure of a detonation wave under the influence of a transverse magnetic field, *Z. Angew. Math. Phys.*, **19** (1968), 600.

56. Y. Lynn: Transition solution for MHD shocks in arbitrary fluids, *Phys. Fluids*, **11** (1968), 441; Magnetogasdynamic shock polar for aligned fields, *J. Plasma Phys.*, **6** (1971), 283.

57. J. C. Pant: Oblique regular reflection of a plane shock in presence of a transverse magnetic field, *Z. Angew. Math. Phys.*, **48** (1968), 73; Some aspects of unsteady curved shock-waves, *Int. J. Engng. Sci.*, **7** (1969), 235.

58. R. A. Gross, L. S. Levine, R. J. Mechar, B. Miller: Normal and transverse ionizing shock studies, *Phys. Fluids*, **12** (1969), 65.

59. J. Tendys: Study of the structure of switch-on ionizing shock waves, *Plasma Phys.*, **11** (1969), 223.

60. K. O. Westzhal, J. F. McKenzie: Interaction of magnetoacoustic and entropy waves with normal MHD shock waves, *Phys. Fluids*, **12** (1969), 1228; Interaction of hydromagnetic waves with hydromagnetic shocks, *Idem*, **13** (1970), 630.

61. L. C. Woods: Critical Alfvén-Mach numbers for transverse field MHD shocks, *Plasma Phys.*, 11 (1969), 25; Jump conditions for a two-fluid magneto-plasma, *Idem*, p. 967.

62. F. V. Coroniti: Dissipation discontinuities in hydromagnetic shock waves, *J. Plasma Phys.*, **4** (1970), 265.

63. M. N. Kogan, A. G. Kulikovsky, G. A. Liubimov: Mechanics of rarefied gases and MHD, in *Mechanics in SSSR at the 50th Anniversary*, Moscow, Nauka, 1970 (in Russian).

64. M. P. Rauga Rao, D. Chaturani: Pressure shocks in dissipative MHD, *Ind. J. Pure Appl. Math.*, (1970), 258.

65. R. J. Bickerton, L. Lenamon, R. V. Murphy: The structure of MHD shock waves, *J. Plasma Phys.*, **5** (1971), 177.

66. V. V. Gogosov, V. A. Poljandkii: Discontinuities in electrohydrodynamics, *Prikl. Mat. Meh.*, **35** (1971), 761.

67. V. D. Sharma, Rishi Ram: The vorticity jump across a stationary magnetogasdynamic, *Z. Angew. Math. Phys.*, **22** (1971), 1126.

68. L. C. Woods: Generalized theory of the stability of shock waves in magnetogasdynamics, *J. Plasma Phys.*, **6** (1971), 615.

69. S. I. Emenalo: Interaction of weak shock waves with MHD boundary layers, *J. Inst. Mat· Appl.*, **9** (1972), 234.

70. D. J. Tembharey, S. K. Sachdeva: Derivation of the discontinuity conditions across a magnetogasdynamic shock surface, *Tensor*, **23** (1972), 240.

71. K. S. UPADHYAYA, R. K. PANDEY: The jump in vortivity and current density across a surface of discontinuity, *Ann. Soc. Sci., Bruxelles, Sér. I* **86** (1972), 291.
72. R. TAUSSING: Magnetohydrodynamic shock structure in strong transverse magnetic fields, *Phys. Fluids*, **16** (1973), 384.
73. C. C. CONLEY, J. A. SMOLLER: Sur l'existence et la structure des ondes de choc en magnétohydrodynamique, *C. R. Acad. Sci. Paris, Sér A* (1973), 387.
74. J. SKIEPKO: Sur l'existence et l'unicité des ondes de choc en magnétodynamique des fluides, *C. R Acad. Sci. Paris*, Sér. A (1973), 391.
75. P. GERMAIN: Sur l'existence et la non unicité de la structure du choc lent en magnétodynamique des fluides. Remarques sur les Notes de C. C. Conley et J. A. Smoller et de J. Skiepko, *C. R. Acad. Sci. Paris, Sér A* (1973), 395.
76. L. DINU: Studies on the shock layer theory in a plasma, *Ph. D. thesis, Bucharest* (1974).

The piston problem

77. G. J. LIUBARSKI, R. V. POLOVIN: The problem of a piston in MHD, *Dokl. Akad. Nauk. S.S.S.R.* **128** (1959), 684 (in Russian).
78. V. V. GOGOSOV, A. A. BARMIN: The piston problem in MHD, *Dokl. Akad. Nauk. S.S.S.R.*, **134** (1960), 1041 (in Russian).
79. R. V. POLOVIN: The motion of a conducting piston in a MHD medium, *Zh. Eksp. Teor. Fiz.*, **11** (1960), 1113 (in Russian).
80. D. L. TURCOTTE, C. K. CHU: On the structure of Alfvén shocks, *Z. Angew. Math. Phys.*, **17** (1966), 528.
81. A. A. BARMIN, A. G. KULIKOVSKY: The piston problem in ionizing gases, *Prikl. Mat. Meh.*, **32** (1968), 495 (in Russian).

Chapter 12

1. L. D. LANDAU, E. M. LIFCHITZ: *Théorie des champs*, Ed. Moscou (1970).
2. L. A. PARS: *A Treatise on Analytical Dynamics*, Heinemann, London 1964.
3. H. WEYL: *The Classical Groups*, Princeton, (1946).
4. W. P. ALLIS: Motion of ions and electrons, *Handbuch der Phys.*, **21** (1956).
5. B. LEHNERT: *Dynamics of charged particles*, North Holland Publ. Co., Amsterdam (1964).
6. M. BORN, L. INFELD: Fondaments de la nouvelle théorie du champ, *Proc. Roy. Soc.*, **A144** (1934), 425.
7. I. E. TAMM: *Fundamentals of electromagnetic theory* (in Russian), Moscow.
8. S. R. DE GROOT, J. VILIEYER: On the derivation of Maxwell's equations, *Nuovo Cimento*, **33** (1964), 1225.

Chapter 13

1. S. CHAPMAN, T. G. COWLING: *The mathematical theory of non-uniform gases*, Cambridge Univ. Press (1970).
2. H. GRAD: Principles of the kinetic theory of gases, *Handbuch der Physik*, **12** (1958), 205.
3. J. YVON: *Eléments de mécanique statistique*, Saclay (1952).
4. H. CABANNES: *Cours de mécanique génerale*, Dunod, Paris (1960).

5. J. Delcroix: *Physique des Plasmas*, Dunod, Paris (1963).
6. J. L. Synge: Classical Dynamics, *Handbuch der Physik*, III/1 (1960).
7. R. Jancel, Th. Kahan: *Electrodynamique des plasmas*, Dunod, Paris (1963).
8. H. Kennard: *Kinetic theory of gases*, McGraw Hill, New York (1938).
9. H. Grad: On the kinetic theory of rarefied gases, *Comm. Pure Appl. Math.*, **2** (1949), 331.
10. C. Cercignani: *Mathematical methods in kinetic theory*, New York, Plenum Press (1969).
11. J. O. Hirschfelder, C. F. Curtiss, R. B. Bird: *Molecular theory of gases and liquids*, John Wiley, New York (1954).
12. A. Sommerfeld: *Thermodynamics and statistical mechanics*, Academic Press, New York (1956).
13. E. P. Gross, E. A. Jackson, S. Ziering: Boundary value problems in kinetic theory of gases, *Ann. Phys.*, **1** (1957), 141.
14. M. Krook: Continuum equations in the dynamics of rarefied gases, *J. Fluid Mech.*, *6* (1959), 523.
15. M. N. Kogan: *Dynamics of rarefied gases* (in Russian), Izd. Nauka, Moscow (1967).
16. W. Wuest: Boundary layers in rarefied gas flow, *Progr. in Aeron. Sci.*, **8** (1967), 295.
17. M. R. Williams: Rarefied gas flow between parallel plates, *Proc. Cambridge Philos. Soc.*, **66** (1969), 189.

Chapter 14

1. J. L. Delcroix: *Physique des plasmas*, Dunod, Paris (1963).
2. R. Jancel, Th. Kahan: *Electrodynamique des plasmas*, Dunod, Paris (1963).
3. S. Chapmann, T. G. Cowling: *The mathematical theory of non-uniform gases*, Cambridge University Press (1970).
4. A. Schlüter: Dynamik der plasmes, I, II, *Z. Naturforchung*, **5a**(1950), 72; **6a** (1951), 73; Plasma in Magnetfeld, *Ann. Phys.*, **6–10** (1952), 422.
5. H. S. Green: Ionic theory of plasmas and magnetohydrodynamics, *Phys. Fluids*, **2** (1959), 341.
6. L. Spitzer, (Jr.): *Physics of fully ionized gases*, Interscience Publishers, New York (1962).
7. L. Dragoş: On the conductivity tensor in fully ionized gases (unpublished).
8. R. Lüst and A. Schlüter: Kraftfreie Magnetfelder, *Z. Astrophys.*, **34**, (1954), 263.
9. L. Spitzer, (Jr.): Equations of motion for an ideal plasma, *Astrophys., J.*, **116** (1952), 299.
10. T. Kihara: Macroscopic foundation of plasma dynamics, *Proc. Phys. Soc. Japan* (1954).
11. B. Lehnert: Plasma physics on cosmical and laboratory scale, *Nuovo Cimento Ser. X suppl.* **13** (1959), 1.
12. E. Covert: Microscopic analysis of magnetohydrodynamics, *Proc. Third Bien. Gas Dyn. Symp.*, Illinois (1959).
13. R. Lüst: Uber die Ausbreitung Zur Wellen in einem Plasma, *Fortschr. der Phys.*, **7** (1959), 503.
14. S. Chandrasekhar: *Plasma Physics*, The University of Chicago Press, Chicago (1960).
15. H. Clauser (gen. ed): *Symposium of plasma dynamics*, Addison-Wesley Publ. Co., Reading, Mass. (1960).
16. V. L. Ginzburg: *The propagation of electromagnetic waves in plasmas*, Pergamon Press, Oxford (1964), (in Russian) (1969).

17. L. KRAUS, H. YOSHIHARA: Electrogasdynamic motion of a charged body in a plasma, *J. Aero-Space Sci.*, **27** (1960), 229.

18. L. OSTER: Linearized theory of plasma oscillations, *Rev. Mod. Phys.*, **32** (1960), 141.

19. H. CABANNES: Théorie cinétique des gas ionisés, *Publ. ONERA Nr.* 101 (1961).

20. J. F. DENISSE, J. L. DELCROIX: *Théorie des ondes dans les plasmas*, Dunod, Paris (1961).

21. H. GRAD: Microscopic and macroscopic models in plasma physics, *Courant Inst. Publ. MF*-19 1961;
Electromagnetic and fluid dynamics of gaseous plasma (ed. J. Fox), Polytechnic Press, Brooklyn—New York, (1961).

22. H. YOSHIHARA: Motion of thin bodies in a highly rarefied plasma, *Phys. Fluids*, **4** (1961), 100.

23. JACQUELINE NAZE: Sur l'équation de Boltzmann des gas faiblement ionisés, *Publications Sci. et Tech. du Min. de l'Air* (1962) 387.

24. T. H. STIX: *The theory of plasma waves*, McGraw — Hill (1962).

25. R. BALESCU: *Statistical mechanics of charged particles*, Interscience Publishers, London (1963).

26. M. A. LEONTOVICH (ed.): *Problems on plasma theory* **1—5** (in Russian), Moscow 1963—1965.

27. H. E. WILHELM: Zur stabilität dissipativer plasmen I, II, *Beiträge aus der Plasma Physik*, **3** (1963), 25.

28. D. K. MONTGOMERY, D.A. TIDMAN: *Plasma kinetic theory*, McGraw — Hill, New York (1964).

29. E. H. HOLT, R. E. HASKELL: *Foundations of plasma dynamics*, Macmillan, New York (1965).

30. E. L. RESLER (JR.) and F. T. HARRIS: Plasma waves, *AIAA Journ.*, **3** (1965), 2033.

31. D. A. TIDMAN, N. A. KRALL: *Shock waves in collision less plasmas*, Wiley, New York (1971).

32. N. A. KRALL, A. W. TRIVELPIECE: *Principles of plasma physics*, McGraw-Hill, New York (1973).

PRINTED IN ROMANIA